# Elementary
# Algebra

## SIXTH EDITION

## Charles P. McKeague

*Cuesta College*

HARCOURT COLLEGE PUBLISHERS

FORT WORTH   PHILADELPHIA   SAN DIEGO   NEW YORK   AUSTIN   ORLANDO   SAN ANTONIO

TORONTO   MONTREAL   LONDON   SYDNEY   TOKYO

Publisher: Emily Barrosse
Executive Editor: Angus McDonald
Marketing Strategist: Julia Downs-Conover
Developmental Editor: Jay Campbell
Project Editor: Robin C. Bonner
Production Manager: Alicia Jackson
Art Director: Lisa Adamitis
Text Designer: Lisa Adamitis
Cover Credit: Vienna, Austria (©Murat Ayranci/SuperStock)

*Elementary Algebra* 6/e, Charles P. McKeague

ISBN: 0-03-026258-5
Library of Congress Catalog Card Number: 99-067248

*Address for domestic orders:*
Harcourt College Publishers, 6277 Sea Harbor Drive, Orlando, FL 32887-6777
1-800-782-4479
email: collegesales@harcourt.com

*Address for international orders:*
International Customer Service, Harcourt, Inc.
6277 Sea Harbor Drive, Orlando, FL 32887-6777
(407) 345-3800
Fax (407) 345-4060
email: hbintl@harcourt.com

*Address for editorial correspondence:*
Harcourt College Publishers, Public Ledger Building, Suite 1250,
150 S. Independence Mall West, Philadelphia, PA 19106-3412

*Web Site Address*
http://www.harcourtcollege.com

Printed in the United States of America

12345678   032   10 987654

# CONTENTS

# Linear Equations and Inequalities in Two Variables    *156*

# Systems of Linear Equations    *216*

# Exponents and Polynomials    *254*

# Factoring  *316*

# Rational Expressions  *364*

# PREFACE TO THE INSTRUCTOR

This sixth edition of *Elementary Algebra* retains the same basic format and style as the fifth edition. The book is intended to be used in a lecture-format class. Each section of the book can be discussed in a forty-five- to fifty-minute class session.

## FEATURES OF THE BOOK

*Increased Visualization of Topics*    This edition contains many more diagrams, charts, and graphs than did the previous edition. The purpose of this is to give students additional information, in visual form, to help them understand the topics we cover. As an example, look in Chapter 1, Section 2, at the Applying the Concepts Problems.

*Tables, Bar Charts, Scatter Diagrams, and Line Graphs*    Beginning in Chapter 1 and then continuing in Chapters 3 and 4, students are required to analyze information from tables. In addition to simply reading tables, they practice converting data in tabular form to data in graphical form that includes histograms and scatter diagrams. From there they move on to graphing ordered pairs and linear equations on a rectangular coordinate system.

*Early Coverage of Graphing*    The material on graphing equations in two variables begins in Chapter 3.

*Blueprint for Problem Solving*    The *Blueprint for Problem Solving* is a detailed outline of the steps needed to successfully attempt application problems. Intended as a guide to problem solving in general, the *Blueprint* overlays the solution process to all the application problems in the first few chapters of the book. As students become more familiar with problem solving, the steps in the *Blueprint* are streamlined.

*Facts From Geometry*    Many of the important facts from geometry are listed under this heading. In most cases, an example or two accompanies each of the facts to give students a chance to see how topics from geometry are related to the algebra they are learning.

*Number Sequences*    An introductory coverage of number sequences is integrated throughout Chapter 1. I find that there are many interesting topics I can cover if students have some experience with number sequences. It is also the easiest way to demonstrate inductive reasoning.

*Unit Analysis*    Chapter 7 contains problems requiring students to convert from one unit of measure to another. The method used to accomplish the conversions is the method they will use if they take a chemistry class. Since this is similar to the method used to multiply rational expressions, unit analysis is covered in Section 7.2 with multiplication and division of rational expressions.

***Getting Ready for Class***   Just before each problem set is a list of four problems under the heading *Getting Ready for Class.* These problems require written responses from students and can be answered by reading the preceding section. They are to be done before students go to class.

***Chapter Openers***   Each chapter begins with an introduction in which a real-world application is used to stimulate interest in the chapter. These opening applications are expanded on later in the chapter. Most of them are explained using the rule of four: verbally, numerically, graphically, and algebraically.

***Study Skills***   Found in the first six chapter openings after the opening application is a list of study skills intended to help students become organized and efficient with their time. The study skills point students in the direction of success.

***Organization of the Problem Sets***   Five main ideas are incorporated into the problem sets.

1. **Drill**   There are enough problems in each set to ensure student proficiency in the material.

2. **Progressive Difficulty**   The problems increase in difficulty as the problem set progresses.

3. **Odd-Even Similarities**   Each pair of consecutive problems is similar. Since the answers to the odd-numbered problems are listed in the back of the book, the similarity of the odd-even pairs of problems allows your students to check their work on an odd-numbered problem and then to try the similar even-numbered problem.

4. **Applying the Concepts Problems**   Students are always curious about how the algebra they are learning can be applied, but at the same time many of them are apprehensive about attempting application problems. I have found that they are more likely to put some time and effort into trying application problems if they do not have to work an overwhelming number of them at one time and if they do not have to work on them every day. For these reasons, I have placed a few application problems toward the end of almost every problem set in the book.

5. **Review Problems**   As was the case in the fifth edition, each problem set, beginning with Chapter 2, contains a few review problems. Where appropriate, the review problems cover material that will be needed in the next section. Otherwise, they cover material from the previous chapter. That is, the review problems in Chapter 5 cover the important points in Chapter 4. Likewise, the review problems in Chapter 6 review the important material from Chapter 5. If you give tests on two chapters at a time, you will find this to be a timesaving feature. Your students will review one chapter as they study the next chapter.

***End-of-Chapter Retrospective***   Each chapter ends with the following items that together give a comprehensive reexamination of the chapter and some of the important problems from previous chapters.

***Chapter Summary***   These list all main points from the chapter. In the margin, next to each topic being reviewed, is an example that illustrates the type of problem associated with the topic being reviewed.

***Chapter Review***   An extensive set of problems that review all the main topics in the chapter, the *Chapter Reviews* contain more problems than do the *Chapter Tests*. Numbers in brackets refer to the section(s) in the text where similar problems can be found.

***Chapter Test***   This set of problems is representative of all the main points of the chapter.

***Cumulative Review***   Starting in Chapter 2, at the end of the *Chapter Test* in each chapter, is a set of problems that reviews material from all preceding chapters. This keeps students current with past topics and helps them retain the information they study.

## CHANGES IN THE SIXTH EDITION

Although there are many more changes in this edition than are listed here, these changes are the more noticeable ones. To start, we have moved most of the material previously in Chapter 9 into the other chapters of the book. The section on compound inequalities is now at the end of Chapter 2, the section on linear inequalities in two variables has been integrated into Chapter 3, and variation is at the end of Chapter 7 on rational expressions. Next, we split Chapter 3 into two chapters so that systems of equations is handled separately. As we mentioned earlier, each chapter now contains a cumulative review that reviews problems from the current chapter and all previous chapters. Finally, in each chapter in this edition, one problem set ends with a set of review problems that reviews all of the previous chapters.

## SUPPLEMENTS TO THE TEXTBOOK

This sixth edition of *Elementary Algebra* is accompanied by a number of useful supplements.

### For the Instructor

• ***Instructor's Edition***   An instructor's version of the text is available. This version contains everything in the student's edition as well as an additional appendix containing answers to the even-numbered problems in the *Problem Sets*.

• ***MathCue Course Management for Windows***   This computer software component is an all-in-one learning and assessment tool. It contains Syllabus, Grade Book, Testing, and E-mail functions. Tests can be administered on-line, grades can be automatically recorded in the gradebook, and professors can facilitate communication through e-mail. Preparing, administering, and taking quizzes and tests, as well as tracking and evaluating student performance, has never been so easy!

- *Printed Test Bank and Prepared Tests*    The test bank consists of multiple-choice and short-answer test items organized by chapter and section. It also includes 15 sample problems per section of the text, labeled by chapter, section, and level of difficulty. The prepared tests include 10 sets of ready-to-copy tests: one set for each chapter and one set for the entire book. Each set comprises 4 multiple-choice and 4 show-your-work tests. Answers for every test item are provided.

- *ESATEST 2000™ Computerized Test Bank*    A flexible, powerful computerized testing system, the ESATEST 2000™ Computerized Test Bank contains all the test bank questions and allows instructors to prepare quizzes and examinations quickly and easily. It offers teachers the ability to select, edit, and create not only test items but algorithms for test items as well. Teachers can tailor tests according to a variety of criteria, scramble the order of test items, and administer tests on-line, either over a network or via the Web. ESATEST 2000™ also includes full-function tradebook and graphing features. Available in Macintosh or Windows format.

- *Instructor's Manual*    The Instructor's Manual includes two distinct parts. The first part consists of the manual itself, which contains complete annotated solutions to all even-numbered problems and to every other odd-numbered problem, beginning with problem 3, in the problem sets. The second part of the manual consists of additional practice problems based upon approximately 50 video clips created by Lori Palmer of Utah Valley State College. These two- to three-minute video clips cover a wide variety of real-world situations and are housed on a CD-ROM that is packaged with the manual. These practice problems are suitable for either individual or group work.

### For the Student

- *Videotape Package*    Free to adopters, the videotape package consists of nine VHS videotapes, each correlating to a different chapter of the book. Each chapter tape is an hour to an hour and a half in length and is divided into lessons that correspond to each section of the chapter. In the videotapes I work selected problems from the text.

- *Core Concepts Video*    This single videotape is over four hours in length and contains more than 50 problem-solving sessions covering most sections of the text. Tailor-made as a take-home tutorial for students with a demanding schedule, this video can be used as a preview of what is to be covered in class, as an aid to completing homework assignments, or as a tool to review for a test.

- *Student Solutions Manual*    Like the Instructor's Manual, this is composed of two distinct parts. The first part is the manual itself, which contains complete annotated solutions to every other odd-numbered problem in the problem sets and to all chapter review and chapter test exercises. The second part of the manual consists of additional practice problems based upon approximately 50 video clips created by Lori Palmer of Utah Valley State College. These two- to three-minute video clips cover a wide variety of real-world situations and are housed on a CD-ROM that is packaged with the manual. These practice problems are suitable for either individual or group work.

- *MathCue Interactive Tutorial Software for Windows*    This unique computer software is an all-in-one learning tool. It contains Tutorial and Practice modes, enabling students to work any number of problems keyed to chapters of the McKeague text. The Solution Finder tool, an algorithm-based problem generator, provides step-by-step feedback.

- *MathCue Tutorial*  This computer software package of tutorials has problems that correspond to every section in the series. The software presents problems to solve and tutors students by displaying annotated, step-by-step solutions. Students may view partial solutions to get started on a problem, see a continuous record of their progress, and back up to review missed problems. Student scores can also be printed. Available in Macintosh format only.

- *MathCue Solution Finder*  This software allows students to enter their own problems into the computer and get annotated, step-by-step solutions in return. This unique program simulates working with a tutor, tracks student progress, refers students to specific sections in the text when appropriate, and prints student scores. Available in Macintosh format only.

- *MathCue Practice*  This algorithm-based software allows students to generate large numbers of practice problems keyed to problem types from each section of the book. Practice scores students' performance and saves students' scores session to session. Available in Macintosh format only.

- *Excursions in Real-World Mathematics, by Gina Shankland and Sara Williams of Mt. Hood Community College*  This exciting Web-based activity program will activate the minds of students as they solve problems that they may encounter in their everyday lives. Real data obtained from actual business, industry, and other academic disciplines serve as the raw material used in solving problems. There are more than 80 application situations to increase students' awareness of the power of mathematics; these correspond to topics in McKeague's *Elementary Algebra 6/e* and *Intermediate Algebra 6/e*. The applications are organized by level, length, and concept. Professors access an easy-to-use, menu-based system to order exactly the problems they wish to teach. Professors receive an Instructor's Manual with tips and solutions, and students receive the problems in the form of a workbook, which includes a guide to solving these applications and is available in the bookstore. Check out the activities at

*http://www.harcourtcollege.com*

and contact your sales representative for details and passwords.

---

Saunders College Publishing, a division of Harcourt College Publishers, may provide complimentary instructional aids and supplements or supplement packages to those adopters qualified under our adoption policy. Please contact your sales representative for more information. If as an adopter or potential user you receive supplements you do not need, please return them to your sales representative or send them to

Attn: Returns Department
Troy Warehouse
465 South Lincoln Drive
Troy, MO 63379

## ACKNOWLEDGMENTS

A project of this size cannot be completed without help from many people. Angus McDonald, my editor at Saunders College Publishing, has been very helpful and encouraging in all parts of the revision process. Julia Downs-Conover, the Senior Math Marketing Strategist at Saunders, has provided important feedback from various math departments across the country. Jay Campbell, my developmental editor, worked especially hard to get the book ready for production. Anne Scanlan-Rohrer and Robin Bonner handled the production part of this revision. They did an outstanding job, and I am very lucky to have these two on our team. Dudley Brooks, Patrick Dawson, Laura Cornell, Alexia Nielsen, and Warren Hawley suggested many new applications and exercises for the problem sets, and these suggestions have strengthened this revision. My son, Patrick, assisted me with this revision, and his influence has made this a better book. Ann Ostberg and Nancy Matthews did an excellent job of accuracy checking and proofreading. My thanks to these people; this book would not have been possible without them.

Thanks also to Diane McKeague, for her encouragement with all my writing endeavors, and to Amy Jacobs, for all her advice and encouragement.

Finally, I am grateful to the following instructors for their suggestions and comments on this revision. Some reviewed the entire manuscript, while others were asked to evaluate the development of specific topics or the overall sequence of topics. My thanks go to the people listed below:

Susan F. Akers, Northeast State Technical Community College
Gerald Allen, Angelo State University
Carol Atnip, University of Louisville
Jon Becker, Indiana University Northwest
Sandra Beken, Horry-Georgetown Technical College
Jacqueline "Monty" Briley, Guilford Technical Community College
Beverly Broomell, Suffolk Community College
Karen Sue Cain, Eastern Kentucky University
Sally Copeland, Johnson County Community College
Jim Corbett, Alvin Community College
Jorge Cossio, Miami-Dade Community College
Barry Cuell, Utah Valley Community College
Margaret F. Donaldson, East Tennessee State University
Elizabeth Farber, Bucks County Community College
Dorothy Gotway, University of Missouri — St. Louis
Carol Hay, Northern Essex Community College
Betsy Huttenlock, Penn State University — Abington
Mike Iannone, College of New Jersey
Jamie King-Blair, Orange Coast College
Harriet Kiser, Floyd College
Jillian M. Knowles, Northern Essex Community College
Molly Krajewski, Daytona Beach Community College
William Livingston, Missouri Southern State College
Gene W. Majors, Fullerton College
Mary Lou McDowell, Widener University

Judy Meckley, Joliet Junior College
Beverly Michael, University of Pittsburgh
Ellen Milosheff, Triton College
Peter Moore, Northern Kentucky State University
Linda J. Murphy, Northern Essex Community College
Nancy Nickerson, Northern Essex Community College
Gene Reid, Central Piedmont Community College
Dennis Riessig, Suffolk Community College
Gladys Rockind, Oakland Community College
Geoffrey Schulz, Community College of Philadelphia
Mark Serebransky, Camden County College
Barbara Jane Sparks, Camden County College
Radu Teodorescu, Western Michigan State University
Shirley M. Thompson, North Lake College
Kathryn C. Wetzel, Amarillo College
Claude Williams, Central Piedmont Community College
Pete Witt, Glendale Community College
Robert Wynegar, University of Tennessee at Chattanooga

**Charles P. McKeague**
**December 1999**

# PREFACE TO THE STUDENT

I often find my students asking themselves the question "Why can't I understand this stuff the first time?" The answer is "You're not expected to." Learning a topic in algebra isn't always accomplished the first time around. There are many instances when you will find yourself reading over new material a number of times before you can begin to work problems. That's just the way things are in algebra. If you don't understand a topic the first time you see it, that doesn't mean there is something wrong with you. Understanding algebra takes time. The process of understanding requires reading the book, studying the examples, working problems, and getting your questions answered.

Here are some questions that are often asked by students starting a beginning algebra class.

*How much math do I need to know before taking algebra?* You should be able to do the four basic operations (addition, subtraction, multiplication, and division) with whole numbers, fractions, and decimals. Most important is your ability to work with whole numbers. If you are a bit weak at working with fractions because you haven't worked with them in a while, don't be too concerned; we will review fractions as we progress through the book. I have had students who eventually did very well in algebra, even though they were initially unsure of themselves when working with fractions.

*What is the best way to study?* The best way to study is to study consistently. You must work problems every day. A number of my students spend an hour or so in the morning working problems and reading over new material and then spend another hour in the evening working problems. The students of mine who are the most successful in algebra are the ones who find a schedule that works for them and then stick to it. They work problems every day.

*If I understand everything that goes on in class, can I take it easy on my homework?* Not necessarily. There is a big difference between understanding a problem someone else is working and working the same problem yourself. There is no substitute for working problems yourself. The concepts and properties are understandable to you only if you yourself work problems involving them.

## HOW TO BE SUCCESSFUL IN ALGEBRA

If you have decided to be successful in algebra, then the following list will be important to you:

1. **If you are in a lecture class, be sure to attend all class sessions on time.** You cannot know exactly what goes on in class unless you are there. Missing class and then expecting to find out what went on from someone else is not the same as being there yourself.

2. **Read the book.**  It is best to read the section that will be covered in class beforehand. Reading in advance, even if you do not understand everything you read, is still better than going to class with no idea of what will be discussed.

3. **Work problems every day, and check your answers.**  The key to success in mathematics is working problems. The more problems you work, the better you will become at working them. The answers to the odd-numbered problems are given in the back of the book. When you have finished an assignment, be sure to compare your answers with those in the book. If you have made a mistake, find out what it is, and correct it.

4. **Do it on your own.**  Don't be misled into thinking someone else's work is your own. Having someone else show you how to work a problem is not the same as working the same problem yourself. It is okay to get help when you are stuck. As a matter of fact, it is a good idea. Just be sure you do the work yourself.

5. **Review every day.**  After you have finished the problems your instructor has assigned, take another fifteen minutes and review a section you have already completed. The more you review, the longer you will retain the material you have learned.

6. **Don't expect to understand every new topic the first time you see it.**  Sometimes you will understand everything you are doing, and sometimes you won't. That's just the way things are in mathematics. Expecting to understand each new topic the first time you see it can lead to disappointment and frustration. The process of understanding algebra takes time. It requires that you read the book, work problems, and get your questions answered.

7. **Spend as much time as it takes for you to master the material.**  No set formula exists for the exact amount of time you need to spend on algebra to master it. You will find out as you go along what is or isn't enough time for you. If you end up spending two or more hours on each section in order to master the material there, then that's how much time it takes; trying to get by with less will not work.

8. **Relax.**  It's probably not as difficult as you think.

**Charles P. McKeague**
**December 1999**

# FEATURES OF THE NEW EDITION

## New Design and Art Program

The newly revamped look of this edition is more open and inviting. Featuring numerous new mathematical and situational figures, the enhanced art program will assist students in visualizing the concepts.

## Linear Equations and Inequalities

*(Index Stock Imagery)*

### INTRODUCTION

Just before starting work on this edition of your text, I flew to Europe for vacation. From time to time the television screens on the plane displayed statistics about the flight. At one point during the flight the temperature outside the plane was $-60°F$. When I returned home, I did some research and found that the relationship between temperature $T$ and altitude $A$ can be described with the formula

$$T = -0.0035A + 70$$

when the temperature on the ground is 70°F. Table 1 and line graph below also describe this relationship.

| Temperature Altitude | |
|---|---|
| Temperature (°F) | |
| 70 | |
| 35 | |
| 0 | |
| −35 | |
| −70 | |

... will start our work with formulas, and you will see how we use ... tables and line graphs like the ones above.

---

Section 1.4   Subtraction of Real Numbers

31

| Professions | Salaries |
|---|---|
| Engineering | |
| Computer Science | $40,920 |
| | $40,523 |
| | $36,036 |
| Business Admin. | |
| Accounting | $33,702 |
| Sales/Marketing | |
| Teaching | $25,735 |

**Starting Salaries in 1998**

(b) Find the total digital camera sales from 1997 through 1999.

| Year | Sales ($ millions) |
|---|---|
| 1996 | |
| 1997 | |
| 1998 | |
| | 819 |

**Digital Camera Sales**

**78. Digital Camera Sales** The following bar chart shows the sales of digital cameras from 1996–1999.
(a) Use the information in the bar chart to fill in the missing entries in the table.

**1600 × 1200 Resolution Digital Camera**
**Your Price: $599⁹⁹**

---

### 1.4 Subtraction of Real Numbers

Suppose that the temperature at noon is 20° Fahrenheit and 12 hours later, at midnight, it has dropped to −15° Fahrenheit. What is the difference between the temperature at noon and the temperature at midnight? Intuitively, we know the difference in the two temperatures is 35°. We also know that the word *difference* indicates subtraction. The difference between 20 and −15 is written

$$20 - (-15)$$

## Increased Visualization of Topics

This edition contains many more diagrams, charts, and graphs than did the previous edition. The purpose of this is to give students additional information, in visual form, to help them understand the topics covered.

**87. Reading Tables and Charts** The following bar chart gives the record low temperatures for each month of the year for Lake Placid, New York. The accompanying table should display the same information. Use the bar chart to complete the table.

Record Low Temperatures for Lake Placid, New York

Daily Lows for Salt Lake City

| Record Low Temperatures in Lake Placid, New York | |
|---|---|
| **Month** | **Temperature (°F)** |
| January | −36 |
| February | |
| March | |
| April | |
| May | |
| June | |
| July | |
| August | |
| September | |
| October | |
| November | |
| December | |

| Daily Low Temperatures in Salt Lake City | |
|---|---|
| **Day** | **Temperature (°F)** |
| Monday | |
| Tuesday | |
| Wednesday | |
| Thursday | |
| Friday | |
| Saturday | |
| Sunday | |

**89. Scuba Diving** Steve is scuba diving near his home in Maui. At one point he is 100 feet below the surface. Represent this number with a negative number. If he descends another 5 feet, what negative number will represent his new position?

**90. Checking Account Balance** Pete and Heidi are married and have a joint checking account that allows them to withdraw money from the account at an automatic teller machine. With each withdrawal from the automatic teller, the bank charges an extra $2 to their account. Suppose the balance in the checkbook is $75. If Heidi writes a check for $55 and Pete withdraws $20 from the automatic teller machine, what will their new balance be?

**91. Geometry** Find the area and perimeter of an $8\frac{1}{2}$-by-11-inch piece of notebook paper.

**88. Reading Tables and Charts** The following bar chart gives the low temperature for each day of one week in Salt Lake City, Utah. The accompanying

FIGURE 2          FIGURE 3

**7.** Here is the sequence of odd numbers written so that the position of each term is noted:

Position    1, 2, 3, 4, 5, . . .
Value       1, 3, 5, 7, 9, . . .

**EXAMPLE 1**   Tables 2 and 3 give the first five terms of the sequence of odd numbers and the sequence of squares as paired data. In each case construct a scatter diagram.

| TABLE 2  Odd Numbers | |
|---|---|
| **Position** | **Value** |
| 1 | 1 |
| 2 | 3 |
| 3 | 5 |
| 4 | 7 |
| 5 | 9 |

| TABLE 3  Squares | |
|---|---|
| **Position** | **Value** |
| 1 | 1 |
| 2 | 4 |
| 3 | 9 |
| 4 | 16 |
| 5 | 25 |

**Solution**   The two scatter diagrams are based on the data from Tables 2 and 3 shown here. Notice how the dots in Figure 4 seem to line up in a straight line, while

FIGURE 4          FIGURE 5

## Tables, Bar Charts, Scatter Diagrams, and Line Graphs

Beginning in Chapter 1 and then continuing in Chapters 3 and 4, students are required to analyze information from tables. In addition to simply reading tables, they practice converting data in tabular form to data in graphical form that includes histograms and scatter diagrams. From there they move on to graphing ordered pairs and linear equations on a rectangular coordinate system.

## Linear Equations and Inequalities

*(Index Stock Imagery)*

**INTRODUCTION**

Just before starting work on this edition of your text, I flew to Europe for vacation. From time to time the television screens on the plane displayed statistics about the flight. At one point during the flight the temperature outside the plane was $-60°F$. When I returned home, I did some research and found that the relationship between temperature $T$ and altitude $A$ can be described with the formula

$$T = -0.0035A + 70$$

when the temperature on the ground is $70°F$. Table 1 and line graph below also describe this relationship.

| TABLE 1  Air Temperature and Altitude | |
|---|---|
| **Altitude (feet)** | **Temperature (°F)** |
| 0 | 70 |
| 10,000 | 35 |
| 20,000 | 0 |
| 30,000 | $-35$ |
| 40,000 | $-70$ |

In this chapter we will start our work with formulas, and you will see how we use formulas to produce tables and line graphs like the ones above.

**Chapter Openers**  Each chapter begins with an introduction in which a real-world application is used to stimulate interest in the chapter. These opening applications are expanded on later in the chapter. Most of them are explained using the rule of four: verbally, numerically, graphically, and algebraically.

**STUDY SKILLS**

If you have successfully completed Chapter 1, then you have made a good start at developing the study skills necessary to succeed in all math classes. Here is the list of study skills for this chapter. Some are a continuation of the skills from Chapter 1, while others are new to this chapter.

1. **Continue to Set and Keep a Schedule**  Sometimes I find students do well in Chapter 1 and then become overconfident. They begin to put in less time with their homework. Don't do it. Keep to the same schedule.

2. **Increase Effectiveness**  You want to become more and more effective with the time you spend on your homework. You want to increase the amount of learning you obtain in the time you have set aside. Increase those activities that you feel are the most beneficial, and decrease those that have not given you the results you want.

3. **List Difficult Problems**  Begin to make lists of problems that give you the most difficulty. These are the problems in which you are repeatedly making mistakes.

4. **Begin to Develop Confidence With Word Problems**  It seems that the major difference between those people who are good at working word problems and those who are not is confidence. The people with confidence know that no matter how long it takes them, they will eventually be able to solve the problem. Those without confidence begin by saying to themselves, "I'll never be able to work this problem." Are you like that? If you are, what you need to do is put your old ideas about you and word problems aside for a while and make a decision to be successful. Sometimes that's all it takes. Instead of telling yourself that you can't do word problems, that you don't like them, or that they're not good for anything anyway, decide to do whatever it takes to master them.

Many of my students keep a notebook that contains everything that they need for the course: class notes, homework, quizzes, tests, and research projects. A three-ring binder with tabs is ideal. Organize your notebook so that you can easily get to any item you wish to look at.

**Study Skills**  Immediately following the chapter opener in Chapters 1 through 6 are a list of study skills intended to help students become organized and efficient with their time. The study skills point students in the direction of success.

**Facts From Geometry** Many of the important facts from geometry are listed under this heading. In most cases, an example or two accompanies each of the facts to give students a chance to see how topics from geometry are related to the algebra they are learning.

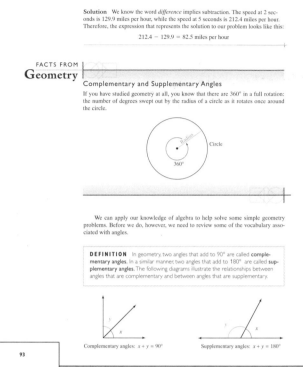

**Solution** We know the word *difference* implies subtraction. The speed at 2 seconds is 129.9 miles per hour, while the speed at 5 seconds is 212.4 miles per hour. Therefore, the expression that represents the solution to our problem looks like this:

$$212.4 - 129.9 = 82.5 \text{ miles per hour}$$

FACTS FROM
**Geometry**

**Complementary and Supplementary Angles**

If you have studied geometry at all, you know that there are 360° in a full rotation: the number of degrees swept out by the radius of a circle as it rotates once around the circle.

We can apply our knowledge of algebra to help solve some simple geometry problems. Before we do, however, we need to review some of the vocabulary associated with angles.

**DEFINITION** In geometry, two angles that add to 90° are called **complementary angles**. In a similar manner, two angles that add to 180° are called **supplementary angles**. The following diagrams illustrate the relationships between angles that are complementary and between angles that are supplementary.

Complementary angles: $x + y = 90°$      Supplementary angles: $x + y = 180°$

**Note:** Again, we place a question mark over the equal sign because we don't know yet whether the expressions on the left and right side of the equal sign will be equal.

So far in this section we have used the addition property of equality to add only numbers to both sides of an equation. It is often necessary to add a term involving a variable to both sides of an equation, as the following example indicates.

**EXAMPLE 9** Solve $3x - 5 = 2x + 7$.

**Solution** We can solve this equation in two steps. First, we add $-2x$ to both sides of the equation. When this has been done, $x$ appears on the left side only. Second, we add 5 to both sides:

$$3x + (-2x) - 5 = 2x + (-2x) + 7 \qquad \text{Add } -2x \text{ to both sides.}$$
$$x - 5 = 7 \qquad \text{Simplify each side.}$$
$$x - 5 + 5 = 7 + 5 \qquad \text{Add } 5 \text{ to both sides.}$$
$$x = 12 \qquad \text{Simplify each side.}$$

**A NOTE ON SUBTRACTION**

Although the addition property of equality is stated for addition only, we can subtract the same number from both sides of an equation as well. Because subtraction is defined as addition of the opposite, subtracting the same quantity from both sides of an equation will not change the solution. If we were to solve the equation in Example 7 using subtraction instead of addition, the steps would look like this:

$$2 - x + 2 = 12 - 2 \qquad \text{Original equation}$$
$$x + 2 - 2 = 12 - 2 \qquad \text{Subtract } 2 \text{ from each side.}$$
$$2 + 2 - x = 10 - 2 \qquad \text{Subtraction}$$

In my experience teaching algebra, I find that students make fewer mistakes if they think in terms of addition rather than subtraction. So, you are probably better off if you continue to use the addition property just the way we have used it in the examples in this section. But, if you are curious as to whether you can subtract the same number from both sides of an equation, the answer is yes.

**Getting Ready for Class**

*After reading through the preceding section, respond in your own words and in complete sentences.*

A. What is a solution to an equation?
B. What are equivalent equations?
C. Explain in words the addition property of equality.
D. How do you check a solution to an equation?

**Getting Ready for Class** Just before each problem set is a list of four problems under the heading *Getting Ready for Class*. These problems require written responses from students and can be answered by reading the preceding section. They are to be done before students go to class.

## PROBLEM SET 2.1

Simplify the following expressions.

1. $3x - 6x$
2. $7x - 5x$
3. $-2a + a$
4. $3a - a$
5. $7x + 3x + 2x$
6. $8x - 2x - x$
7. $3a - 2a + 5a$
8. $7a - a + 2a$
9. $4x - 3 + 2x$
10. $5x + 6 - 3x$
11. $3a + 4a + 5$
12. $6a + 7a + 8$
13. $2x - 3 + 3x - 2$
14. $6x + 5 - 2x + 3$
15. $3a - 1 + a + 3$
16. $-a + 2 + 8a - 7$
17. $-4x + 8 - 5x - 10$
18. $-9x - 1 + x - 4$
19. $7a + 3 + 2a + 3a$
20. $8a - 2 + a + 5a$
21. $5(2x - 1) + 4$
22. $2(4x - 3) + 2$
23. $7(3y + 2) - 8$
24. $6(4y + 2) - 7$
25. $-3(2x - 1) + 5$
26. $-4(3x - 2) - 6$
27. $5 - 2(a + 1)$
28. $7 - 8(2a + 3)$

Simplify the following expressions.

29. $6 - 4(x - 5)$
30. $12 - 3(4x - 2)$
31. $-9 - 4(2 - y) + 1$
32. $-10 - 3(2 - y) + 3$
33. $-6 + 2(2 - 3x) + 1$
34. $-7 - 4(3 - x) + 1$
35. $(4x - 7) - (2x + 5)$
36. $(7x - 3) - (4x + 2)$
37. $8(2a + 4) - (6a - 1)$
38. $9(3a + 5) - (8a - 7)$
39. $3(x - 2) + (x - 3)$
40. $2(2x + 1) - (x + 4)$
41. $4(2y - 8) - (y + 7)$
42. $5(y - 3) - (y - 4)$
43. $-9(2x + 1) - (x + 5)$
44. $-3(3x - 2) - (2x + 3)$

Evaluate the following expressions when $x$ is 2. (Find the value of the expressions if $x$ is 2.)

45. $3x - 1$
46. $4x + 3$
47. $-2x - 5$
48. $-3x + 6$

49. $x^2 - 8x + 16$
50. $x^2 - 10x + 25$
51. $(x - 4)^2$
52. $(x - 5)^2$

Find the value of the following expressions when $x$ is $-5$. Then simplify the expression, and check to see that it has the same value for $x = -5$.

53. $7x - 4 - x - 3$
54. $3x + 4 + 7x - 6$
55. $5(2x + 1) + 4$
56. $2(3x - 10) + 5$

Find the value of each expression when $x$ is $-3$ and $y$ is 5.

57. $x^2 - 2xy + y^2$
58. $x^2 + 2xy + y^2$
59. $(x - y)^2$
60. $(x + y)^2$
61. $x^2 + 6xy + 9y^2$
62. $x^2 + 10xy + 25y^2$
63. $(x + 3y)^2$
64. $(x + 5y)^2$

Find the value of $12x - 3$ for each of the following values of $x$.

65. $\dfrac{1}{2}$
66. $\dfrac{1}{3}$
67. $\dfrac{1}{4}$
68. $\dfrac{1}{6}$
69. $\dfrac{3}{2}$
70. $\dfrac{2}{3}$
71. $\dfrac{3}{4}$
72. $\dfrac{5}{6}$

Substitute 1, 2, 3, and 4 for $n$ in each of the following expressions.

73. $2n + 3$
74. $5n + 2$
75. $n^2 + 1$
76. $(n + 1)^2$
77. Fill in the tables below to find the sequences formed by substituting the first four counting numbers into the expressions $3n$ and $n^3$.

(a)

| $n$ | 1 | 2 | 3 | 4 |
|-----|---|---|---|---|
| $3n$ | | | | |

(b)

| $n$ | 1 | 2 | 3 | 4 |
|-----|---|---|---|---|
| $n^3$ | | | | |

---

**Drill Problems** There are enough problems in each set to ensure student proficiency in the material. The problems are progressive in level of difficulty, and each pair of consecutive problems is similar.

---

## Applying the Concepts Problems

These problems are designed to show students how the algebra they are learning can be applied to real-life situations. More than 200 new application problems have been added, and, in keeping with AMATYC standards, many emphasize reading and interpreting graphical and tabular data.

---

55. $(4 + 5)^2$
56. $(6 + 3)^2$
57. $4^2 + 5^2$
58. $6^2 + 3^2$
59. $3 \cdot 10^2 + 4 \cdot 10 + 5$
60. $6 \cdot 10^2 + 5 \cdot 10 + 4$
61. $2 \cdot 10^3 + 3 \cdot 10^2 + 4 \cdot 10 + 5$
62. $5 \cdot 10^3 + 6 \cdot 10^2 + 7 \cdot 10 + 8$
63. $10 - 2(4 \cdot 5 - 16)$
64. $15 - 5(3 \cdot 2 - 4)$
65. $4[7 + 3(2 \cdot 9 - 8)]$
66. $5[10 + 2(3 \cdot 6 - 10)]$
67. $5(7 - 3) + 8(6 - 4)$
68. $3(10 - 4) + 6(12 - 10)$
69. $3(4 \cdot 5 - 12) + 6(7 \cdot 6 - 40)$
70. $6(8 \cdot 3 - 4) + 5(7 \cdot 3 - 1)$
71. $3^4 + 4^2 \div 2^3 - 5^2$
72. $2^5 + 6^2 \div 2^2 - 3^2$
73. $5^2 + 3^4 \div 9^2 + 6^2$
74. $6^2 + 2^5 \div 4^2 + 7^2$

Find the next number in each sequence.

75. 1, 2, 3, 4, . . . . (The sequence of counting numbers)
76. 0, 1, 2, 3, . . . (The sequence of whole numbers)
77. 2, 4, 6, 8, . . . (The sequence of even numbers)
78. 1, 3, 5, 7, . . . (The sequence of odd numbers)
79. 1, 4, 9, 16, . . . (The sequence of squares)
80. 1, 8, 27, 64, . . . (The sequence of cubes)
81. 2, 2, 4, 6, . . . (A Fibonacci-like sequence)
82. 5, 5, 10, 15, . . . (A Fibonacci-like sequence)

### Applying the Concepts

83. **Reading Tables and Charts** The following table and bar chart give the amount of caffeine in five different soft drinks. How much caffeine is in each of the following?
(a) A 6-pack of Jolt
(b) 2 Coca-Colas plus 3 Tabs

| Caffeine Content in Soft Drinks | |
|---|---|
| Drink | Caffeine (milligrams) |
| Jolt | 100 |
| Tab | 47 |
| Coca-Cola | 45 |
| Diet Pepsi | 36 |
| 7UP | 0 |

Caffeine (mg)

84. **Reading Tables and Charts** The following table and bar chart give the amount of caffeine in five different nonprescription drugs. How much caffeine is in each of the following?
(a) A box of 12 Excedrin
(b) 1 Dexatrim plus 4 Excedrin

| Caffeine Content in Nonprescription Drugs | |
|---|---|
| Nonprescription Drug | Caffeine (milligrams) |
| Dexatrim | 200 |
| NoDoz | 100 |
| Excedrin | 65 |
| Triaminicin tablets | 30 |
| Dristan tablets | 16 |

Caffeine (mg)

85. **Reading Tables and Charts** The following bar chart gives the number of calories burned by a 150-

**Chapter Summary** These list all main points from the chapter. In the margin, next to each topic, is an example that illustrates the type of problem associated with the topic being reviewed.

C H A P T E R **2** **SUMMARY**

**Examples**

**1.** The terms $2x$, $5x$, and $-7x$ are all similar since their variable parts are the same.

**2.** Simplify $3x + 4x$.

$$3x + 4x = (3 + 4)x$$
$$= 7x$$

**3.** The solution set for the equation $x + 2 = 5$ is $\{3\}$ because when $x$ is 3 the equation is $3 + 2 = 5$, or $5 = 5$.

**4.** The equations $a - 4 = 3$ and $a - 2 = 5$ are equivalent since both have solution set $\{7\}$.

**5.** Solve $x - 5 = 12$.

$$x - 5 + 5 = 12 + 5$$
$$x + 0 = 17$$
$$x = 17$$

**6.** Solve $3x = 18$.

$$\frac{1}{3}(3x) = \frac{1}{3}(18)$$
$$x = 6$$

**7.** Solve $2(x + 3) = 10$.

$$2x + 6 = 10$$
$$2x + 6 + (-6) = 10 + (-6)$$
$$2x = 4$$
$$\frac{1}{2}(2x) = \frac{1}{2}(4)$$
$$x = 2$$

**Similar Terms [2.1]**
A *term* is a number or a number and one or more variables multiplied together. *Similar terms* are terms with the same variable part.

**Simplifying Expressions [2.1]**
In this chapter we simplified expressions that contained variables by using the distributive property to combine similar terms.

**Solution Set [2.2]**
The *solution set* for an equation (or inequality) is all the numbers that, when used in place of the variable, make the equation (or inequality) a true statement.

**Equivalent Equations [2.2]**
Two equations are called *equivalent* if they have the same solution set.

**Addition Property of Equality [2.2]**
When the same quantity is added to both sides of an equation, the solution set for the equation is unchanged. Adding the same amount to both sides of an equation produces an equivalent equation.

**Multiplication Property of Equality [2.3]**
If both sides of an equation are multiplied by the same nonzero number, the solution set is unchanged. Multiplying both sides of an equation by a nonzero quantity produces an equivalent equation.

**Strategy for Solving Linear Equations in One Variable [2.4]**

*Step 1a:* Use the distributive property to separate terms, if necessary.

*1b:* If fractions are present, consider multiplying both sides by the LCD to eliminate the fractions. If decimals are present, consider multiplying both sides by a power of 10 to clear the equation of decimals.

*1c:* Combine similar terms on each side of the equation.

*Step 2:* Use the addition property of equality to get all variable terms on one side of the equation and all constant terms on the other side. A variable term is a term that contains the variable (for example, $5x$). A constant term is a term that

does not contain the variable (the number 3, for example).

*Step 3:* Use the multiplication property of equality to get $x$ (that is, $1x$) by itself on one side of the equation.

*Step 4:* Check your solution in the original equation to be sure that you have not made a mistake in the solution process.

**8.** Solving $P = 2l + 2w$ for $l$, we have

$$P - 2w = 2l$$
$$\frac{P - 2w}{2} = l$$

**Formulas [2.5]**
A *formula* is an equation with more than one variable. To solve a formula for one of its variables, we use the addition and multiplication properties of equality to move everything except the variable in question to one side of the equal sign so that the variable in question is alone on the other side.

**Blueprint for Problem Solving [2.6, 2.7]**

*Step 1: Read* the problem, and then mentally *list* the items that are known and the items that are unknown.

*Step 2: Assign a variable* to one of the unknown items. (In most cases this will amount to letting $x$ = the item that is asked for in the problem.) Then *translate* the other *information* in the problem to expressions involving the variable.

*Step 3: Reread* the problem, and then *write an equation*, using the items and variables listed in steps 1 and 2, that describes the situation.

*Step 4: Solve the equation* found in step 3.

*Step 5: Write* your *answer* using a complete sentence.

*Step 6: Reread* the problem, and *check* your solution with the original words in the problem.

**9.** Solve $x + 5 < 7$.

$$x + 5 + (-5) < 7 + (-5)$$
$$x < 2$$

**Addition Property for Inequalities [2.8]**
Adding the same quantity to both sides of an inequality produces an equivalent inequality, one with the same solution set.

**10.** Solve $-3a \le 18$.

$$-\frac{1}{3}(-3a) \ge -\frac{1}{3}(18)$$
$$a \ge -6$$

**Multiplication Property for Inequalities [2.8]**
Multiplying both sides of an inequality by a positive number never changes the solution set. If both sides are multiplied by a negative number, the sense of the inequality must be reversed to produce an equivalent inequality.

**11.** Solve $3(x - 4) \ge -2$.

**Strategy for Solving Linear Inequalities in One Variable [2.8]**

*Step 1a:* Use the distributive property to separate terms, if necessary.

**Blueprint for Problem Solving**
Included in the *Chapter Summary*, the *Blueprint for Problem Solving* is a detailed outline of the steps needed to successfully attempt application problems. Intended as a guide to problem solving in general, the *Blueprint* overlays the solution process to all the application problems in the first few chapters of the book. As students become more familiar with problem solving, the steps in the *Blueprint* are streamlined.

$$3x - 12 \geq -2$$
$$3x - 12 + 12 \geq -2 + 12$$
$$3x \geq 10$$
$$\frac{1}{3}(3x) \geq \frac{1}{3}(10)$$
$$x \geq \frac{10}{3}$$

**12.** $x < -3$ or $x > 1$

$-2 \leq x \leq 3$

*1b:* If fractions are present, consider multiplying both sides by the LCD to eliminate the fractions. If decimals are present, consider multiplying both sides by a power of 10 to clear the inequality of decimals.

*1c:* Combine similar terms on each side of the inequality.

**Step 2:** Use the addition property for inequalities to get all variable terms on one side of the inequality and all constant terms on the other side.

**Step 3:** Use the multiplication property for inequalities to get $x$ by itself on one side of the inequality.

**Step 4:** Graph the solution set.

**Compound Inequalities [2.9]**
Two inequalities connected by the word *and* or *or* form a compound inequality. If the connecting word is *or*, we graph all points that are on either graph. If the connecting word is *and*, we graph only those points that are common to both graphs. The inequality $-2 \leq x \leq 3$ is equivalent to the compound inequality $-2 \leq x$ and $x \leq 3$.

### COMMON MISTAKES

1. Trying to subtract away coefficients (the number in front of variables) when solving equations. For example:

$$4x = 12$$
$$4x - 4 = 12 - 4$$
$$x = 8 \leftarrow \text{Mistake}$$

It is not incorrect to add $(-4)$ to both sides; it's just that $4x - 4$ is not equal to $x$. Both sides should be multiplied by $\frac{1}{4}$ to solve for $x$.

2. Forgetting to reverse the direction of the inequality symbol when multiplying both sides of an inequality by a negative number. For instance:

$$-3x < 12$$
$$-\frac{1}{3}(-3x) < -\frac{1}{3}(12) \leftarrow \text{Mistake}$$
$$x < -4$$

It is not incorrect to multiply both sides by $-\frac{1}{3}$. But if we do, we must also reverse the sense of the inequality.

**Common Mistakes**  Located at the end of each *Chapter Summary*, this section alerts students to problems they are likely to encounter and provides helpful hints for avoiding these pitfalls.

---

**Chapter Review**  These exercises are representative of all the different problem types within a given chapter. The *Chapter Reviews* are longer and more extensive than the *Chapter Tests*.

---

### ⊕ CHAPTER 2    REVIEW

The numbers in brackets refer to the sections of the text in which similar problems can be found.

Simplify each expression as much as possible. [2.1]

**1.** $5x - 8x$      **2.** $6x - 3 - 8x$

**3.** $-a + 2 + 5a - 9$

**4.** $5(2a - 1) - 4(3a - 2)$

**5.** $6 - 2(3y + 1) - 4$    **6.** $4 - 2(3x - 1) - 5$

Find the value of each expression when $x$ is 3. [2.1]

**7.** $7x - 2$      **8.** $-4x - 5 + 2x$

**9.** $-x - 2x - 3x$

Find the value of each expression when $x$ is $-2$. [2.1]

**10.** $5x - 3$    **11.** $-3x + 2$    **12.** $7 - x - 3$

Solve each equation. [2.2, 2.3]

**13.** $x + 2 = -6$    **14.** $x - \frac{1}{2} = \frac{4}{7}$

**15.** $10 - 3y + 4y = 12$

**16.** $-3 - 4 = -y - 2 + 2y$

**17.** $2x = -10$    **18.** $3x = 0$

**19.** $\frac{x}{3} = 4$    **20.** $-\frac{x}{4} = 2$

**21.** $3a - 2 = 5a$    **22.** $\frac{7}{10}a = \frac{1}{5}a + \frac{1}{2}$

**23.** $3x + 2 = 5x - 8$    **24.** $6x - 3 = 2x + 7$

**25.** $0.7x - 0.1 = 0.5x - 0.1$

**26.** $0.2x - 0.3 = 0.8x - 0.3$

Solve each equation. Be sure to simplify each side first. [2.4]

**27.** $2(x - 5) = 10$    **28.** $12 = 2(5x - 4)$

**29.** $\frac{1}{2}(3t - 2) + \frac{1}{2} = \frac{5}{2}$

**30.** $\frac{3}{5}(5x - 10) = \frac{2}{3}(9x + 3)$

**31.** $2(3x + 7) = 4(5x - 1) + 18$

**32.** $7 - 3(y + 4) = 10$

Use the formula $4x - 5y = 20$ to find $y$ if [2.5]

**33.** $x$ is 5      **34.** $x$ is 0

**35.** $x$ is $-5$      **36.** $x$ is 10

Solve each of the following formulas for the indicated variable. [2.5]

**37.** $2x - 5y = 10$ for $y$    **38.** $5x - 2y = 10$ for $y$

**39.** $V = \pi r^2 h$ for $h$    **40.** $P = 2l + 2w$ for $w$

**41.** What number is 86% of 240? [2.5]

**42.** What percent of 2,000 is 180? [2.5]

Solve each of the following word problems. In each case, be sure to show the equation that describes the situation. [2.6, 2.7]

**43. Number Problem**  The sum of twice a number and 6 is 28. Find the number.

**44. Geometry**  The length of a rectangle is 5 times as long as the width. If the perimeter is 60 meters, find the length and the width.

**45. Investing**  A man invests a certain amount of money in an account that pays 9% annual interest. He invests $300 more than that in an account that pays 10% annual interest. If his total interest after a year is $125, how much does he have invested in each account?

**46. Coin Problem**  A collection of 15 coins is worth $1.00. If the coins are dimes and nickels, how many of each coin are there?

Solve each inequality. [2.8]

**47.** $-2x < 4$    **48.** $-5x > -10$

**49.** $-\frac{a}{2} \leq -3$    **50.** $-\frac{a}{3} > 5$

Solve each inequality, and graph the solution. [2.8]

**51.** $-4x + 5 > 37$    **52.** $2x + 10 < 5x - 11$

**53.** $2(3t + 1) + 6 \geq 5(2t + 4)$

Graph the solution to each of the following compound inequalities. [2.9]

**54.** $x < -2$ or $x > 5$

**55.** $-5x \geq 25$ or $2x - 3 \geq 9$

**56.** $-1 < 2x + 1 < 9$

**Chapter Test** This set of problems tests the student's understanding of all key points in the chapter.

C H A P T E R   2   **TEST**

Simplify each of the following expressions. [2.1]

1. $3x + 2 - 7x + 3$    2. $4a - 5 - a + 1$
3. $7 - 3(y + 5) - 4$
4. $8(2x + 1) - 5(x - 4)$
5. Find the value of $2x - 3 - 7x$ when $x = -5$. [2.1]
6. Find the value of $x^2 + 2xy + y^2$ when $x = 2$ and $y = 3$. [2.1]

Solve the following equations. [2.2, 2.3, 2.4]

7. $2x - 5 = 7$    8. $2y + 4 = 5y$
9. $\frac{1}{2}x - \frac{1}{10} = \frac{1}{5}x + \frac{1}{2}$
10. $\frac{2}{5}(5x - 10) = -5$
11. $-5(2x + 1) - 6 = 19$
12. $0.04x + 0.06(100 - x) = 4.6$
13. $2(t - 4) + 3(t + 5) = 2t - 2$
14. $2x - 4(5x + 1) = 3x + 17$
15. What number is 15% of 38? [2.5]
16. 240 is 12% of what number? [2.5]
17. If $2x - 3y = 12$, find $x$ when $y = -2$. [2.5]
18. The formula for the volume of a cone is $V = \frac{1}{3}\pi r^2 h$. Find $h$ if $V = 88$ cubic inches, $\pi = \frac{22}{7}$, and $r = 3$ inches. [2.5]

19. Solve $2x + 5y = 20$ for $y$. [2.5]
20. Solve $h = x + vt + 16t^2$ for $v$. [2.5]

Solve each word problem. [2.6, 2.7]

21. **Age Problem** Dave is twice as old as Rick. Ten years ago the sum of their ages was 40. How old are they now?
22. **Geometry** A rectangle is twice as long as it is wide. The perimeter is 60 inches. What are the length and width?
23. **Coin Problem** A man has a collection of dimes and quarters with a total value of $3.50. If he has 7 more dimes than quarters, how many of each coin does he have?
24. **Investing** A woman has money in two accounts. One account pays 7% annual interest, while the other pays 9% annual interest. If she has $600 more invested at 9% than she does at 7% and her total interest for a year is $182, how much does she have in each account?

Solve each inequality, and graph the solution. [2.8]

25. $2x + 3 < 5$    26. $-5a > 20$
27. $0.4 - 0.2x \geq 1$    28. $4 - 5(m + 1) \leq 9$

Solve each inequality, and graph the solution. [2.9]

29. $3 - 4x \geq -5$ or $2x \geq 10$
30. $-7 < 2x - 1 < 9$

C H A P T E R S   1 - 2   **CUMULATIVE REVIEW**

Simplify.

1. $6 + 3(6 + 2)$    2. $-11 + 17 + (-13)$
3. $7 - 9 - 12$    4. $8 - 4 \cdot 5$    5. $\frac{1}{5}(10x)$
6. $-6(5 - 11) - 4(13 - 6)$
7. $\left(-\frac{2}{3}\right)^3$    8. $-8(9x)$    9. $-\frac{3}{4} \div \frac{15}{16}$
10. $\frac{2}{5} \cdot \frac{4}{7}$    11. $\frac{-4(-6)}{-9}$
12. $\frac{2(-9) + 3(6)}{-7 - 8}$    13. $\frac{(5 - 3)^2}{5^2 - 3^2}$
14. $\frac{5}{9} + \frac{1}{3}$    15. $\frac{4}{21} - \frac{9}{35}$
16. $3(2x - 1) + 5(x + 2)$

Solve each equation.

17. $7x = 6x + 4$    18. $x + 12 = -4$
19. $-\frac{3}{5}x = 30$    20. $3x - 4 = 11$
21. $5x - 7 = x - 1$
22. $4(3x - 8) + 5(2x + 7) = 25$
23. $15 - 3(2t + 4) = 1$
24. $3(2a - 7) - 4(a - 3) = 15$
25. $\frac{1}{3}(x - 6) = \frac{1}{4}(x + 8)$
26. Solve $P = a + b + c$ for $c$.
27. Solve $3x + 4y = 12$ for $y$.

Solve each inequality, and graph the solution.

28. $3(x - 4) \leq 6$    29. $-5x + 9 < -6$
30. $x - 1 < 1$ or $2x + 1 > 7$
31. $-2 < x + 1 < 5$
32. Translate into symbols: The difference of $x$ and 5 is 12.
33. Give the opposite, reciprocal, and absolute value of $-\frac{2}{3}$.
34. Find the perimeter and area:

3 in.

3 in.

35. Find the next number in the sequence $4, 1, -2, -5, \dots$.
36. State the property or properties that justify the following: $5 + (-5) = 0$
37. Use the distributive property: $\frac{1}{4}(8x - 4)$
38. List all the rational numbers from the set $\{-5, -3, -1.7, 2.3, \frac{12}{7}, \pi\}$
39. Reduce to lowest terms: $\frac{234}{312}$
40. Evaluate $x^2 - 8x - 9$ when $x = -2$.
41. Evaluate $a^2 - 2ab + b^2$ when $a = 3, b = -2$.
42. What is 30% of 50?
43. **Number Problem** Twice a number increased by 7 is 31. Find the number.
44. **Gambling** A gambler won $5, then lost $8, won $1, and finally lost $7. How much did she win or lose?
45. **Geometry** A right triangle has one 42° angle. Find the other two angles.
46. **Geometry** The length of a rectangle is 4 inches more than the width. The perimeter is 28. Find the length and width.
47. **Geometry** Two angles are complementary. If one is 25°, find the other one.
48. **Coin Problem** Arbi has $2.20 in dimes and quarters. If he has 8 more dimes than quarters, how many coins of each does he have?
49. **Investing** I have invested money in two accounts. One account pays 5% annual interest, and the other pays 6%. I have $200 more in the 6% account than I have in the 5% account. If the total amount of interest was $56, how much do I have in each account?

**Cumulative Review** Starting in Chapter 2, at the end of the *Chapter Test* in each chapter, is a set of problems that reviews material from all preceding chapters. This keeps students current with past topics and helps them retain information they study.

# The Basics

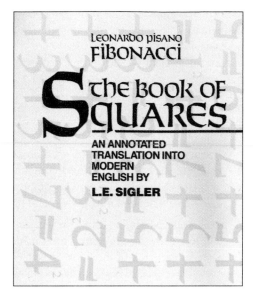

LEONARDO PISANO
FIBONACCI

## The Book of Squares

AN ANNOTATED
TRANSLATION INTO
MODERN
ENGLISH BY
**L.E. SIGLER**

## INTRODUCTION

Much of what we do in mathematics is concerned with recognizing patterns. If you recognize the patterns in the following two sequences, then you can easily extend each sequence.

Sequence of odd numbers = 1, 3, 5, 7, 9, . . .
Sequence of squares = 1, 4, 9, 16, 25, . . .

Once we have classified groups of numbers as to the characteristics they share, we sometimes discover that a relationship exists between the groups. Although it may not be obvious at first, there is a relationship that exists *between* the two sequences shown. The introduction to *The Book of Squares,* written in 1225 by the mathematician known as Fibonacci, begins this way:

> I thought about the origin of all square numbers and discovered that they arise out of the increasing sequence of odd numbers.

The relationship that Fibonacci refers to is shown visually in Figure 1.

**FIGURE I**

Many times we describe a relationship or pattern in a number of different ways. Figure 1 is a visual description of a relationship. In this chapter we will work on describing relationships numerically and verbally, in writing.

## STUDY SKILLS

Some of the students enrolled in my beginning algebra classes develop difficulties early in the course. Their difficulties are not associated with their ability to learn mathematics; they all have the potential to pass the course. Students who get off to a poor start do so because they have not developed the study skills necessary to be successful in algebra; they do not put themselves on an effective homework schedule, and when they work problems, they do it their way, not my way. Here is a list of things you can do to begin to develop effective study skills.

1. **Put Yourself on a Schedule**   The general rule is that you spend 2 hours on homework for every hour you are in class. Make a schedule for yourself in which you set aside 2 hours each day to work on algebra. Once you make the schedule, stick to it. Don't just complete your assignments and stop. Use all the time you have set aside. If you complete an assignment and have time left over, read the next section in the book, and then work more problems.

2. **Find Your Mistakes and Correct Them**   There is more to studying algebra than just working problems. You must always check your answers with the answers in the back of the book. When you have made a mistake, find out what it is and correct it. Making mistakes is part of the process of learning mathematics. In the prologue to *The Book of Squares,* Leonardo Fibonacci (ca. 1170– ca. 1250) had this to say about the content of his book:

   > I have come to request indulgence if in any place it contains something more or less than right or necessary; for to remember everything and be mistaken in nothing is divine rather than human  . . .

   Fibonacci knew, as you know, that human beings make mistakes. You cannot learn algebra without making mistakes.

3. **Imitate Success**   Your work should look like the work you see in this book and the work your instructor shows. The steps shown in solving problems in this book were written by someone who has been successful in mathematics. The same is true of your instructor. Your work should imitate the work of people who have been successful in mathematics.

Suppose you have a checking account that costs you $15 a month, plus $0.05 for each check you write. If you write 10 checks in a month, then the monthly charge for your checking account will be

$$15 + 10(0.05)$$

Do you add 15 and 10 first and then multiply by 0.05? Or do you multiply 10 and 0.05 first and then add 15? If you don't know the answer to this question, you will after you have read through this section.

Since much of what we do in algebra involves comparison of quantities, we will begin by listing some symbols used to compare mathematical quantities. The comparison symbols fall into two major groups: equality symbols and inequality symbols.

We will let the letters $a$ and $b$ stand for (represent) any two mathematical quantities. When we use letters to represent numbers, as we are doing here, we call the letters *variables*.

## VARIABLES: AN INTUITIVE LOOK

When you filled out the application for the school you are attending, there was a space to fill in your first name. "First Name" is a variable quantity, because the value it takes depends on who is filling out the application. For example, if your first name is Manuel, then the value of "First Name" is Manuel. On the other hand, if your first name is Christa, then the value of "First Name" is Christa.

If we denote "First Name" as $FN$, "Last Name" as $LN$, and "Whole Name" as $WN$, then we take the concept of a variable further and write the relationship between the names this way:

$$FN + LN = WN$$

(We use the $+$ symbol loosely here to represent writing the names together with a space between them.) This relationship we have written holds for all people who have only a first name and a last name. For those people who have a middle name, the relationship between the names is

$$FN + MN + LN = WN$$

A similar situation exists in algebra when we let a letter stand for a number or a group of numbers. For instance, if we say "let $a$ and $b$ represent numbers," then $a$ and $b$ are called **variables** because the values they take on vary. We use the variables $a$ and $b$ in the following lists so that the relationships shown there are true for all numbers that we will encounter in this book. By using variables, the following statements are general statements about all numbers, rather than specific statements about only a few numbers.

## COMPARISON SYMBOLS

| | | |
|---|---|---|
| *Equality:* | $a = b$ | $a$ is equal to $b$ ($a$ and $b$ represent the same number) |
| | $a \neq b$ | $a$ is not equal to $b$ |
| *Inequality:* | $a < b$ | $a$ is less than $b$ |
| | $a \not< b$ | $a$ is not less than $b$ |
| | $a > b$ | $a$ is greater than $b$ |
| | $a \not> b$ | $a$ is not greater than $b$ |
| | $a \geq b$ | $a$ is greater than or equal to $b$ |
| | $a \leq b$ | $a$ is less than or equal to $b$ |

The symbols for inequality, $<$ and $>$, always point to the smaller of the two quantities being compared. For example, $3 < x$ means 3 is smaller than $x$. In this case we can say "3 is less than $x$" or "$x$ is greater than 3"; both statements are correct. Similarly, the expression $5 > y$ can be read as "5 is greater than $y$" or as "$y$ is less than 5" because the inequality symbol is pointing to $y$, meaning $y$ is the smaller of the two quantities.

Next, we consider the symbols used to represent the four basic operations: addition, subtraction, multiplication, and division.

## OPERATION SYMBOLS

| | | |
|---|---|---|
| *Addition:* | $a + b$ | The *sum* of $a$ and $b$ |
| *Subtraction:* | $a - b$ | The *difference* of $a$ and $b$ |
| *Multiplication:* | $a \cdot b, (a)(b), a(b), (a)b, ab$ | The *product* of $a$ and $b$ |
| *Division:* | $a \div b, a/b, \dfrac{a}{b}, b\overline{)a}$ | The *quotient* of $a$ and $b$ |

When we encounter the word *sum,* the implied operation is addition. To find the sum of two numbers, we simply add them. *Difference* implies subtraction, *product* implies multiplication, and *quotient* implies division. Notice also that there is more than one way to write the product or quotient of two numbers.

**Note:** In the past you may have used the notation $3 \times 5$ to denote multiplication. In algebra it is best to avoid this notation if possible, since the multiplication symbol $\times$ can be confused with the variable $x$ when written by hand.

## GROUPING SYMBOLS

Parentheses ( ) and brackets [ ] are the symbols used for grouping numbers together. (Occasionally, braces { } are also used for grouping, although they are usually reserved for set notation, as we shall see.)

The following examples show the written language equivalents of the symbols for comparing, operating, and grouping.

**EXAMPLES**

| *Mathematical Expression* | *Written Equivalent* |
|---|---|
| **1.** $4 + 1 = 5$ | The sum of 4 and 1 is 5. |
| **2.** $8 - 1 < 10$ | The difference of 8 and 1 is less than 10. |
| **3.** $2(3 + 4) = 14$ | Twice the sum of 3 and 4 is 14. |
| **4.** $3x \geq 15$ | The product of 3 and $x$ is greater than or equal to 15. |
| **5.** $\dfrac{y}{2} = y - 2$ | The quotient of $y$ and 2 is equal to the difference of $y$ and 2. |

## EXPONENTS

The last type of notation we need to discuss is the notation that allows us to write repeated multiplications in a more compact form — *exponents.* In the expression $2^3$, the 2 is called the *base,* and the 3 is called the *exponent.* The exponent 3 tells us the number of times the base appears in the product. That is,

$$2^3 = 2 \cdot 2 \cdot 2 = 8$$

The expression $2^3$ is said to be in exponential form, while $2 \cdot 2 \cdot 2$ is said to be in expanded form. Here are some additional examples of expressions involving exponents.

**EXAMPLES**  Expand and multiply.

| | |
|---|---|
| **6.** $5^2 = 5 \cdot 5 = 25$ | Base 5, exponent 2 |
| **7.** $2^5 = 2 \cdot 2 \cdot 2 \cdot 2 \cdot 2 = 32$ | Base 2, exponent 5 |
| **8.** $10^3 = 10 \cdot 10 \cdot 10 = 1,000$ | Base 10, exponent 3 |

*Notation and Vocabulary*  Here is how we read expressions containing exponents.

| Mathematical Expression | Written Equivalent |
|---|---|
| $5^2$ | five to the second power |
| $5^3$ | five to the third power |
| $5^4$ | five to the fourth power |
| $5^5$ | five to the fifth power |
| $5^6$ | five to the sixth power |

We have a shorthand vocabulary for second and third powers, because the area of a square with a side of 5 is $5^2$, and the volume of a cube with a side of 5 is $5^3$.

$5^2$ can be read "five squared."        $5^3$ can be read "five cubed."

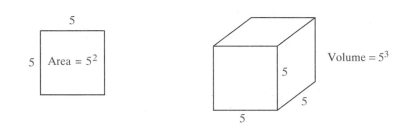

The symbols for comparing, operating, and grouping are to mathematics what punctuation symbols are to English. These symbols are the punctuation symbols for mathematics.

Consider the following sentence:

Paul said John is tall.

It can have two different meanings, depending on how it is punctuated.

1. "Paul," said John, "is tall."
2. Paul said, "John is tall."

Let's take a look at a similar situation in mathematics. Consider the following mathematical statement:

$$5 + 2 \cdot 7$$

If we add the 5 and 2 first and then multiply by 7, we get an answer of 49. On the other hand, if we multiply the 2 and the 7 first and then add 5, we are left with 19. We have a problem that seems to have two different answers, depending on whether we add first or multiply first. We would like to avoid this type of situation. That is, every problem like $5 + 2 \cdot 7$ should have only one answer. Therefore, we have developed the following rule for the order of operations.

### RULE (ORDER OF OPERATIONS)

When evaluating a mathematical expression, we will perform the operations in the following order, beginning with the expression in the innermost parentheses or brackets first and working our way out.

1. Simplify all numbers with exponents, working from left to right if more than one of these expressions is present.
2. Then do all multiplications and divisions left to right.
3. Perform all additions and subtractions left to right.

**E X A M P L E S**   Simplify each expression using the rule for order of operations.

**9.** $5 + 8 \cdot 2 = 5 + 16$               Multiply $8 \cdot 2$ first.

$\qquad = 21$

**10.** $12 \div 4 \cdot 2 = 3 \cdot 2$               Work left to right.

$\qquad = 6$

**11.** $2[5 + 2(6 + 3 \cdot 4)] = 2[5 + 2(6 + 12)]$ ⎫   Simplify within the
$\qquad = 2[5 + 2(18)]$ ⎭       innermost grouping
$\qquad\qquad\qquad\qquad\qquad\qquad\qquad$        symbols first.
$\qquad = 2[5 + 36]$         Next, simplify inside
$\qquad = 2[41]$            the brackets.
$\qquad = 82$            Multiply.

**12.** $10 + 12 \div 4 + 2 \cdot 3 = 10 + 3 + 6$       Multiply and divide left to
$\qquad\qquad\qquad\qquad\qquad\qquad\qquad$         right.
$\qquad = 19$            Add left to right.

**13.** $2^4 + 3^3 \div 9 - 4^2 = 16 + 27 \div 9 - 16$     Simplify numbers with
$\qquad\qquad\qquad\qquad\qquad\qquad\qquad$         exponents.
$\qquad = 16 + 3 - 16$         Then, divide.
$\qquad = 19 - 16$ ⎫        Finally, add and subtract
$\qquad = 3$ ⎭            left to right.

## READING TABLES AND BAR CHARTS

The table below shows the average amount of caffeine in a number of beverages. The diagram in Figure 1 is a bar chart. It is a visual presentation of the information in Table 1. The table gives information in numerical form, while the chart gives the same information in a geometric way. In mathematics, it is important to be able to move back and forth between the two forms.

**TABLE 1  Caffeine Content of Hot Drinks**

| Drink (6-ounce cup) | Caffeine (milligrams) |
|---|---|
| Brewed coffee | 100 |
| Instant coffee | 70 |
| Tea | 50 |
| Cocoa | 5 |
| Decaffeinated coffee | 4 |

**FIGURE 1**

**EXAMPLE   14**    Referring to Table 1 and Figure 1, suppose you have 3 cups of coffee, 1 cup of tea, and 2 cups of decaf in one day. Write an expression that will give the total amount of caffeine in these six drinks, then simplify the expression.

**Solution**  From the table or the bar chart, we find the number of milligrams of caffeine in each drink; then we write an expression for the total amount of caffeine:

$$3(100) + 50 + 2(4)$$

Using the rule for order of operations, we get 358 total milligrams of caffeine.

## NUMBER SEQUENCES AND INDUCTIVE REASONING

Suppose someone asks you to give the next number in the sequence of numbers below. (The dots mean that the sequence continues in the same pattern forever.)

$$2, 5, 8, 11, \ldots$$

If you notice that each number is 3 more than the number before it, you would say the next number in the sequence is 14 because $11 + 3 = 14$. When we reason in this way, we are using what is called *inductive reasoning*. In mathematics we use inductive reasoning when we notice a pattern to a sequence of numbers and then use the pattern to extend the sequence.

**EXAMPLE   15**   Find the next number in each sequence.

**(a)** 3, 8, 13, 18, . . .
**(b)** 2, 10, 50, 250, . . .
**(c)** 2, 4, 7, 11, . . .

**Solution**  In order to find the next number in each sequence, we need to look for a pattern or relationship.

**(a)** For the first sequence, each number is 5 more than the number before it; therefore, the next number will be $18 + 5 = 23$.
**(b)** For the sequence in part (b), each number is 5 times the number before it; therefore, the next number in the sequence will be $5 \cdot 250 = 1{,}250$.
**(c)** For the sequence in part (c), there is no number to add each time or multiply by each time. However, the pattern becomes apparent when we look at the differences between consecutive numbers:

Proceeding in the same manner, we would add 5 to the next term, giving us $11 + 5 = 16$.

In the introduction to this chapter, we mentioned the mathematician known as Fibonacci. There is a special sequence in mathematics named for Fibonacci. Here it is:

Fibonacci sequence = 1, 1, 2, 3, 5, 8, . . .

Can you see the relationship among the numbers in this sequence? Start with two 1's, then add two consecutive members of the sequence to get the next number. Here is a diagram.

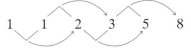

Sometimes we refer to the numbers in a sequence as *terms* of the sequence.

**E X A M P L E    1 6**    Write the first ten terms of the Fibonacci sequence.

**Solution**   The first six terms are given above. We extend the sequence by adding 5 and 8 to obtain the seventh term, 13. Then we add 8 and 13 to obtain 21. Continuing in this manner, the first ten terms in the Fibonacci sequence are

1, 1, 2, 3, 5, 8, 13, 21, 34, 55

# PROBLEM SET 1.1

For each sentence below, write an equivalent expression in symbols.

**1.** The sum of $x$ and 5 is 14.

**2.** The difference of $x$ and 4 is 8.

**3.** The product of 5 and $y$ is less than 30.

**4.** The product of 8 and $y$ is greater than 16.

**5.** The product of 3 and $y$ is less than or equal to the sum of $y$ and 6.

**6.** The product of 5 and $y$ is greater than or equal to the difference of $y$ and 16.

**7.** The quotient of $x$ and 3 is equal to the sum of $x$ and 2.

**8.** The quotient of $x$ and 2 is equal to the difference of $x$ and 4.

Expand and multiply.

**9.** $3^2$        **10.** $4^2$        **11.** $7^2$

**12.** $9^2$        **13.** $2^3$        **14.** $3^3$

**15.** $4^3$        **16.** $5^3$        **17.** $2^4$

**18.** $3^4$        **19.** $10^2$        **20.** $10^4$

**21.** $11^2$        **22.** $111^2$

Use the rule for order of operations to simplify each expression as much as possible.

**23.** $2 \cdot 3 + 5$        **24.** $8 \cdot 7 + 1$

**25.** $2(3 + 5)$        **26.** $8(7 + 1)$

**27.** $5 + 2 \cdot 6$        **28.** $8 + 9 \cdot 4$

**29.** $(5 + 2) \cdot 6$        **30.** $(8 + 9) \cdot 4$

**31.** $5 \cdot 4 + 5 \cdot 2$        **32.** $6 \cdot 8 + 6 \cdot 3$

**33.** $5(4 + 2)$        **34.** $6(8 + 3)$

**35.** $8 + 2(5 + 3)$        **36.** $7 + 3(8 - 2)$

**37.** $(8 + 2)(5 + 3)$        **38.** $(7 + 3)(8 - 2)$

**39.** $20 + 2(8 - 5) + 1$        **40.** $10 + 3(7 + 1) + 2$

**41.** $5 + 2(3 \cdot 4 - 1) + 8$

**42.** $11 - 2(5 \cdot 3 - 10) + 2$

**43.** $8 + 10 \div 2$        **44.** $16 - 8 \div 4$

**45.** $4 + 8 \div 4 - 2$        **46.** $6 + 9 \div 3 + 2$

**47.** $3 + 12 \div 3 + 6 \cdot 5$        **48.** $18 + 6 \div 2 + 3 \cdot 4$

**49.** $3 \cdot 8 + 10 \div 2 + 4 \cdot 2$

**50.** $5 \cdot 9 + 10 \div 2 + 3 \cdot 3$

**51.** $(5 + 3)(5 - 3)$        **52.** $(7 + 2)(7 - 2)$

**53.** $5^2 - 3^2$        **54.** $7^2 - 2^2$

**55.** $(4 + 5)^2$          **56.** $(6 + 3)^2$

**57.** $4^2 + 5^2$          **58.** $6^2 + 3^2$

**59.** $3 \cdot 10^2 + 4 \cdot 10 + 5$     **60.** $6 \cdot 10^2 + 5 \cdot 10 + 4$

**61.** $2 \cdot 10^3 + 3 \cdot 10^2 + 4 \cdot 10 + 5$

**62.** $5 \cdot 10^3 + 6 \cdot 10^2 + 7 \cdot 10 + 8$

**63.** $10 - 2(4 \cdot 5 - 16)$     **64.** $15 - 5(3 \cdot 2 - 4)$

**65.** $4[7 + 3(2 \cdot 9 - 8)]$

**66.** $5[10 + 2(3 \cdot 6 - 10)]$

**67.** $5(7 - 3) + 8(6 - 4)$

**68.** $3(10 - 4) + 6(12 - 10)$

**69.** $3(4 \cdot 5 - 12) + 6(7 \cdot 6 - 40)$

**70.** $6(8 \cdot 3 - 4) + 5(7 \cdot 3 - 1)$

**71.** $3^4 + 4^2 \div 2^3 - 5^2$     **72.** $2^5 + 6^2 \div 2^2 - 3^2$

**73.** $5^2 + 3^4 \div 9^2 + 6^2$     **74.** $6^2 + 2^5 \div 4^2 + 7^2$

Find the next number in each sequence.

**75.** 1, 2, 3, 4, . . . (The sequence of counting numbers)

**76.** 0, 1, 2, 3, . . . (The sequence of whole numbers)

**77.** 2, 4, 6, 8, . . . (The sequence of even numbers)

**78.** 1, 3, 5, 7, . . . (The sequence of odd numbers)

**79.** 1, 4, 9, 16, . . . (The sequence of squares)

**80.** 1, 8, 27, 64, . . . (The sequence of cubes)

**81.** 2, 2, 4, 6, . . . (A Fibonacci-like sequence)

**82.** 5, 5, 10, 15, . . . (A Fibonacci-like sequence)

## Applying the Concepts

**83. Reading Tables and Charts**  The following table and bar chart give the amount of caffeine in five different soft drinks. How much caffeine is in each of the following?
(a)  A 6-pack of Jolt
(b)  2 Coca-Colas plus 3 Tabs

| Caffeine Content in Soft Drinks | |
| --- | --- |
| **Drink** | **Caffeine (milligrams)** |
| Jolt | 100 |
| Tab | 47 |
| Coca-Cola | 45 |
| Diet Pepsi | 36 |
| 7UP | 0 |

Caffeine (mg)

**84. Reading Tables and Charts**  The following table and bar chart give the amount of caffeine in five different nonprescription drugs. How much caffeine is in each of the following?
(a)  A box of 12 Excedrin
(b)  1 Dexatrim plus 4 Excedrin

| Caffeine Content in Nonprescription Drugs | |
| --- | --- |
| **Nonprescription Drug** | **Caffeine (milligrams)** |
| Dexatrim | 200 |
| NoDoz | 100 |
| Excedrin | 65 |
| Triaminicin tablets | 30 |
| Dristan tablets | 16 |

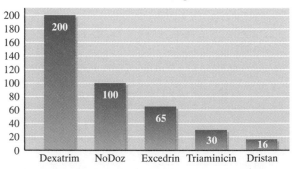

Caffeine (mg)

**85. Reading Tables and Charts**  The following bar chart gives the number of calories burned by a 150-

pound person during 1 hour of various exercises. The accompanying table should display the same information. Use the bar chart to complete the table.

| Calories Burned by 150-Pound Person | |
|---|---|
| **Activity** | **Calories Burned in 1 Hour** |
| Bicycling | 374 |
| Bowling | |
| Handball | |
| Jogging | |
| Skiing | |

**Calories Burned in 1 Hour by a 150-Pound Person**

**86. Reading Tables and Charts** The following bar chart gives the number of calories consumed by eating some popular fast foods. The accompanying table should display the same information. Use the bar chart to complete the table.

| Calories in Fast Food | |
|---|---|
| **Food** | **Calories** |
| McDonald's Hamburger | 270 |
| Burger King Hamburger | |
| Jack in the Box Hamburger | |
| McDonald's Big Mac | |
| Burger King Whopper | |

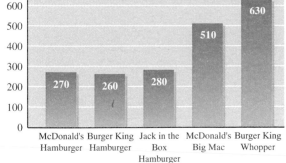

**Calories in Fast Food**

**87. Stock Market** On Monday Maria buys 10 shares of a certain stock. On Tuesday she buys 4 more shares of the same stock. If the stock splits Wednesday, then she has twice the number of shares she had on Tuesday. Write an expression using parentheses and the numbers 2, 4, and 10 to describe this situation.

**88. Baseball Cards** Patrick has a collection of 25 baseball cards. He then buys 3 packs of gum, and each pack contains 5 baseball cards. Write an expression using the numbers 25, 3, and 5 to describe this situation.

**89. Gambling** A gambler begins an evening in Las Vegas with $50. After an hour he has tripled her money. The next hour she loses $14. Write an expression using the numbers 3, 50, and 14 to describe this situation.

**90. Number of Passengers** A flight from Los Angeles to New York has 128 passengers. The plane stops in Denver, where 50 of the passengers get off and 21 new passengers get on. Write an expression containing the numbers 128, 50, and 21 to describe this situation.

**Food Labels** In 1993 the government standardized the way in which nutrition information was presented on the labels of most packaged food products. The food label shown below is from a package of cookies that I ate

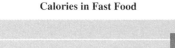

**SANDWICH CREMES**

| Nutrition Facts | | | | |
|---|---|---|---|---|
| Serving Size 5 Cookies (about 43 g) Servings Per Container 2 | | | | |
| | Amount/serving | %DV* | Amount/serving | %DV* |
| | **Total Fat** 9 g | **15%** | **Total Carb.** 30 g | **10%** |
| **Calories** 210 | Sat. Fat 2.5 g | **12%** | Fiber 1 g | **2%** |
| **Fat Calories** 90 | **Cholest.** less than 5 mg | **2%** | Sugars 14 g | |
| | **Sodium** 110 mg | **5%** | **Protein** 3 g | |
| * Percent Daily Values (DV) are based on a 2,000 calorie diet. | Vitamin A 0% • Vitamin C 0% • Calcium 2% • Iron 8% | | | |

at lunch the day I was writing this problem set. Use the information on the food label to answer Problems 91–94.

**91.** How many cookies are in the package?

**92.** If I paid $0.50 for the package of cookies, how much did each cookie cost?

**93.** If the "calories" category stands for calories per serving, how many calories did I consume by eating the whole package of cookies?

**94.** Suppose that, while swimming, I burn 11 calories each minute. If I swim for 20 minutes, will I burn enough calories to cancel out the calories I added by eating 5 cookies?

**Food Labels**  The food label shown below was taken from a bag of corn chips. Use the information on the food label to answer Problems 95–98.

**95.** Approximately how many chips are in the bag?

**96.** If the bag of chips costs $0.99, approximately how much does 1 serving of chips cost?

**97.** The table toward the bottom of the label gives the recommended amount of total fat in grams (g) that should be consumed by a person eating 2,000 calories per day and by a person eating 2,500 calories per day. Use the numbers in the table to estimate the recommended fat intake for a person eating 3,000 calories per day.

**98.** Deirdre burns 256 calories per hour by trotting her horse at a constant rate. How long must she ride in order to burn the calories consumed by eating 4 servings of these chips?

## Nutrition Facts

Serving Size 1 oz. (28 g/About 32 chips)
Servings Per Container 7

**Amount Per Serving**

**Calories** 160        Calories from Fat 90

|  | % Daily Value* |
|---|---|
| **Total Fat** 10 g | **15%** |
| Saturated Fat 1.5 g | **8%** |
| **Cholesterol** 0 mg | **0%** |
| **Sodium** 160 mg | **7%** |
| **Total Carbohydrate** 15 g | **5%** |
| Dietary Fiber 1 g | **4%** |
| Sugars 0 g | |
| **Protein** 2 g | |

Vitamin A 0%    •    Vitamin C 0%

Calcium 2%    •    Iron 0%

* Percent Daily Values are based on a 2,000 calorie diet. Your daily values may be higher or lower depending on your calorie needs:

| | Calories: | 2,000 | 2,500 |
|---|---|---|---|
| Total Fat | Less than | 65 g | 80 g |
| Sat Fat | Less than | 20 g | 25 g |
| Cholesterol | Less than | 300 mg | 300 mg |
| Sodium | Less than | 2,400 mg | 2,400 mg |
| Total Carbohydrate | | 300 g | 375 g |
| Dietary Fiber | | 25 g | 30 g |

Calories per gram:
Fat 9    •    Carbohydrate 4    •    Protein 4

## 1.2  Real Numbers

Table 1 and Figure 1 give the record low temperature, in degrees Fahrenheit, for each month of the year in the city of Jackson, Wyoming. Notice that some of these temperatures are represented by negative numbers.

| TABLE 1 | Record Low Temperatures for Jackson, Wyoming |
|---|---|
| **Month** | **Temperature (degrees Fahrenheit)** |
| January | −50 |
| February | −44 |
| March | −32 |
| April | −5 |
| May | 12 |
| June | 19 |
| July | 24 |
| August | 18 |
| September | 14 |
| October | 2 |
| November | −27 |
| December | −49 |

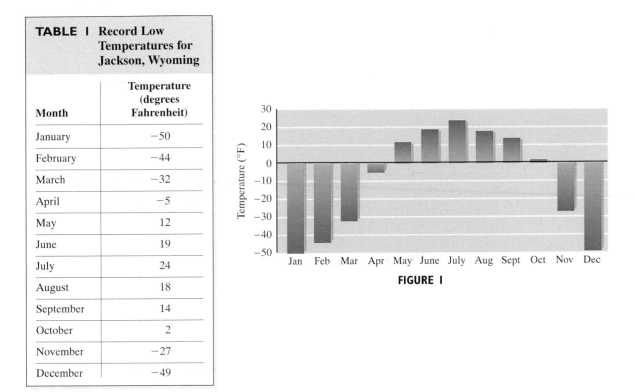

**FIGURE 1**

In this section we start our work with negative numbers. To represent negative numbers in algebra, we use what is called the **real number line.** Here is how we construct a real number line. We first draw a straight line and label a convenient point on the line with 0. Then we mark off equally spaced distances in both directions from 0. Label the points to the right of 0 with the numbers 1, 2, 3, . . . (the dots mean "and so on"). The points to the left of 0 we label in order, −1, −2, −3, . . . . Here is what it looks like.

The numbers increase in value going from left to right. If we move to the right, we are moving in the positive direction. If we move to the left, we are moving in the negative direction. When we compare two numbers on the number line, the number on the left is always smaller than the number on the right. For instance, −3 is smaller than −1 since it is to the left of −1 on the number line.

**Note:** If there is no sign (+ or −) in front of a number, the number is assumed to be positive (+).

**EXAMPLE I**   Locate and label the points on the real number line associated with the numbers $-3.5$, $-1\frac{1}{4}$, $\frac{1}{2}$, $\frac{3}{4}$, 2.5.

**Solution**   We draw a real number line from $-4$ to 4 and label the points in question.

---

> **DEFINITION**   The number associated with a point on the real number line is called the **coordinate** of that point.

In the preceding example, the numbers $\frac{1}{2}$, $\frac{3}{4}$, 2.5, $-3.5$, and $-1\frac{1}{4}$ are the coordinates of the points they represent.

> **DEFINITION**   The numbers that can be represented with points on the real number line are called **real numbers**.

Real numbers include whole numbers, fractions, decimals, and other numbers that are not as familiar to us as these.

## FRACTIONS ON THE NUMBER LINE

As we proceed through Chapter 1, from time to time we will review some of the major concepts associated with fractions. To begin, here is the formal definition of a fraction.

> **DEFINITION**   If $a$ and $b$ are real numbers, then the expression
>
> $$\frac{a}{b}$$
>
> is called a **fraction.** The top number $a$ is called the **numerator,** and the bottom number $b$ is called the **denominator.** The restriction $b \neq 0$ keeps us from writing an expression that is undefined. (As you will see, division by zero is not allowed.)

The number line can be used to visualize fractions. Recall that for the fraction $\frac{a}{b}$, $a$ is called the numerator and $b$ is called the denominator. The denominator indicates the number of equal parts in the interval from 0 to 1 on the number line. The numerator indicates how many of those parts we have. If we take that part of the number line from 0 to 1 and divide it into *three equal parts,* we say that we have divided it into *thirds* (see Figure 2). Each of the three segments is $\frac{1}{3}$ (one third) of the whole segment from 0 to 1.

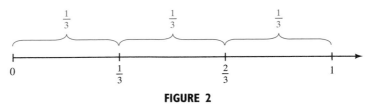

**FIGURE 2**

Two of these smaller segments together are $\frac{2}{3}$ (two thirds) of the whole segment. And three of them would be $\frac{3}{3}$ (three thirds), or the whole segment.

Let's do the same thing again with six equal divisions of the segment from 0 to 1 (see Figure 3). In this case we say each of the smaller segments has a length of $\frac{1}{6}$ (one sixth).

**FIGURE 3**

The same point we labeled with $\frac{1}{3}$ in Figure 2 is now labeled with $\frac{2}{6}$. Likewise, the point we labeled earlier with $\frac{2}{3}$ is now labeled $\frac{4}{6}$. It must be true then that

$$\frac{2}{6} = \frac{1}{3} \qquad \text{and} \qquad \frac{4}{6} = \frac{2}{3}$$

Actually, there are many fractions that name the same point as $\frac{1}{3}$. If we were to divide the segment between 0 and 1 into 12 equal parts, 4 of these 12 equal parts $\left(\frac{4}{12}\right)$ would be the same as $\frac{2}{6}$ or $\frac{1}{3}$. That is,

$$\frac{4}{12} = \frac{2}{6} = \frac{1}{3}$$

Even though these three fractions look different, each names the same point on the number line, as shown in Figure 4. All three fractions have the same *value* because they all represent the same number.

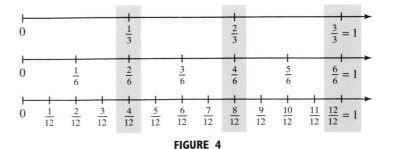

**FIGURE 4**

> **DEFINITION**   Fractions that represent the same number are said to be **equivalent.** Equivalent fractions may look different, but they must have the same value.

It is apparent that every fraction has many different representations, each of which is equivalent to the original fraction. The next two properties give us a way of changing the terms of a fraction without changing its value.

### PROPERTY 1

Multiplying the numerator and denominator of a fraction by the same nonzero number never changes the value of the fraction.

### PROPERTY 2

Dividing the numerator and denominator of a fraction by the same nonzero number never changes the value of the fraction.

**EXAMPLE 2**   Write $\frac{3}{4}$ as an equivalent fraction with denominator 20.

**Solution**   The denominator of the original fraction is 4. The fraction we are trying to find must have a denominator of 20. We know that if we multiply 4 by 5, we get 20. Property 1 indicates that we are free to multiply the denominator by 5 as long as we do the same to the numerator.

$$\frac{3}{4} = \frac{3 \cdot 5}{4 \cdot 5} = \frac{15}{20}$$

The fraction $\frac{15}{20}$ is equivalent to the fraction $\frac{3}{4}$.

## ABSOLUTE VALUES AND OPPOSITES

Representing numbers on the number line lets us give each number two important properties: a direction from zero and a distance from zero. The direction from zero is represented by the sign in front of the number. (A number without a sign is understood to be positive.) The distance from zero is called the absolute value of the number, as the following definition indicates.

> **DEFINITION**   The **absolute value** of a real number is its distance from zero on the number line. If $x$ represents a real number, then the absolute value of $x$ is written $|x|$.

**EXAMPLES**   Write each expression without absolute value symbols.

**3.** $|5| = 5$          The number 5 is 5 units from zero.

**4.** $|-5| = 5$      The number $-5$ is 5 units from zero.

**5.** $\left|-\dfrac{1}{2}\right| = \dfrac{1}{2}$      The number $-\dfrac{1}{2}$ is $\dfrac{1}{2}$ unit from zero.

The absolute value of a number is *never* negative. It is the distance the number is from zero without regard to which direction it is from zero. When working with the absolute value of sums and differences, we must simplify the expression inside the absolute value symbols first, and then find the absolute value of the simplified expression.

**EXAMPLES**  Simplify each expression.

**6.** $|8 - 3| = |5| = 5$

**7.** $|3 \cdot 2^3 + 2 \cdot 3^2| = |3 \cdot 8 + 2 \cdot 9| = |24 + 18| = |42| = 42$

**8.** $|9 - 2| - |8 - 6| = |7| - |2| = 7 - 2 = 5$

Another important concept associated with numbers on the number line is that of opposites. Here is the definition.

**DEFINITION**  Numbers the same distance from zero but in opposite directions from zero are called **opposites**.

**EXAMPLES**  Give the opposite of each number.

| | *Number* | *Opposite* | |
|---|---|---|---|
| **9.** | 5 | $-5$ | 5 and $-5$ are opposites. |
| **10.** | $-3$ | 3 | $-3$ and 3 are opposites. |
| **11.** | $\dfrac{1}{4}$ | $-\dfrac{1}{4}$ | $\dfrac{1}{4}$ and $-\dfrac{1}{4}$ are opposites. |
| **12.** | $-2.3$ | 2.3 | $-2.3$ and 2.3 are opposites. |

Each negative number is the opposite of some positive number, and each positive number is the opposite of some negative number. The opposite of a negative number is a positive number. In symbols, if $a$ represents a positive number, then

$$-(-a) = a$$

Opposites always have the same absolute value. When you add any two opposites, the result is always zero:

$$a + (-a) = 0$$

## RECIPROCALS AND MULTIPLICATION WITH FRACTIONS

The last concept we want to cover in this section is the concept of reciprocals. Understanding reciprocals requires some knowledge of multiplication with fractions. To multiply two fractions, we simply multiply numerators and multiply denominators.

**EXAMPLE  13**   Multiply $\frac{3}{4} \cdot \frac{5}{7}$.

**Solution**   The product of the numerators is 15, and the product of the denominators is 28:

$$\frac{3}{4} \cdot \frac{5}{7} = \frac{3 \cdot 5}{4 \cdot 7} = \frac{15}{28}$$

**EXAMPLE  14**   Multiply $7(\frac{1}{3})$.

**Solution**   The number 7 can be thought of as the fraction $\frac{7}{1}$:

$$7\left(\frac{1}{3}\right) = \frac{7}{1}\left(\frac{1}{3}\right) = \frac{7 \cdot 1}{1 \cdot 3} = \frac{7}{3}$$

**Note:**   In past math classes you may have written fractions like $\frac{7}{3}$ (improper fractions) as mixed numbers, such as $2\frac{1}{3}$. In algebra it is usually better to write them as improper fractions rather than mixed numbers.

**EXAMPLE  15**   Expand and multiply $(\frac{2}{3})^3$.

**Solution**   Using the definition of exponents from the previous section, we have

$$\left(\frac{2}{3}\right)^3 = \frac{2}{3} \cdot \frac{2}{3} \cdot \frac{2}{3} = \frac{8}{27}$$

We are now ready for the definition of reciprocals.

---

**DEFINITION**   Two numbers whose product is 1 are called **reciprocals.**

---

**EXAMPLES**   Give the reciprocal of each number.

|  | *Number* | *Reciprocal* |  |
|---|---|---|---|
| **16.** | 5 | $\frac{1}{5}$ | Because $5(\frac{1}{5}) = \frac{5}{1}(\frac{1}{5}) = \frac{5}{5} = 1$ |
| **17.** | 2 | $\frac{1}{2}$ | Because $2(\frac{1}{2}) = \frac{2}{1}(\frac{1}{2}) = \frac{2}{2} = 1$ |
| **18.** | $\frac{1}{3}$ | 3 | Because $\frac{1}{3}(3) = \frac{1}{3}(\frac{3}{1}) = \frac{3}{3} = 1$ |

**19.**   $\dfrac{3}{4}$          $\dfrac{4}{3}$          Because $\dfrac{3}{4}\left(\dfrac{4}{3}\right) = \dfrac{12}{12} = 1$

**20.**   $\dfrac{2}{5}$          $\dfrac{5}{2}$          Because $\dfrac{2}{5}\left(\dfrac{5}{2}\right) = \dfrac{10}{10} = 1$

Although we will not develop multiplication with negative numbers until later in the chapter, you should know that the reciprocal of a negative number is also a negative number. For example, the reciprocal of $-4$ is $-\dfrac{1}{4}$.

**EXAMPLE 21**   Find the next number in each sequence.

**(a)** $1, \dfrac{1}{2}, \dfrac{1}{3}, \dfrac{1}{4}, \cdots$          **(b)** $1, 1, \dfrac{1}{2}, \dfrac{1}{3}, \dfrac{1}{5}, \cdots$

**Solution**   As we did with the sequences in Section 1.1, we look for a pattern that defines the sequence.

**(a)** In the first case, the sequence is simply the reciprocals of the counting numbers. The next number in the sequence will be the reciprocal of 5, which is $\dfrac{1}{5}$.

**(b)** This sequence is formed by taking the reciprocals of the numbers in the Fibonacci sequence: $1, 1, 2, 3, 5, \cdots$

The next number in the Fibonacci sequence will be 8. Therefore, the next number in the sequence in question is $\dfrac{1}{8}$.

# FACTS FROM
# Geometry

## Formulas for Area and Perimeter

A square, rectangle, and triangle are shown in the following figures. Note that we have labeled the dimensions of each with variables. The formulas for the perimeter and area of each object are given in terms of its dimensions.

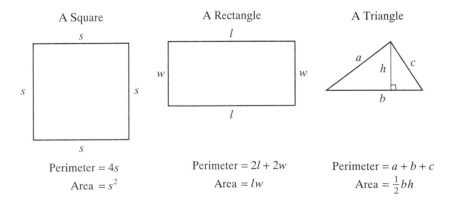

A Square

Perimeter $= 4s$

Area $= s^2$

A Rectangle

Perimeter $= 2l + 2w$

Area $= lw$

A Triangle

Perimeter $= a + b + c$

Area $= \dfrac{1}{2}bh$

The formula for perimeter gives us the distance around the outside of the object along its sides, while the formula for area gives us a measure of the amount of surface the object has.

**Note:** The vertical line labeled $h$ in the triangle is its height, or altitude. It extends from the top of the triangle down to the base, meeting the base at an angle of 90°. The altitude of a triangle is always perpendicular to the base. The small square shown where the altitude meets the base is used to indicate that the angle formed is 90°.

**E X A M P L E   2 2**   Find the perimeter and area of each figure.

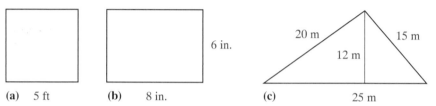

**(a)**   5 ft          **(b)**   8 in.          **(c)**          25 m

**Solution**   We use the preceding formulas to find the perimeter and the area. In each case, the units for perimeter are linear units, while the units for area are square units.

**(a)** Perimeter $= 4s = 4 \cdot 5$ feet $= 20$ feet
Area $= s^2 = (5 \text{ feet})^2 = 25$ square feet

**(b)** Perimeter $= 2l + 2w = 2(8 \text{ inches}) + 2(6 \text{ inches}) = 28$ inches
Area $= lw = (8 \text{ inches})(6 \text{ inches}) = 48$ square inches

**(c)** Perimeter $= a + b + c = (20 \text{ meters}) + (25 \text{ meters}) + (15 \text{ meters})$
$= 60$ meters

Area $= \dfrac{1}{2} bh = \dfrac{1}{2} (25 \text{ meters})(12 \text{ meters}) = 150$ square meters

# PROBLEM SET 1.2

Draw a number line that extends from $-5$ to $+5$. Label the points with the following coordinates.

**1.** 5

**2.** $-2$

**3.** $-4$

**4.** $-3$

**5.** 1.5

**6.** $-1.5$

**7.** $\dfrac{9}{4}$

**8.** $\dfrac{8}{3}$

Write each of the following fractions as an equivalent fraction with denominator 24.

**9.** $\dfrac{3}{4}$

**10.** $\dfrac{5}{6}$

**11.** $\dfrac{1}{2}$

**12.** $\dfrac{1}{8}$

**13.** $\dfrac{5}{8}$

**14.** $\dfrac{7}{12}$

Write each fraction as an equivalent fraction with denominator 60.

**15.** $\dfrac{3}{5}$   **16.** $\dfrac{5}{12}$   **17.** $\dfrac{11}{30}$

**18.** $\dfrac{9}{10}$

For each of the following numbers, give the opposite, the reciprocal, and the absolute value. (Assume all variables are nonzero.)

**19.** 10   **20.** 8   **21.** $\dfrac{3}{4}$

**22.** $\dfrac{5}{7}$   **23.** $\dfrac{11}{2}$   **24.** $\dfrac{16}{3}$

**25.** $-3$   **26.** $-5$   **27.** $-\dfrac{2}{5}$

**28.** $-\dfrac{3}{8}$   **29.** $x$   **30.** $a$

Place one of the symbols $<$ or $>$ between each of the following to make the resulting statement true.

**31.** $-5$   $-3$   **32.** $-8$   $-1$   **33.** $-3$   $-7$
**34.** $-6$   $-5$   **35.** $|-4|$   $-|-4|$
**36.** $3$   $-|-3|$   **37.** $7$   $-|-7|$

**38.** $-7$   $|-7|$   **39.** $-\dfrac{3}{4}$   $-\dfrac{1}{4}$   **40.** $-\dfrac{2}{3}$   $-\dfrac{1}{3}$

**41.** $-\dfrac{3}{2}$   $-\dfrac{3}{4}$   **42.** $-\dfrac{8}{3}$   $-\dfrac{17}{3}$

Simplify each expression.

**43.** $|8 - 2|$   **44.** $|6 - 1|$
**45.** $|5 \cdot 2^3 - 2 \cdot 3^2|$   **46.** $|2 \cdot 10^2 + 3 \cdot 10|$
**47.** $|7 - 2| - |4 - 2|$   **48.** $|10 - 3| - |4 - 1|$
**49.** $10 - |7 - 2(5 - 3)|$   **50.** $12 - |9 - 3(7 - 5)|$
**51.** $15 - |8 - 2(3 \cdot 4 - 9)| - 10$
**52.** $25 - |9 - 3(4 \cdot 5 - 18)| - 20$

Multiply the following.

**53.** $\dfrac{2}{3} \cdot \dfrac{4}{5}$   **54.** $\dfrac{1}{4} \cdot \dfrac{3}{5}$   **55.** $\dfrac{1}{2}\,(3)$

**56.** $\dfrac{1}{3}\,(2)$   **57.** $\dfrac{1}{4}\,(5)$   **58.** $\dfrac{1}{5}\,(4)$

**59.** $\dfrac{4}{3} \cdot \dfrac{3}{4}$   **60.** $\dfrac{5}{7} \cdot \dfrac{7}{5}$   **61.** $6\left(\dfrac{1}{6}\right)$

**62.** $8\left(\dfrac{1}{8}\right)$   **63.** $3 \cdot \dfrac{1}{3}$   **64.** $4 \cdot \dfrac{1}{4}$

Expand and multiply.

**65.** $\left(\dfrac{3}{4}\right)^2$   **66.** $\left(\dfrac{5}{6}\right)^2$   **67.** $\left(\dfrac{2}{3}\right)^3$

**68.** $\left(\dfrac{1}{2}\right)^3$   **69.** $\left(\dfrac{1}{10}\right)^4$   **70.** $\left(\dfrac{1}{10}\right)^5$

Find the next number in each sequence.

**71.** $1, \dfrac{1}{3}, \dfrac{1}{5}, \dfrac{1}{7}, \ldots$ (Reciprocals of odd numbers)
**72.** $\dfrac{1}{2}, \dfrac{1}{4}, \dfrac{1}{6}, \dfrac{1}{8}, \ldots$ (Reciprocals of even numbers)
**73.** $1, \dfrac{1}{4}, \dfrac{1}{9}, \dfrac{1}{16}, \ldots$ (Reciprocals of squares)
**74.** $1, \dfrac{1}{8}, \dfrac{1}{27}, \dfrac{1}{64}, \ldots$ (Reciprocals of cubes)

Find the perimeter and area of each figure.

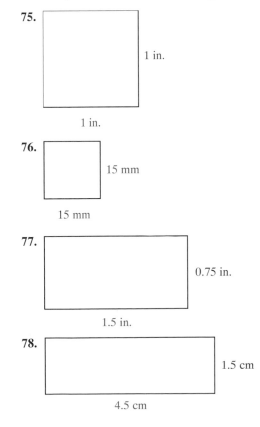

**75.**   1 in.
1 in.

**76.**   15 mm
15 mm

**77.**   0.75 in.
1.5 in.

**78.**   1.5 cm
4.5 cm

**79.**

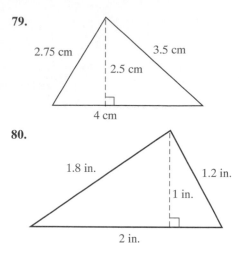

**80.**

Fahrenheit scale, we write

32°F, which is read "32 degrees Fahrenheit"

Use this information for Problems 83 and 84.

**83. Temperature and Altitude**    Marilyn is flying from Seattle to San Francisco on a Boeing 737 jet. When the plane reaches an altitude of 35,000 feet, the temperature outside the plane is 64 degrees below zero Fahrenheit. Represent the temperature with a negative number. If the temperature outside the plane gets warmer by 10 degrees, what will the new temperature be?

**84. Temperature Change**    At 10:00 in the morning in White Bear Lake, Minnesota, John notices the temperature outside is 10 degrees below zero Fahrenheit. Write the temperature as a negative number. An hour later it has warmed up by 6 degrees. What is the temperature at 11:00 that morning?

## Applying the Concepts

**81. Football Yardage**    A football team gains 6 yards on one play and then loses 8 yards on the next play. To what number on the number line does a loss of 8 yards correspond? The total yards gained or lost on the two plays corresponds to what negative number?

**82. Checking Account Balance**    A woman has a balance of $20 in her checking account. If she writes a check for $30, what negative number can be used to represent the new balance in her checking account?

**Temperature**    In the United States, temperature is measured on the Fahrenheit temperature scale. On this scale, water boils at 212 degrees and freezes at 32 degrees. To denote a temperature of 32 degrees on the

**Wind Chill**    Table 2 is a table of wind chill temperatures. The top row gives the air temperature, while the first column is wind speed in miles per hour. The numbers within the table indicate how cold the weather will feel. For example, if the thermometer reads 30°F, and the wind is blowing at 15 miles per hour, the wind chill temperature is 9°F. Use Table 2 to answer Questions 85 and 86.

**85. Reading Tables**    Find the wind chill temperature if the thermometer reads 20°F and the wind is blowing at 25 miles per hour.

**86. Reading Tables**    Which will feel colder: a day with an air temperature of 10°F with a 25-mile-per-hour wind, or a day with an air temperature of −5° F and a 10-mile-per-hour wind?

| **TABLE 2  Wind Chill Temperatures** | | | | | | | | |
|---|---|---|---|---|---|---|---|---|
| | **Air Temperature (°F)** | | | | | | | |
| **Wind Speed** | *30°* | *25°* | *20°* | *15°* | *10°* | *5°* | *0°* | *−5°* |
| 10 mph | 16° | 10° | 3° | −3° | −9° | −15° | −22° | −27° |
| 15 mph | 9° | 2° | −5° | −11° | −18° | −25° | −31° | −38° |
| 20 mph | 4° | −3° | −10° | −17° | −24° | −31° | −39° | −46° |
| 25 mph | 1° | −7° | −15° | −22° | −29° | −36° | −44° | −51° |
| 30 mph | −2° | −10° | −18° | −25° | −33° | −41° | −49° | −56° |

87. **Reading Tables and Charts**   The following bar chart gives the record low temperatures for each month of the year for Lake Placid, New York. The accompanying table should display the same information. Use the bar chart to complete the table.

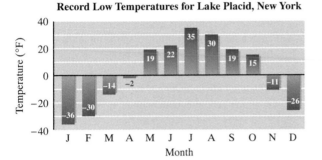

**Record Low Temperatures for Lake Placid, New York**

| Record Low Temperatures in Lake Placid, New York | |
|---|---|
| **Month** | **Temperature (°F)** |
| January | −36 |
| February | |
| March | |
| April | |
| May | |
| June | |
| July | |
| August | |
| September | |
| October | |
| November | |
| December | |

88. **Reading Tables and Charts**   The following bar chart gives the low temperature for each day of one week in Salt Lake City, Utah. The accompanying table should display the same information. Use the bar chart to complete the table.

**Daily Lows for Salt Lake City**

| Daily Low Temperatures in Salt Lake City | |
|---|---|
| **Day** | **Temperature (°F)** |
| Monday | |
| Tuesday | |
| Wednesday | |
| Thursday | |
| Friday | |
| Saturday | |
| Sunday | |

89. **Scuba Diving**   Steve is scuba diving near his home in Maui. At one point he is 100 feet below the surface. Represent this number with a negative number. If he descends another 5 feet, what negative number will represent his new position?

90. **Checking Account Balance**   Pete and Heidi are married and have a joint checking account that allows them to withdraw money from the account at an automatic teller machine. With each withdrawal from the automatic teller, the bank charges an extra $2 to their account. Suppose the balance in the checkbook is $75. If Heidi writes a check for $55 and Pete withdraws $20 from the automatic teller machine, what will their new balance be?

91. **Geometry**   Find the area and perimeter of an $8\frac{1}{2}$-by-11-inch piece of notebook paper.

92. **Geometry**   Find the area and perimeter of an $8\frac{1}{2}$-by-$5\frac{1}{2}$-inch piece of paper.

**Calories and Exercise**  Table 3 gives the amount of energy expended per hour for various activities, for a person weighing 120, 150, or 180 pounds. Use Table 3 to answer Questions 93–96.

| TABLE 2 | Energy Expended from Exercising | | |
|---|---|---|---|
| | **Calories per Hour** | | |
| **Activity** | **120 lb** | **150 lb** | **180 lb** |
| Bicycling | 299 | 374 | 449 |
| Bowling | 212 | 265 | 318 |
| Handball | 544 | 680 | 816 |
| Horseback trotting | 278 | 347 | 416 |
| Jazzercise | 272 | 340 | 408 |
| Jogging | 544 | 680 | 816 |
| Skiing (downhill) | 435 | 544 | 653 |

**93. Reading Tables**  Suppose you weigh 120 pounds. How many calories will you burn if you play handball for 2 hours and then ride your bicycle for an hour?

**94. Reading Tables**  How many calories are burned by a person weighing 150 pounds who jogs for $\frac{1}{2}$ hour and then goes bicycling for 2 hours?

**95. Reading Tables**  Two people go skiing. One weighs 180 pounds, and the other weighs 120 pounds. If they ski for 3 hours, how many more calories are burned by the person weighing 180 pounds?

**96. Reading Tables**  Two people spend 3 hours bowling. If one weighs 120 pounds and the other weighs 150 pounds, how many more calories are burned during the evening by the person weighing 150 pounds?

## 1.3  Addition of Real Numbers

Suppose that you are playing a friendly game of poker with some friends, and you lose $3 on the first hand and $4 on the second hand. If you represent winning with positive numbers and losing with negative numbers, how can you translate this situation into symbols? Since you lost $3 and $4 for a total of $7, one way to represent this situation is with addition of negative numbers:

$$(-\$3) + (-\$4) = -\$7$$

From this equation, we see that the sum of two negative numbers is a negative number. To generalize addition with positive and negative numbers, we use the number line.

Since real numbers have both a distance from zero (absolute value) and a direction from zero (sign), we can think of addition of two numbers in terms of distance and direction from zero.

Let's look at a problem for which we know the answer. Suppose we want to add the numbers 3 and 4. The problem is written  3 + 4.  To put it on the number line, we read the problem as follows:

**1.** The 3 tells us to "start at the origin and move 3 units in the positive direction."

**2.** The + sign is read "and then move."

**3.** The 4 means "4 units in the positive direction."

To summarize,  3 + 4  means to start at the origin, move 3 units in the *positive* direction and then 4 units in the *positive* direction.

We end up at 7, which is the answer to our problem: $3 + 4 = 7$.
  Let's try other combinations of positive and negative 3 and 4 on the number line.

**EXAMPLE  1**  Add $3 + (-4)$.

**Solution**  Starting at the origin, move 3 units in the *positive* direction and then 4 units in the *negative* direction.

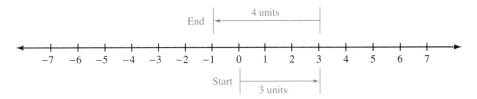

We end up at $-1$; therefore, $3 + (-4) = -1$.

**EXAMPLE  2**  Add $-3 + 4$.

**Solution**  Starting at the origin, move 3 units in the *negative* direction and then 4 units in the *positive* direction.

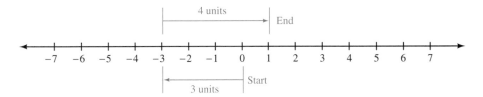

We end up at $+1$; therefore, $-3 + 4 = 1$.

**EXAMPLE  3**  Add $-3 + (-4)$.

**Solution**  Starting at the origin, move 3 units in the *negative* direction and then 4 units in the *negative* direction.

We end up at $-7$; therefore, $-3 + (-4) = -7$. Here is a summary of what we have just completed:

$$3 + 4 = 7$$

$$3 + (-4) = -1$$

$$-3 + 4 = 1$$

$$-3 + (-4) = -7$$

Let's do four more problems on the number line and then summarize our results in a rule we can use to add any two real numbers.

**EXAMPLE 4**   Add $5 + 7 = 12$.

**Solution**

**EXAMPLE 5**   Add $5 + (-7) = -2$.

**Solution**

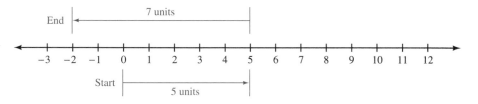

**EXAMPLE 6**   Add $-5 + 7 = 2$.

**Solution**

**E X A M P L E   7**   Add $-5 + (-7) = -12$.

**Solution**

End                                                7 units                              5 units                    Start

If we look closely at the results of the preceding addition problems, we can see that they support (or justify) the following rule.

**RULE**

To add two real numbers with

1. The *same* sign: Simply add their absolute values and use the common sign. (If both numbers are positive, the answer is positive. If both numbers are negative, the answer is negative.)
2. *Different* signs: Subtract the smaller absolute value from the larger. The answer will have the sign of the number with the larger absolute value.

This rule covers all possible combinations of addition with real numbers. You must memorize it. After you have worked a number of problems, it will seem almost automatic.

**E X A M P L E   8**   Add all combinations of positive and negative 10 and 13.

**Solution**   Rather than work these problems on the number line, we use the rule for adding positive and negative numbers to obtain our answers:

$$10 + 13 = 23$$

$$10 + (-13) = -3$$

$$-10 + 13 = 3$$

$$-10 + (-13) = -23$$

**E X A M P L E   9**   Add all possible combinations of positive and negative 12 and 17.

**Solution**   Applying the rule for adding positive and negative numbers, we have

$$12 + 17 = 29$$

$$12 + (-17) = -5$$

$$-12 + 17 = 5$$

$$-12 + (-17) = -29$$

**EXAMPLE 10** Add $-3 + 2 + (-4)$.

**Solution** Applying the rule for order of operations, we add left to right:

$$-3 + 2 + (-4) = -1 + (-4)$$
$$= -5$$

**EXAMPLE 11** Add $-8 + [2 + (-5)] + (-1)$.

**Solution** Adding inside the brackets first, and then left to right, we have

$$-8 + [2 + (-5)] + (-1) = -8 + (-3) + (-1)$$
$$= -11 + (-1)$$
$$= -12$$

**EXAMPLE 12** Simplify $-10 + 2(-8 + 11) + (-4)$.

**Solution** First, we simplify inside the parentheses. Then, we multiply. Finally, we add left to right:

$$-10 + 2(-8 + 11) + (-4) = -10 + 2(3) + (-4)$$
$$= -10 + 6 + (-4)$$
$$= -4 + (-4)$$
$$= -8$$

## ARITHMETIC SEQUENCES

The pattern in a sequence of numbers is easy to identify when each number in the sequence comes from the preceding number by adding the same amount each time. This leads us to our next level of classification, in which we classify together groups of sequences with a common characteristic.

> **DEFINITION** An **arithmetic sequence** is a sequence of numbers in which each number (after the first number) comes from adding the same amount to the number before it.

Here is an example of an arithmetic sequence:

$$2, 5, 8, 11, \ldots$$

Each number is obtained by adding 3 to the number before it.

**EXAMPLE 13** Each of the following is an arithmetic sequence. Find the next two numbers in each sequence.

(a) 7, 10, 13, . . .

(b) 9.5, 10, 10.5, . . .

(c) 5, 0, −5, . . .

**Solution**  Since we know that each sequence is arithmetic, we know to look for the number that is added to each term to produce the next consecutive term.

(a) 7, 10, 13, . . . : Each term is found by adding 3 to the term before it. Therefore, the next two terms will be 16 and 19.

(b) 9.5, 10, 10.5, . . . : Each term comes from adding 0.5 to the term before it. Therefore, the next two terms will be 11 and 11.5.

(c) 5, 0, −5, . . . : Each term comes from adding −5 to the term before it. Therefore, the next two terms will be −5 + (−5) = −10 and −10 + (−5) = −15.

# PROBLEM SET 1.3

**1.** Add all positive and negative combinations of 3 and 5. (Look back to Examples 8 and 9.)

**2.** Add all positive and negative combinations of 6 and 4.

**3.** Add all positive and negative combinations of 15 and 20.

**4.** Add all positive and negative combinations of 18 and 12.

Work the following problems. You may want to begin by doing a few on the number line.

**5.** 6 + (−3)

**6.** 7 + (−8)

**7.** 13 + (−20)

**8.** 15 + (−25)

**9.** 18 + (−32)

**10.** 6 + (−9)

**11.** −6 + 3

**12.** −8 + 7

**13.** −30 + 5

**14.** −18 + 6

**15.** −6 + (−6)

**16.** −5 + (−5)

**17.** −9 + (−10)

**18.** −8 + (−6)

**19.** −10 + (−15)

**20.** −18 + (−30)

Work the following problems using the rule for addition of real numbers. You may want to refer back to the rule for order of operations.

**21.** 5 + (−6) + (−7)

**22.** 6 + (−8) + (−10)

**23.** −7 + 8 + (−5)

**24.** −6 + 9 + (−3)

**25.** 5 + [6 + (−2)] + (−3)

**26.** 10 + [8 + (−5)] + (−20)

**27.** [6 + (−2)] + [3 + (−1)]

**28.** [18 + (−5)] + [9 + (−10)]

**29.** 20 + (−6) + [3 + (−9)]

**30.** 18 + (−2) + [9 + (−13)]

**31.** −3 + (−2) + [5 + (−4)]

**32.** −6 + (−5) + [−4 + (−1)]

**33.** (−9 + 2) + [5 + (−8)] + (−4)

**34.** (−7 + 3) + [9 + (−6)] + (−5)

**35.** [−6 + (−4)] + [7 + (−5)] + (−9)

**36.** [−8 + (−1)] + [8 + (−6)] + (−6)

**37.** (−6 + 9) + (−5) + (−4 + 3) + 7

**38.** (−10 + 4) + (−3) + (−3 + 8) + 6

The problems that follow involve some multiplication. Be sure that you work inside the parentheses first, then multiply, and, finally, add left to right.

**39.** −5 + 2(−3 + 7)

**40.** −3 + 4(−2 + 7)

**41.** 9 + 3(−8 + 10)

**42.** 4 + 5(−2 + 6)

**43.** −10 + 2(−6 + 8) + (−2)

**44.** −20 + 3(−7 + 10) + (−4)

**45.** 2(−4 + 7) + 3(−6 + 8)

**46.** 5(−2 + 5) + 7(−1 + 6)

Each sequence below is an arithmetic sequence. In each case, find the next two numbers in the sequence.

**47.** 3, 8, 13, 18, . . .       **48.** 1, 5, 9, 13, . . .

**49.** 10, 15, 20, 25, . . .       **50.** 10, 16, 22, 28, . . .

**51.** 20, 15, 10, 5, . . .       **52.** 24, 20, 16, 12, . . .

**53.** 6, 0, −6, . . .       **54.** 1, 0, −1, . . .

**55.** 8, 4, 0, . . .       **56.** 5, 2, −1, . . .

**57.** Is the sequence of odd numbers an arithmetic sequence?

**58.** Is the sequence of squares an arithmetic sequence?

Recall that the word *sum* indicates addition. Write the numerical expression that is equivalent to each of the following phrases and then simplify.

**59.** The sum of 5 and 9       **60.** The sum of 6 and −3

**61.** Four added to the sum of −7 and −5

**62.** Six added to the sum of −9 and 1

**63.** The sum of −2 and −3 increased by 10

**64.** The sum of −4 and −12 increased by 2

Answer the following questions.

**65.** What number do you add to −8 to get −5?

**66.** What number do you add to 10 to get 4?

**67.** The sum of what number and −6 is −9?

**68.** The sum of what number and −12 is 8?

## Applying the Concepts

**69. Temperature Change** The temperature at noon is 12 degrees below zero Fahrenheit. By 1:00 it has risen 4 degrees. Write an expression using the numbers −12 and 4 to describe this situation.

**70. Stock Value** On Monday a certain stock gains 2 points. On Tuesday it loses 3 points. Write an expression using positive and negative numbers with addition to describe this situation and then simplify.

**71. Gambling** On three consecutive hands of draw poker, a gambler wins $10, loses $6, and then loses another $8. Write an expression using positive and negative numbers and addition to describe this situation and then simplify.

**72. Number Problem** You know from your past experience with numbers that subtracting 5 from 8 results in 3: 8 − 5 = 3. What addition problem that starts with the number 8 gives the same result?

**73. Checkbook Balance** Suppose that you balance your checkbook and find that you are overdrawn by $30. That is, your balance is −$30. Then you go to the bank and deposit $40. Translate this situation into an addition problem, the answer to which gives the new balance in your checkbook.

**74. Checkbook Balance** The balance in your checkbook is −$25. If you make a deposit of $75 and then write a check for $18, what is the new balance?

**75. Salary Increase** Colleen has a job for which her starting wage is $6.50 per hour. If she gets a raise of $0.25 every 6 months, write a sequence of numbers that gives her wage every 6 months for the first 3 years. Is the sequence an arithmetic sequence?

**76. Coin Collection** Benja has a coin collection that she purchased for $650. Each year the collection increases $50 in value. Write a sequence of numbers that gives the value of the coin collection every year for the first 5 years. Is this sequence an arithmetic sequence?

**77. Starting Salaries** The table and bar chart on p. 31 show the salaries for new teachers compared with starting salaries in other professions in 1998.
(a) Use the information in the bar chart to fill in the missing entries in the table.
(b) What is the average family income of a married couple graduating in 1998 if one majored in chemistry and the other in engineering?

| Professions | Salaries |
|---|---|
| Engineering | |
| Computer Science | $40,920 |
| | $40,523 |
| | $36,036 |
| Business Admin. | |
| Accounting | $33,702 |
| Sales/Marketing | |
| Teaching | $25,735 |

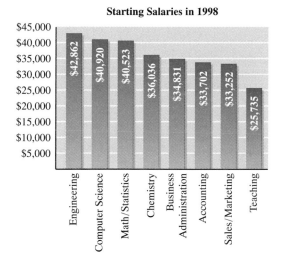

Starting Salaries in 1998

78. **Digital Camera Sales** The following bar chart shows the sales of digital cameras from 1996–1999.
   (a) Use the information in the bar chart to fill in the missing entries in the table.

(b) Find the total digital camera sales from 1997 through 1999.

| Year | Sales ($ millions) |
|---|---|
| 1996 | |
| 1997 | |
| 1998 | |
| | 819 |

Digital Camera Sales

**1600 × 1200 Resolution Digital Camera**
**Your Price: $599⁹⁹**

---

**1.4** ▶ **Subtraction of Real Numbers**

Suppose that the temperature at noon is 20° Fahrenheit and 12 hours later, at midnight, it has dropped to −15° Fahrenheit. What is the difference between the temperature at noon and the temperature at midnight? Intuitively, we know the difference in the two temperatures is 35°. We also know that the word *difference* indicates subtraction. The difference between 20 and −15 is written

$$20 - (-15)$$

It must be true that $20 - (-15) = 35$. In this section we will see how our definition for subtraction confirms that this last statement is in fact correct.

In the previous section we spent some time developing the rule for addition of real numbers. Since we want to make as few rules as possible, we can define subtraction in terms of addition. By doing so, we can then use the rule for addition to solve our subtraction problems.

**RULE**

To subtract one real number from another, simply add its opposite.

Algebraically, the rule is written like this: If $a$ and $b$ represent two real numbers, then it is always true that

$$\underbrace{a - b}_{\text{To subtract } b} = \underbrace{a + (-b)}_{\text{add the opposite of } b}$$

This is how subtraction is defined in algebra. This definition of subtraction will not conflict with what you already know about subtraction, but it will allow you to do subtraction using negative numbers.

**EXAMPLE 1** Subtract all possible combinations of positive and negative 7 and 2.

**Solution**

$$\left.\begin{array}{l} 7 - 2 = 7 + (-2) = 5 \\ -7 - 2 = -7 + (-2) = -9 \end{array}\right\} \quad \begin{array}{l}\text{Subtracting } +2 \text{ is the same} \\ \text{as adding } -2.\end{array}$$

$$\left.\begin{array}{l} 7 - (-2) = 7 + 2 = 9 \\ -7 - (-2) = -7 + 2 = -5 \end{array}\right\} \quad \begin{array}{l}\text{Subtracting } -2 \text{ is the same} \\ \text{as adding } +2.\end{array}$$

Notice that each subtraction problem is first changed to an addition problem. The rule for addition is then used to arrive at the answer.

We have defined subtraction in terms of addition, and we still obtain answers consistent with the answers we are used to getting with subtraction. Moreover, we can now do subtraction problems involving both positive and negative numbers.

As you proceed through the following examples and the problem set, you will begin to notice shortcuts you can use in working the problems. You will not always have to change subtraction to addition of the opposite to be able to get answers quickly. Use all the shortcuts you wish as long as you consistently get the correct answers.

**EXAMPLE 2** Subtract all combinations of positive and negative 8 and 13.

**Solution**

$$\left.\begin{array}{l} 8 - 13 = 8 + (-13) = -5 \\ -8 - 13 = -8 + (-13) = -21 \end{array}\right\} \quad \begin{array}{l}\text{Subtracting } +13 \text{ is the} \\ \text{same as adding } -13.\end{array}$$

$$8 - (-13) = \quad 8 + 13 = 21$$
$$-8 - (-13) = -8 + 13 = 5$$

Subtracting $-13$ is the same as adding $+13$.

---

**E X A M P L E S**   Simplify each expression as much as possible.

**3.** $7 + (-3) - 5 = 7 + (-3) + (-5)$          Begin by changing all
$$= 4 + (-5)$$                subtractions to additions.
$$= -1$$                 Then add left to right.

**4.** $8 - (-2) - 6 = 8 + 2 + (-6)$          Begin by changing all
$$= 10 + (-6)$$                subtractions to additions.
$$= 4$$                 Then add left to right.

**5.** $-2 - (-3 + 1) - 5 = -2 - (-2) - 5$          Do what is in the parentheses
$$= -2 + 2 + (-5)$$          first.
$$= -5$$

---

The next two examples involve multiplication and exponents as well as subtraction. Remember, according to the rule for order of operations, we evaluate the numbers containing exponents and multiply before we subtract.

---

**E X A M P L E   6**   Simplify $2 \cdot 5 - 3 \cdot 8 - 4 \cdot 9$.

**Solution**   First, we multiply left to right, and then we subtract.

$$2 \cdot 5 - 3 \cdot 8 - 4 \cdot 9 = 10 - 24 - 36$$
$$= -14 - 36$$
$$= -50$$

---

**E X A M P L E   7**   Simplify $3 \cdot 2^3 - 2 \cdot 4^2$.

**Solution**   We begin by evaluating each number that contains an exponent. Then we multiply before we subtract:

$$3 \cdot 2^3 - 2 \cdot 4^2 = 3 \cdot 8 - 2 \cdot 16$$
$$= 24 - 32$$
$$= -8$$

---

**E X A M P L E   8**   Subtract 7 from $-3$.

**Solution**   First, we write the problem in terms of subtraction. Then we change to addition of the opposite:

$$-3 - 7 = -3 + (-7)$$
$$= -10$$

**EXAMPLE 9**  Subtract −5 from 2.

**Solution**  Subtracting −5 is the same as adding +5:

$$2 - (-5) = 2 + 5$$
$$= 7$$

**EXAMPLE 10**  Find the difference of 9 and 2.

**Solution**  Written in symbols, the problem looks like this:

$$9 - 2 = 7$$

The difference of 9 and 2 is 7.

**EXAMPLE 11**  Find the difference of 3 and −5.

**Solution**  Subtracting −5 from 3 we have

$$3 - (-5) = 3 + 5$$
$$= 8$$

In the sport of drag racing, two cars at the starting line race to the finish line $\frac{1}{4}$ mile away. The car that crosses the finish line first wins the race.

Jim Rizzoli owns and races an alcohol dragster. On board the dragster is a computer that records data during each of Jim's races. Table 1 gives some of the data from a race Jim was in during the 1993 Winternationals. Figure 1 gives the same information visually.

| TABLE 1 | Speed of a Race Car |
|---|---|
| Time in Seconds | Speed in Miles/Hour |
| 0 | 0 |
| 1 | 72.7 |
| 2 | 129.9 |
| 3 | 162.8 |
| 4 | 192.2 |
| 5 | 212.4 |
| 6 | 228.1 |

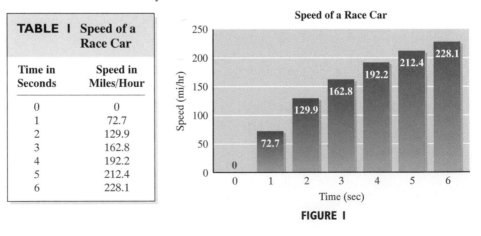

FIGURE 1

**EXAMPLE 12**  Use Table 1 to find the difference in speed after 5 seconds and after 2 seconds have elapsed during the race.

**Solution**   We know the word *difference* implies subtraction. The speed at 2 seconds is 129.9 miles per hour, while the speed at 5 seconds is 212.4 miles per hour. Therefore, the expression that represents the solution to our problem looks like this:

$$212.4 - 129.9 = 82.5 \text{ miles per hour}$$

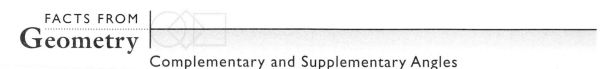

## FACTS FROM
# Geometry

### Complementary and Supplementary Angles

If you have studied geometry at all, you know that there are 360° in a full rotation: the number of degrees swept out by the radius of a circle as it rotates once around the circle.

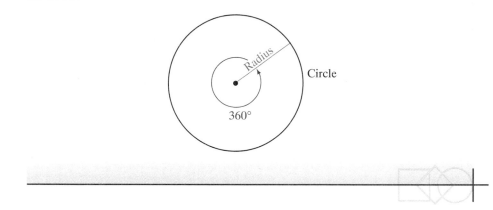

We can apply our knowledge of algebra to help solve some simple geometry problems. Before we do, however, we need to review some of the vocabulary associated with angles.

> **DEFINITION**   In geometry, two angles that add to 90° are called **complementary angles**. In a similar manner, two angles that add to 180° are called **supplementary angles**. The following diagrams illustrate the relationships between angles that are complementary and between angles that are supplementary.

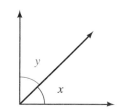

Complementary angles: $x + y = 90°$

Supplementary angles: $x + y = 180°$

**EXAMPLE 13**    Find *x* in each of the following diagrams.

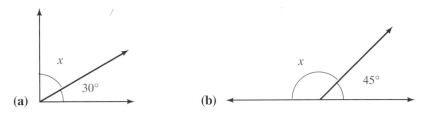

**(a)**                                        **(b)**

**Solution**    We use subtraction to find each angle.

(a) Since the two angles are complementary, we can find *x* by subtracting 30° from 90°:

$$x = 90° - 30° = 60°$$

We say 30° and 60° are complementary angles. The complement of 30° is 60°.

(b) The two angles in the diagram are supplementary. To find *x*, we subtract 45° from 180°:

$$x = 180° - 45° = 135°$$

We say 45° and 135° are supplementary angles. The supplement of 45° is 135°.

## SUBTRACTION AND TAKING AWAY

For some people taking algebra for the first time, subtraction of positive and negative numbers can be a problem. These people may believe that $-5 - 9$ should be $-4$ or 4, not $-14$. If this is happening to you, you are probably thinking of subtraction in terms of taking one number away from another. Thinking of subtraction in this way works well with positive numbers if you always subtract the smaller number from the larger. In algebra, however, we encounter many situations other than this. The definition of subtraction, that $a - b = a + (-b)$, clearly indicates the correct way to use subtraction. That is, when working subtraction problems, you should think "addition of the opposite," not "take one number away from another." To be successful in algebra, you need to apply properties and definitions exactly as they are presented here.

# PROBLEM SET 1.4

The following problems are intended to give you practice with subtraction of positive and negative numbers. Remember, in algebra subtraction is not taking one number away from another. Instead, subtracting a number is equivalent to adding its opposite.

Subtract.

1. $5 - 8$          2. $6 - 7$          3. $3 - 9$
4. $2 - 7$          5. $5 - 5$          6. $8 - 8$
7. $-8 - 2$         8. $-6 - 3$         9. $-4 - 12$
10. $-3 - 15$       11. $-6 - 6$        12. $-3 - 3$

**13.** $-8 - (-1)$    **14.** $-6 - (-2)$    **15.** $15 - (-20)$

**16.** $20 - (-5)$    **17.** $-4 - (-4)$    **18.** $-5 - (-5)$

Simplify each expression by applying the rule for order of operations.

**19.** $3 - 2 - 5$               **20.** $4 - 8 - 6$

**21.** $9 - 2 - 3$               **22.** $8 - 7 - 12$

**23.** $-6 - 8 - 10$            **24.** $-5 - 7 - 9$

**25.** $-22 + 4 - 10$          **26.** $-13 + 6 - 5$

**27.** $10 - (-20) - 5$         **28.** $15 - (-3) - 20$

**29.** $8 - (2 - 3) - 5$         **30.** $10 - (4 - 6) - 8$

**31.** $7 - (3 - 9) - 6$         **32.** $4 - (3 - 7) - 8$

**33.** $5 - (-8 - 6) - 2$        **34.** $4 - (-3 - 2) - 1$

**35.** $-(5 - 7) - (2 - 8)$      **36.** $-(4 - 8) - (2 - 5)$

**37.** $-(3 - 10) - (6 - 3)$     **38.** $-(3 - 7) - (1 - 2)$

**39.** $16 - [(4 - 5) - 1]$      **40.** $15 - [(4 - 2) - 3]$

**41.** $5 - [(2 - 3) - 4]$       **42.** $6 - [(4 - 1) - 9]$

**43.** $21 - [-(3 - 4) - 2] - 5$

**44.** $30 - [-(10 - 5) - 15] - 25$

The following problems involve multiplication and exponents. Use the rule for order of operations to simplify each expression as much as possible.

**45.** $2 \cdot 8 - 3 \cdot 5$          **46.** $3 \cdot 4 - 6 \cdot 7$

**47.** $3 \cdot 5 - 2 \cdot 7$          **48.** $6 \cdot 10 - 5 \cdot 20$

**49.** $5 \cdot 9 - 2 \cdot 3 - 6 \cdot 2$    **50.** $4 \cdot 3 - 7 \cdot 1 - 9 \cdot 4$

**51.** $3 \cdot 8 - 2 \cdot 4 - 6 \cdot 7$    **52.** $5 \cdot 9 - 3 \cdot 8 - 4 \cdot 5$

**53.** $2 \cdot 3^2 - 5 \cdot 2^2$       **54.** $3 \cdot 7^2 - 2 \cdot 8^2$

**55.** $4 \cdot 3^3 - 5 \cdot 2^3$       **56.** $3 \cdot 6^2 - 2 \cdot 3^2 - 8 \cdot 6^2$

Rewrite each of the following phrases as an equivalent expression in symbols and then simplify.

**57.** Subtract 4 from $-7$.       **58.** Subtract 5 from $-19$.

**59.** Subtract $-8$ from 12.      **60.** Subtract $-2$ from 10.

**61.** Subtract $-7$ from $-5$.     **62.** Subtract $-9$ from $-3$.

**63.** Subtract 17 from the sum of 4 and $-5$.

**64.** Subtract $-6$ from the sum of 6 and $-3$.

Recall that the word *difference* indicates subtraction. The difference of $a$ and $b$ is $a - b$, in that order. Write a numerical expression that is equivalent to each of the following phrases and then simplify.

**65.** The difference of 8 and 5.

**66.** The difference of 5 and 8.

**67.** The difference of $-8$ and 5.

**68.** The difference of $-5$ and 8.

**69.** The difference of 8 and $-5$.

**70.** The difference of 5 and $-8$.

Answer the following questions.

**71.** What number do you subtract from 8 to get $-2$?

**72.** What number do you subtract from 1 to get $-5$?

**73.** What number do you subtract from 8 to get 10?

**74.** What number do you subtract from 1 to get 5?

## Applying the Concepts

**75. Savings Account Balance**   A man with $1,500 in a savings account makes a withdrawal of $730. Write an expression using subtraction that describes this situation.

| First Bank Account No. 12345 | | | |
|---|---|---|---|
| Date | Withdrawals | Deposits | Balance |
| 1/1/99 | | | 1,500 |
| 2/2/99 | 730 | | |

**76. Temperature Change**   The temperature inside a space shuttle is 73°F before reentry. During reentry, the temperature inside the craft increases 10°. Upon landing it drops 8°F. Write an expression using the numbers 73, 10, and 8 to describe this situation. What is the temperature inside the shuttle upon landing?

**77. Gambling**   A man who has lost $35 playing roulette in Las Vegas wins $15 playing blackjack. He then loses $20 playing the wheel of fortune. Write an expression using the numbers $-35$, 15, and 20 to describe this situation and then simplify it.

**78. Altitude Change**   An airplane flying at 10,000 feet lowers its altitude by 1,500 feet to avoid other air traffic. Then it increases its altitude by 3,000 feet to clear a mountain range. Write an expression that describes this situation and then simplify it.

**79. Checkbook Balance**   Bob has $98 in his checking account when he writes a check for $65 and then another check for $53. Write a subtraction problem that gives the new balance in Bob's checkbook. What is his new balance?

**80. Temperature Change**   The temperature at noon is 23°F. Six hours later it has dropped 19°F, and by midnight it has dropped another 10°F. Write a subtraction problem that gives the temperature at midnight. What is the temperature at midnight?

**81. Depreciation** Stacey buys a used car for $4,500. With each year that passes, the car drops $550 in value. Write a sequence of numbers that gives the value of the car at the beginning of each of the first 5 years she owns it. Can this sequence be considered an arithmetic sequence?

**82. Depreciation** Wade buys a computer system for $6,575. Each year after that he finds that the system is worth $1,250 less than it was the year before. Write a sequence of numbers that gives the value of the computer system at the beginning of each of the first 4 years he owns it. Can this sequence be considered an arithmetic sequence?

**Drag Racing** Table 2 extends the information given in Table 1 of this section. In addition to showing the time and speed of Jim Rizzoli's dragster during a race, it also shows the distance past the starting line that his dragster has traveled. Use the information in Table 2 to answer Problems 83–88.

| TABLE 2 Speed and Distance for a Race Car | | |
|---|---|---|
| **Time in Seconds** | **Speed in Miles/Hour** | **Distance Traveled in Feet** |
| 0 | 0 | 0 |
| 1 | 72.7 | 69 |
| 2 | 129.9 | 231 |
| 3 | 162.8 | 439 |
| 4 | 192.2 | 728 |
| 5 | 212.4 | 1,000 |
| 6 | 228.1 | 1,373 |

**83.** Find the difference in the distance traveled by the dragster after 5 seconds and after 2 seconds.

**84.** How much faster is he traveling after 4 seconds than he is after 2 seconds?

**85.** How far from the starting line is he after 3 seconds?

**86.** How far from the starting line is he when his speed is 192.2 miles per hour?

**87.** How many seconds have gone by between the time his speed is 162.8 miles per hour and the time at which he has traveled 1,000 feet?

**88.** How many seconds have gone by between the time

at which he has traveled 231 feet and the time at which his speed is 228.1 miles per hour?

Look carefully at the relationship between the numbers in Table 1 of this section and the diagram in Figure 1 that appears next to it, then work the following problems.

**89. Garbage Production** The bar chart below shows the annual production of garbage in the United States for some specific years.
(a) Use the information in the bar chart to fill in the missing entries in the table.
(b) How much more garbage was there in 1990 than in 1970?

| Year | Garbage (millions of tons) |
|---|---|
| 1960 | |
| 1970 | |
| 1980 | |
| | 205 |
| 1997 | |

**The Growing Garbage Problem**

© *Tony Stone Images /Lonnie Duka*

**90. Grass Growth**   The bar chart below shows the growth of a certain species of grass over a period of 10 days.

(a) Use the chart to fill in the missing entries in the table.

| Day | Plant Height (inches) |
|-----|-----------------------|
| 0   | 0                     |
| 1   | 0.5                   |
| 2   |                       |
|     | 1.5                   |
| 4   |                       |
| 5   | 4                     |
| 6   |                       |
| 7   |                       |
|     | 13                    |
| 9   | 18                    |
| 10  |                       |

**Overall Plant Height**

(b) How much higher is the grass after 8 days than after 3 days?

**91. Wireless Phone Costs**   The bar chart below shows the projected cost of wireless phone use through 2003.

(a) Use the chart to fill in the missing entries in the table.

(b) What is the difference in cost between 1998 and 1999?

| Year | Cents/Minute |
|------|--------------|
| 1998 | 33           |
| 1999 |              |
| 2000 |              |
|      | 23           |
| 2002 |              |
|      | 20           |

**Wireless Talk**

92. **Temperature**  The following table and bar chart below both give the record low for each month of the year for Lake Placid, New York.
    (a) Find the difference of the April record low and the May record low.
    (b) Find the difference of the October record low and the November record low.

**Record Low Temperatures for Lake Placid, New York**

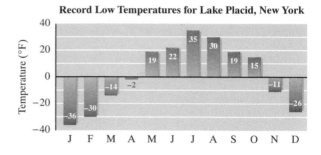

| Month | Temperature (°F) |
|---|---|
| January | −36 |
| February | −30 |
| March | −14 |
| April | −2 |
| May | 19 |
| June | 22 |
| July | 35 |
| August | 30 |
| September | 19 |
| October | 15 |
| November | −11 |
| December | −26 |

Find $x$ in each of the following diagrams.

93.  94.  95.  96.

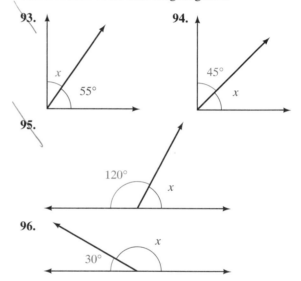

---

### 1.5  Properties of Real Numbers

In this section we will list all the facts (properties) that you know from past experience are true about numbers in general. We will give each property a name so we can refer to it later in the book. Mathematics is very much like a game. The game involves numbers. The rules of the game are the properties and rules we are developing in this chapter. The goal of the game is to extend the basic rules to as many new situations as possible.

You know from past experience with numbers that it makes no difference in which order you add two numbers. That is, $3 + 5$ is the same as $5 + 3$. This fact about numbers is called the *commutative property of addition*. We say addition is a commutative operation. Changing the order of the numbers does not change the answer.

There is one other basic operation that is commutative. Since $3(5)$ is the same as $5(3)$, we say multiplication is a commutative operation. Changing the order of the two numbers you are multiplying does not change the answer.

For all properties listed in this section, $a$, $b$, and $c$ represent real numbers.

### COMMUTATIVE PROPERTY OF ADDITION

*In symbols:*   $a + b = b + a$
*In words:*     Changing the *order* of the numbers in a sum will not change the result.

### COMMUTATIVE PROPERTY OF MULTIPLICATION

*In symbols:*   $a \cdot b = b \cdot a$
*In words:*     Changing the *order* of the numbers in a product will not change the result.

### EXAMPLES

1. The statement $5 + 8 = 8 + 5$ is an example of the commutative property of addition.
2. The statement $2 \cdot y = y \cdot 2$ is an example of the commutative property of multiplication.
3. The expression $5 + x + 3$ can be simplified using the commutative property of addition:

$$5 + x + 3 = x + 5 + 3 \qquad \text{Commutative property of addition}$$
$$= x + 8 \qquad \text{Addition}$$

The other two basic operations, subtraction and division, are not commutative. The order in which we subtract or divide two numbers makes a difference in the answer.

Another property of numbers that you have used many times has to do with grouping. You know that when we add three numbers it makes no difference which two we add first. When adding $3 + 5 + 7$, we can add the 3 and 5 first and then the 7, or we can add the 5 and 7 first and then the 3. Mathematically, it looks like this: $(3 + 5) + 7 = 3 + (5 + 7)$. This property is true of multiplication as well. Operations that behave in this manner are called *associative* operations. The answer will not change when we change the association (or grouping) of the numbers.

### ASSOCIATIVE PROPERTY OF ADDITION

*In symbols:*   $a + (b + c) = (a + b) + c$
*In words:*     Changing the *grouping* of the numbers in a sum will not change the result.

### ASSOCIATIVE PROPERTY OF MULTIPLICATION

*In symbols:*   $a(bc) = (ab)c$
*In words:*     Changing the *grouping* of the numbers in a product will not change the result.

The following examples illustrate how the associative properties can be used to simplify expressions that involve both numbers and variables.

**E X A M P L E S**   Simplify.

**4.** $4 + (5 + x) = (4 + 5) + x$      Associative property of addition

$\qquad\qquad = 9 + x$      Addition

**5.** $5(2x) = (5 \cdot 2)x$      Associative property of multiplication

$\qquad = 10x$      Multiplication

**6.** $\dfrac{1}{5}(5x) = \left(\dfrac{1}{5} \cdot 5\right)x$      Associative property of multiplication

$\qquad\quad = 1x$      Multiplication

$\qquad\quad = x$

The associative and commutative properties apply to problems that are either all multiplication or all addition. There is a third basic property that involves both addition and multiplication. It is called the *distributive property* and looks like this.

**DISTRIBUTIVE PROPERTY**

*In symbols:*      $a(b + c) = ab + ac$
*In words:*      Multiplication *distributes* over addition.

**Note:**  Since subtraction is defined in terms of addition, it is also true that the distributive property applies to subtraction as well as addition. That is, $a(b - c) = ab - ac$ for any three real numbers $a$, $b$, and $c$.

You will see as we progress through the book that the distributive property is used very frequently in algebra. We can give a visual justification to the distributive property by finding the areas of rectangles. Figure 1 shows a large rectangle that is made up of two smaller rectangles.

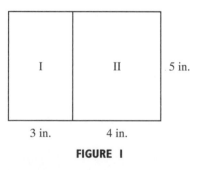

**FIGURE I**

We can find the area of the large rectangle two different ways.

**Method 1**  We can calculate the area of the large rectangle directly by finding its length and width. The width is 5 inches, and the length is (3 + 4) inches.

$$\text{Area of large rectangle} = 5(3 + 4)$$
$$= 5(7)$$
$$= 35 \text{ square inches}$$

***Method 2***   Since the area of the large rectangle is the sum of the areas of the two smaller rectangles, we find the area of each small rectangle and then add to find the area of the large rectangle.

| Area of large rectangle | = | Area of rectangle I | + | Area of rectangle II |
|---|---|---|---|---|
| | = | 5(3) | + | 5(4) |
| | = | 15 | + | 20 |
| | = | | 35 square inches | |

In both cases the result is 35 square inches. Since the results are the same, the two original expressions must be equal. Stated mathematically, $5(3 + 4) = 5(3) + 5(4)$. We can either add the 3 and the 4 first and then multiply that sum by 5, or we can multiply the 3 and the 4 separately by 5 and then add the results. In either case we get the same answer.

Here are some examples that illustrate how we use the distributive property.

**EXAMPLES**   Apply the distributive property to each expression and then simplify the result.

**7.** $2(x + 3) = 2(x) + 2(3)$        Distributive property
$\phantom{2(x + 3)} = 2x + 6$             Multiplication

**8.** $5(2x - 8) = 5(2x) - 5(8)$      Distributive property
$\phantom{5(2x - 8)} = 10x - 40$         Multiplication

Notice in this example that multiplication distributes over subtraction as well as addition.

**9.** $4(x + y) = 4x + 4y$                    Distributive property

**10.** $5(2x + 4y) = 5(2x) + 5(4y)$       Distributive property
$\phantom{5(2x + 4y)} = 10x + 20y$         Multiplication

**11.** $\dfrac{1}{2}(3x + 6) = \dfrac{1}{2}(3x) + \dfrac{1}{2}(6)$     Distributive property

$\phantom{\dfrac{1}{2}(3x + 6)} = \dfrac{3}{2}x + 3$               Multiplication

**12.** $4(2a + 3) + 8 = 4(2a) + 4(3) + 8$    Distributive property
$\phantom{4(2a + 3) + 8} = 8a + 12 + 8$       Multiplication
$\phantom{4(2a + 3) + 8} = 8a + 20$           Addition

## SPECIAL NUMBERS

In addition to the three properties mentioned so far, we want to include in our list two special numbers that have unique properties. They are the numbers zero and one.

## ADDITIVE IDENTITY PROPERTY

There exists a unique number 0 such that
*In symbols:*    $a + 0 = a$   and   $0 + a = a$
*In words:*    Zero preserves identities under addition. (The identity of the number is unchanged after addition with 0.)

## MULTIPLICATIVE IDENTITY PROPERTY

There exists a unique number 1 such that
*In symbols:*    $a(1) = a$   and   $1(a) = a$
*In words:*    The number 1 preserves identities under multiplication. (The identity of the number is unchanged after multiplication by 1.)

## ADDITIVE INVERSE PROPERTY

For each real number $a$, there exists a unique number $-a$ such that
*In symbols:*    $a + (-a) = 0$
*In words:*    Opposites add to 0.

## MULTIPLICATIVE INVERSE PROPERTY

For every real number $a$, except 0, there exists a unique real number $\dfrac{1}{a}$ such that

*In symbols:*    $a\left(\dfrac{1}{a}\right) = 1$

*In words:*    Reciprocals multiply to 1.

Of all the basic properties listed, the commutative, associative, and distributive properties are the ones we will use most often. They are important because they will be used as justifications or reasons for many of the things we will do in the future.

The following examples illustrate how we use the preceding properties. Each one contains an algebraic expression that has been changed in some way. The property that justifies the change is written to the right.

**EXAMPLES**   State the property that justifies the given statement.

**13.** $x + 5 = 5 + x$             Commutative property of addition

**14.** $(2 + x) + y = 2 + (x + y)$     Associative property of addition

**15.** $6(x + 3) = 6x + 18$         Distributive property

**16.** $2 + (-2) = 0$              Additive inverse property

**17.** $3\left(\dfrac{1}{3}\right) = 1$            Multiplicative inverse property

**18.** $(2 + 0) + 3 = 2 + 3$      0 is the identity element for addition.

**19.** $(2 + 3) + 4 = 3 + (2 + 4)$      Commutative and associative properties of addition

**20.** $(x + 2) + y = (x + y) + 2$      Commutative and associative properties of addition

As a final note on the properties of real numbers, we should mention that although some of the properties are stated for only two or three real numbers, they hold for as many numbers as needed. For example, the distributive property holds for expressions like $3(x + y + z + 5 + 2)$. That is,

$$3(x + y + z + 5 + 2) = 3x + 3y + 3z + 15 + 6$$

It is not important how many numbers are contained in the sum, only that it is a sum. Multiplication, you see, distributes over addition, whether there are 2 numbers in the sum or 200.

## PROBLEM SET 1.5

State the property or properties that justify the following.

**1.** $3 + 2 = 2 + 3$

**2.** $5 + 0 = 5$

**3.** $4\left(\dfrac{1}{4}\right) = 1$

**4.** $10(0.1) = 1$

**5.** $4 + x = x + 4$

**6.** $3(x - 10) = 3x - 30$

**7.** $2(y + 8) = 2y + 16$

**8.** $3 + (4 + 5) = (3 + 4) + 5$

**9.** $(3 + 1) + 2 = 1 + (3 + 2)$

**10.** $(5 + 2) + 9 = (2 + 5) + 9$

**11.** $(8 + 9) + 10 = (8 + 10) + 9$

**12.** $(7 + 6) + 5 = (5 + 6) + 7$

**13.** $3(x + 2) = 3(2 + x)$    **14.** $2(7y) = (7 \cdot 2)y$

**15.** $x(3y) = 3(xy)$    **16.** $a(5b) = 5(ab)$

**17.** $4(xy) = 4(yx)$    **18.** $3[2 + (-2)] = 3(0)$

**19.** $8[7 + (-7)] = 8(0)$    **20.** $7(1) = 7$

Each of the following problems has a mistake in it. Correct the right-hand side.

**21.** $3(x + 2) = 3x + 2$    **22.** $5(4 + x) = 4 + 5x$

**23.** $9(a + b) = 9a + b$    **24.** $2(y + 1) = 2y + 1$

**25.** $3(0) = 3$    **26.** $5\left(\dfrac{1}{5}\right) = 5$

**27.** $3 + (-3) = 1$    **28.** $8(0) = 8$

**29.** $10(1) = 0$    **30.** $3 \cdot \dfrac{1}{3} = 0$

Use the associative property to rewrite each of the following expressions and then simplify the result. (See Examples 4, 5, and 6.)

**31.** $4 + (2 + x)$    **32.** $5 + (6 + x)$

**33.** $(x + 2) + 7$    **34.** $(x + 8) + 2$

**35.** $3(5x)$    **36.** $5(3x)$

**37.** $9(6y)$    **38.** $6(9y)$

**39.** $\dfrac{1}{2}(3a)$    **40.** $\dfrac{1}{3}(2a)$

**41.** $\dfrac{1}{3}(3x)$    **42.** $\dfrac{1}{4}(4x)$

**43.** $\dfrac{1}{2}(2y)$    **44.** $\dfrac{1}{7}(7y)$

**45.** $\dfrac{3}{4}\left(\dfrac{4}{3}x\right)$    **46.** $\dfrac{3}{2}\left(\dfrac{2}{3}x\right)$

**47.** $\dfrac{6}{5}\left(\dfrac{5}{6}a\right)$    **48.** $\dfrac{2}{5}\left(\dfrac{5}{2}a\right)$

Apply the distributive property to each of the following expressions. Simplify when possible.

**49.** $8(x + 2)$    **50.** $5(x + 3)$

**51.** $8(x - 2)$    **52.** $5(x - 3)$

**53.** $4(y + 1)$    **54.** $4(y - 1)$

**55.** $3(6x + 5)$    **56.** $3(5x + 6)$

**57.** $2(3a + 7)$    **58.** $5(3a + 2)$

**59.** $9(6y - 8)$

**60.** $2(7y - 4)$

**61.** $\frac{1}{2}(3x - 6)$

**62.** $\frac{1}{3}(2x - 6)$

**63.** $\frac{1}{3}(3x + 6)$

**64.** $\frac{1}{2}(2x + 4)$

**65.** $3(x + y)$

**66.** $2(x - y)$

**67.** $8(a - b)$

**68.** $7(a + b)$

**69.** $6(2x + 3y)$

**70.** $8(3x + 2y)$

**71.** $4(3a - 2b)$

**72.** $5(4a - 8b)$

**73.** $\frac{1}{2}(6x + 4y)$

**74.** $\frac{1}{3}(6x + 9y)$

**75.** $4(a + 4) + 9$

**76.** $6(a + 2) + 8$

**77.** $2(3x + 5) + 2$

**78.** $7(2x + 1) + 3$

**79.** $7(2x + 4) + 10$

**80.** $3(5x + 6) + 20$

## Applying the Concepts

**81. Getting Dressed**   While getting dressed for work, a man puts on his socks and puts on his shoes. Are the two statements "put on your socks" and "put on your shoes" commutative?

**82. Getting Dressed**   Are the statements "put on your left shoe" and "put on your right shoe" commutative?

**83. Skydiving**   A skydiver flying over the jump area is about to do two things: jump out of the plane and pull the rip cord. Are the two events "jump out of the plane" and "pull the rip cord" commutative? That is, will changing the order of the events always produce the same result?

**84. Commutative Property**   Give an example of two events in your daily life that are commutative.

**85. Division**   Give an example that shows that division is not a commutative operation. That is, find two numbers for which changing the order of division gives two different answers.

**86. Subtraction**   Simplify the expression $10 - (5 - 2)$ and the expression $(10 - 5) - 2$ to show that subtraction is not an associative operation.

**87. Take-Home Pay**   Jose works at a winery. His monthly salary is $2,400. To cover his taxes and retirement, the winery withholds $480 from each check. Calculate his yearly "take-home" pay using the numbers 2,400, 480, and 12. Do the calculation two different ways so that the results give further justification for the distributive property.

**88. Hours Worked**   Carlo works as a waiter. He works 4 days a week. The lunch shift is 2 hours, and the dinner shift is 3 hours. Find the total number of hours he works per week, using the numbers 2, 3, and 4. Do the calculation two different ways so that the results give further justification for the distributive property.

**89. Distributive Property**   Use the distributive property to rewrite the formula for the perimeter of a rectangle: $P = 2l + 2w$.

**90. College Expenses**   Maria is estimating her expenses for attending college for a year. Tuition is $650 per academic quarter. She estimates she will spend $225 on books each quarter. If she plans on attending 3 academic quarters during the year, how much can she expect to spend? Do the calculation two different ways so that the results give further justification for the distributive property.

## 1.6  Multiplication of Real Numbers

Suppose that you own 5 shares of a stock and the price per share drops $3. How much money have you lost? Intuitively, we know the loss is $15. Since it is a loss, we can express it as $-$15. To describe this situation with numbers, we would write

5 shares   each loses $3   for a total of $15

$$5(-3) = -15$$

Reasoning in this manner, we conclude that the product of a positive number with a negative number is a negative number. Let's look at multiplication in more detail.

From our experience with counting numbers, we know that multiplication is simply repeated addition. That is, $3(5) = 5 + 5 + 5$. We will use this fact, along with our knowledge of negative numbers, to develop the rule for multiplication of any two real numbers. The following examples illustrate multiplication with three of the possible combinations of positive and negative numbers.

**E X A M P L E S**    Multiply.

**1.** Two positives:     $3(5) = 5 + 5 + 5$
$$= 15 \qquad \text{Positive answer}$$

**2.** One positive:     $3(-5) = -5 + (-5) + (-5)$
$$= -15 \qquad \text{Negative answer}$$

**3.** One negative:     $-3(5) = 5(-3) \qquad \text{Commutative property}$
$$= -3 + (-3) + (-3) + (-3) + (-3)$$
$$= -15 \qquad \text{Negative answer}$$

**4.** Two negatives:     $-3(-5) = ?$

With two negatives, $-3(-5)$, it is not possible to work the problem in terms of repeated addition. (It doesn't "make sense" to write $-5$ down a $-3$ number of times.) The answer is probably $+15$ (that's just a guess), but we need some justification for saying so. We will solve a different problem and, in so doing, get the answer to the problem $(-3)(-5)$.

Here is a problem to which we know the answer. We will work it two different ways.

$$-3[5 + (-5)] = -3(0) = 0$$

The answer is zero. We can also work the problem using the distributive property.

$$-3[5 + (-5)] = -3(5) + (-3)(-5) \qquad \text{Distributive property}$$
$$= -15 \ + \qquad ?$$

Since the answer to the problem is 0, our ? must be $+15$. (What else could we add to $-15$ to get 0? Only $+15$.)

**Note:**  You may have to read the explanation for Example 4 several times before you understand it completely. The purpose of the explanation in Example 4 is simply to justify the fact that the product of two negative numbers is a positive number. If you have no trouble believing that, then it is not so important that you understand everything in the explanation.

Here is a summary of the results we have obtained from the first four examples:

| Original Numbers Have | | The Answer Is |
|---|---|---|
| The same sign | $3(5) = 15$ | Positive |
| Different signs | $3(-5) = -15$ | Negative |
| Different signs | $-3(5) = -15$ | Negative |
| The same sign | $-3(-5) = 15$ | Positive |

By examining Examples 1 through 4 and the preceding table, we can use the information there to write the following rule. This rule tells us how to multiply any two real numbers.

### RULE

To multiply any two real numbers, simply multiply their absolute values. The sign of the answer is

**1.** *Positive* if both numbers have the same sign (both + or both −).
**2.** *Negative* if the numbers have opposite signs (one +, the other −).

The following examples illustrate how we use the preceding rule to multiply real numbers.

**EXAMPLES**  Multiply.

**5.** $-8(-3) = 24$
**6.** $-10(-5) = 50$    If the two numbers in the product have the same sign, the answer is positive.
**7.** $-4(-7) = 28$

**8.** $5(-7) = -35$
**9.** $-4(8) = -32$    If the two numbers in the product have different signs, the answer is negative.
**10.** $-6(10) = -60$

*Note:*  Some students have trouble with the expression $-8(-3)$ because they want to subtract rather than multiply. Because we are very precise with the notation we use in algebra, the expression $-8(-3)$ has only one meaning — multiplication. A subtraction problem that uses the same numbers is $-8 - 3$. Compare the two following lists.

| *All Multiplication* | *No Multiplication* |
|:---:|:---:|
| $5(4)$ | $5 + 4$ |
| $-5(4)$ | $-5 + 4$ |
| $5(-4)$ | $5 - 4$ |
| $-5(-4)$ | $-5 - 4$ |

In the following examples, we combine the rule for order of operations with the rule for multiplication to simplify expressions. Remember, the rule for order of operations specifies that we are to work inside the parentheses first and then simplify numbers containing exponents. After this, we multiply and divide, left to right. The last step is to add and subtract, left to right.

**EXAMPLES**  Simplify as much as possible.

**11.** $-5(-3)(-4) = 15(-4)$
$$= -60$$

**12.** $4(-3) + 6(-5) - 10 = -12 + (-30) - 10$      Multiply.

$$= -42 - 10 \qquad \text{Add.}$$

$$= -52 \qquad \text{Subtract.}$$

**13.** $(-2)^3 = (-2)(-2)(-2)$      Definition of exponents

$$= -8 \qquad \text{Multiply, left to right.}$$

**14.** $-3(-2)^3 - 5(-4)^2 = -3(-8) - 5(16)$      Exponents first

$$= 24 - 80 \qquad \text{Multiply.}$$

$$= -56 \qquad \text{Subtract.}$$

**15.** $6 - 4(7 - 2) = 6 - 4(5)$      Inside parentheses first

$$= 6 - 20 \qquad \text{Multiply.}$$

$$= -14 \qquad \text{Subtract.}$$

## MULTIPLYING FRACTIONS

Previously, we mentioned that to multiply two fractions we multiply numerators and multiply denominators. We can apply the rule for multiplication of positive and negative numbers to fractions in the same way we apply it to other numbers. We multiply absolute values: The product is positive if both fractions have the same sign and negative if they have different signs. Here are some examples.

**EXAMPLES**  Multiply.

**16.** $-\dfrac{3}{4}\left(\dfrac{5}{7}\right) = -\dfrac{3 \cdot 5}{4 \cdot 7}$      Different signs give a negative answer.

$$= -\dfrac{15}{28}$$

**17.** $-6\left(\dfrac{1}{2}\right) = -\dfrac{6}{1}\left(\dfrac{1}{2}\right)$      Different signs give a negative answer.

$$= -\dfrac{6}{2}$$

$$= -3$$

**18.** $-\dfrac{2}{3}\left(-\dfrac{3}{2}\right) = \dfrac{2 \cdot 3}{3 \cdot 2}$      Same signs give a positive answer.

$$= \dfrac{6}{6}$$

$$= 1$$

**EXAMPLE 19**   Figure 1 gives the calories that are burned in 1 hour for a variety of forms of exercise by a person weighing 150 pounds. Figure 2 gives the calories that are consumed by eating some popular fast foods. Find the net change in calories for a 150-pound person playing handball for 2 hours and then eating a Whopper.

**FIGURE 1**

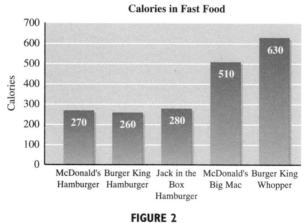

**FIGURE 2**

**Solution**    The net change in calories will be the difference of the calories gained from eating and the calories lost from exercise.

$$\text{Net change in calories} = 630 - 2(680) = -730 \text{ calories}$$

We can use the rule for multiplication of real numbers, along with the associative property, to multiply expressions that contain numbers and variables.

**E X A M P L E S**    Apply the associative property and then multiply.

**20.** $-3(2x) = (-3 \cdot 2)x$          Associative property
$\quad\quad\quad\quad = -6x$                   Multiplication

**21.** $6(-5y) = [6(-5)]y$            Associative property
$\quad\quad\quad\quad = -30y$                 Multiplication

**22.** $-2\left(-\dfrac{1}{2}x\right) = \left[(-2)\left(-\dfrac{1}{2}\right)\right]x$     Associative property

$\qquad\qquad = 1x$     Multiplication

$\qquad\qquad = x$     Multiplication

The following examples show how we can use both the distributive property and multiplication with real numbers.

**EXAMPLES**   Apply the distributive property to each expression.

**23.** $-2(a + 3) = -2a + (-2)(3)$     Distributive property

$\qquad\qquad = -2a + (-6)$     Multiplication

$\qquad\qquad = -2a - 6$

**24.** $-3(2x + 1) = -3(2x) + (-3)(1)$     Distributive property

$\qquad\qquad = -6x + (-3)$     Multiplication

$\qquad\qquad = -6x - 3$

**25.** $-\dfrac{1}{3}(2x - 6) = -\dfrac{1}{3}(2x) - \left(-\dfrac{1}{3}\right)(6)$     Distributive property

$\qquad\qquad = -\dfrac{2}{3}x - (-2)$     Multiplication

$\qquad\qquad = -\dfrac{2}{3}x + 2$

**26.** $-4(3x - 5) - 8 = -4(3x) - (-4)(5) - 8$     Distributive property

$\qquad\qquad = -12x - (-20) - 8$     Multiplication

$\qquad\qquad = -12x + 20 - 8$     Definition of subtraction

$\qquad\qquad = -12x + 12$     Subtraction

## GEOMETRIC SEQUENCES

A *geometric sequence* is a sequence of numbers in which each number (after the first number) comes from multiplying the number before it by the same amount each time. For example, the sequence

$$2, 6, 18, 54, \ldots$$

is a geometric sequence because each number is obtained by multiplying the number before it by 3.

**EXAMPLE  27**   Each of the following is a geometric sequence. Find the next number in each sequence.

(a) 5, 10, 20, . . .

(b) 3, −15, 75, . . .

(c) $\dfrac{1}{8}, \dfrac{1}{4}, \dfrac{1}{2}, \ldots$

**Solution** Since each sequence is a geometric sequence, we know that each term is obtained from the previous term by multiplying by the same number each time.

(a) 5, 10, 20, . . . : Starting with 5, each number is obtained from the previous number by multiplying by 2 each time, The next number will be $20 \cdot 2 = 40$.

(b) 3, −15, 75, . . . : The sequence starts with 3. After that, each number is obtained by multiplying by −5 each time. The next number will be $75(-5) = -375$.

(c) $\dfrac{1}{8}, \dfrac{1}{4}, \dfrac{1}{2}, \ldots$ : This sequence starts with $\dfrac{1}{8}$. Multiplying each number in the sequence by 2 produces the next number in the sequence. To extend the sequence, we multiply $\dfrac{1}{2}$ by 2:

$$\frac{1}{2} \cdot 2 = 1$$

The next number in the sequence is 1.

# PROBLEM SET 1.6

Use the rule for multiplying two real numbers to find each of the following products.

**1.** $7(-6)$   **2.** $8(-4)$   **3.** $-7(3)$

**4.** $-5(4)$   **5.** $-8(2)$   **6.** $-16(3)$

**7.** $-3(-1)$   **8.** $-7(-1)$   **9.** $-11(-11)$

**10.** $-12(-12)$

Use the rule for order of operations to simplify each expression as much as possible.

**11.** $-3(2)(-1)$   **12.** $-2(3)(-4)$

**13.** $-3(-4)(-5)$   **14.** $-5(-6)(-7)$

**15.** $-2(-4)(-3)(-1)$   **16.** $-1(-3)(-2)(-1)$

**17.** $(-7)^2$   **18.** $(-8)^2$

**19.** $(-3)^3$   **20.** $(-2)^4$

**21.** $-2(2 - 5)$   **22.** $-3(3 - 7)$

**23.** $-5(8 - 10)$   **24.** $-4(6 - 12)$

**25.** $(4 - 7)(6 - 9)$   **26.** $(3 - 10)(2 - 6)$

**27.** $(-3 - 2)(-5 - 4)$   **28.** $(-3 - 6)(-2 - 8)$

**29.** $-3(-6) + 4(-1)$   **30.** $-4(-5) + 8(-2)$

**31.** $2(3) - 3(-4) + 4(-5)$

**32.** $5(4) - 2(-1) + 5(6)$   **33.** $4(-3)^2 + 5(-6)^2$

**34.** $2(-5)^2 + 4(-3)^2$   **35.** $7(-2)^3 - 2(-3)^3$

**36.** $10(-2)^3 - 5(-2)^4$   **37.** $6 - 4(8 - 2)$

**38.** $7 - 2(6 - 3)$   **39.** $9 - 4(3 - 8)$

**40.** $8 - 5(2 - 7)$

**41.** $-4(3 - 8) - 6(2 - 5)$

**42.** $-8(2 - 7) - 9(3 - 5)$   **43.** $7 - 2[-6 - 4(-3)]$

**44.** $6 - 3[-5 - 3(-1)]$

**45.** $7 - 3[2(-4 - 4) - 3(-1 - 1)]$

**46.** $5 - 3[7(-2 - 2) - 3(-3 + 1)]$

**47.** $8 - 6[-2(-3 - 1) + 4(-2 - 3)]$

**48.** $4 - 2[-3(-1 + 8) + 5(-5 + 7)]$

Multiply the following fractions. (See Examples 16–19.)

**49.** $-\dfrac{2}{3} \cdot \dfrac{5}{7}$   **50.** $-\dfrac{6}{5} \cdot \dfrac{2}{7}$

**51.** $-8\left(\dfrac{1}{2}\right)$

**52.** $-12\left(\dfrac{1}{3}\right)$

**53.** $-\dfrac{3}{4}\left(-\dfrac{4}{3}\right)$

**54.** $-\dfrac{5}{8}\left(-\dfrac{8}{5}\right)$

**55.** $\left(-\dfrac{3}{4}\right)^2$

**56.** $\left(-\dfrac{2}{5}\right)^2$

**57.** $\left(-\dfrac{2}{3}\right)^3$

**58.** $\left(-\dfrac{1}{2}\right)^3$

Find the following products. (See Examples 20, 21, and 22.)

**59.** $-2(4x)$    **60.** $-8(7x)$    **61.** $-7(-6x)$

**62.** $-8(-9x)$    **63.** $-\dfrac{1}{3}(-3x)$    **64.** $-\dfrac{1}{5}(-5x)$

**65.** $-4\left(-\dfrac{1}{4}x\right)$    **66.** $-2\left(-\dfrac{1}{2}x\right)$

Apply the distributive property to each expression; then simplify the result. (See Examples 23–26.)

**67.** $-4(a+2)$    **68.** $-7(a+6)$

**69.** $-\dfrac{1}{2}(3x-6)$    **70.** $-\dfrac{1}{4}(2x-4)$

**71.** $-3(2x-5)-7$    **72.** $-4(3x-1)-8$

**73.** $-5(3x+4)-10$    **74.** $-3(4x+5)-20$

**75.** Five added to the product of 3 and $-10$ is what number?

**76.** If the product of $-8$ and $-2$ is decreased by 4, what number results?

**77.** Write an expression for twice the product of $-4$ and $x$, and then simplify it.

**78.** Write an expression for twice the product of $-2$ and $3x$, and then simplify it.

**79.** What number results if 8 is subtracted from the product of $-9$ and 2?

**80.** What number results if $-8$ is subtracted from the product of $-9$ and 2?

Each of the following is a geometric sequence. In each case, find the next number in the sequence.

**81.** 1, 2, 4, . . .    **82.** 1, 5, 25, . . .

**83.** 10, $-20$, 40, . . .    **84.** 10, $-30$, 90, . . .

**85.** 1, $\dfrac{1}{2}$, $\dfrac{1}{4}$, . . .    **86.** 1, $\dfrac{1}{3}$, $\dfrac{1}{9}$, . . .

**87.** 8, 4, 2, . . .    **88.** 20, 10, 5, . . .

**89.** 3, $-6$, 12, . . .    **90.** $-3$, 6, $-12$, . . .

## Applying the Concepts

**91. Stock Value**  Suppose you own 20 shares of a stock. If the price per share drops $3, how much money have you lost?

**92. Stock Value**  Imagine that you purchase 50 shares of a stock at a price of $18 per share. If the stock is selling for $11 a share a week after you purchased it, how much money have you lost?

**93. Temperature Change**  The temperature is 25°F at 5:00 in the afternoon. If the temperature drops 6°F every hour after that, what is the temperature at 9:00 in the evening?

**94. Temperature Change**  The temperature is $-5$°F at 6:00 in the evening. If the temperature drops 3°F every hour after that, what is the temperature at midnight?

**95. Investment Value**  Suppose you purchase $500 worth of a mutual fund and find that the value of your purchase doubles every 2 years. Write a sequence of numbers that gives the value of your purchase every 2 years for the first 10 years you own it. Is this sequence a geometric sequence?

**96. Number of Bacteria**  A colony of 150 bacteria doubles in size every 8 hours. Starting with the original 150 bacteria, write a sequence of numbers that gives the number of bacteria in the colony every 8 hours during a 40-hour time span. Is this sequence a geometric sequence?

**97. Reading Charts**  Refer to Figures 1 and 2 in Example 19 in this section to find the net change in calories for a 150-pound person who bowls for 3 hours and then eats 2 Whoppers.

**98. Reading Charts**  Refer to Figures 1 and 2 in Example 19 in this section to find the net change in calories for a 150-pound person who eats a Big Mac and then skis for 3 hours.

Suppose that you and four friends bought equal shares of an investment for a total of $15,000 and then sold it later for only $13,000. How much did each person lose? Since the total amount of money that was lost can be represented by −$2,000 and there are 5 people with equal shares, we can represent each person's loss with division:

$$\frac{-\$2,000}{5} = -\$400$$

From this discussion it seems reasonable to say that a negative number divided by a positive number is a negative number. Here is a more detailed discussion of division with positive and negative numbers.

The last of the four basic operations is division. We will use the same approach to define division as we used for subtraction. That is, we will define division in terms of rules we already know.

Recall that we developed the rule for subtraction of real numbers by defining subtraction in terms of addition. We changed our subtraction problems to addition problems and then added to get our answers. Since we already have a rule for multiplication of real numbers and division is the inverse operation of multiplication, we will simply define division in terms of multiplication.

We know that division by the number 2 is the same as multiplication by $\frac{1}{2}$. That is, 6 divided by 2 is 3, which is the same as 6 times $\frac{1}{2}$. Similarly, dividing a number by 5 gives the same result as multiplying by $\frac{1}{5}$. We can extend this idea to all real numbers with the following rule.

> **RULE**
>
> If $a$ and $b$ represent any two real numbers ($b$ cannot be 0), then it is always true that
>
> $$a \div b = \frac{a}{b} = a\left(\frac{1}{b}\right)$$

Division by a number is the same as multiplication by its reciprocal. Since every division problem can be written as a multiplication problem and since we already know the rule for multiplication of two real numbers, we do not have to write a new rule for division of real numbers. We will simply replace our division problem with multiplication and use the rule we already have.

**EXAMPLES** Write each division problem as an equivalent multiplication problem and then multiply.

**1.** $\dfrac{6}{2} = 6\left(\dfrac{1}{2}\right) = 3$ The product of two positives is positive.

**2.** $\dfrac{6}{-2} = 6\left(-\dfrac{1}{2}\right) = -3$

**3.** $\dfrac{-6}{2} = -6\left(\dfrac{1}{2}\right) = -3$

The product of a positive and a negative is a negative.

**4.** $\dfrac{-6}{-2} = -6\left(-\dfrac{1}{2}\right) = 3$      The product of two negatives is positive.

---

The second step in these examples is used only to show that we *can* write division in terms of multiplication. (In actual practice we wouldn't write $\frac{6}{2}$ as $6(\frac{1}{2})$.) The answers, therefore, follow from the rule for multiplication. That is, like signs produce a positive answer, and unlike signs produce a negative answer.

Here are some examples. This time we will not show division as multiplication by the reciprocal. We will simply divide. If the original numbers have the same signs, the answer will be positive. If the original numbers have different signs, the answer will be negative.

**EXAMPLES**   Divide.

**5.** $\dfrac{12}{6} = 2$            Like signs give a positive answer.

**6.** $\dfrac{12}{-6} = -2$           Unlike signs give a negative answer.

**7.** $\dfrac{-12}{6} = -2$           Unlike signs give a negative answer.

**8.** $\dfrac{-12}{-6} = 2$           Like signs give a positive answer.

**9.** $\dfrac{15}{-3} = -5$           Unlike signs give a negative answer.

**10.** $\dfrac{-40}{-5} = 8$           Like signs give a positive answer.

**11.** $\dfrac{-14}{2} = -7$           Unlike signs give a negative answer.

---

## DIVISION WITH FRACTIONS

We can apply the definition of division to fractions. Since dividing by a fraction is equivalent to multiplying by its reciprocal, we can divide a number by the fraction $\frac{3}{4}$ by multiplying it by the reciprocal of $\frac{3}{4}$, which is $\frac{4}{3}$. For example,

$$\frac{2}{5} \div \frac{3}{4} = \frac{2}{5} \cdot \frac{4}{3} = \frac{8}{15}$$

You may have learned this rule in previous math classes. In some math classes, multiplication by the reciprocal is referred to as "inverting the divisor and multiplying." No matter how you say it, division by any number (except 0) is always equivalent to multiplication by its reciprocal. Here are additional examples that involve division by fractions.

**EXAMPLES** Divide.

**12.** $\dfrac{2}{3} \div \dfrac{5}{7} = \dfrac{2}{3} \cdot \dfrac{7}{5}$  Rewrite as multiplication by the reciprocal.

$= \dfrac{14}{15}$  Multiply.

**13.** $-\dfrac{3}{4} \div \dfrac{7}{9} = -\dfrac{3}{4} \cdot \dfrac{9}{7}$  Rewrite as multiplication by the reciprocal.

$= -\dfrac{27}{28}$  Multiply.

**14.** $8 \div \left(-\dfrac{4}{5}\right) = \dfrac{8}{1}\left(-\dfrac{5}{4}\right)$  Rewrite as multiplication by the reciprocal.

$= -\dfrac{40}{4}$  Multiply.

$= -10$  Divide 40 by 4.

The last step in each of the following examples involves reducing a fraction to lowest terms. To reduce a fraction to lowest terms, we divide the numerator and denominator by the largest number that divides each of them exactly. For example, to reduce $\frac{15}{20}$ to lowest terms, we divide 15 and 20 by 5 to get $\frac{3}{4}$.

**EXAMPLES** Simplify as much as possible.

**15.** $\dfrac{-4(5)}{6} = \dfrac{-20}{6}$  Simplify the numerator.

$= -\dfrac{10}{3}$  Reduce to lowest terms by dividing the numerator and denominator by 2.

**16.** $\dfrac{30}{-4-5} = \dfrac{30}{-9}$  Simplify the denominator.

$= -\dfrac{10}{3}$  Reduce to lowest terms by dividing the numerator and denominator by 3.

In the examples that follow, the numerators and denominators contain expressions that are somewhat more complicated than those we have seen thus far. To apply the rule for order of operations to these examples, we treat fraction bars the same way we treat grouping symbols. That is, fraction bars separate numerators and denominators so that each will be simplified separately.

**E X A M P L E S**   Simplify.

**17.** $\dfrac{2(-3) + 4}{12} = \dfrac{-6 + 4}{12}$      In the numerator, we multiply
                                                    before we add.

$= \dfrac{-2}{12}$      Addition

$= -\dfrac{1}{6}$      Reduce to lowest terms by dividing the
                        numerator and denominator by 2.

**18.** $\dfrac{5(-4) + 6(-1)}{2(3) - 4(1)} = \dfrac{-20 + (-6)}{6 - 4}$      Multiplication before addition

$= \dfrac{-26}{2}$      Simplify the numerator and
                        denominator.

$= -13$      Divide $-26$ by 2.

We must be careful when we are working with expressions such as $(-5)^2$ and $-5^2$, that we include the negative sign with the base only when parentheses indicate we are to do so.

Unless there are parentheses to indicate otherwise, we consider the base to be only the number directly below and to the left of the exponent. If we want to include a negative sign with the base, we must use parentheses.

To simplify a more complicated expression, we follow the same rule. For example,

$7^2 - 3^2 = 49 - 9$      The bases are 7 and 3; the sign between the two terms
                          is a subtraction sign.

For another example,

$5^3 - 3^4 = 125 - 81$      We simplify exponents first, and then subtract.

**E X A M P L E S**   Simplify.

**19.** $\dfrac{5^2 - 3^2}{-5 + 3} = \dfrac{25 - 9}{-2}$      Simplify numerator and denominator
                                                        separately.

$= \dfrac{16}{-2}$

$= -8$

**20.** $\dfrac{(3 + 2)^2}{-3^2 - 2^2} = \dfrac{5^2}{-9 - 4}$      Simplify numerator and denominator
                                                              separately.

$= \dfrac{25}{-13}$

$= -\dfrac{25}{13}$

## DIVISION WITH THE NUMBER 0

For every division problem, there is an associated multiplication problem involving the same numbers. For example, the following two problems say the same thing about the numbers 2, 3, and 6:

*Division*  *Multiplication*

$$\frac{6}{3} = 2 \qquad 6 = 2(3)$$

We can use this relationship between division and multiplication to clarify division involving the number 0.

First, dividing 0 by a number other than 0 is allowed and always results in 0. To see this, consider dividing 0 by 5. We know the answer is 0 because of the relationship between multiplication and division. This is how we write it:

$$\frac{0}{5} = 0 \qquad \text{because} \qquad 0 = 0(5)$$

On the other hand, dividing a nonzero number by 0 is not allowed in the real numbers. Suppose we were attempting to divide 5 by 0. We don't know if there is an answer to this problem, but if there is, let's say the answer is a number that we can represent with the letter $n$. If 5 divided by 0 is a number $n$, then

$$\frac{5}{0} = n \qquad \text{and} \qquad 5 = n(0)$$

This is impossible, however, because no matter what number $n$ is, when we multiply it by 0 the answer must be 0. It can never be 5. In algebra, we say expressions like $\frac{5}{0}$ are undefined, because there is no answer to them. That is, division by 0 is not allowed in the real numbers.

The only other possibility for division involving the number 0 is 0 divided by 0. We will treat problems like $\frac{0}{0}$ as if they were undefined also.

## PROBLEM SET 1.7

Find the following quotients (divide).

**1.** $\dfrac{8}{-4}$    **2.** $\dfrac{10}{-5}$    **3.** $\dfrac{-48}{16}$

**4.** $\dfrac{-32}{4}$    **5.** $\dfrac{-7}{21}$    **6.** $\dfrac{-25}{100}$

**7.** $\dfrac{-39}{-13}$    **8.** $\dfrac{-18}{-6}$    **9.** $\dfrac{-6}{-42}$

**10.** $\dfrac{-4}{-28}$    **11.** $\dfrac{0}{-32}$    **12.** $\dfrac{0}{17}$

The following problems review all four operations with positive and negative numbers. Perform the indicated operations.

**13.** $-3 + 12$    **14.** $5 + (-10)$

**15.** $-3 - 12$    **16.** $5 - (-10)$

**17.** $-3(12)$    **18.** $5(-10)$

**19.** $-3 \div 12$    **20.** $5 \div (-10)$

Divide and reduce all answers to lowest terms.

**21.** $\dfrac{4}{5} \div \dfrac{3}{4}$    **22.** $\dfrac{6}{8} \div \dfrac{3}{4}$

**23.** $-\dfrac{5}{6} \div \left(-\dfrac{5}{8}\right)$

**24.** $-\dfrac{7}{9} \div \left(-\dfrac{1}{6}\right)$

**25.** $\dfrac{10}{13} \div \left(-\dfrac{5}{4}\right)$

**26.** $\dfrac{5}{12} \div \left(-\dfrac{10}{3}\right)$

**27.** $-\dfrac{5}{6} \div \dfrac{5}{6}$

**28.** $-\dfrac{8}{9} \div \dfrac{8}{9}$

**29.** $-\dfrac{3}{4} \div \left(-\dfrac{3}{4}\right)$

**30.** $-\dfrac{6}{7} \div \left(-\dfrac{6}{7}\right)$

The following problems involve more than one operation. Simplify as much as possible.

**31.** $\dfrac{3(-2)}{-10}$

**32.** $\dfrac{4(-3)}{24}$

**33.** $\dfrac{-5(-5)}{-15}$

**34.** $\dfrac{-7(-3)}{-35}$

**35.** $\dfrac{-8(-7)}{-28}$

**36.** $\dfrac{-3(-9)}{-6}$

**37.** $\dfrac{27}{4-13}$

**38.** $\dfrac{27}{13-4}$

**39.** $\dfrac{20-6}{5-5}$

**40.** $\dfrac{10-12}{3-3}$

**41.** $\dfrac{-3+9}{2 \cdot 5 - 10}$

**42.** $\dfrac{-4+8}{2 \cdot 4 - 8}$

**43.** $\dfrac{15(-5)-25}{2(-10)}$

**44.** $\dfrac{10(-3)-20}{5(-2)}$

**45.** $\dfrac{27-2(-4)}{-3(5)}$

**46.** $\dfrac{20-5(-3)}{10(-3)}$

**47.** $\dfrac{12-6(-2)}{12(-2)}$

**48.** $\dfrac{3(-4)+5(-6)}{10-6}$

**49.** $\dfrac{5^2-2^2}{-5+2}$

**50.** $\dfrac{7^2-4^2}{-7+4}$

**51.** $\dfrac{8^2-2^2}{8^2+2^2}$

**52.** $\dfrac{4^2-6^2}{4^2+6^2}$

**53.** $\dfrac{(5+3)^2}{-5^2-3^2}$

**54.** $\dfrac{(7+2)^2}{-7^2-2^2}$

**55.** $\dfrac{(8-4)^2}{8^2-4^2}$

**56.** $\dfrac{(6-2)^2}{6^2-2^2}$

**57.** $\dfrac{-4 \cdot 3^2 - 5 \cdot 2^2}{-8(7)}$

**58.** $\dfrac{-2 \cdot 5^2 + 3 \cdot 2^3}{-3(13)}$

**59.** $\dfrac{3 \cdot 10^2 + 4 \cdot 10 + 5}{345}$

**60.** $\dfrac{5 \cdot 10^2 + 6 \cdot 10 + 7}{567}$

**61.** $\dfrac{7-[(2-3)-4]}{-1-2-3}$

**62.** $\dfrac{2-[(3-5)-8]}{-3-4-5}$

**63.** $\dfrac{6(-4)-2(5-8)}{-6-3-5}$

**64.** $\dfrac{3(-4)-5(9-11)}{-9-2-3}$

**65.** $\dfrac{3(-5-3)+4(7-9)}{5(-2)+3(-4)}$

**66.** $\dfrac{-2(6-10)-3(8-5)}{6(-3)-6(-2)}$

**67.** $\dfrac{|3-9|}{3-9}$

**68.** $\dfrac{|4-7|}{4-7}$

Answer the following questions.

**69.** What is the quotient of $-12$ and $-4$?

**70.** The quotient of $-4$ and $-12$ is what number?

**71.** What number do we divide by $-5$ to get 2?

**72.** What number do we divide by $-3$ to get 4?

**73.** Twenty-seven divided by what number is $-9$?

**74.** Fifteen divided by what number is $-3$?

**75.** If the quotient of $-20$ and 4 is decreased by 3, what number results?

**76.** If $-4$ is added to the quotient of 24 and $-8$, what number results?

## Applying the Concepts

**77. Investment**   Suppose that you and 3 friends bought equal shares of an investment for a total of $15,000 and then sold it later for only $13,600. How much did each person lose?

**78. Investment**   If 8 people invest $500 each in a stamp collection and after a year the collection is worth $3,800, how much did each person lose?

**79. Temperature Change**   Suppose that the temperature outside is dropping at a constant rate. If the temperature is 75°F at noon and drops to 61°F by 4:00 in the afternoon, by how much did the temperature change each hour?

**80. Temperature Change**   In a chemistry class, a thermometer is placed in a beaker of hot water. The initial temperature of the water is 165°F. After 10 minutes, the water has cooled to 72°F. If the water temperature drops at a constant rate, by how much does the water temperature change each minute?

# Subsets of the Real Numbers

In Section 1.2 we introduced the real numbers and defined them as the numbers associated with points on the real number line. At that time we said the real numbers include whole numbers, fractions, and decimals, as well as other numbers that are not as familiar to us as these numbers. In this section we take a more detailed look at the kinds of numbers that make up the set of real numbers.

The numbers that make up the set of real numbers can be classified as *counting numbers, whole numbers, integers, rational numbers,* and *irrational numbers;* each is said to be a *subset* of the real numbers.

> **DEFINITION**  Set *A* is called a **subset** of set *B* if set *A* is contained in set *B*, that is, if each and every element in set *A* is also a member of set *B*.

Here is a detailed description of the major subsets of the real numbers.

The counting numbers are the numbers with which we count. They are the numbers 1, 2, 3, and so on. The notation we use to specify a group of numbers like this is *set notation.* We use the symbols { and } to enclose the members of the set.

$$\text{Counting numbers} = \{1, 2, 3, \ldots\}$$

**EXAMPLE 1**  Which of the numbers in the following set are not counting numbers?

$$\left\{-3, 0, \frac{1}{2}, 1, 1.5, 3\right\}$$

**Solution**  The numbers $-3$, $0$, $\frac{1}{2}$, and $1.5$ are not counting numbers.

The whole numbers include the counting numbers and the number 0.

$$\text{Whole numbers} = \{0, 1, 2, \ldots\}$$

The set of integers includes the whole numbers and the opposites of all the counting numbers.

$$\text{Integers} = \{\ldots, -3, -2, -1, 0, 1, 2, 3, \ldots\}$$

When we refer to positive integers, we are referring to the numbers 1, 2, 3, . . . . Likewise, the negative integers are $-1$, $-2$, $-3$, . . . . The number 0 is neither positive nor negative.

**EXAMPLE 2**  Which of the numbers in the following set are not integers?

$$\left\{-5, -1.75, 0, \frac{2}{3}, 1, \pi, 3\right\}$$

**Solution**  The only numbers in the set that are not integers are $-1.75$, $\frac{2}{3}$, and $\pi$.

The set of *rational numbers* is the set of numbers commonly called "fractions" together with the integers. The set of rational numbers is difficult to list in the same way we have listed the other sets, so we will use a different kind of notation:

$$\text{Rational numbers} = \left\{ \frac{a}{b} \,\middle|\, a \text{ and } b \text{ are integers } (b \neq 0) \right\}$$

This notation is read "The set of elements $\frac{a}{b}$ such that $a$ and $b$ are integers (and $b$ is not 0)." If a number can be put in the form $\frac{a}{b}$, where $a$ and $b$ are both from the set of integers, then it is called a rational number.

Rational numbers include any number that can be written as the ratio of two integers. That is, rational numbers are numbers that can be put in the form

$$\frac{\text{integer}}{\text{integer}}$$

---

**E X A M P L E   3**    Show why each of the numbers in the following set is a rational number.

$$\left\{ -3, -\frac{2}{3}, 0, 0.333 \ldots , 0.75 \right\}$$

**Solution**    The number $-3$ is a rational number because it can be written as the ratio of $-3$ to 1. That is,

$$-3 = \frac{-3}{1}$$

Similarly, the number $-\frac{2}{3}$ can be thought of as the ratio of $-2$ to 3, while the number 0 can be thought of as the ratio of 0 to 1.

Any repeating decimal, such as $0.333 \ldots$ (the dots indicate that the 3's repeat forever), can be written as the ratio of two integers. In this case $0.333 \ldots$ is the same as the fraction $\frac{1}{3}$.

Finally, any decimal that terminates after a certain number of digits can be written as the ratio of two integers. The number 0.75 is equal to the fraction $\frac{3}{4}$ and is therefore a rational number.

---

Still other numbers exist, each of which is associated with a point on the real number line, that cannot be written as the ratio of two integers. In decimal form they never terminate and never repeat a sequence of digits indefinitely. They are called *irrational numbers* (because they are not rational):

$$\text{Irrational numbers} = \{\text{nonrational numbers; nonrepeating,}$$
$$\text{nonterminating decimals}\}$$

We cannot write any irrational number in a form that is familiar to us, because they are all nonterminating, nonrepeating decimals. Since they are not rational, they cannot be written as the ratio of two integers. They have to be represented in other ways. One irrational number you have probably seen before is $\pi$. It is not 3.14. Rather, 3.14 is an approximation to $\pi$. It cannot be written as a decimal number.

Other representations for irrational numbers are $\sqrt{2}$, $\sqrt{3}$, $\sqrt{5}$, $\sqrt{6}$, and, in general, the square root of any number that is not itself a perfect square. (If you are not familiar with square roots, you will be after Chapter 8.) Right now it is enough to know that some numbers on the number line cannot be written as the ratio of two integers or in decimal form. We call them irrational numbers.

The set of real numbers is the set of numbers that are either rational or irrational. That is, a real number is either rational or irrational.

$$\text{Real numbers} = \{\text{all rational numbers and all irrational numbers}\}$$

## PRIME NUMBERS AND FACTORING

The following diagram shows the relationship between multiplication and factoring:

Multiplication

Factors $\longrightarrow$ 3 $\cdot$ 4 $=$ 12 $\longleftarrow$ Product

Factoring

When we read the problem from left to right, we say the product of 3 and 4 is 12. Or we multiply 3 and 4 to get 12. When we read the problem in the other direction, from right to left, we say we have *factored* 12 into 3 times 4, or 3 and 4 are *factors* of 12.

The number 12 can be factored still further:

$$12 = 4 \cdot 3$$
$$= 2 \cdot 2 \cdot 3$$
$$= 2^2 \cdot 3$$

The numbers 2 and 3 are called *prime factors* of 12, because neither of them can be factored any further.

---

**DEFINITION**   If $a$ and $b$ represent integers, then $a$ is said to be a **factor** (or divisor) of $b$ if $a$ divides $b$ evenly—that is, if $a$ divides $b$ with no remainder.

---

**DEFINITION**   A **prime number** is any positive integer larger than 1 whose only positive factors (divisors) are itself and 1.

---

**Note:**   The number 15 is not a prime number since it has factors of 3 and 5. That is, $15 = 3 \cdot 5$. When a whole number larger than 1 is not prime, it is said to be *composite*.

Here is a list of the first few prime numbers.

$$\text{Prime numbers} = \{2, 3, 5, 7, 11, 13, 17, 19, 23, 29, 31, 37, 41, \ldots\}$$

When a number is not prime, we can factor it into the product of prime numbers. To factor a number into the product of primes, we simply factor it until it cannot be factored further.

**E X A M P L E   4**    Factor the number 60 into the product of prime numbers.

**Solution**    We begin by writing 60 as the product of any two positive integers whose product is 60, like 6 and 10:

$$60 = 6 \cdot 10$$

We then factor these numbers:

$$
\begin{aligned}
60 &= 6 \cdot 10 \\
&= (2 \cdot 3) \cdot (2 \cdot 5) \\
&= 2 \cdot 2 \cdot 3 \cdot 5 \\
&= 2^2 \cdot 3 \cdot 5
\end{aligned}
$$

**Note:**    It is customary to write the prime factors in order from smallest to largest.

**E X A M P L E   5**    Factor the number 630 into the product of primes.

**Solution**    Let's begin by writing 630 as the product of 63 and 10:

$$
\begin{aligned}
630 &= 63 \cdot 10 \\
&= (7 \cdot 9) \cdot (2 \cdot 5) \\
&= 7 \cdot 3 \cdot 3 \cdot 2 \cdot 5 \\
&= 2 \cdot 3^2 \cdot 5 \cdot 7
\end{aligned}
$$

It makes no difference which two numbers we start with, as long as their product is 630. We will always get the same result because a number has only one set of prime factors.

$$
\begin{aligned}
630 &= 18 \cdot 35 \\
&= 3 \cdot 6 \cdot 5 \cdot 7 \\
&= 3 \cdot 2 \cdot 3 \cdot 5 \cdot 7 \\
&= 2 \cdot 3^2 \cdot 5 \cdot 7
\end{aligned}
$$

**Note:**    There are some "tricks" to finding the divisors of a number. For instance, if a number ends in 0 or 5, then it is divisible by 5. If a number ends in an even number (0, 2, 4, 6, or 8), then it is divisible by 2. A number is divisible by 3 if the sum of its digits is divisible by 3. For example, 921 is divisible by 3 because the sum of its digits is $9 + 2 + 1 = 12$, which is divisible by 3.

When we have factored a number into the product of its prime factors, we not only know what prime numbers divide the original number, but we also know all of the other numbers that divide it as well. For instance, if we were to factor 210 into its prime factors, we would have $210 = 2 \cdot 3 \cdot 5 \cdot 7$, which means that 2, 3, 5, and 7 divide 210, as well as any combination of products of 2, 3, 5, and 7. That is, since 3 and 7 divide 210, then so does their product 21. Since 3, 5, and 7 each divide 210, then so does their product 105:

$$210 = 2 \cdot 3 \cdot 5 \cdot 7$$

Although there are many ways in which factoring is used in arithmetic and algebra, one simple application is in reducing fractions to lowest terms.

Recall that we reduce fractions to lowest terms by dividing the numerator and denominator by the same number. We can use the prime factorization of numbers to help us reduce fractions with large numerators and denominators.

**EXAMPLE 6** Reduce $\frac{210}{231}$ to lowest terms.

**Solution** First, we factor 210 and 231 into the product of prime factors. Then we reduce to lowest terms by dividing the numerator and denominator by any factors they have in common.

$$\frac{210}{231} = \frac{2 \cdot 3 \cdot 5 \cdot 7}{3 \cdot 7 \cdot 11} \qquad \text{Factor the numerator and denominator completely.}$$

$$= \frac{2 \cdot \cancel{3} \cdot 5 \cdot \cancel{7}}{\cancel{3} \cdot \cancel{7} \cdot 11} \qquad \text{Divide the numerator and denominator by } 3 \cdot 7.$$

$$= \frac{2 \cdot 5}{11}$$

$$= \frac{10}{11}$$

**Note:** The small lines we have drawn through the factors that are common to the numerator and denominator are used to indicate that we have divided the numerator and denominator by those factors.

# PROBLEM SET 1.8

Given the numbers in the set $\{-3, -2.5, 0, 1, \frac{3}{2}, \sqrt{15}\}$:

**1.** List all the whole numbers.

**2.** List all the integers.

**3.** List all the rational numbers.

**4.** List all the irrational numbers.

**5.** List all the real numbers.

Given the numbers in the set $\{-10, -8, -0.333 \ldots, -2, 9, \frac{25}{3}, \pi\}$:

**6.** List all the whole numbers.

**7.** List all the integers.

**8.** List all the rational numbers.

**9.** List all the irrational numbers.

**10.** List all the real numbers.

Identify the following statements as either true or false.

**11.** Every whole number is also an integer.

**12.** The set of whole numbers is a subset of the set of integers.

**13.** A number can be both rational and irrational.

**14.** The set of rational numbers and the set of irrational numbers have some elements in common.

**15.** Some whole numbers are also negative integers.

**16.** Every rational number is also a real number.

**17.** All integers are also rational numbers.

**18.** The set of integers is a subset of the set of rational numbers.

Label each of the following numbers as *prime* or *composite*. If a number is composite, then factor it completely.

**19.** 48    **20.** 72    **21.** 37

**22.** 23    **23.** 1,023    **24.** 543

Factor the following into the product of primes. When the number has been factored completely, write its prime factors from smallest to largest.

**25.** 144    **26.** 288    **27.** 38

**28.** 63    **29.** 105    **30.** 210

**31.** 180    **32.** 900    **33.** 385

**34.** 1,925    **35.** 121    **36.** 546

**37.** 420    **38.** 598    **39.** 620

**40.** 2,310

Reduce each fraction to lowest terms by first factoring the numerator and denominator into the product of prime factors and then dividing out any factors they have in common.

**41.** $\dfrac{105}{165}$    **42.** $\dfrac{165}{385}$    **43.** $\dfrac{525}{735}$

**44.** $\dfrac{550}{735}$    **45.** $\dfrac{385}{455}$    **46.** $\dfrac{385}{735}$

**47.** $\dfrac{322}{345}$    **48.** $\dfrac{266}{285}$    **49.** $\dfrac{205}{369}$

**50.** $\dfrac{111}{185}$    **51.** $\dfrac{215}{344}$    **52.** $\dfrac{279}{310}$

**53.** Factor $6^3$ into the product of prime factors by first factoring 6 and then raising each of its factors to the third power.

**54.** Factor $12^2$ into the product of prime factors by first factoring 12 and then raising each of its factors to the second power.

**55.** Factor $9^4 \cdot 16^2$ into the product of prime factors by first factoring 9 and 16 completely.

**56.** Factor $10^2 \cdot 12^3$ into the product of prime factors by first factoring 10 and 12 completely.

**57.** Simplify the expression $3 \cdot 8 + 3 \cdot 7 + 3 \cdot 5$ and then factor the result into the product of primes. (Notice one of the factors of the answer is 3.)

**58.** Simplify the expression $5 \cdot 4 + 5 \cdot 9 + 5 \cdot 3$ and then factor the result into the product of primes.

Recall the Fibonacci sequence we introduced earlier in this chapter.

Fibonacci sequence $= 1, 1, 2, 3, 5, 8, \ldots$

Any number in the Fibonacci sequence is a *Fibonacci number.*

**59.** The Fibonacci numbers are not a subset of which of the following sets: real numbers, rational numbers, irrational numbers, whole numbers?

**60.** Name three Fibonacci numbers that are prime numbers.

**61.** Name three Fibonacci numbers that are composite numbers.

**62.** Is the sequence of odd numbers a subset of the Fibonacci numbers?

---

**1.9** **Addition and Subtraction With Fractions**

You may recall from previous math classes that to add two fractions with the same denominator, you simply add their numerators and put the result over the common denominator:

$$\frac{3}{4} + \frac{2}{4} = \frac{3+2}{4} = \frac{5}{4}$$

The reason we add numerators but do not add denominators is that we must follow the distributive property. To see this, you first have to recall that $\frac{3}{4}$ can be written as $3 \cdot \frac{1}{4}$, and $\frac{2}{4}$ can be written as $2 \cdot \frac{1}{4}$ (dividing by 4 is equivalent to multiplying by $\frac{1}{4}$). Here is the addition problem again, this time showing the use of the distributive property:

$$\frac{3}{4} + \frac{2}{4} = 3 \cdot \frac{1}{4} + 2 \cdot \frac{1}{4}$$

$$= (3 + 2) \cdot \frac{1}{4} \qquad \text{Distributive property}$$

$$= 5 \cdot \frac{1}{4}$$

$$= \frac{5}{4}$$

What we have here is the sum of the numerators placed over the *common denominator*. In symbols we have the following.

## ADDITION AND SUBTRACTION OF FRACTIONS

If $a$, $b$, and $c$ are integers and $c$ is not equal to 0, then

$$\frac{a}{c} + \frac{b}{c} = \frac{a + b}{c}$$

This rule holds for subtraction as well. That is,

$$\frac{a}{c} - \frac{b}{c} = \frac{a - b}{c}$$

In Examples 1–4, find the sum or difference. (Add or subtract as indicated.) Reduce all answers to lowest terms. (Assume all variables represent nonzero numbers.)

## EXAMPLES

**1.** $\dfrac{3}{8} + \dfrac{1}{8} = \dfrac{3 + 1}{8}$     Add numerators; keep the same denominator.

$\qquad\qquad = \dfrac{4}{8}$     The sum of 3 and 1 is 4.

$\qquad\qquad = \dfrac{1}{2}$     Reduce to lowest terms.

**2.** $\dfrac{a + 5}{8} - \dfrac{3}{8} = \dfrac{a + 5 - 3}{8}$     Combine numerators; keep the same denominator.

$\qquad\qquad = \dfrac{a + 2}{8}$

**3.** $\dfrac{9}{x} - \dfrac{3}{x} = \dfrac{9-3}{x}$    Subtract numerators; keep the same denominator.

$\qquad = \dfrac{6}{x}$    The difference of 9 and 3 is 6.

**4.** $\dfrac{3}{7} + \dfrac{2}{7} - \dfrac{9}{7} = \dfrac{3+2-9}{7}$

$\qquad = \dfrac{-4}{7}$

$\qquad = -\dfrac{4}{7}$    Unlike signs give a negative answer.

As Examples 1–4 indicate, addition and subtraction are simple, straightforward processes when all the fractions have the same denominator. We will now turn our attention to the process of adding fractions that have different denominators. In order to get started, we need the following definition.

> **DEFINITION**   The **least common denominator (LCD)** for a set of denominators is the smallest number that is exactly divisible by each denominator. (Note that in some books the least common denominator is also called the *least common multiple*.)
>
>   In other words, all the denominators of the fractions involved in a problem must divide into the least common denominator exactly. That is, they divide it without giving a remainder.

**EXAMPLE 5**   Find the LCD for the fractions $\dfrac{5}{12}$ and $\dfrac{7}{18}$.

**Solution**   The least common denominator for the denominators 12 and 18 must be the smallest number divisible by both 12 and 18. We can factor 12 and 18 completely and then build the LCD from these factors. Factoring 12 and 18 completely gives us

$$12 = 2\cdot2\cdot3 \qquad 18 = 2\cdot3\cdot3$$

Now, if 12 is going to divide the LCD exactly, then the LCD must have factors of $2\cdot2\cdot3$. If 18 is to divide it exactly, it must have factors of $2\cdot3\cdot3$. We don't need to repeat the factors that 12 and 18 have in common:

12 divides the LCD

$$\left.\begin{array}{l}12 = 2\cdot2\cdot3\\18 = 2\cdot3\cdot3\end{array}\right\} \text{LCD} = 2\cdot2\cdot3\cdot3 = 36$$

18 divides the LCD

In other words, first we write down the factors of 12, then we attach the factors of 18 that do not already appear as factors of 12. We start with $2 \cdot 2 \cdot 3$ because those are the factors of 12. Then we look at the first factor of 18. It is 2. Since 2 already appears in the expression $2 \cdot 2 \cdot 3$, we don't need to attach another one. Next, we look at the factors $3 \cdot 3$. The expression $2 \cdot 2 \cdot 3$ has one 3. In order for it to contain the expression $3 \cdot 3$, we attach another 3. The final expression, our LCD, is $2 \cdot 2 \cdot 3 \cdot 3$.

The LCD for 12 and 18 is 36. It is the smallest number that is divisible by both 12 and 18; 12 divides it exactly three times, and 18 divides it exactly two times.

We can use the results of Example 5 to find the sum of the fractions $\frac{5}{12}$ and $\frac{7}{18}$.

**EXAMPLE 6**   Add $\frac{5}{12} + \frac{7}{18}$.

**Solution**   We can add fractions only when they have the same denominators. In Example 5 we found the LCD for $\frac{5}{12}$ and $\frac{7}{18}$ to be 36. We change $\frac{5}{12}$ and $\frac{7}{18}$ to equivalent fractions that each have 36 for a denominator by applying Property 1 (see p. 16) for fractions:

$$\frac{5}{12} = \frac{5 \cdot \mathbf{3}}{12 \cdot \mathbf{3}} = \frac{15}{36}$$

$$\frac{7}{18} = \frac{7 \cdot \mathbf{2}}{18 \cdot \mathbf{2}} = \frac{14}{36}$$

The fraction $\frac{15}{36}$ is equivalent to $\frac{5}{12}$, since it was obtained by multiplying both the numerator and denominator by 3. Likewise, $\frac{14}{36}$ is equivalent to $\frac{7}{18}$, since it was obtained by multiplying the numerator and denominator by 2. All we have left to do is to add numerators:

$$\frac{15}{36} + \frac{14}{36} = \frac{29}{36}$$

The sum of $\frac{5}{12}$ and $\frac{7}{18}$ is the fraction $\frac{29}{36}$. Let's write the complete problem again step-by-step.

$$\frac{5}{12} + \frac{7}{18} = \frac{5 \cdot \mathbf{3}}{12 \cdot \mathbf{3}} + \frac{7 \cdot \mathbf{2}}{18 \cdot \mathbf{2}} \qquad \text{Rewrite each fraction as an equivalent}$$
$$\text{fraction with denominator 36.}$$

$$= \frac{15}{36} + \frac{14}{36}$$

$$= \frac{29}{36} \qquad \text{Add numerators; keep the common}$$
$$\text{denominator.}$$

**EXAMPLE 7**   Find the LCD for $\frac{3}{4}$ and $\frac{1}{6}$.

**Solution**   We factor 4 and 6 into products of prime factors and build the LCD from these factors:

$$\left.\begin{array}{l} 4 = 2 \cdot 2 \\ 6 = 2 \cdot 3 \end{array}\right\} \quad \text{LCD} = 2 \cdot 2 \cdot 3 = 12$$

The LCD is 12. Both denominators divide it exactly; 4 divides 12 exactly three times, and 6 divides 12 exactly two times.

---

**EXAMPLE  8**   Add $\frac{3}{4} + \frac{1}{6}$.

**Solution**   In Example 7 we found that the LCD for these two fractions is 12. We begin by changing $\frac{3}{4}$ and $\frac{1}{6}$ to equivalent fractions with denominator 12:

$$\frac{3}{4} = \frac{3 \cdot \mathbf{3}}{4 \cdot \mathbf{3}} = \frac{9}{12}$$

$$\frac{1}{6} = \frac{1 \cdot \mathbf{2}}{6 \cdot \mathbf{2}} = \frac{2}{12}$$

The fraction $\frac{9}{12}$ is equal to the fraction $\frac{3}{4}$, since it was obtained by multiplying the numerator and denominator of $\frac{3}{4}$ by 3. Likewise, $\frac{2}{12}$ is equivalent to $\frac{1}{6}$, since it was obtained by multiplying the numerator and denominator of $\frac{1}{6}$ by 2. To complete the problem, we add numerators:

$$\frac{9}{12} + \frac{2}{12} = \frac{11}{12}$$

The sum of $\frac{3}{4}$ and $\frac{1}{6}$ is $\frac{11}{12}$. Here is how the complete problem looks:

$$\frac{3}{4} + \frac{1}{6} = \frac{3 \cdot \mathbf{3}}{4 \cdot \mathbf{3}} + \frac{1 \cdot \mathbf{2}}{6 \cdot \mathbf{2}} \qquad \text{Rewrite each fraction as an equivalent fraction with denominator 12.}$$

$$= \frac{9}{12} + \frac{2}{12}$$

$$= \frac{11}{12} \qquad \text{Add numerators; keep the same denominator.}$$

---

**EXAMPLE  9**   Subtract $\frac{7}{15} - \frac{3}{10}$.

**Solution**   Let's factor 15 and 10 completely and use these factors to build the LCD:

15 divides the LCD

$$\left.\begin{array}{l} 15 = 3 \cdot 5 \\ 10 = 2 \cdot 5 \end{array}\right\} \text{LCD} = 2 \cdot 3 \cdot 5 = 30$$

10 divides the LCD

Changing to equivalent fractions and subtracting, we have

$$\frac{7}{15} - \frac{3}{10} = \frac{7 \cdot \mathbf{2}}{15 \cdot \mathbf{2}} - \frac{3 \cdot \mathbf{3}}{10 \cdot \mathbf{3}}$$    Rewrite as equivalent fractions with the LCD for denominator.

$$= \frac{14}{30} - \frac{9}{30}$$

$$= \frac{5}{30}$$    Subtract numerators; keep the LCD.

$$= \frac{1}{6}$$    Reduce to lowest terms.

As a summary of what we have done so far and as a guide to working other problems, we will now list the steps involved in adding and subtracting fractions with different denominators.

### TO ADD OR SUBTRACT ANY TWO FRACTIONS

***Step 1:*** Factor each denominator completely and use the factors to build the LCD. (Remember, the LCD is the smallest number divisible by each of the denominators in the problem.)

***Step 2:*** Rewrite each fraction as an equivalent fraction that has the LCD for its denominator. This is done by multiplying both the numerator and denominator of the fraction in question by the appropriate whole number.

***Step 3:*** Add or subtract the numerators of the fractions produced in step 2. This is the numerator of the sum or difference. The denominator of the sum or difference is the LCD.

***Step 4:*** Reduce the fraction produced in step 3 to lowest terms if it is not already in lowest terms.

The idea behind adding or subtracting fractions is really very simple. We can add or subtract only fractions that have the same denominators. If the fractions we are trying to add or subtract do not have the same denominators, we rewrite each of them as an equivalent fraction with the LCD for a denominator.

Here are some further examples of sums and differences of fractions.

**EXAMPLE 10**   Add $\frac{1}{6} + \frac{1}{8} + \frac{1}{4}$.

**Solution**   We begin by factoring the denominators completely and building the LCD from the factors that result:

We then change to equivalent fractions and add as usual:

$$\frac{1}{6} + \frac{1}{8} + \frac{1}{4} = \frac{1 \cdot 4}{6 \cdot 4} + \frac{1 \cdot 3}{8 \cdot 3} + \frac{1 \cdot 6}{4 \cdot 6}$$

$$= \frac{4}{24} + \frac{3}{24} + \frac{6}{24}$$

$$= \frac{13}{24}$$

**EXAMPLE  11**   Subtract $3 - \frac{5}{6}$.

**Solution**   The denominators are 1 (because $3 = \frac{3}{1}$) and 6. The smallest number divisible by both 1 and 6 is 6.

$$3 - \frac{5}{6} = \frac{3}{1} - \frac{5}{6}$$

$$= \frac{3 \cdot 6}{1 \cdot 6} - \frac{5}{6}$$

$$= \frac{18}{6} - \frac{5}{6}$$

$$= \frac{13}{6}$$

**EXAMPLE  12**   Find the next number in each sequence.

**(a)** $\frac{1}{2}, 0, -\frac{1}{2}, \ldots$     **(b)** $\frac{1}{2}, 1, \frac{3}{2}, \ldots$     **(c)** $\frac{1}{2}, \frac{1}{4}, \frac{1}{8}, \ldots$

**Solution**

**(a)** $\frac{1}{2}, 0, -\frac{1}{2}, \ldots$ : Adding $-\frac{1}{2}$ to each term produces the next term. The fourth term will be $-\frac{1}{2} + (-\frac{1}{2}) = -1$. This is an arithmetic sequence.

**(b)** $\frac{1}{2}, 1, \frac{3}{2}, \ldots$ : Each term comes from the term before it by adding $\frac{1}{2}$. The fourth term will be $\frac{3}{2} + \frac{1}{2} = 2$. This sequence is also an arithmetic sequence.

**(c)** $\frac{1}{2}, \frac{1}{4}, \frac{1}{8}, \ldots$ : This is a geometric sequence in which each term comes from the term before it by multiplying by $\frac{1}{2}$ each time. The next term will be $\frac{1}{8} \cdot \frac{1}{2} = \frac{1}{16}$.

# PROBLEM SET 1.9

Find the following sums and differences and reduce to lowest terms. Add and subtract as indicated. Assume all variables represent nonzero numbers.

1. $\dfrac{3}{6} + \dfrac{1}{6}$

2. $\dfrac{2}{5} + \dfrac{3}{5}$

3. $\dfrac{3}{8} - \dfrac{5}{8}$

4. $\dfrac{1}{7} - \dfrac{6}{7}$

5. $-\dfrac{1}{4} + \dfrac{3}{4}$

6. $-\dfrac{4}{9} + \dfrac{7}{9}$

7. $\dfrac{x}{3} - \dfrac{1}{3}$

8. $\dfrac{x}{8} - \dfrac{1}{8}$

9. $\dfrac{1}{4} + \dfrac{2}{4} + \dfrac{3}{4}$

10. $\dfrac{2}{5} + \dfrac{3}{5} + \dfrac{4}{5}$

11. $\dfrac{x+7}{2} - \dfrac{1}{2}$

12. $\dfrac{x+5}{4} - \dfrac{3}{4}$

13. $\dfrac{1}{10} - \dfrac{3}{10} - \dfrac{4}{10}$

14. $\dfrac{3}{20} - \dfrac{1}{20} - \dfrac{4}{20}$

15. $\dfrac{1}{a} + \dfrac{4}{a} + \dfrac{5}{a}$

16. $\dfrac{5}{a} + \dfrac{4}{a} + \dfrac{3}{a}$

Find the LCD for each of the following; then use the methods developed in this section to add and subtract as indicated. Assume all variables represent nonzero numbers.

17. $\dfrac{1}{8} + \dfrac{3}{4}$

18. $\dfrac{1}{6} + \dfrac{2}{3}$

19. $\dfrac{3}{10} - \dfrac{1}{5}$

20. $\dfrac{5}{6} - \dfrac{1}{12}$

21. $\dfrac{4}{9} + \dfrac{1}{3}$

22. $\dfrac{1}{2} + \dfrac{1}{4}$

23. $2 + \dfrac{1}{3}$

24. $3 + \dfrac{1}{2}$

25. $-\dfrac{3}{4} + 1$

26. $-\dfrac{3}{4} + 2$

27. $\dfrac{1}{2} + \dfrac{2}{3}$

28. $\dfrac{2}{3} + \dfrac{1}{4}$

29. $\dfrac{5}{12} - \left(-\dfrac{3}{8}\right)$

30. $\dfrac{9}{16} - \left(-\dfrac{7}{12}\right)$

31. $-\dfrac{1}{20} + \dfrac{8}{30}$

32. $-\dfrac{1}{30} + \dfrac{9}{40}$

33. $\dfrac{17}{30} + \dfrac{11}{42}$

34. $\dfrac{19}{42} + \dfrac{13}{70}$

35. $\dfrac{25}{84} + \dfrac{41}{90}$

36. $\dfrac{23}{70} + \dfrac{29}{84}$

37. $\dfrac{13}{126} - \dfrac{13}{180}$

38. $\dfrac{17}{84} - \dfrac{17}{90}$

39. $\dfrac{3}{4} + \dfrac{1}{8} + \dfrac{5}{6}$

40. $\dfrac{3}{8} + \dfrac{2}{5} + \dfrac{1}{4}$

41. $\dfrac{1}{2} + \dfrac{1}{3} + \dfrac{1}{4} + \dfrac{1}{6}$

42. $\dfrac{1}{8} + \dfrac{1}{4} + \dfrac{1}{5} + \dfrac{1}{10}$

43. Find the sum of $\dfrac{3}{7}$, 2, and $\dfrac{1}{9}$.

44. Find the sum of 6, $\dfrac{6}{11}$, and 11.

45. Give the difference of $\dfrac{7}{8}$ and $\dfrac{1}{4}$.

46. Give the difference of $\dfrac{9}{10}$ and $\dfrac{1}{100}$.

Find the fourth term in each sequence.

47. $\dfrac{1}{3}$, 0, $-\dfrac{1}{3}$, . . .

48. $\dfrac{2}{3}$, 0, $-\dfrac{2}{3}$, . . .

49. $\dfrac{1}{3}$, 1, $\dfrac{5}{3}$, . . .

50. 1, $\dfrac{3}{2}$, 2, . . .

51. 1, $\dfrac{1}{5}$, $\dfrac{1}{25}$, . . .

52. 1, $-\dfrac{1}{2}$, $\dfrac{1}{4}$, . . .

# CHAPTER 1    SUMMARY

The number(s) in brackets next to each heading indicates the section(s) in which that topic is discussed.

*Note:* We will use the margins in the chapter summaries to give examples that correspond to the topic being reviewed whenever it is appropriate.

## Symbols [1.1]

| | |
|---|---|
| $a = b$ | $a$ is equal to $b$ |
| $a \neq b$ | $a$ is not equal to $b$ |
| $a < b$ | $a$ is less than $b$ |
| $a \not< b$ | $a$ is not less than $b$ |
| $a > b$ | $a$ is greater than $b$ |
| $a \not> b$ | $a$ is not greater than $b$ |

**Examples**

**1.** $2^5 = 2 \cdot 2 \cdot 2 \cdot 2 \cdot 2 = 32$

$5^2 = 5 \cdot 5 = 25$

$10^3 = 10 \cdot 10 \cdot 10 = 1,000$

$1^4 = 1 \cdot 1 \cdot 1 \cdot 1 = 1$

## Exponents [1.1]

Exponents are notation used to indicate repeated multiplication. In the expression $3^4$, 3 is the *base* and 4 is the *exponent*.

$$3^4 = 3 \cdot 3 \cdot 3 \cdot 3 = 81$$

**2.** $10 + (2 \cdot 3^2 - 4 \cdot 2)$

$\quad = 10 + (2 \cdot 9 - 4 \cdot 2)$

$\quad = 10 + (18 - 8)$

$\quad = 10 + 10$

$\quad = 20$

## Order of Operations [1.1]

When evaluating a mathematical expression, we will perform the operations in the following order, beginning with the expression in the innermost parentheses or brackets and working our way out.

**1.** Simplify all numbers with exponents, working from left to right if more than one of these numbers is present.

**2.** Then do all multiplications and divisions left to right.

**3.** Finally, perform all additions and subtractions left to right.

**3.** Add all combinations of positive and negative 10 and 13:

$10 + 13 = 23$

$10 + (-13) = -3$

$-10 + 13 = 3$

$-10 + (-13) = -23$

## Addition of Real Numbers [1.3]

To add two real numbers with

**1.** The same sign: Simply add their absolute values and use the common sign.

**2.** Different signs: Subtract the smaller absolute value from the larger absolute value. The answer has the same sign as the number with the larger absolute value.

**4.** Subtracting 2 is the same as adding $-2$:

$7 - 2 = 7 + (-2) = 5$

## Subtraction of Real Numbers [1.4]

To subtract one number from another, simply add the opposite of the number you are subtracting. That is, if $a$ and $b$ represent real numbers, then

$$a - b = a + (-b)$$

**5.** $\quad 3(5) = 15$

$\quad\quad 3(-5) = -15$

$\quad\quad -3(5) = -15$

$\quad\quad -3(-5) = 15$

**Multiplication of Real Numbers [1.6]**
To multiply two real numbers, simply multiply their absolute values. Like signs give a positive answer. Unlike signs give a negative answer.

**6.** $\frac{-6}{2} = -6\left(\frac{1}{2}\right) = -3$

$\quad \frac{-6}{-2} = -6\left(-\frac{1}{2}\right) = 3$

**Division of Real Numbers [1.7]**
Division by a number is the same as multiplication by its reciprocal. Like signs give a positive answer. Unlike signs give a negative answer.

**7.** $|5| = 5$

$\quad |-5| = 5$

**Absolute Value [1.2]**
The absolute value of a real number is its distance from zero on the real number line. Absolute value is never negative.

**8.** The numbers 3 and $-3$ are opposites; their sum is 0:

$\quad\quad 3 + (-3) = 0$

**Opposites [1.2, 1.5]**
Any two real numbers the same distance from zero on the number line but in opposite directions from zero are called opposites. Opposites always add to zero.

**9.** The numbers 2 and $\frac{1}{2}$ are reciprocals; their product is 1:

$\quad\quad 2\left(\frac{1}{2}\right) = 1$

**Reciprocals [1.2, 1.5]**
Any two real numbers whose product is 1 are called reciprocals. Every real number has a reciprocal except 0.

**Properties of Real Numbers [1.5]**

|  | *For Addition* | *For Multiplication* |
|---|---|---|
| *Commutative:* | $a + b = b + a$ | $a \cdot b = b \cdot a$ |
| *Associative:* | $a + (b + c) = (a + b) + c$ | $a \cdot (b \cdot c) = (a \cdot b) \cdot c$ |
| *Identity:* | $a + 0 = a$ | $a \cdot 1 = a$ |
| *Inverse:* | $a + (-a) = 0$ | $a\left(\dfrac{1}{a}\right) = 1$ |
| *Distributive:* | $a(b + c) = ab + ac$ | |

**10. (a)** 7 and 100 are counting numbers, but 0 and $-2$ are not.

**(b)** 0 and 241 are whole numbers, but $-4$ and $\frac{1}{2}$ are not.

**(c)** $-15$, 0, and 20 are integers.

**(d)** $-4$, $-\frac{1}{2}$, 0.75, and 0.666 . . . are rational numbers.

**(e)** $-\pi$, $\sqrt{3}$, and $\pi$ are irrational numbers.

**(f)** All the numbers listed above are real numbers.

**Subsets of the Real Numbers [1.8]**

*Counting numbers:*      $\{1, 2, 3, \ldots\}$

*Whole numbers:*      $\{0, 1, 2, 3, \ldots\}$

*Integers:*      $\{\ldots, -3, -2, -1, 0, 1, 2, 3, \ldots\}$

*Rational numbers:*      {all numbers that can be expressed as the ratio of two integers}

*Irrational numbers:*      {all numbers on the number line that cannot be expressed as the ratio of two integers}

*Real numbers:*      {all numbers that are either rational or irrational}

**11.** The number 150 can be factored into the product of prime numbers:

$$150 = 15 \cdot 10$$
$$= (3 \cdot 5)(2 \cdot 5)$$
$$= 2 \cdot 3 \cdot 5^2$$

**12.** The LCD for $\frac{5}{12}$ and $\frac{7}{18}$ is 36.

**13.** $\dfrac{5}{12} + \dfrac{7}{18} = \dfrac{5}{12} \cdot \dfrac{3}{3} + \dfrac{7}{18} \cdot \dfrac{2}{2}$

$$= \dfrac{15}{36} + \dfrac{14}{36}$$

$$= \dfrac{29}{36}$$

## Factoring [1.8]

Factoring is the reverse of multiplication.

Multiplication

Factoring

## Least Common Denominator (LCD) [1.9]

The *least common denominator* (LCD) for a set of denominators is the smallest number that is exactly divisible by each denominator.

## Addition and Subtraction of Fractions [1.9]

To add (or subtract) two fractions with a common denominator, add (or subtract) numerators and use the common denominator.

$$\frac{a}{c} + \frac{b}{c} = \frac{a+b}{c} \qquad \text{and} \qquad \frac{a}{c} - \frac{b}{c} = \frac{a-b}{c}$$

### COMMON MISTAKES

**1.** Interpreting absolute value as changing the sign of the number inside the absolute value symbols. $|-5| = +5, |+5| = -5$. (The first expression is correct, the second one is not.) To avoid this mistake, remember: Absolute value is a distance, and distance is always measured in positive units.

**2.** Using the phrase "two negatives make a positive." This works only with multiplication and division. With addition, two negative numbers produce a negative answer. It is best not to use the phrase "two negatives make a positive" at all.

## CHAPTER 1    REVIEW

The numbers in brackets refer to the sections of the text in which similar problems can be found.

Write the numerical expression that is equivalent to each phrase, and then simplify. [1.3, 1.4, 1.6, 1.7]

**1.** The sum of $-7$ and $-10$

**2.** Five added to the sum of $-7$ and 4

**3.** The sum of $-3$ and 12 increased by 5

**4.** The difference of 4 and 9

**5.** The difference of 9 and $-3$

**6.** The difference of $-7$ and $-9$

**7.** The product of $-3$ and $-7$ decreased by 6

**8.** Ten added to the product of 5 and $-6$

**9.** Twice the product of $-8$ and $3x$

**10.** The quotient of $-25$ and $-5$

**11.** The quotient of $-40$ and 8 decreased by 7

**12.** The quotient of $-45$ and 15 increased by 9

Locate and label the points on the number line with the following coordinates. [1.2]

**13.** 5       **14.** $-4$       **15.** $\dfrac{24}{8}$

**16.** $-\dfrac{11}{2}$       **17.** $1.25$       **18.** $-1.5$

For each number give the absolute value. [1.2]

**19.** 12       **20.** $-3$       **21.** $-\dfrac{4}{5}$

**22.** $-\dfrac{7}{10}$

Simplify. [1.2]

**23.** $|-1.8|$       **24.** $-|-10|$

For each number, give the opposite and the reciprocal. [1.2]

**25.** 6       **26.** $\dfrac{3}{10}$       **27.** $-9$

**28.** $-\dfrac{12}{5}$

Multiply. [1.2, 1.6]

**29.** $\left(\dfrac{2}{5}\right)\left(\dfrac{3}{7}\right)$       **30.** $\dfrac{1}{2}(-10)$

**31.** $\left(-\dfrac{4}{5}\right)\left(\dfrac{25}{16}\right)$

Add. [1.3]

**32.** $-9 + 12$       **33.** $-18 + (-20)$

**34.** $-3 + 8 + (-5)$

**35.** $(-5) + (-10) + (-7)$

**36.** $-2 + (-8) + [-9 + (-6)]$

**37.** $(-21) + 40 + (-23) + 5$

Subtract. [1.4]

**38.** $6 - 9$       **39.** $14 - (-8)$

**40.** $-12 - (-8)$       **41.** $4 - 9 - 15$

Simplify. [1.4]

**42.** $-14 + 7 - 8$       **43.** $5 - (-10 - 2) - 3$

**44.** $6 - [(3 - 4) - 5]$

**45.** $20 - [-(10 - 3) - 8] - 7$

Find the products. [1.6]

**46.** $(-5)(6)$       **47.** $4(-3)$

**48.** $-2(3)(4)$       **49.** $(-1)(-3)(-1)(-4)$

Find the following quotients. [1.7]

**50.** $\dfrac{12}{-3}$       **51.** $\dfrac{-9}{36}$       **52.** $-\dfrac{8}{9} \div \dfrac{4}{3}$

Simplify. [1.1, 1.6, 1.7]

**53.** $4 \cdot 5 + 3$       **54.** $9 \cdot 3 + 4 \cdot 5$

**55.** $2^3 - 4 \cdot 3^2 + 5^2$

**56.** $12 - 3(2 \cdot 5 + 7) + 4$

**57.** $20 + 8 \div 4 + 2 \cdot 5$       **58.** $2(3 - 5) - (2 - 8)$

**59.** $-4(-5) + 10$

**60.** $(-2)(3) - (4)(-3) - 9$

**61.** $3(4 - 7)^2 - 5(3 - 8)^2$

**62.** $(-5 - 2)(-3 - 7)$

**63.** $\dfrac{4(-3)}{-6}$       **64.** $\dfrac{3^2 + 5^2}{(3 - 5)^2}$

**65.** $\dfrac{15 - 10}{6 - 6}$       **66.** $\dfrac{8(-5) - 24}{4(-2)}$

**67.** $\dfrac{2(-7) + (-11)(-4)}{7 - (-3)}$

State the property or properties that justify the following. [1.5]

**68.** $9(3y) = (9 \cdot 3)y$       **69.** $8(1) = 8$

**70.** $(4 + y) + 2 = (y + 4) + 2$

**71.** $5 + (-5) = 0$       **72.** $6\left(\dfrac{1}{6}\right) = 1$

**73.** $8 + 0 = 8$

**74.** $(4 + 2) + y = (4 + y) + 2$

**75.** $5(w - 6) = 5w - 30$

Use the associative property to rewrite each expression and then simplify the result. [1.5]

**76.** $7 + (5 + x)$       **77.** $4(7a)$

**78.** $\dfrac{1}{9}(9x)$       **79.** $\dfrac{4}{5}\left(\dfrac{5}{4}y\right)$

Apply the distributive property to each of the following expressions. Simplify when possible. [1.5, 1.6]

**80.** $7(2x + 3)$

**81.** $3(2a - 4)$

**82.** $\dfrac{1}{2}(5x - 6)$

**83.** $-\dfrac{1}{2}(3x - 6)$

For the set $\{\sqrt{7}, -\frac{1}{3}, 0, 5, -4.5, \frac{2}{5}, \pi, -3\}$ list all the [1.8]

**84.** Rational numbers

**85.** Whole numbers

**86.** Irrational numbers

**87.** Integers

Factor into the product of primes. [1.8]

**88.** 90

**89.** 840

Combine. [1.9]

**90.** $\dfrac{18}{35} + \dfrac{13}{42}$

**91.** $\dfrac{9}{70} + \dfrac{11}{84}$

Find the next number in each sequence. [1.1, 1.2, 1.3, 1.6, 1.9]

**92.** $10, 7, 4, 1, \ldots$

**93.** $10, -30, 90, -270, \ldots$

**94.** $1, 1, 2, 3, 5, \ldots$

**95.** $4, 6, 8, 10, \ldots$

**96.** $1, \dfrac{1}{2}, 0, -\dfrac{1}{2}, \ldots$

**97.** $1, -\dfrac{1}{2}, \dfrac{1}{4}, -\dfrac{1}{8}, \ldots$

---

# CHAPTER 1 TEST

Translate into symbols. [1.1]

**1.** The sum of $x$ and 3 is 8.

**2.** The product of 5 and $y$ is 15.

Simplify according to the rule for order of operations. [1.1]

**3.** $5^2 + 3(9 - 7) + 3^2$

**4.** $10 - 6 \div 3 + 2^3$

For each number, name the opposite, reciprocal, and absolute value. [1.2]

**5.** $-4$

**6.** $\dfrac{3}{4}$

Add. [1.3]

**7.** $3 + (-7)$

**8.** $|-9 + (-6)| + |-3 + 5|$

Subtract. [1.4]

**9.** $-4 - 8$

**10.** $9 - (7 - 2) - 4$

Match each expression below with the letter of the property that justifies it. [1.5]

**11.** $(x + y) + z = x + (y + z)$

**12.** $3(x + 5) = 3x + 15$

**13.** $5(3x) = (5 \cdot 3)x$

**14.** $(x + 5) + 7 = 7 + (x + 5)$

   **a.** Commutative property of addition

   **b.** Commutative property of multiplication

   **c.** Associative property of addition

   **d.** Associative property of multiplication

   **e.** Distributive property

Multiply. [1.6]

**15.** $-3(7)$

**16.** $-4(8)(-2)$

**17.** $8\left(-\dfrac{1}{4}\right)$

**18.** $\left(-\dfrac{2}{3}\right)^3$

Simplify using the rule for order of operations. [1.6]

**19.** $-3(-4) - 8$

**20.** $5(-6)^2 - 3(-2)^3$

Simplify as much as possible. [1.7]

**21.** $7 - 3(2 - 8)$

**22.** $4 - 2[-3(-1 + 5) + 4(-3)]$

**23.** $\dfrac{4(-5) - 2(7)}{-10 - 7}$

**24.** $\dfrac{2(-3 - 1) + 4(-5 + 2)}{-3(2) - 4}$

Apply the associative property and then simplify. [1.5, 1.6]

**25.** $3 + (5 + 2x)$       **26.** $-2(-5x)$

Multiply by applying the distributive property. [1.5, 1.6]

**27.** $2(3x + 5)$       **28.** $-\dfrac{1}{2}(4x - 2)$

From the set of numbers $\{1, 1.5, \sqrt{2}, \frac{3}{4}, -8\}$, list [1.8]

**29.** All the integers

**30.** All the rational numbers

**31.** All the irrational numbers

**32.** All the real numbers

Factor into the product of primes. [1.8]

**33.** 592       **34.** 1,340

Combine. [1.9]

**35.** $\dfrac{5}{15} + \dfrac{11}{42}$       **36.** $\dfrac{5}{x} + \dfrac{3}{x}$

Write an expression in symbols that is equivalent to each written phrase and then simplify it.

**37.** The sum of 8 and $-3$ [1.1, 1.3]

**38.** The difference of $-24$ and 2 [1.1, 1.4]

**39.** The product of $-5$ and $-4$ [1.1, 1.6]

**40.** The quotient of $-24$ and $-2$ [1.1, 1.7]

Find the next number in each sequence. [1.1, 1.2, 1.3, 1.6, 1.9]

**41.** $-8, -3, 2, 7, \ldots$       **42.** $8, -4, 2, -1, \ldots$

# Linear Equations and Inequalities

*(Index Stock Imagery)*

## INTRODUCTION

Just before starting work on this edition of your text, I flew to Europe for vacation. From time to time the television screens on the plane displayed statistics about the flight. At one point during the flight the temperature outside the plane was $-60°F$. When I returned home, I did some research and found that the relationship between temperature $T$ and altitude $A$ can be described with the formula

$$T = -0.0035A + 70$$

when the temperature on the ground is 70°F. Table 1 and the line graph below also describe this relationship.

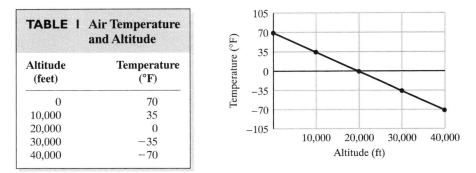

| TABLE 1 | Air Temperature and Altitude |
|---|---|
| **Altitude (feet)** | **Temperature (°F)** |
| 0 | 70 |
| 10,000 | 35 |
| 20,000 | 0 |
| 30,000 | −35 |
| 40,000 | −70 |

In this chapter we will start our work with formulas, and you will see how we use formulas to produce tables and line graphs like the ones above.

## STUDY SKILLS

If you have successfully completed Chapter 1, then you have made a good start at developing the study skills necessary to succeed in all math classes. Here is the list of study skills for this chapter. Some are a continuation of the skills from Chapter 1, while others are new to this chapter.

1. **Continue to Set and Keep a Schedule**  Sometimes I find students do well in Chapter 1 and then become overconfident. They begin to put in less time with their homework. Don't do it. Keep to the same schedule.

2. **Increase Effectiveness**  You want to become more and more effective with the time you spend on your homework. You want to increase the amount of learning you obtain in the time you have set aside. Increase those activities that you feel are the most beneficial, and decrease those that have not given you the results you want.

3. **List Difficult Problems**  Begin to make lists of problems that give you the most difficulty. These are the problems in which you are repeatedly making mistakes.

4. **Begin to Develop Confidence With Word Problems**  It seems that the major difference between those people who are good at working word problems and those who are not is confidence. The people with confidence know that no matter how long it takes them, they will eventually be able to solve the problem. Those without confidence begin by saying to themselves, "I'll never be able to work this problem." Are you like that? If you are, what you need to do is put your old ideas about you and word problems aside for a while and make a decision to be successful. Sometimes that's all it takes. Instead of telling yourself that you can't do word problems, that you don't like them, or that they're not good for anything anyway, decide to do whatever it takes to master them.

Many of my students keep a notebook that contains everything that they need for the course: class notes, homework, quizzes, tests, and research projects. A three-ring binder with tabs is ideal. Organize your notebook so that you can easily get to any item you wish to look at.

## 2.1 Simplifying Expressions

If a cellular phone company charges $35 per month plus $0.25 for each minute, or fraction of a minute, that you use one of their cellular phones, then the amount of your monthly bill is given by the expression $35 + 0.25t$. To find the amount you will pay for using that phone 30 minutes in one month, you substitute 30 for $t$ and simplify the resulting expression. This process is one of the topics we will study in this section.

As you will see in the next few sections, the first step in solving an equation is to simplify both sides as much as possible. In the first part of this section, we will practice simplifying expressions by combining what are called *similar* (or like) terms.

For our immediate purposes, a term is a number or a number and one or more variables multiplied together. For example, the number 5 is a term, as are the expressions $3x$, $-7y$, and $15xy$.

---

**DEFINITION**  Two or more terms with the same variable part are called **similar** (or **like**) terms.

---

The terms $3x$ and $4x$ are similar, since their variable parts are identical. Likewise, the terms $18y$, $-10y$, and $6y$ are similar terms.

To simplify an algebraic expression, we simply reduce the number of terms in the expression. We accomplish this by applying the distributive property along with our knowledge of addition and subtraction of positive and negative real numbers. The following examples illustrate the procedure.

**EXAMPLES**  Simplify by combining similar terms.

1. $3x + 4x = (3 + 4)x$      Distributive property
    $= 7x$         Addition of 3 and 4
2. $7a - 10a = (7 - 10)a$     Distributive property
     $= -3a$        Addition of 7 and $-10$
3. $18y - 10y + 6y = (18 - 10 + 6)y$   Distributive property
       $= 14y$       Addition of 18, $-10$, and 6

When the expression we intend to simplify is more complicated, we use the commutative and associative properties first.

**EXAMPLES**  Simplify each expression.

4. $3x + 5 + 2x - 3 = 3x + 2x + 5 - 3$    Commutative property
      $= (3x + 2x) + (5 - 3)$    Associative property
      $= (3 + 2)x + (5 - 3)$    Distributive property
      $= 5x + 2$        Addition

**5.** $4a - 7 - 2a + 3 = (4a - 2a) + (-7 + 3)$     Commutative and
                                                                       associative properties

$= (4 - 2)a + (-7 + 3)$     Distributive property

$= 2a - 4$     Addition

**6.** $5x + 8 - x - 6 = (5x - x) + (8 - 6)$     Commutative and
                                                                    associative properties

$= (5 - 1)x + (8 - 6)$     Distributive property

$= 4x + 2$     Addition

Notice that in each case the result has fewer terms than the original expression. Since there are fewer terms, the resulting expression is said to be simpler than the original expression.

## SIMPLIFYING EXPRESSIONS CONTAINING PARENTHESES

If an expression contains parentheses, it is often necessary to apply the distributive property to remove the parentheses before combining similar terms.

**EXAMPLE 7**  Simplify the expression $5(2x - 8) - 3$.

**Solution**  We begin by distributing the 5 across $2x - 8$. We then combine similar terms:

$$5(2x - 8) - 3 = 10x - 40 - 3 \quad \text{Distributive property}$$
$$= 10x - 43$$

**EXAMPLE 8**  Simplify $7 - 3(2y + 1)$.

**Solution**  By the rule for order of operations, we must multiply before we add or subtract. For that reason, it would be incorrect to subtract 3 from 7 first. Instead, we multiply $-3$ and $2y + 1$ to remove the parentheses and then combine similar terms:

$$7 - 3(2y + 1) = 7 - 6y - 3 \quad \text{Distributive property}$$
$$= -6y + 4$$

**EXAMPLE 9**  Simplify $5(x - 2) - (3x + 4)$.

**Solution**  We begin by applying the distributive property to remove the parentheses. The expression $-(3x + 4)$ can be thought of as $-1(3x + 4)$. Thinking of it in this way allows us to apply the distributive property:

$$-1(3x + 4) = -1(3x) + (-1)(4)$$
$$= -3x - 4$$

The complete solution looks like this:

$$5(x - 2) - (3x + 4) = 5x - 10 - 3x - 4 \qquad \text{Distributive property}$$
$$= 2x - 14 \qquad \text{Combine similar terms.}$$

As you can see from the explanation in Example 9, we use the distributive property to simplify expressions in which parentheses are preceded by a negative sign. In general we can write

$$-(a + b) = -1(a + b)$$
$$= -a + (-b)$$
$$= -a - b$$

The negative sign outside the parentheses ends up changing the sign of each term within the parentheses. In words, we say "the opposite of a sum is the sum of the opposites."

## THE VALUE OF AN EXPRESSION

An expression like $3x + 2$ has a certain value depending on what number we assign to $x$. For instance, when $x$ is 4, $3x + 2$ becomes $3(4) + 2$, or 14. When $x$ is $-8$, $3x + 2$ becomes $3(-8) + 2$, or $-22$. The value of an expression is found by replacing the variable with a given number.

**E X A M P L E S**   Find the value of the following expressions by replacing the variable with the given number.

| Expression | Value of the Variable | Value of the Expression |
|---|---|---|
| **10.** $3x - 1$ | $x = 2$ | $3(2) - 1 = 6 - 1 = 5$ |
| **11.** $7a + 4$ | $a = -3$ | $7(-3) + 4 = -21 + 4 = -17$ |
| **12.** $2x - 3 + 4x$ | $x = -1$ | $2(-1) - 3 + 4(-1)$ |
| | | $\quad = -2 - 3 + (-4) = -9$ |
| **13.** $2x - 5 - 8x$ | $x = 5$ | $2(5) - 5 - 8(5)$ |
| | | $\quad = 10 - 5 - 40 = -35$ |
| **14.** $y^2 - 6y + 9$ | $y = 4$ | $4^2 - 6(4) + 9 = 16 - 24 + 9 = 1$ |

Simplifying an expression should not change its value. That is, if an expression has a certain value when $x$ is 5, then it will always have that value no matter how much it has been simplified, as long as $x$ is 5. If we were to simplify the expression in Example 13 first, it would look like

$$2x - 5 - 8x = -6x - 5$$

When $x$ is 5, the simplified expression $-6x - 5$ is

$$-6(5) - 5 = -30 - 5 = -35$$

It has the same value as the original expression when $x$ is 5.

We can also find the value of an expression that contains two variables if we know the values for both variables.

**EXAMPLE   15**   Find the value of the expression $2x - 3y + 4$ when $x$ is $-5$ and $y$ is 6.

**Solution**   Substituting $-5$ for $x$ and 6 for $y$, the expression becomes

$$2(-5) - 3(6) + 4 = -10 - 18 + 4$$
$$= -28 + 4$$
$$= -24$$

**EXAMPLE   16**   Find the value of the expression $x^2 - 2xy + y^2$ when $x$ is 3 and $y$ is $-4$.

**Solution**   Replacing each $x$ in the expression with the number 3 and each $y$ in the expression with the number $-4$ gives us

$$3^2 - 2(3)(-4) + (-4)^2 = 9 - 2(3)(-4) + 16$$
$$= 9 - (-24) + 16$$
$$= 33 + 16$$
$$= 49$$

## MORE ABOUT SEQUENCES

As the next example indicates, when we substitute the counting numbers, in order, into algebraic expressions, we form some of the sequences of numbers that we studied in Chapter 1. To review, recall that the sequence of counting numbers (also called the sequence of positive integers) is

$$\text{Counting numbers} = 1, 2, 3, \ldots$$

**EXAMPLE   17**   Substitute 1, 2, 3, and 4 for $n$ in the expression $2n - 1$.

**Solution**   Substituting as indicated, we have

When $n = 1, 2n - 1 = 2 \cdot 1 - 1 = 1.$

When $n = 2, 2n - 1 = 2 \cdot 2 - 1 = 3.$

When $n = 3, 2n - 1 = 2 \cdot 3 - 1 = 5.$

When $n = 4, 2n - 1 = 2 \cdot 4 - 1 = 7.$

As you can see, substituting the first four counting numbers into the formula $2n - 1$ produces the first four numbers in the sequence of odd numbers.

The next example is similar to Example 17 but uses tables to display the information.

**E X A M P L E   1 8**    Fill in the tables below to find the sequences formed by substituting the first four counting numbers into the expressions $2n$ and $n^2$.

**(a)**

| $n$ | 1 | 2 | 3 | 4 |
|-----|---|---|---|---|
| $2n$ |   |   |   |   |

**(b)**

| $n$ | 1 | 2 | 3 | 4 |
|-----|---|---|---|---|
| $n^2$ |   |   |   |   |

**Solution**    Proceeding as we did in the previous example, we substitute the numbers 1, 2, 3, and 4 into the given expressions.

**(a)**  When $n = 1$, $2n = 2 \cdot 1 = 2$.
   When $n = 2$, $2n = 2 \cdot 2 = 4$.
   When $n = 3$, $2n = 2 \cdot 3 = 6$.
   When $n = 4$, $2n = 2 \cdot 4 = 8$.

As you can see, the expression $2n$ produces the sequence of even numbers when $n$ is replaced by the counting numbers. Placing these results into our first table gives us

| $n$ | 1 | 2 | 3 | 4 |
|-----|---|---|---|---|
| $2n$ | 2 | 4 | 6 | 8 |

**(b)**  The expression $n^2$ produces the sequence of squares when $n$ is replaced by 1, 2, 3, and 4. In table form we have

| $n$ | 1 | 2 | 3 | 4 |
|-----|---|---|---|---|
| $n^2$ | 1 | 4 | 9 | 16 |

# Getting Ready for Class

*After reading through the preceding section, respond in your own words and in complete sentences.*

**A.** What are similar terms?

**B.** Explain how the distributive property is used to combine similar terms.

**C.** What is wrong with writing $3x + 4x = 7x^2$?

**D.** Explain how you would find the value of $5x + 3$ when $x$ is 6.

# PROBLEM SET 2.1

Simplify the following expressions.

**1.** $3x - 6x$

**2.** $7x - 5x$

**3.** $-2a + a$

**4.** $3a - a$

**5.** $7x + 3x + 2x$

**6.** $8x - 2x - x$

**7.** $3a - 2a + 5a$

**8.** $7a - a + 2a$

**9.** $4x - 3 + 2x$

**10.** $5x + 6 - 3x$

**11.** $3a + 4a + 5$

**12.** $6a + 7a + 8$

**13.** $2x - 3 + 3x - 2$

**14.** $6x + 5 - 2x + 3$

**15.** $3a - 1 + a + 3$

**16.** $-a + 2 + 8a - 7$

**17.** $-4x + 8 - 5x - 10$

**18.** $-9x - 1 + x - 4$

**19.** $7a + 3 + 2a + 3a$

**20.** $8a - 2 + a + 5a$

**21.** $5(2x - 1) + 4$

**22.** $2(4x - 3) + 2$

**23.** $7(3y + 2) - 8$

**24.** $6(4y + 2) - 7$

**25.** $-3(2x - 1) + 5$

**26.** $-4(3x - 2) - 6$

**27.** $5 - 2(a + 1)$

**28.** $7 - 8(2a + 3)$

Simplify the following expressions.

**29.** $6 - 4(x - 5)$

**30.** $12 - 3(4x - 2)$

**31.** $-9 - 4(2 - y) + 1$

**32.** $-10 - 3(2 - y) + 3$

**33.** $-6 + 2(2 - 3x) + 1$

**34.** $-7 - 4(3 - x) + 1$

**35.** $(4x - 7) - (2x + 5)$

**36.** $(7x - 3) - (4x + 2)$

**37.** $8(2a + 4) - (6a - 1)$

**38.** $9(3a + 5) - (8a - 7)$

**39.** $3(x - 2) + (x - 3)$

**40.** $2(2x + 1) - (x + 4)$

**41.** $4(2y - 8) - (y + 7)$

**42.** $5(y - 3) - (y - 4)$

**43.** $-9(2x + 1) - (x + 5)$

**44.** $-3(3x - 2) - (2x + 3)$

Evaluate the following expressions when $x$ is 2. (Find the value of the expressions if $x$ is 2.)

**45.** $3x - 1$

**46.** $4x + 3$

**47.** $-2x - 5$

**48.** $-3x + 6$

**49.** $x^2 - 8x + 16$

**50.** $x^2 - 10x + 25$

**51.** $(x - 4)^2$

**52.** $(x - 5)^2$

Find the value of the following expressions when $x$ is $-5$. Then simplify the expression, and check to see that it has the same value for $x = -5$.

**53.** $7x - 4 - x - 3$

**54.** $3x + 4 + 7x - 6$

**55.** $5(2x + 1) + 4$

**56.** $2(3x - 10) + 5$

Find the value of each expression when $x$ is $-3$ and $y$ is 5.

**57.** $x^2 - 2xy + y^2$

**58.** $x^2 + 2xy + y^2$

**59.** $(x - y)^2$

**60.** $(x + y)^2$

**61.** $x^2 + 6xy + 9y^2$

**62.** $x^2 + 10xy + 25y^2$

**63.** $(x + 3y)^2$

**64.** $(x + 5y)^2$

Find the value of $12x - 3$ for each of the following values of $x$.

**65.** $\dfrac{1}{2}$

**66.** $\dfrac{1}{3}$

**67.** $\dfrac{1}{4}$

**68.** $\dfrac{1}{6}$

**69.** $\dfrac{3}{2}$

**70.** $\dfrac{2}{3}$

**71.** $\dfrac{3}{4}$

**72.** $\dfrac{5}{6}$

Substitute 1, 2, 3, and 4 for $n$ in each of the following expressions.

**73.** $2n + 3$

**74.** $5n + 2$

**75.** $n^2 + 1$

**76.** $(n + 1)^2$

**77.** Fill in the tables below to find the sequences formed by substituting the first four counting numbers into the expressions $3n$ and $n^3$.

(a)

| $n$ | 1 | 2 | 3 | 4 |
|-----|---|---|---|---|
| $3n$ | | | | |

(b)

| $n$ | 1 | 2 | 3 | 4 |
|-----|---|---|---|---|
| $n^3$ | | | | |

**78.** Fill in the tables below to find the sequences formed by substituting the first four counting numbers into the expressions $2n - 1$ and $2n + 1$.

(a)

| $n$ | 1 | 2 | 3 | 4 |
|---|---|---|---|---|
| $2n - 1$ | | | | |

(b)

| $n$ | 1 | 2 | 3 | 4 |
|---|---|---|---|---|
| $2n + 1$ | | | | |

## Applying the Concepts

**79. Temperature and Altitude**   If the temperature on the ground is 70°F, then the temperature at $A$ feet above the ground can be found from the expression $-0.0035A + 70$. Find the temperature at the following altitudes.

(a) 8,000 feet    (b) 12,000 feet    (c) 24,000 feet

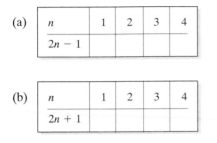

*(Index Stock Imagery)*

**80. Perimeter of a Rectangle**   The expression $2l + 2w$ gives the perimeter of a rectangle with length $l$

5 m

8 m

and width $w$. Find the perimeter of the rectangles with the following lengths and widths.

(a) Length = 8 meters, width = 5 meters
(b) Length = 10 feet, width = 3 feet

3 ft

10 ft

**81. Cellular Phone Rates**   A cellular phone company charges $35 per month plus $0.25 for each minute, or fraction of a minute, that you use one of their cellular phones. The expression $35 + 0.25t$ gives the amount of money you will pay for using one of their phones for $t$ minutes a month. Find the monthly bill for using one of their phones.

(a) 10 minutes in a month
(b) 20 minutes in a month
(c) 30 minutes in a month

**82. Cost of Bottled Water**   A water bottling company charges $7.00 per month for their water dispenser and $1.10 for each gallon of water delivered. If you have $g$ gallons of water delivered in a month, then the expression $7 + 1.1g$ gives the amount of your bill for that month. Find the monthly bill for each of the following deliveries.

(a) 10 gallons   (b) 20 gallons   (c) 30 gallons

*Cool, Refreshing Spring Water*

for only **$7.00** per month and
**$1.10** per gallon!

**83. Taxes**   We all have to pay taxes. Suppose that 21% of your monthly pay is withheld for federal income taxes and another 8% is withheld for Social Security, state income tax, and other miscellaneous items. If $G$ is your monthly pay before any money is deducted (your gross pay), then the amount of money that you take home each month is given by the expression $G - 0.21G - 0.08G$. Simplify this expression, and then find your take-home pay if your gross pay is $1,250 per month.

**84. Taxes**  If you work $H$ hours a day and you pay 21% of your income for federal income tax and another 8% for miscellaneous deductions, then the expression $0.21H + 0.08H$ tells you how many hours a day you are working to pay for those taxes and miscellaneous deductions. Simplify this expression. If you work 8 hours a day under these conditions, how many hours do you work to pay for your taxes and miscellaneous deductions?

## Translating Words Into Symbols

Translate each phrase into an algebraic expression. Then find the value of the expression when $x$ is $-2$.

**85.** The difference of $x$ and 5

**86.** The difference of 5 and $x$

**87.** Twice the sum of $x$ and 10

**88.** The sum of twice $x$ and 10

**89.** The quotient of 10 and $x$

**90.** The quotient of $x$ and 10

**91.** The sum of $3x$ and $-2$, decreased by 5

**92.** The product of $3x$ and $-2$, decreased by 5

## Review Problems

From here on, each problem set will end with some review problems. In mathematics it is very important to review work you have done previously. The more you review, the better you will understand the topics we cover and the longer you will remember them. Also, there are times when material that seemed confusing earlier becomes less confusing the second time around.

The following problems review material we covered previously in Section 1.9. Reviewing these problems will help you with the material in the next section.

Add and subtract as indicated.

**93.** $-3 - \dfrac{1}{2}$

**94.** $-5 - \dfrac{1}{3}$

**95.** $\dfrac{4}{5} + \dfrac{1}{10} + \dfrac{3}{8}$

**96.** $\dfrac{3}{10} + \dfrac{7}{25} + \dfrac{3}{4}$

---

## 2.2    Addition Property of Equality

When light comes into contact with any object, it is reflected, absorbed, and transmitted, as shown in Figure 1.

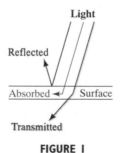

**FIGURE I**

For a certain type of glass, 88% of the light hitting the glass is transmitted through to the other side, while 6% of the light is absorbed into the glass. To find the percent of light that is reflected by the glass, we can solve the equation

$$88 + R + 6 = 100$$

Solving equations of this type is what we study in this section. To solve an equation we must find all replacements for the variable that make the equation a true statement.

> **DEFINITION**   The **solution set** for an equation is the set of all numbers that when used in place of the variable make the equation a true statement.

For example, the equation $x + 2 = 5$ has solution set $\{3\}$ because when $x$ is 3 the equation becomes the true statement $3 + 2 = 5$, or $5 = 5$.

**EXAMPLE  1**   Is 5 a solution to $2x - 3 = 7$?

**Solution**   We substitute 5 for $x$ in the equation and then simplify to see if a true statement results. A true statement means we have a solution; a false statement indicates the number we are using is not a solution.

$$\text{When} \qquad\qquad x = 5$$
$$\text{the equation} \qquad 2x - 3 = 7$$
$$\text{becomes} \qquad 2(5) - 3 \overset{?}{=} 7$$
$$10 - 3 \overset{?}{=} 7$$
$$7 = 7 \qquad \text{A true statement}$$

Since $x = 5$ turns the equation into the true statement $7 = 7$, we know 5 is a solution to the equation.

**Note:**   We can use a question mark over the equal signs to show that we don't know yet whether the two sides of the equation are equal.

**EXAMPLE  2**   Is $-2$ a solution to $8 = 3x + 4$?

**Solution**   Substituting $-2$ for $x$ in the equation, we have

$$8 \overset{?}{=} 3(-2) + 4$$
$$8 \overset{?}{=} -6 + 4$$
$$8 = -2 \qquad\qquad \text{A false statement}$$

Substituting $-2$ for $x$ in the equation produces a false statement. Therefore, $x = -2$ is not a solution to the equation.

The important thing about an equation is its solution set. We therefore make the following definition in order to classify together all equations with the same solution set.

> **DEFINITION**  Two or more equations with the same solution set are said to be **equivalent equations.**

Equivalent equations may look different but must have the same solution set.

## EXAMPLE 3

(a) $x + 2 = 5$ and $x = 3$ are equivalent equations, since both have solution set $\{3\}$.

(b) $a - 4 = 3$, $a - 2 = 5$, and $a = 7$ are equivalent equations, since they all have solution set $\{7\}$.

(c) $y + 3 = 4$, $y - 8 = -7$, and $y = 1$ are equivalent equations, since they all have solution set $\{1\}$.

If two numbers are equal and we increase (or decrease) both of them by the same amount, the resulting quantities are also equal. We can apply this concept to equations. Adding the same amount to both sides of an equation always produces an equivalent equation — one with the same solution set. This fact about equations is called the *addition property of equality* and can be stated more formally as follows.

### ADDITION PROPERTY OF EQUALITY

For any three algebraic expressions $A$, $B$, and $C$,

$$\text{if} \qquad A = B$$

$$\text{then} \qquad A + C = B + C$$

*In words:*  Adding the same quantity to both sides of an equation will not change the solution set.

This property is just as simple as it seems. We can add any amount to both sides of an equation and always be sure we have not changed the solution set.

Consider the equation $x + 6 = 5$. We want to solve this equation for the value of $x$ that makes it a true statement. We want to end up with $x$ on one side of the equal sign and a number on the other side. Since we want $x$ by itself, we will add $-6$ to both sides:

$$x + 6 + (\mathbf{-6}) = 5 + (\mathbf{-6}) \qquad \text{Addition property of equality}$$

$$x + 0 = -1 \qquad \text{Addition}$$

$$x = -1$$

All three equations say the same thing about $x$. They all say that $x$ is $-1$. All three equations are equivalent. The last one is just easier to read.

Here are some further examples of how the addition property of equality can be used to solve equations.

**E X A M P L E   4**   Solve the equation $x - 5 = 12$ for $x$.

**Solution**   Since we want $x$ alone on the left side, we choose to add $+5$ to both sides:

$$x - 5 + \mathbf{5} = 12 + \mathbf{5} \qquad \text{Addition property of equality}$$

$$x + 0 = 17$$

$$x = 17$$

To check our solution, we substitute 17 for $x$ in the original equation:

$$\text{When} \qquad x = 17$$

$$\text{the equation} \qquad x - 5 = 12$$

$$\text{becomes} \qquad 17 - 5 \stackrel{?}{=} 12$$

$$12 = 12 \qquad \text{A true statement}$$

As you can see, our solution checks. The purpose for checking a solution to an equation is to catch any mistakes we may have made in the process of solving the equation.

**E X A M P L E   5**   Solve for $a$: $a + \frac{3}{4} = -\frac{1}{2}$

**Solution**   Since we want $a$ by itself on the left side of the equal sign, we add the opposite of $\frac{3}{4}$ to each side of the equation.

$$a + \frac{3}{4} + \left(-\frac{\mathbf{3}}{\mathbf{4}}\right) = -\frac{1}{2} + \left(-\frac{\mathbf{3}}{\mathbf{4}}\right) \qquad \text{Addition property of equality}$$

$$a + 0 = -\frac{1}{2} \cdot \frac{\mathbf{2}}{\mathbf{2}} + \left(-\frac{3}{4}\right) \qquad \text{LCD on the right side is 4.}$$

$$a = -\frac{2}{4} + \left(-\frac{3}{4}\right) \qquad \tfrac{2}{4} \text{ is equivalent to } \tfrac{1}{2}.$$

$$a = -\frac{5}{4} \qquad \text{Add fractions.}$$

The solution is $a = -\frac{5}{4}$. To check our result, we replace $a$ with $-\frac{5}{4}$ in the original equation. The left side then becomes $-\frac{5}{4} + \frac{3}{4}$, which reduces to $-\frac{1}{2}$, so our solution checks.

**E X A M P L E   6**   Solve for $x$: $7.3 + x = -2.4$

**Solution**  Again, we want to isolate $x$, so we add the opposite of 7.3 to both sides:

$$7.3 + (-\textbf{7.3}) + x = -2.4 + (-\textbf{7.3}) \qquad \text{Addition property of equality}$$

$$0 + x = -9.7$$

$$x = -9.7$$

Sometimes it is necessary to simplify each side of an equation before using the addition property of equality. The reason we simplify both sides first is that we want as few terms as possible on each side of the equation before we use the addition property of equality. The following examples illustrate this procedure.

**EXAMPLE 7**  Solve for $x$: $-x + 2 + 2x = 7 + 5$

**Solution**  We begin by combining similar terms on each side of the equation. Then we use the addition property to solve the simplified equation.

$$x + 2 = 12 \qquad \text{Simplify both sides first.}$$

$$x + 2 + (-\textbf{2}) = 12 + (-\textbf{2}) \qquad \text{Addition property of equality}$$

$$x + 0 = 10$$

$$x = 10$$

**EXAMPLE 8**  Solve $4(2a - 3) - 7a = 2 - 5$.

**Solution**  We must begin by applying the distributive property to separate terms on the left side of the equation. Following that, we combine similar terms and then apply the addition property of equality.

$$4(2a - 3) - 7a = 2 - 5 \qquad \text{Original equation}$$

$$8a - 12 - 7a = 2 - 5 \qquad \text{Distributive property}$$

$$a - 12 = -3 \qquad \text{Simplify each side.}$$

$$a - 12 + \textbf{12} = -3 + \textbf{12} \qquad \text{Add } \textbf{12} \text{ to each side.}$$

$$a = 9 \qquad \text{Addition}$$

To check our solution, we replace $a$ with 9 in the original equation.

$$4(2 \cdot 9 - 3) - 7 \cdot 9 \stackrel{?}{=} 2 - 5$$

$$4(18 - 3) - 63 \stackrel{?}{=} -3$$

$$4(15) - 63 \stackrel{?}{=} -3$$

$$60 - 63 \stackrel{?}{=} -3$$

$$-3 = -3 \qquad \text{A true statement}$$

Our solution checks.

**Note:**  Again, we place a question mark over the equal sign because we don't know yet whether the expressions on the left and right side of the equal sign will be equal.

So far in this section we have used the addition property of equality to add only numbers to both sides of an equation. It is often necessary to add a term involving a variable to both sides of an equation, as the following example indicates.

**E X A M P L E   9**    Solve $3x - 5 = 2x + 7$.

**Solution**   We can solve this equation in two steps. First, we add $-2x$ to both sides of the equation. When this has been done, $x$ appears on the left side only. Second, we add 5 to both sides:

$$3x + (-2x) - 5 = 2x + (-2x) + 7 \qquad \text{Add } -2x \text{ to both sides.}$$
$$x - 5 = 7 \qquad \text{Simplify each side.}$$
$$x - 5 + 5 = 7 + 5 \qquad \text{Add } 5 \text{ to both sides.}$$
$$x = 12 \qquad \text{Simplify each side.}$$

## A NOTE ON SUBTRACTION

Although the addition property of equality is stated for addition only, we can subtract the same number from both sides of an equation as well. Because subtraction is defined as addition of the opposite, subtracting the same quantity from both sides of an equation will not change the solution. If we were to solve the equation in Example 7 using subtraction instead of addition, the steps would look like this:

$$x + 2 = 12 \qquad \text{Original equation}$$
$$x + 2 - 2 = 12 - 2 \qquad \text{Subtract } 2 \text{ from each side.}$$
$$x = 10 \qquad \text{Subtraction}$$

In my experience teaching algebra, I find that students make fewer mistakes if they think in terms of addition rather than subtraction. So, you are probably better off if you continue to use the addition property just the way we have used it in the examples in this section. But, if you are curious as to whether you can subtract the same number from both sides of an equation, the answer is yes.

## Getting Ready for Class

*After reading through the preceding section, respond in your own words and in complete sentences.*

**A.**  What is a solution to an equation?

**B.**  What are equivalent equations?

**C.**  Explain in words the addition property of equality.

**D.**  How do you check a solution to an equation?

# PROBLEM SET 2.2

Find the solution set for the following equations. Be sure to show when you have used the addition property of equality.

**1.** $x - 3 = 8$
**2.** $x - 2 = 7$
**3.** $x + 2 = 6$
**4.** $x + 5 = 4$
**5.** $a + \dfrac{1}{2} = -\dfrac{1}{4}$
**6.** $a + \dfrac{1}{3} = -\dfrac{5}{6}$
**7.** $x + 2.3 = -3.5$
**8.** $x + 7.9 = -3.4$
**9.** $y + 11 = -6$
**10.** $y - 3 = -1$
**11.** $x - \dfrac{5}{8} = -\dfrac{3}{4}$
**12.** $x - \dfrac{2}{5} = -\dfrac{1}{10}$
**13.** $m - 6 = -10$
**14.** $m - 10 = -6$
**15.** $6.9 + x = 3.3$
**16.** $7.5 + x = 2.2$
**17.** $5 = a + 4$
**18.** $12 = a - 3$
**19.** $-\dfrac{5}{9} = x - \dfrac{2}{5}$
**20.** $-\dfrac{7}{8} = x - \dfrac{4}{5}$

Simplify both sides of the following equations as much as possible, and then solve.

**21.** $4x + 2 - 3x = 4 + 1$
**22.** $5x + 2 - 4x = 7 - 3$
**23.** $8a - \dfrac{1}{2} - 7a = \dfrac{3}{4} + \dfrac{1}{8}$
**24.** $9a - \dfrac{4}{5} - 8a = \dfrac{3}{10} - \dfrac{1}{5}$
**25.** $-3 - 4x + 5x = 18$
**26.** $10 - 3x + 4x = 20$
**27.** $-11x + 2 + 10x + 2x = 9$
**28.** $-10x + 5 - 4x + 15x = 0$
**29.** $-2.5 + 4.8 = 8x - 1.2 - 7x$
**30.** $-4.8 + 6.3 = 7x - 2.7 - 6x$
**31.** $2y - 10 + 3y - 4y = 18 - 6$
**32.** $3y - 20 + 6y - 8y = 21$
**33.** $15 - 21 = 8x + 3x - 10x$
**34.** $23 - 17 = -7x - x + 9x$
**35.** $24 - 3 + 8a - 5a - 2a = 21$
**36.** $30 - 4 + 7a - 2 - 6a = 30$

The following equations contain parentheses. Apply the distributive property to remove the parentheses; then simplify each side before using the addition property of equality.

**37.** $2(x + 3) - x = 4$
**38.** $5(x + 1) - 4x = 2$
**39.** $-3(x - 4) + 4x = 3 - 7$
**40.** $-2(x - 5) + 3x = 4 - 9$
**41.** $5(2a + 1) - 9a = 8 - 6$
**42.** $4(2a - 1) - 7a = 9 - 5$
**43.** $-(x + 3) + 2x - 1 = 6$
**44.** $-(x - 7) + 2x - 8 = 4$
**45.** $4y - 3(y - 6) + 2 = 8$
**46.** $7y - 6(y - 1) + 3 = 9$
**47.** $2(3x + 1) - 5(x + 2) = 1 - 10$
**48.** $4(2x + 1) - 7(x - 1) = 2 - 6$
**49.** $-3(2m - 9) + 7(m - 4) = 12 - 9$
**50.** $-5(m - 3) + 2(3m + 1) = 15 - 8$

Solve the following equations by the method used in Example 9 in this section. Check each solution in the original equation.

**51.** $4x = 3x + 2$
**52.** $6x = 5x - 4$
**53.** $8a = 7a - 5$
**54.** $9a = 8a - 3$
**55.** $2x = 3x + 1$
**56.** $4x = 3x + 5$
**57.** $3y + 4 = 2y + 1$
**58.** $5y + 6 = 4y + 2$
**59.** $2m - 3 = m + 5$
**60.** $8m - 1 = 7m - 3$
**61.** $4x - 7 = 5x + 1$
**62.** $3x - 7 = 4x - 6$
**63.** $5x - \dfrac{2}{3} = 4x + \dfrac{4}{3}$
**64.** $3x - \dfrac{5}{4} = 2x + \dfrac{1}{4}$
**65.** $8a - 7.1 = 7a + 3.9$
**66.** $10a - 4.3 = 9a + 4.7$

## Applying the Concepts

**67. Magic Square**  The sum of the numbers in each row, each column, and each diagonal of the following magic square is 6. Use this fact, along with the information in the first column of the magic square, to write an equation containing the variable $x$, and then solve the equation to find $x$. Next, write and solve equations that will give you $y$ and $z$.

| | | |
|---|---|---|
| $x$ | 4 | $y$ |
| $-2$ | 2 | 6 |
| 5 | $z$ | 1 |

**68. Magic Square**   The sum of the numbers in each row and each column of the magic square below is 0. Use this fact, along with the information in the second row of the magic square, to write an equation containing the variable $a$, and then solve the equation to find $a$. Next, write and solve an equation that will allow you to find the value of $b$. Next, write and solve equations that will give you $c$ and $d$.

| | | |
|---|---|---|
| $-3$ | $d$ | $b$ |
| $a$ | $-1$ | $-6$ |
| $-4$ | $c$ | 1 |

**69. Light**   When light comes into contact with any object, it is reflected, absorbed, and transmitted, as shown in the following figure. If $T$ represents the percent of light transmitted, $R$ the percent of light reflected, and $A$ the percent of light absorbed by a surface, then the equation $T + R + A = 100$ shows one way these quantities are related.

(a) For glass, $T = 88$ and $A = 6$, meaning that 88% of the light hitting the glass is transmitted

and 6% is absorbed. Substitute $T = 88$ and $A = 6$ into the equation $T + R + A = 100$ and solve for $R$ to find the percent of light that is reflected.

(b) For flat black paint, $A = 95$ and no light is transmitted, meaning that $T = 0$. What percent of light is reflected by flat black paint?

(c) A pure white surface can reflect 98% of light, so $R = 98$. If no light is transmitted, what percent of light is absorbed by the pure white surface?

(d) Typical, shiny gray metals reflect 70–80% of light. Suppose a thick sheet of aluminum absorbs 25% of light. What percent of light is reflected by this shiny gray metal? (Assume no light is transmitted.)

**70. Movie Tickets**   A movie theater has a total of 300 seats. For a special Saturday night preview, they reserve 20 seats to give away to their VIP guests at no charge. If $x$ represents the number of tickets they can sell for the preview, then $x$ can be found by solving the equation $x + 20 = 300$.

(a) Solve the equation for $x$.

(b) If tickets for the preview are $7.50 each, what is the maximum amount of money they can make from ticket sales?

**71. Geometry**   The three angles shown in the triangle at the front of the tent in the following figure add up to 180°. Use this fact to write an equation containing $x$, and then solve the equation to find the number of degrees in the angle at the top of the triangle.

**72. Geometry** The following figure shows part of a room. From a point on the floor, the angle of elevation to the top of the door is 47°, while the angle of elevation to the ceiling above the door is 59°. Use this diagram to write an equation involving $x$, and then solve the equation to find the number of degrees in the angle that extends from the top of the door to the ceiling.

**Translations** Translate each sentence below into an equation, and then solve the equation.

**73.** The difference of $x$ and 3 is 8.

**74.** The difference of 3 and 8 is $x$.

**75.** The sum of $2y$ and 3 is equal to the sum of $y$ and 5.

**76.** The sum of 3 and $2y$ is equal to the sum of 5 and $y$.

## Review Problems

The following problems review material we covered in Section 1.5. Reviewing this material will help you in Section 2.3.

Apply the associative property to each expression and then simplify the result.

**77.** $3(6x)$    **78.** $5(4x)$    **79.** $\frac{1}{5}(5x)$

**80.** $\frac{1}{3}(3x)$    **81.** $8\left(\frac{1}{8}y\right)$    **82.** $6\left(\frac{1}{6}y\right)$

**83.** $-2\left(-\frac{1}{2}x\right)$    **84.** $-4\left(-\frac{1}{4}x\right)$

**85.** $-\frac{4}{3}\left(-\frac{3}{4}a\right)$    **86.** $-\frac{5}{2}\left(-\frac{2}{5}a\right)$

---

**2.3**   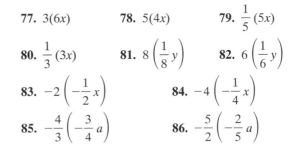 Multiplication Property of Equality

As we have mentioned before, we all have to pay taxes. According to Figure 1, people have been paying taxes for quite a long time.

**FIGURE 1** Collection of taxes, ca. 3000 B.C. Clerks and scribes appear at the right, with pen and papyrus, and officials and taxpayers appear at the left.

If 21% of your monthly pay is withheld for federal income taxes and another 8% is withheld for Social Security, state income tax, and other miscellaneous items, leaving you with $987.50 a month in take-home pay, then the amount you earned before the deductions were removed from your check is given by the equation

$$G - 0.21G - 0.08G = 987.5$$

In this section we will learn how to solve equations of this type.

In the previous section, we found that adding the same number to both sides of an equation never changed the solution set. The same idea holds for multiplication by numbers other than zero. We can multiply both sides of an equation by the same nonzero number and always be sure we have not changed the solution set. (The reason we cannot multiply both sides by zero will become apparent later.) This fact about equations is called the *multiplication property of equality,* which can be stated formally as follows.

## MULTIPLICATION PROPERTY OF EQUALITY

For any three algebraic expressions $A$, $B$, and $C$, where $C \neq 0$,

$$\text{if} \qquad A = B$$

$$\text{then} \qquad AC = BC$$

*In words:*  Multiplying both sides of an equation by the same nonzero number will not change the solution set.

Suppose we want to solve the equation $5x = 30$. We have $5x$ on the left side but would like to have just $x$. We choose to multiply both sides by $\frac{1}{5}$ since $(\frac{1}{5})(5) = 1$. Here is the solution:

$$5x = 30$$

$$\frac{1}{5}(5x) = \frac{1}{5}(30) \qquad \text{Multiplication property of equality}$$

$$\left(\frac{1}{5} \cdot 5\right)x = \frac{1}{5}(30) \qquad \text{Associative property of multiplication}$$

$$1x = 6$$

$$x = 6$$

We chose to multiply by $\frac{1}{5}$ because it is the reciprocal of 5. We can see that multiplication by any number except zero will not change the solution set. If, however, we were to multiply both sides by zero, the result would always be $0 = 0$, since multiplication by zero always results in zero. Although the statement $0 = 0$ is true, we have lost our variable and cannot solve the equation. This is the only restriction of the multiplication property of equality. We are free to multiply both sides of an equation by any number except zero.

Here are some more examples that use the multiplication property of equality.

**EXAMPLE 1**  Solve for $a$: $-4a = 24$

**Solution**  Since we want $a$ alone on the left side, we choose to multiply both sides by $-\frac{1}{4}$:

$$-\frac{1}{4}(-4a) = -\frac{1}{4}(24) \qquad \text{Multiplication property of equality}$$

$$\left[-\frac{1}{4}(-4)\right]a = -\frac{1}{4}(24) \qquad \text{Associative property}$$

$$a = -6$$

**EXAMPLE 2**  Solve for $t$: $-\frac{t}{3} = 5$

**Solution**  Since division by 3 is the same as multiplication by $\frac{1}{3}$, we can write $-\frac{t}{3}$ as $-\frac{1}{3}t$. To solve the equation, we multiply each side by the reciprocal of $-\frac{1}{3}$, which is $-3$.

$$-\frac{t}{3} = 5 \qquad \text{Original equation}$$

$$-\frac{1}{3}t = 5 \qquad \text{Dividing by 3 is equivalent to multiplying by } \frac{1}{3}.$$

$$-3\left(-\frac{1}{3}t\right) = -3(5) \qquad \text{Multiply each side by } -3.$$

$$\left[-3\left(-\frac{1}{3}\right)\right]t = -3(5) \qquad \text{Associative property}$$

$$t = -15 \qquad \text{Multiplication}$$

**EXAMPLE 3**  Solve $\frac{2}{3}y = 4$.

**Solution**  We can multiply both sides by $\frac{3}{2}$ and have $1y$ on the left side:

$$\frac{3}{2}\left(\frac{2}{3}y\right) = \frac{3}{2}(4) \qquad \text{Multiplication property of equality}$$

$$\left(\frac{3}{2}\cdot\frac{2}{3}\right)y = \frac{3}{2}(4) \qquad \text{Associative property}$$

$$y = 6 \qquad \text{Simplify } \frac{3}{2}(4) = \frac{3}{2}(\frac{4}{1}) = \frac{12}{2} = 6.$$

Notice in Examples 1 through 3 that if the variable is being multiplied by a number like $-4$ or $\frac{2}{3}$, we always multiply by the number's reciprocal, $-\frac{1}{4}$ or $\frac{3}{2}$, to end up with just the variable on one side of the equation.

---

**E X A M P L E   4**    Solve $5 + 8 = 10x + 20x - 4x$.

**Solution**   Our first step will be to simplify each side of the equation:

$$13 = 26x \qquad \text{Simplify both sides first.}$$

$$\frac{1}{26}(13) = \frac{1}{26}(26x) \qquad \text{Multiplication property of equality}$$

$$\frac{13}{26} = x \qquad \text{Multiplication}$$

$$\frac{1}{2} = x \qquad \text{Reduce to lowest terms.}$$

---

In the next three examples, we will use both the addition property of equality and the multiplication property of equality.

---

**E X A M P L E   5**    Solve for $x$: $6x + 5 = -13$

**Solution**   We begin by adding $-5$ to both sides of the equation:

$$6x + 5 + (-5) = -13 + (-5) \qquad \text{Add } -5 \text{ to both sides.}$$

$$6x = -18 \qquad \text{Simplify.}$$

$$\frac{1}{6}(6x) = \frac{1}{6}(-18) \qquad \text{Multiply both sides by } \tfrac{1}{6}.$$

$$x = -3$$

---

**E X A M P L E   6**    Solve for $x$: $5x = 2x + 12$

**Solution**   We begin by adding $-2x$ to both sides of the equation:

$$5x + (-2x) = 2x + (-2x) + 12 \qquad \text{Add } -2x \text{ to both sides.}$$

$$3x = 12 \qquad \text{Simplify.}$$

$$\frac{1}{3}(3x) = \frac{1}{3}(12) \qquad \text{Multiply both sides by } \tfrac{1}{3}.$$

$$x = 4 \qquad \text{Simplify.}$$

---

Notice that in Example 6 we used the addition property of equality first in order to combine all the terms containing $x$ on the left side of the equation. Once this had been done, we used the multiplication property to isolate $x$ on the left side.

---

**EXAMPLE 7**   Solve for $x$: $3x - 4 = -2x + 6$

**Solution**   We begin by adding $2x$ to both sides:

$$3x + \mathbf{2x} - 4 = -2x + \mathbf{2x} + 6 \qquad \text{Add } \mathbf{2x} \text{ to both sides.}$$

$$5x - 4 = 6 \qquad\qquad\qquad \text{Simplify.}$$

Now we add 4 to both sides:

$$5x - 4 + \mathbf{4} = 6 + \mathbf{4} \qquad \text{Add } \mathbf{4} \text{ to both sides.}$$

$$5x = 10 \qquad\qquad\qquad \text{Simplify.}$$

$$\frac{\mathbf{1}}{\mathbf{5}}(5x) = \frac{\mathbf{1}}{\mathbf{5}}(10) \qquad \text{Multiply by } \tfrac{\mathbf{1}}{\mathbf{5}}.$$

$$x = 2 \qquad\qquad\qquad \text{Simplify.}$$

---

The next example involves fractions. You will see that the properties we use to solve equations containing fractions are the same as the properties we used to solve the previous equations. Also, the LCD that we used previously to add fractions can be used with the multiplication property of equality to simplify equations containing fractions.

---

**EXAMPLE 8**   Solve $\frac{2}{3}x + \frac{1}{2} = -\frac{3}{4}$.

**Solution**   We can solve this equation by applying our properties and working with the fractions, or we can begin by eliminating the fractions. Let's use both methods.

*Method 1*   Working with the fractions.

$$\frac{2}{3}x + \frac{1}{2} + \left(-\frac{\mathbf{1}}{\mathbf{2}}\right) = -\frac{3}{4} + \left(-\frac{\mathbf{1}}{\mathbf{2}}\right) \qquad \text{Add } -\tfrac{1}{2} \text{ to each side.}$$

$$\frac{2}{3}x = -\frac{5}{4} \qquad\qquad \text{Note that } -\tfrac{3}{4} + \left(-\tfrac{1}{2}\right) = -\tfrac{3}{4} + \left(-\tfrac{2}{4}\right).$$

$$\frac{\mathbf{3}}{\mathbf{2}}\left(\frac{2}{3}x\right) = \frac{\mathbf{3}}{\mathbf{2}}\left(-\frac{5}{4}\right) \qquad \text{Multiply each side by } \tfrac{\mathbf{3}}{\mathbf{2}}.$$

$$x = -\frac{15}{8}$$

*Method 2*   Eliminating the fractions in the beginning.

Our original equation has denominators of 3, 2, and 4. The LCD for these three denominators is 12, and it has the property that all three denominators will divide it evenly. Therefore, if we multiply both sides of our equation by 12, each denominator will divide into 12, and we will be left with an equation that does not contain any denominators other than 1.

$$12\left(\frac{2}{3}x + \frac{1}{2}\right) = 12\left(-\frac{3}{4}\right) \qquad \text{Multiply each side by the LCD } \textbf{12.}$$

$$12\left(\frac{2}{3}x\right) + 12\left(\frac{1}{2}\right) = 12\left(-\frac{3}{4}\right) \qquad \text{Distributive property on the left side}$$

$$8x + 6 = -9 \qquad \text{Multiply.}$$

$$8x = -15 \qquad \text{Add } -6 \text{ to each side.}$$

$$x = -\frac{15}{8} \qquad \text{Multiply each side by } \frac{1}{8}.$$

As the third line in the previous expression indicates, multiplying each side of the equation by the LCD eliminates all the fractions from the equation.

As you can see, both methods yield the same solution.

## A NOTE ON DIVISION

Since division is defined as multiplication by the reciprocal, multiplying both sides of an equation by the same number is equivalent to dividing both sides of the equation by the reciprocal of that number. That is, multiplying each side of an equation by $\frac{1}{3}$ and dividing each side of the equation by 3 are equivalent operations. If we were to solve the equation $3x = 18$ using division instead of multiplication, the steps would look like this:

$$3x = 18 \qquad \text{Original equation}$$

$$\frac{3x}{3} = \frac{18}{3} \qquad \text{Divide each side by } \textbf{3.}$$

$$x = 6 \qquad \text{Division}$$

Using division instead of multiplication on a problem like this may save you some writing. On the other hand, with multiplication, it is easier to explain "why" we end up with just one $x$ on the left side of the equation. (The "why" has to do with the associative property of multiplication.) My suggestion is that you continue to use multiplication to solve equations like this one until you understand the process completely. Then, if you find it more convenient, you can use division instead of multiplication.

# Getting Ready for Class

*After reading through the preceding section, respond in your own words and in complete sentences.*

**A.** Explain in words the multiplication property of equality.

**B.** If an equation contains fractions, how do you use the multiplication property of equality to clear the equation of fractions?

**C.** Why is it okay to divide both sides of an equation by the same number?

**D.** Explain in words how you would solve the equation $3x = 7$, using the multiplication property of equality.

## PROBLEM SET 2.3

Solve the following equations. Be sure to show your work.

**1.** $5x = 10$

**2.** $6x = 12$

**3.** $7a = 28$

**4.** $4a = 36$

**5.** $-8x = 4$

**6.** $-6x = 2$

**7.** $8m = -16$

**8.** $5m = -25$

**9.** $-3x = -9$

**10.** $-9x = -36$

**11.** $-7y = -28$

**12.** $-15y = -30$

**13.** $2x = 0$

**14.** $7x = 0$

**15.** $-5x = 0$

**16.** $-3x = 0$

**17.** $\dfrac{x}{3} = 2$

**18.** $\dfrac{x}{4} = 3$

**19.** $-\dfrac{m}{5} = 10$

**20.** $-\dfrac{m}{7} = 1$

**21.** $-\dfrac{x}{2} = -\dfrac{3}{4}$

**22.** $-\dfrac{x}{3} = \dfrac{5}{6}$

**23.** $\dfrac{2}{3}a = 8$

**24.** $\dfrac{3}{4}a = 6$

**25.** $-\dfrac{3}{5}x = \dfrac{9}{5}$

**26.** $-\dfrac{2}{5}x = \dfrac{6}{15}$

**27.** $-\dfrac{5}{8}y = -20$

**28.** $-\dfrac{7}{2}y = -14$

Simplify both sides as much as possible, and then solve.

**29.** $-4x - 2x + 3x = 24$

**30.** $7x - 5x + 8x = 20$

**31.** $4x + 8x - 2x = 15 - 10$

**32.** $5x + 4x + 3x = 4 + 8$

**33.** $-3 - 5 = 3x + 5x - 10x$

**34.** $10 - 16 = 12x - 6x - 3x$

**35.** $18 - 13 = \dfrac{1}{2}a + \dfrac{3}{4}a - \dfrac{5}{8}a$

**36.** $20 - 14 = \dfrac{1}{3}a + \dfrac{5}{6}a - \dfrac{2}{3}a$

Solve the following equations by multiplying both sides by $-1$.

**37.** $-x = 4$

**38.** $-x = -3$

**39.** $-x = -4$

**40.** $-x = 3$

**41.** $15 = -a$

**42.** $-15 = -a$

**43.** $-y = \dfrac{1}{2}$

**44.** $-y = -\dfrac{3}{4}$

Solve each of the following equations using the method shown in Examples 5–8 in this section.

**45.** $3x - 2 = 7$

**46.** $2x - 3 = 9$

**47.** $2a + 1 = 3$

**48.** $5a - 3 = 7$

**49.** $\dfrac{1}{8} + \dfrac{1}{2}x = \dfrac{1}{4}$

**50.** $\dfrac{1}{3} + \dfrac{1}{7}x = -\dfrac{8}{21}$

**51.** $6x = 2x - 12$

**52.** $8x = 3x - 10$

**53.** $2y = -4y + 18$

**54.** $3y = -2y - 15$

**55.** $-7x = -3x - 8$

**56.** $-5x = -2x - 12$

**57.** $8x + 4 = 2x - 5$

**58.** $5x + 6 = 3x - 6$

**59.** $x + \dfrac{1}{2} = \dfrac{1}{4}x - \dfrac{5}{8}$

**60.** $\dfrac{1}{3}x + \dfrac{2}{5} = \dfrac{1}{5}x - \dfrac{2}{5}$

Solve.

**61.** $6m - 3 = m + 2$    **62.** $6m - 5 = m + 5$

**63.** $\dfrac{1}{2}m - \dfrac{1}{4} = \dfrac{1}{12}m + \dfrac{1}{6}$

**64.** $\dfrac{1}{2}m - \dfrac{5}{12} = \dfrac{1}{12}m + \dfrac{5}{12}$

**65.** $9y + 2 = 6y - 4$    **66.** $6y + 14 = 2y - 2$

**67.** $\dfrac{3}{2}y + \dfrac{1}{3} = y - \dfrac{2}{3}$    **68.** $\dfrac{3}{2}y + \dfrac{7}{2} = \dfrac{1}{2}y - \dfrac{1}{2}$

## Applying the Concepts

**69. Break-Even Point**   Movie theaters pay a certain price for the movies that you and I see. Suppose a theater pays \$1,500 for each showing of a popular movie. If they charge \$7.50 for each ticket they sell, then the equation $7.5x = 1,500$ gives the number of tickets they must sell to equal the \$1,500 cost of showing the movie. This number is called the break-even point. Solve the equation for $x$ to find the break-even point.

**70. Stock Sales**   Suppose you purchase $x$ shares of a stock at \$12 per share. After 6 months you decide to sell all your shares at \$20 per share. Your broker charges you \$15 for the trade, which gives you a profit of \$3,985. In this situation, solving the equation

$$20x - 12x - 15 = 3,985$$

will tell you how many shares of the stock you purchased originally. How many shares did you purchase originally?

**71. Basketball**   Laura plays basketball for her community college. In one game she scored 13 points total, with a combination of free throws, field goals, and three-pointers. Each free throw is worth 1 point, each field goal is 2 points, and each 3-pointer is worth 3 points. If she made 1 free throw and 3 field goals, then solving the equation

$$1 + 3(2) + 3x = 13$$

will give us the number of three-pointers she made. Solve the equation to find the number of three-point shots Laura made.

**72. Taxes**   If 21% of your monthly pay is withheld for federal income taxes and another 8% is withheld for Social Security, state income tax, and other miscellaneous items, leaving you with \$987.50 a month in take-home pay, then the amount you earned before the deductions were removed from your check is given by the equation

$$G - 0.21G - 0.08G = 987.5$$

Solve this equation to find your gross income.

**73. Similar Triangles**   If two triangles are similar, their corresponding sides are in proportion.

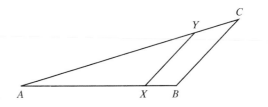

Triangle $ABC$ is similar to triangle $AXY$ means that

$$\frac{AX}{AB} = \frac{AY}{AC} = \frac{XY}{BC}$$

(a) If $AB = 12$, $AY = 15$, and $AC = 20$, find $AX$.
(b) If $XY = 8$, $BC = 10$, and $AX = 12$, find $AB$.
(c) If $YC = 6$, $XB = 4$, and $AX = 8$, find $AC$.

**74. Rhind Papyrus**   The *Rhind Papyrus* is an ancient Egyptian document that contains some mathematical riddles. One problem on the *Rhind Papyrus* asks the reader to find a quantity such that when it is added to one-fourth of itself the sum is 15. The equation that describes this situation is

$$x + \frac{1}{4}x = 15$$

Solve this equation.

*(© The British Museum)*

**Translations**   Translate each sentence below into an equation, and then solve the equation.

**75.** The sum of $3x$ and 2 is 19.

**76.** The difference of 4 and $6x$ is 20.

**77.** Twice the sum of $x$ and 10 is 40.

**78.** The sum of twice $x$ and 10 is 40.

## Review Problems

The following problems review material we covered in Section 2.1. Reviewing this material will help you in the next section.

Simplify each expression.

**79.** $5(2x - 8) - 3$

**80.** $4(3x - 1) + 7$

**81.** $-2(3x + 5) + 3(x - 1)$

**82.** $6(x + 3) - 2(2x + 4)$

**83.** $7 - 3(2y + 1)$

**84.** $8 - 5(3y - 4)$

**85.** $4x - (9x - 3) + 4$

**86.** $x - (5x + 2) - 3$

---

**2.4**   **Solving Linear Equations**

We will now use the material we have developed in the first three sections of this chapter to build a method for solving any linear equation.

> **DEFINITION**   A **linear equation** in one variable is any equation that can be put in the form $ax + b = 0$, where $a$ and $b$ are real numbers and $a$ is not zero.

Each of the equations we will solve in this section is a linear equation in one variable. The steps we use to solve a linear equation in one variable are listed here.

### STRATEGY FOR SOLVING LINEAR EQUATIONS IN ONE VARIABLE

*Step 1a:* Use the distributive property to separate terms, if necessary.

*1b:* If fractions are present, consider multiplying both sides by the LCD to eliminate the fractions. If decimals are present, consider multiplying both sides by a power of 10 to clear the equation of decimals.

*1c:* Combine similar terms on each side of the equation.

*Step 2:* Use the addition property of equality to get all variable terms on one side of the equation and all constant terms on the other side. A variable term is a term that contains the variable (e.g., $5x$). A constant term is a term that does not contain the variable (e.g., the number 3).

*Step 3:* Use the multiplication property of equality to get $x$ (that is, $1x$) by itself on one side of the equation.

*Step 4:* Check your solution in the original equation to be sure that you have not made a mistake in the solution process.

As you will see as you work through the examples in this section, it is not always necessary to use all four steps when solving equations. The number of steps used depends upon the equation. In Example 1 there are no fractions or decimals in the original equation, so step 1b will not be used. Likewise, after applying the distributive property to the left side of the equation in Example 1, there are no similar terms to combine on either side of the equation, making step 1c also unnecessary.

**EXAMPLE  1**  Solve $2(x + 3) = 10$.

**Solution**  To begin, we apply the distributive property to the left side of the equation to separate terms:

*Step 1a:*               $2x + 6 = 10$                    Distributive property

*Step 2:*  $\begin{cases} 2x + 6 + (-6) = 10 + (-6) & \text{Addition property of} \\ \qquad\quad 2x = 4 & \text{equality} \end{cases}$

*Step 3:*  $\begin{cases} \dfrac{1}{2}(2x) = \dfrac{1}{2}(4) & \text{Multiply each side by } \dfrac{1}{2}. \\[2mm] \qquad x = 2 & \text{The solution is 2.} \end{cases}$

The solution to our equation is 2. We check our work (to be sure we have not made either a mistake in applying the properties or an arithmetic mistake) by substituting 2 into our original equation and simplifying each side of the result separately.

**Check:**  $\begin{cases} \text{When} & x = 2 \\ \text{the equation} & 2(x + 3) = 10 \\ \text{becomes} & 2(2 + 3) \stackrel{?}{=} 10 \\ & 2(5) \stackrel{?}{=} 10 \\ & 10 = 10 \qquad \text{A true statement} \end{cases}$

*Step 4:*

Our solution checks.

The general method of solving linear equations is actually very simple. It is based on the properties we developed in Chapter 1 and on two very simple new properties. We can add any number to both sides of the equation and multiply both sides by any nonzero number. The equation may change in form, but the solution set will not. If we look back to Example 1, each equation looks a little different from each preceding equation. What is interesting and useful is that each equation says the same thing about $x$. They all say $x$ is 2. The last equation, of course, is the easiest to read, and that is why our goal is to end up with $x$ by itself.

**EXAMPLE  2**  Solve for $x$: $3(x - 5) + 4 = 13$

**Solution**  Our first step will be to apply the distributive property to the left side of the equation:

*Step 1a:*    $3x - 15 + 4 = 13$    Distributive property

*Step 1c:*    $3x - 11 = 13$    Simplify the left side.

*Step 2:* $\begin{cases} 3x - 11 + \mathbf{11} = 13 + \mathbf{11} & \text{Add } \mathbf{11} \text{ to both sides.} \\ \quad\quad 3x = 24 \end{cases}$

*Step 3:* $\begin{cases} \dfrac{1}{3}(3x) = \dfrac{1}{3}(24) & \text{Multiply both sides by } \frac{1}{3}. \\ \quad\quad x = 8 & \text{The solution is 8.} \end{cases}$

**Check:**

*Step 4:* $\begin{cases} \quad\quad \text{When} \quad\quad\quad\quad\quad x = 8 \\ \text{the equation} \quad\quad 3(x - 5) + 4 = 13 \\ \quad \text{becomes} \quad\quad 3(8 - 5) + 4 \overset{?}{=} 13 \\ \quad\quad\quad\quad\quad\quad\quad 3(3) + 4 \overset{?}{=} 13 \\ \quad\quad\quad\quad\quad\quad\quad 9 + 4 \overset{?}{=} 13 \\ \quad\quad\quad\quad\quad\quad\quad 13 = 13 \quad\quad \text{A true statement} \end{cases}$

---

**E X A M P L E   3**    Solve $5(x - 3) + 2 = 5(2x - 8) - 3$.

**Solution**    In this case we apply the distributive property on each side of the equation:

*Step 1a:*    $5x - 15 + 2 = 10x - 40 - 3$    Distributive property

*Step 1c:*    $5x - 13 = 10x - 43$    Simplify each side.

*Step 2:* $\begin{cases} 5x + (-\mathbf{5x}) - 13 = 10x + (-\mathbf{5x}) - 43 & \text{Add } -\mathbf{5x} \text{ to both sides.} \\ \quad\quad -13 = 5x - 43 \\ \quad -13 + \mathbf{43} = 5x - 43 + \mathbf{43} & \text{Add } \mathbf{43} \text{ to both sides.} \\ \quad\quad 30 = 5x \end{cases}$

*Step 3:* $\begin{cases} \dfrac{1}{5}(30) = \dfrac{1}{5}(5x) & \text{Multiply both sides by } \frac{1}{5}. \\ \quad\quad 6 = x & \text{The solution is 6.} \end{cases}$

**Check:**

*Step 4:* $\begin{cases} \text{Replacing } x \text{ with 6 in the original equation, we have} \\ 5(6 - 3) + 2 \overset{?}{=} 5(2 \cdot 6 - 8) - 3 \\ \quad 5(3) + 2 \overset{?}{=} 5(12 - 8) - 3 \\ \quad 5(3) + 2 \overset{?}{=} 5(4) - 3 \\ \quad 15 + 2 \overset{?}{=} 20 - 3 \\ \quad\quad 17 = 17 \quad\quad\quad\quad\quad \text{A true statement} \end{cases}$

**Note:** It makes no difference on which side of the equal sign $x$ ends up. Most people prefer to have $x$ on the left side because we read from left to right, and it seems to sound better to say $x$ is 6 rather than 6 is $x$. Both expressions, however, have exactly the same meaning.

**EXAMPLE 4**   Solve the equation $0.08x + 0.09(x + 2,000) = 690$.

**Solution**   We can solve the equation in its original form by working with the decimals, or we can eliminate the decimals first by using the multiplication property of equality and solving the resulting equation. Both methods follow.

**Method 1**   Working with the decimals

$$0.08x + 0.09(x + 2,000) = 690 \qquad \text{Original equation}$$

**Step 1a:**  $0.08x + 0.09x + 0.09(2,000) = 690$ \qquad Distributive property

**Step 1c:**  $\qquad 0.17x + 180 = 690$ \qquad Simplify the left side.

**Step 2:**
$$\begin{cases} 0.17x + 180 + (\mathbf{-180}) = 690 + (\mathbf{-180}) \qquad \text{Add } \mathbf{-180} \text{ to each side.} \\ \\ 0.17x = 510 \end{cases}$$

**Step 3:**
$$\begin{cases} \dfrac{0.17x}{\mathbf{0.17}} = \dfrac{510}{\mathbf{0.17}} \qquad \text{Divide each side by } \mathbf{0.17}. \\ \\ x = 3,000 \end{cases}$$

Note that we divided each side of the equation by 0.17 to obtain our solution. This is still an application of the multiplication property of equality, since dividing by 0.17 is equivalent to multiplying by $\frac{1}{0.17}$.

**Method 2**   Eliminating the decimals in the beginning

$$0.08x + 0.09(x + 2,000) = 690 \qquad \text{Original equation}$$

**Step 1a:**  $0.08x + 0.09x + 180 = 690$ \qquad Distributive property

**Step 1b:**
$$\begin{cases} \mathbf{100}(0.08x + 0.09x + 180) = \mathbf{100}(690) \qquad \text{Multiply both sides by } \mathbf{100}. \\ \\ 8x + 9x + 18,000 = 69,000 \end{cases}$$

**Step 1c:**  $\qquad 17x + 18,000 = 69,000$ \qquad Simplify the left side.

**Step 2:**  $\qquad 17x = 51,000$ \qquad Add $-18,000$ to each side.

**Step 3:**
$$\begin{cases} \dfrac{17x}{\mathbf{17}} = \dfrac{51,000}{\mathbf{17}} \qquad \text{Divide each side by } \mathbf{17}. \\ \\ x = 3,000 \end{cases}$$

**Check:**

**Step 4:**

Substituting 3,000 for $x$ in the original equation, we have

$$0.08(3,000) + 0.09(3,000 + 2,000) \stackrel{?}{=} 690$$

$$0.08(3,000) + 0.09(5,000) \stackrel{?}{=} 690$$

$$240 + 450 \stackrel{?}{=} 690$$

$$690 = 690 \qquad \text{A true statement}$$

**E X A M P L E   5**    Solve $7 - 3(2y + 1) = 16$.

**Solution**    We begin by multiplying $-3$ times the sum of $2y$ and 1:

| | | |
|---|---|---|
| *Step 1a:* | $7 - 6y - 3 = 16$ | Distributive property |
| *Step 1c:* | $-6y + 4 = 16$ | Simplify the left side. |
| *Step 2:* | $\begin{cases} -6y + 4 + (-4) = 16 + (-4) \\ -6y = 12 \end{cases}$ | Add $-4$ to both sides. |
| *Step 3:* | $\begin{cases} -\dfrac{1}{6}(-6y) = -\dfrac{1}{6}(12) \\[2mm] y = -2 \end{cases}$ | Multiply both sides by $-\dfrac{1}{6}$. |

*Step 4:*    Replacing $y$ with $-2$ in the original equation yields a true statement.

There are two things to notice about the example that follows: first, the distributive property is used to remove parentheses that are preceded by a negative sign, and, second, the addition property and the multiplication property are not shown in as much detail as in the previous examples.

**E X A M P L E   6**    Solve $3(2x - 5) - (2x - 4) = 6 - (4x + 5)$.

**Solution**    When we apply the distributive property to remove the grouping symbols and separate terms, we have to be careful with the signs. Remember, we can think of $-(2x - 4)$ as $-1(2x - 4)$, so that

$$-(2x - 4) = -1(2x - 4) = -2x + 4$$

It is not uncommon for students to make a mistake with this type of simplification and write the result as $-2x - 4$, which is incorrect. Here is the complete solution to our equation:

| | |
|---|---|
| $3(2x - 5) - (2x - 4) = 6 - (4x + 5)$ | Original equation |
| $6x - 15 - 2x + 4 = 6 - 4x - 5$ | Distributive property |
| $4x - 11 = -4x + 1$ | Simplify each side. |
| $8x - 11 = 1$ | Add $4x$ to each side. |

$$8x = 12 \qquad \text{Add 11 to each side.}$$

$$x = \frac{12}{8} \qquad \text{Multiply each side by } \tfrac{1}{8}.$$

$$x = \frac{3}{2} \qquad \text{Reduce to lowest terms.}$$

The solution, $\frac{3}{2}$, checks when replacing $x$ in the original equation.

## Getting Ready for Class

*After reading through the preceding section, respond in your own words and in complete sentences.*

**A.** What is the first step in solving a linear equation containing parentheses?

**B.** What is the last step in solving a linear equation?

**C.** Explain in words how you would solve the equation $2x - 3 = 8$.

**D.** If an equation contains decimals, what can you do to eliminate the decimals?

## PROBLEM SET 2.4

Solve each of the following equations using the four steps shown in this section.

**1.** $2(x + 3) = 12$

**2.** $3(x - 2) = 6$

**3.** $6(x - 1) = -18$

**4.** $4(x + 5) = 16$

**5.** $2(4a + 1) = -6$

**6.** $3(2a - 4) = 12$

**7.** $14 = 2(5x - 3)$

**8.** $-25 = 5(3x + 4)$

**9.** $-2(3y + 5) = 14$

**10.** $-3(2y - 4) = -6$

**11.** $-5(2a + 4) = 0$

**12.** $-3(3a - 6) = 0$

**13.** $1 = \frac{1}{2}(4x + 2)$

**14.** $1 = \frac{1}{3}(6x + 3)$

**15.** $3(t - 4) + 5 = -4$

**16.** $5(t - 1) + 6 = -9$

Solve each equation.

**17.** $4(2y + 1) - 7 = 1$

**18.** $6(3y + 2) - 8 = -2$

**19.** $\frac{1}{2}(x - 3) = \frac{1}{4}(x + 1)$

**20.** $\frac{1}{3}(x - 4) = \frac{1}{2}(x - 6)$

**21.** $-0.7(2x - 7) = 0.3(11 - 4x)$

**22.** $-0.3(2x - 5) = 0.7(3 - x)$

**23.** $-2(3y + 1) = 3(1 - 6y) - 9$

**24.** $-5(4y - 3) = 2(1 - 8y) + 11$

**25.** $\frac{3}{4}(8x - 4) + 3 = \frac{2}{5}(5x + 10) - 1$

**26.** $\frac{5}{6}(6x + 12) + 1 = \frac{2}{3}(9x - 3) + 5$

**27.** $0.06x + 0.08(100 - x) = 6.5$

**28.** $0.05x + 0.07(100 - x) = 6.2$

**29.** $6 - 5(2a - 3) = 1$

**30.** $-8 - 2(3 - a) = 0$

**31.** $0.2x - 0.5 = 0.5 - 0.2(2x - 13)$

**32.** $0.4x - 0.1 = 0.7 - 0.3(6 - 2x)$

**33.** $2(t - 3) + 3(t - 2) = 28$

**34.** $-3(t - 5) - 2(2t + 1) = -8$

**35.** $5(x - 2) - (3x + 4) = 3(6x - 8) + 10$

**36.** $3(x - 1) - (4x - 5) = 2(5x - 1) - 7$

**37.** $2(5x - 3) - (2x - 4) = 5 - (6x + 1)$

**38.** $3(4x - 2) - (5x - 8) = 8 - (2x + 3)$

**39.** $-(3x + 1) - (4x - 7) = 4 - (3x + 2)$

**40.** $-(6x + 2) - (8x - 3) = 8 - (5x + 1)$

**44.** $\dfrac{3}{2}(4)$    **45.** $\dfrac{5}{9} \cdot \dfrac{9}{5}$    **46.** $\dfrac{3}{7} \cdot \dfrac{7}{3}$

## Review Problems

The problems that follow review material we covered in Sections 1.2, 1.5, and 1.6. Reviewing these problems will help you understand the next section.

Apply the distributive property, and then simplify each expression as much as possible.

**47.** $2(3x - 5)$    **48.** $4(2x - 6)$

**49.** $\dfrac{1}{2}(3x + 6)$    **50.** $\dfrac{1}{4}(2x + 8)$

Multiply.

**41.** $\dfrac{1}{2}(3)$    **42.** $\dfrac{1}{3}(2)$    **43.** $\dfrac{2}{3}(6)$

**51.** $\dfrac{1}{3}(-3x + 6)$    **52.** $\dfrac{1}{2}(-2x + 6)$

## 2.5    Formulas

In this section we continue solving equations by working with formulas. To begin, here is the definition for a formula.

> **DEFINITION**    In mathematics, a **formula** is an equation that contains more than one variable.

The equation $P = 2l + 2w$, which tells us how to find the perimeter of a rectangle, is an example of a formula.

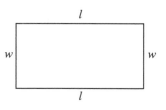

To begin our work with formulas, we will consider some examples in which we are given numerical replacements for all but one of the variables.

**EXAMPLE 1**    The perimeter $P$ of a rectangular livestock pen is 40 feet. If the width $w$ is 6 feet, find the length.

**Solution**  First, we substitute 40 for $P$ and 6 for $w$ in the formula
$P = 2l + 2w$. Then we solve for $l$:

| | | |
|---|---|---|
| When | $P = 40$ and $w = 6$ | |
| the formula | $P = 2l + 2w$ | |
| becomes | $40 = 2l + 2(6)$ | |
| or | $40 = 2l + 12$ | Multiply 2 and 6. |
| | $28 = 2l$ | Add $-12$ to each side. |
| | $14 = l$ | Multiply each side by $\frac{1}{2}$. |

To summarize our results, if a rectangular pen has a perimeter of 40 feet and a
width of 6 feet, then the length must be 14 feet.

**E X A M P L E   2**   Find $y$ when $x = 4$ in the formula $3x + 2y = 6$.

**Solution**  We substitute 4 for $x$ in the formula and then solve for $y$:

| | | |
|---|---|---|
| When | $x = 4$ | |
| the formula | $3x + 2y = 6$ | |
| becomes | $3(4) + 2y = 6$ | |
| or | $12 + 2y = 6$ | Multiply 3 and 4. |
| | $2y = -6$ | Add $-12$ to each side. |
| | $y = -3$ | Multiply each side by $\frac{1}{2}$. |

    In the next examples we will solve a formula for one of its variables, without
being given numerical replacements for the other variables.

Consider the formula for the area of a triangle:

$A = \frac{1}{2}bh$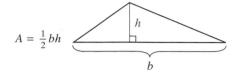

where $A$ = area, $b$ = length of the base, and $h$ = the height of the triangle.

Suppose we want to solve this formula for $h$. What we must do is isolate the variable $h$ on one side of the equal sign. We begin by multiplying both sides by 2:

$$2 \cdot A = 2 \cdot \frac{1}{2}bh$$

$$2A = bh$$

Then we divide both sides by $b$:

$$\frac{2A}{b} = \frac{bh}{b}$$

$$h = \frac{2A}{b}$$

The original formula $A = \frac{1}{2}bh$ and the final formula $h = \frac{2A}{b}$ both give the same relationship among $A$, $b$, and $h$. The first one has been solved for $A$, and the second one has been solved for $h$.

**RULE**

To solve a formula for one of its variables, we must isolate that variable on either side of the equal sign. All other variables and constants will appear on the other side.

**EXAMPLE 3**   Solve $3x + 2y = 6$ for $y$.

**Solution**   To solve for $y$, we must isolate $y$ on the left side of the equation. To begin, we use the addition property of equality to add $-3x$ to each side:

| | |
|---|---|
| $3x + 2y = 6$ | Original formula |
| $3x + (\mathbf{-3x}) + 2y = (\mathbf{-3x}) + 6$ | Add $\mathbf{-3x}$ to each side. |
| $2y = -3x + 6$ | Simplify the left side. |
| $\frac{\mathbf{1}}{\mathbf{2}}(2y) = \frac{\mathbf{1}}{\mathbf{2}}(-3x + 6)$ | Multiply each side by $\frac{\mathbf{1}}{\mathbf{2}}$. |
| $y = -\frac{3}{2}x + 3$ | Multiplication |

**EXAMPLE  4**  Solve $h = vt - 16t^2$ for $v$.

**Solution**  Let's begin by interchanging the left and right sides of the equation. That way, the variable we are solving for, $v$, will be on the left side.

$$vt - 16t^2 = h \qquad \text{Exchange sides.}$$

$$vt - 16t^2 + \mathbf{16t^2} = h + \mathbf{16t^2} \qquad \text{Add } \mathbf{16t^2} \text{ to each side.}$$

$$vt = h + 16t^2$$

$$\frac{vt}{t} = \frac{h + 16t^2}{t} \qquad \text{Divide each side by } \boldsymbol{t.}$$

$$v = \frac{h + 16t^2}{t}$$

We know we are finished because we have isolated the variable we are solving for on the left side of the equation and it does not appear on the other side.

## FACTS FROM
# Geometry

In Chapter 1 we defined complementary angles as angles that add to 90°. That is, if $x$ and $y$ are complementary angles, then

$$x + y = 90°$$

If we solve this formula for $y$, we obtain a formula equivalent to our original formula:

$$y = 90° - x$$

Complementary angles

Since $y$ is the complement of $x$, we can generalize by saying that the complement of angle $x$ is the angle $90° - x$. By a similar reasoning process, we can say that the supplement of angle $x$ is the angle $180° - x$. To summarize, if $x$ is an angle, then

the complement of $x$ is $90° - x$ and

the supplement of $x$ is $180° - x$.

$180° - x$

$x$

Supplementary angles

If you go on to take a trigonometry class, you will see this formula again.

**EXAMPLE  5**  Find the complement and the supplement of 25°.

**Solution**  We can use the formulas above with $x = 25°$.

The complement of 25° is $90° - 25° = 65°$.

The supplement of 25° is $180° - 25° = 155°$.

## BASIC PERCENT PROBLEMS

The last examples in this section show how basic percent problems can be translated directly into equations. To understand these examples, you must recall that percent means "per hundred." That is, 75% is the same as $\frac{75}{100}$, 0.75, and, in reduced fraction form, $\frac{3}{4}$. Likewise, the decimal 0.25 is equivalent to 25%. To change a decimal to a percent, we move the decimal point two places to the right and write the % symbol. To change from a percent to a decimal, we drop the % symbol and move the decimal point two places to the left. The table that follows gives some of the most commonly used fractions and decimals and their equivalent percents.

| Fraction | Decimal | Percent |
|----------|---------|---------|
| $\frac{1}{2}$ | 0.5 | 50% |
| $\frac{1}{4}$ | 0.25 | 25% |
| $\frac{3}{4}$ | 0.75 | 75% |
| $\frac{1}{3}$ | $0.33\frac{1}{3}$ | $33\frac{1}{3}\%$ |
| $\frac{2}{3}$ | $0.66\frac{2}{3}$ | $66\frac{2}{3}\%$ |
| $\frac{1}{5}$ | 0.2 | 20% |
| $\frac{2}{5}$ | 0.4 | 40% |
| $\frac{3}{5}$ | 0.6 | 60% |
| $\frac{4}{5}$ | 0.8 | 80% |

**EXAMPLE 6**   What number is 25% of 60?

**Solution**   To solve a problem like this, we let $x$ = the number in question (that is, the number we are looking for). Then, we translate the sentence directly into an equation by using an equal sign for the word *is* and multiplication for the word *of*. Here is how it is done:

$$\underbrace{\text{What number}}_{x} \;\; \overset{\downarrow}{\text{is}} \;\; \overset{\downarrow}{25\%} \;\; \overset{\downarrow}{\text{of}} \;\; \overset{\downarrow}{60?}$$
$$x \;\; = 0.25 \;\; \cdot \;\; 60$$
$$x = 15$$

Notice that we must write 25% as a decimal in order to do the arithmetic in the problem.

   The number 15 is 25% of 60.

**EXAMPLE 7**   What percent of 24 is 6?

**Solution**   Translating this sentence into an equation, as we did in Example 6, we have:

$$\underbrace{\text{What percent}}\ \underset{\downarrow}{\text{of}}\ \underset{\downarrow}{24}\ \underset{\downarrow}{\text{is}}\ \underset{\downarrow}{6?}$$

$$\underset{x}{\underbrace{\phantom{xxx}}}\quad \cdot\ 24\ =\ 6$$

$$\text{or}\qquad 24x\ =\ 6$$

Next, we multiply each side by $\frac{1}{24}$. (This is the same as dividing each side by 24.)

$$\frac{1}{24}(24x) = \frac{1}{24}(6)$$

$$x = \frac{6}{24}$$

$$= \frac{1}{4}$$

$$= 0.25,\ \text{or } 25\%$$

The number 6 is 25% of 24.

---

**E X A M P L E   8**   45 is 75% of what number?

**Solution**   Again, we translate the sentence directly:

$$\underset{45}{\underset{\downarrow}{45}}\ \underset{=}{\underset{\downarrow}{\text{is}}}\ \underset{0.75}{\underset{\downarrow}{75\%}}\ \underset{\cdot}{\underset{\downarrow}{\text{of}}}\ \underbrace{\text{what number?}}_{x}$$

Next, we multiply each side by $\frac{1}{0.75}$ (which is the same as dividing each side by 0.75):

$$\frac{1}{0.75}(45) = \frac{1}{0.75}(0.75x)$$

$$\frac{45}{0.75} = x$$

$$60 = x$$

The number 45 is 75% of 60.

---

We can solve application problems involving percent by translating each problem into one of the three basic percent problems shown in Examples 6, 7, and 8.

---

**E X A M P L E   9**   The American Dietetics Association (ADA) recommends eating foods in which the calories from fat are less than 30% of the total calories. The nutrition labels from two kinds of granola bars are shown in Figure 1. For each bar, what percent of the total calories comes from fat?

**Solution**  The information needed to solve this problem is located toward the top of each label. Each serving of Bar I contains 210 calories, of which 70 calories come from fat. To find the percent of total calories that come from fat, we must answer this question:

70 is what percent of 210?

Bar I

**Nutrition Facts**
Serving Size 2 bars (47g)
Servings Per Container 6

**Amount Per Serving**

| | |
|---|---|
| **Calories** | 210 |
| Calories from Fat | 70 |

| | % Daily Value* |
|---|---|
| **Total Fat** 8g | 12% |
| Saturated Fat 1g | 5% |
| **Cholesterol** 0mg | 0% |
| **Sodium** 150mg | 6% |
| **Total Carbohydrate** 32g | 11% |
| Dietary Fiber 2g | 10% |
| Sugars 12g | |
| **Protein** 4g | |

* Percent Daily Values are based on a 2,000 calorie diet. Your daily values may be higher or lower depending on your calorie needs.

Bar II

**Nutrition Facts**
Serving Size 1 bar (21g)
Servings Per Container 8

**Amount Per Serving**

| | |
|---|---|
| **Calories** | 80 |
| Calories from Fat | 15 |

| | % Daily Value* |
|---|---|
| **Total Fat** 1.5g | 2% |
| Saturated Fat 0g | 0% |
| **Cholesterol** 0mg | 0% |
| **Sodium** 60mg | 3% |
| **Total Carbohydrate** 16g | 5% |
| Dietary Fiber 1g | 4% |
| Sugars 5g | |
| **Protein** 2g | |

* Percent Daily Values are based on a 2,000 calorie diet. Your daily values may be higher or lower depending on your calorie needs.

**FIGURE I**

For Bar II, one serving contains 80 calories, of which 15 calories come from fat. To find the percent of total calories that come from fat, we must answer this question:

15 is what percent of 80?

Translating each question into symbols, we have

70 is what percent of 210?         15 is what percent of 80?

$70 = x \cdot 210$                              $15 = x \cdot 80$

$x = \dfrac{70}{210}$                            $x = \dfrac{15}{80}$

$x = 0.33$ to the nearest hundredth        $x = 0.19$ to the nearest hundredth

$x = 33\%$                                    $x = 19\%$

Comparing the two bars, 33% of the calories in Bar I are fat calories, while 19% of the calories in Bar II are fat calories. According to the ADA, Bar II is the healthier choice.

## Getting Ready for Class

*After reading through the preceding section, respond in your own words and in complete sentences.*

**A.** What is a formula?

**B.** How do you solve a formula for one of its variables?

**C.** What are complementary angles?

**D.** What does percent mean?

## PROBLEM SET 2.5

Use the formula $P = 2l + 2w$ to find the length $l$ of a rectangular lot if

**1.** The width $w$ is 50 feet and the perimeter $P$ is 300 feet.

**2.** The width $w$ is 75 feet and the perimeter $P$ is 300 feet.

Use the formula $2x + 3y = 6$ to find $y$ if

**3.** $x$ is 3

**4.** $x$ is $-2$

**5.** $x$ is 0

**6.** $x$ is $-3$

Use the formula $2x - 5y = 20$ to find $x$ if

**7.** $y$ is 2

**8.** $y$ is $-4$

**9.** $y$ is 0

**10.** $y$ is $-6$

Use the equation $y = 2x - 1$ to find $x$ when

**11.** $y$ is 7

**12.** $y$ is 9

**13.** $y$ is 3

**14.** $y$ is $-1$

Solve each of the following for the indicated variable.

**15.** $A = lw$ for $l$

**16.** $A = lw$ for $w$

**17.** $d = rt$ for $r$

**18.** $d = rt$ for $t$

**19.** $V = lwh$ for $h$

**20.** $V = lwh$ for $l$

**21.** $PV = nRT$ for $P$

**22.** $PV = nRT$ for $T$

**23.** $P = a + b + c$ for $a$

**24.** $P = a + b + c$ for $b$

**25.** $x - 3y = -1$ for $x$

**26.** $x + 3y = 2$ for $x$

**27.** $-3x + y = 6$ for $y$

**28.** $2x + y = -17$ for $y$

**29.** $2x + 3y = 6$ for $y$

**30.** $4x + 5y = 20$ for $y$

**31.** $6x + 3y = 12$ for $y$

**32.** $3x + 6y = 12$ for $y$

**33.** $5x - 2y = 3$ for $y$

**34.** $7x - 3y = 5$ for $y$

**35.** $P = 2l + 2w$ for $w$

**36.** $P = 2l + 2w$ for $l$

**37.** $h = vt + 16t^2$ for $v$

**38.** $h = vt - 16t^2$ for $v$

**39.** $A = \pi r^2 + 2\pi rh$ for $h$

**40.** $A = 2\pi r^2 + 2\pi rh$ for $h$

Solve each formula for $y$.

**41.** $\dfrac{x}{2} + \dfrac{y}{3} = 1$

**42.** $\dfrac{x}{5} + \dfrac{y}{4} = 1$

**43.** $\dfrac{x}{7} - \dfrac{y}{3} = 1$

**44.** $\dfrac{x}{4} - \dfrac{y}{9} = 1$

**45.** $-\dfrac{1}{4}x + \dfrac{1}{8}y = 1$

**46.** $-\dfrac{1}{9}x + \dfrac{1}{3}y = 1$

Find the complement and the supplement of each angle.

**47.** $30°$

**48.** $60°$

**49.** $45°$

**50.** $15°$

Translate each of the following into an equation, and then solve that equation.

**51.** What number is 25% of 40?

**52.** What number is 75% of 40?

**53.** What number is 12% of 2,000?

**54.** What number is 9% of 3,000?

**55.** What percent of 28 is 7?

**56.** What percent of 28 is 21?

**57.** What percent of 40 is 14?

**58.** What percent of 20 is 14?

**59.** 32 is 50% of what number?

**60.** 16 is 50% of what number?

**61.** 240 is 12% of what number?

**62.** 360 is 12% of what number?

## Applying the Concepts

**63. Percent of People Voting**  In the 60 years between 1932 and 1992, the largest percentage of eligible voters who actually cast votes in a presidential election was 62.8%, when John Kennedy defeated Richard Nixon in 1960. Kennedy and Nixon split a total of nearly 68,840,000 votes.
(a) Let $T$ = the total number of eligible voters and write an equation that relates $T$, 62.8%, and 68,840,000.
(b) Solve the equation you wrote in part (a) to find the number of people who were eligible to vote in the 1960 election.

**64. Women in the Military**  The number of women in the military on active duty in 1981 was 171,418. This number was 76.8% of the number of women on active duty in 1984. How many women were on active duty in 1984?

**65. Jewelry Content**  Silver jewelry is actually a mixture of silver and copper. Suppose you have a silver bracelet that weighs 20 grams and contains 12 grams of silver.
(a) What percent of the bracelet is silver?
(b) What percent of the bracelet is copper?

**66. Treasury Bonds**  If you buy a $5,000 treasury bond that pays 6% interest each year, then in $t$ years its value will be

$$V = 5,000(1.06)^t$$

Find the value of this bond after 2 years.

**More About Temperatures**  As we mentioned in Chapter 1, in the U.S. system, temperature is measured on the Fahrenheit scale. In the metric system, temperature is measured on the Celsius scale. On the Celsius scale, water boils at 100 degrees and freezes at 0 degrees. To denote a temperature of 100 degrees on the Celsius scale, we write

100°C, which is read "100 degrees Celsius"

Table 1 is intended to give you an intuitive idea of the relationship between the two temperature scales. Table 2 gives the formulas, in both symbols and words, that are used to convert between the two scales.

**TABLE 1**

| Situation | Temperature Fahrenheit | Celsius |
|---|---|---|
| Water freezes | 32°F | 0°C |
| Room temperature | 68°F | 20°C |
| Normal body temperature | 98.6°F | 37°C |
| Water boils | 212°F | 100°C |
| Bake cookies | 365°F | 185°C |

**67.** Let $F = 212$ in the formula $C = \frac{5}{9}(F - 32)$, and solve for $C$. Does the value of $C$ agree with the information in Table 1?

**68.** Let $C = 100$ in the formula $F = \frac{9}{5}C + 32$, and solve for $F$. Does the value of $F$ agree with the information in Table 1?

**69.** Let $F = 68$ in the formula $C = \frac{5}{9}(F - 32)$, and solve for $C$. Does the value of $C$ agree with the information in Table 1?

**70.** Let $C = 37$ in the formula $F = \frac{9}{5}C + 32$, and solve for $F$. Does the value of $F$ agree with the information in Table 1?

**TABLE 2**

| To Convert From | Formula in Symbols | Formula in Words |
|---|---|---|
| Fahrenheit to Celsius | $C = \frac{5}{9}(F - 32)$ | Subtract 32, multiply by 5, then divide by 9. |
| Celsius to Fahrenheit | $F = \frac{9}{5}C + 32$ | Multiply by $\frac{9}{5}$, then add 32. |

**71.** Solve the formula $F = \frac{9}{5}C + 32$ for $C$.

**72.** Solve the formula $C = \frac{5}{9}(F - 32)$ for $F$.

**Nutrition Labels**  The nutrition label on the top is from a quart of vanilla ice cream. The label on the bottom is from a pint of vanilla frozen yogurt. Use the information on these labels for Problems 73–76. Round your answers to the nearest tenth of a percent.

Vanilla Ice Cream

**Nutrition Facts**
Serving Size 1/2 cup (65g)
Servings 8

| Amount/Serving | |
| --- | --- |
| **Calories**  150 | Calories from Fat 90 |
| | **% Daily Value*** |
| **Total Fat** 10g | 16% |
| Saturated Fat 6g | 32% |
| **Cholesterol** 35mg | 12% |
| **Sodium** 30mg | 1% |
| **Total Carbohydrate** 14g | 5% |
| Dietary Fiber 0g | 0% |
| Sugars 11g | |
| **Protein** 2g | |

| | | |
| --- | --- | --- |
| Vitamin A 6% | • | Vitamin C 0% |
| Calcium 6% | • | Iron 0% |

* Percent Daily Values are based on a 2,000 calorie diet.

Vanilla Frozen Yogurt

**Nutrition Facts**
Serving Size 1/2 cup (98g)
Servings Per Container 4

| Amount Per Serving | |
| --- | --- |
| **Calories**  160 | Calories from Fat 25 |
| | **% Daily Value*** |
| **Total Fat** 2.5g | 4% |
| Saturated Fat 1.5g | 7% |
| **Cholesterol** 45mg | 15% |
| **Sodium** 55mg | 2% |
| **Total Carbohydrate** 26g | 9% |
| Dietary Fiber 0g | 0% |
| Sugars 19g | |
| **Protein** 8g | |

| | | |
| --- | --- | --- |
| Vitamin A 0% | • | Vitamin C 0% |
| Calcium 25% | • | Iron 0% |

* Percent Daily Values are based on a 2,000 calorie diet.

**73.** What percent of the calories in one serving of the vanilla ice cream are fat calories?

**74.** What percent of the calories in one serving of the frozen yogurt are fat calories?

**75.** One serving of frozen yogurt is 98 grams, of which 26 grams are carbohydrates. What percent of one serving are carbohydrates?

**76.** One serving of vanilla ice cream is 65 grams. Find the percent of one serving that is sugar.

**77. Temperature and Altitude**  As we mentioned in the introduction to this chapter, there is a relationship between air temperature and altitude. The higher above the Earth we are, the lower the temperature. If the temperature on the ground is 56°F, then the temperature $T$ at altitude $A$ is given by the formula

$$T = -0.0035A + 56$$

Use this formula to fill in the table below.

| Altitude (feet) | Temperature (°F) |
| --- | --- |
| 0 | |
| 9,000 | |
| 15,000 | |
| 21,000 | |
| 28,000 | |
| 40,000 | |

**78. Complementary Angles**  The diagram below shows sunlight hitting the ground. Angle $\alpha$ (alpha) is called the angle of inclination, and angle $\theta$ (theta) is called the angle of incidence. As the sun moves across the sky, the values of these angles change. Assume that $\alpha + \theta = 90°$, and fill in the table below.

| $\alpha$ | $\theta$ |
| --- | --- |
| 0° | |
| 15° | |
| 30° | |
| 45° | |
| 60° | |
| 90° | |

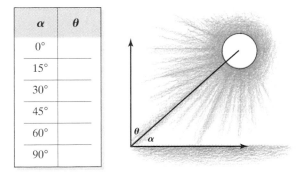

**79. Circumference**  The circumference of a circle is given by the formula $C = 2\pi r$. Find $r$ if
(a) The circumference $C$ is 44 meters and $\pi$ is $\frac{22}{7}$.
(b) The circumference is 9.42 inches and $\pi$ is 3.14.

**80. Volume**  The volume of a cylinder is given by the formula $V = \pi r^2 h$. Find the height $h$ if

(a) The volume $V$ is 42 cubic feet, the radius is $\frac{7}{22}$ feet, and $\pi$ is $\frac{22}{7}$.

(b) The volume is 6.28 cubic centimeters, the radius is 3 centimeters, and $\pi$ is 3.14.

## Review Problems

The following problems review material we covered in Chapter 1. Reviewing these problems will help you in the next section.

Write an equivalent expression in words. Include the words *sum* and *difference* when possible.

**81.** $4 + 1 = 5$

**82.** $7 + 3 = 10$

**83.** $6 - 2 = 4$

**84.** $8 - 1 = 7$

For each of the following expressions, write an equivalent expression with numbers.

**85.** Twice the sum of 6 and 3

**86.** 4 added to the product of 5 and $-1$

**87.** The sum of twice 5 and 3 is 13.

**88.** Twice the difference of 8 and 2 is 12.

## 2.6   Applications

As you begin reading through the examples in this section, you may find yourself asking why some of these problems seem so contrived. The title of the section is "Applications," but many of the problems here don't seem to have much to do with "real life." You are right about that. Example 3 is what we refer to as an "age problem." But imagine a conversation in which you ask someone how old her children are and she replies, "Bill is 6 years older than Tom. Three years ago the sum of their ages was 21. You figure it out." Although many of the "application" problems in this section are contrived, they are also good for practicing the strategy we will use to solve all application problems.

To begin this section, we list the steps used in solving application problems. We call this strategy the *Blueprint for Problem Solving*. It is an outline that will overlay the solution process we use on all application problems.

### Blueprint for Problem Solving

*Step 1:* ***Read*** the problem, and then mentally ***list*** the items that are known and the items that are unknown.

*Step 2:* ***Assign a variable*** to one of the unknown items. (In most cases this will amount to letting $x =$ the item that is asked for in the problem.) Then ***translate*** the other ***information*** in the problem to expressions involving the variable.

*Step 3:* ***Reread*** the problem, and then ***write an equation,*** using the items and variables listed in steps 1 and 2, that describes the situation.

*Step 4:* ***Solve the equation*** found in step 3.

*Step 5:* ***Write*** your ***answer*** using a complete sentence.

*Step 6:* ***Reread*** the problem, and ***check*** your solution with the original words in the problem.

There are a number of substeps within each of the steps in our blueprint. For instance, with steps 1 and 2 it is always a good idea to draw a diagram or picture, if it helps visualize the relationship between the items in the problem. In other cases a table helps organize the information. As you gain more experience using the blueprint to solve application problems, you will find additional techniques that expand the blueprint.

To help with problems of the type shown next in Example 1, here are some common words and phrases and their mathematical translations.

| Words | Algebra |
|---|---|
| The sum of $a$ and $b$ | $a + b$ |
| The difference of $a$ and $b$ | $a - b$ |
| The product of $a$ and $b$ | $a \cdot b$ |
| The quotient of $a$ and $b$ | $\dfrac{a}{b}$ |
| of | $\cdot$ (multiply) |
| is | $=$ (equals) |
| A number | $x$ |
| 4 more than $x$ | $x + 4$ |
| 4 times $x$ | $4x$ |
| 4 less than $x$ | $x - 4$ |

## NUMBER PROBLEMS

**EXAMPLE  1**   The sum of twice a number and 3 is 7. Find the number.

**Solution**   Using the Blueprint for Problem Solving as an outline, we solve the problem as follows:

   *Step 1: **Read** the problem, and then mentally **list** the items that are known and the items that are unknown.*

   *Known items:*      The numbers 3 and 7

   *Unknown item:*      The number in question

   *Step 2: **Assign a variable** to one of the unknown items. Then **translate** the other information in the problem to expressions involving the variable.*

   Let $x$ = the number asked for in the problem,

   then "the sum of twice a number and three" translates to $2x + 3$.

   *Step 3: **Reread** the problem, and then **write an equation,** using the items and variables listed in steps 1 and 2, that describes the situation.*

With all word problems, the word *is* translates to =.

The sum of twice $x$ and 3 is 7

$$2x + 3 \qquad = 7$$

**Step 4:** *Solve the equation* found in step 3.

$$2x + 3 = 7$$
$$2x + 3 + (\mathbf{-3}) = 7 + (\mathbf{-3})$$
$$2x = 4$$
$$\frac{1}{2}(2x) = \frac{1}{2}(4)$$
$$x = 2$$

**Step 5:** *Write* your **answer** using a complete sentence.

The number is 2.

**Step 6:** *Reread* the problem, and **check** your solution with the original words in the problem.

The sum of twice 2 and 3 is 7; a true statement.

You may find some examples and problems in this section, and the problem set that follows, that you can solve without using algebra or our blueprint. It is very important that you solve those problems using the methods we are showing here. The purpose behind these problems is to give you experience using the blueprint as a guide to solving problems written in words. Your answers are much less important than the work that you show to obtain your answer. You will be able to condense the steps in the blueprint later in the course. For now, though, you need to show your work in the same detail that we are showing in the examples in this section.

**EXAMPLE 2**    One number is 3 more than twice another; their sum is 18. Find the numbers.

**Solution**

**Step 1:** *Read and list.*
   *Known items:*        Two numbers that add to 18. One is 3 more than twice the other.
   *Unknown items:*     The numbers in question

**Step 2:** *Assign a variable and translate information.*
   Let $x =$ the first number. The other is $2x + 3$.

**Step 3:** *Reread and write an equation.*

Their sum   is 18

$$x + (2x + 3) = 18$$

*Step 4:  Solve the equation.*

$$x + (2x + 3) = 18$$
$$3x + 3 = 18$$
$$3x + 3 + (\mathbf{-3}) = 18 + (\mathbf{-3})$$
$$3x = 15$$
$$x = 5$$

*Step 5:  Write the answer.*
    The first number is 5. The other is $2 \cdot 5 + 3 = 13$.
*Step 6:  Reread and check.*
    The sum of 5 and 13 is 18, and 13 is 3 more than twice 5.

## AGE PROBLEM

Remember as you read through the steps in the solutions to the examples in this section that step 1 is done mentally. Read the problem, and then mentally list the items that you know and the items that you don't know. The purpose of step 1 is to give you direction as you begin to work application problems. Finding the solution to an application problem is a process; it doesn't happen all at once. The first step is to read the problem with a purpose in mind. That purpose is to mentally note the items that are known and the items that are unknown.

**EXAMPLE 3**    Bill is 6 years older than Tom. Three years ago Bill's age was 4 times Tom's age. Find the age of each boy now.

**Solution**    Applying the Blueprint for Problem Solving, we have

*Step 1:  Read and list.*

| | |
|---|---|
| *Known items:* | Bill is 6 years older than Tom. Three years ago Bill's age was 4 times Tom's age. |
| *Unknown items:* | Bill's age and Tom's age |

*Step 2:  Assign a variable and translate information.*
    Let $x$ = Tom's age now. That makes Bill $x + 6$ years old now. A table like the one shown here can help organize the information in an age problem. Notice how we placed the $x$ in the box that corresponds to Tom's age now.

| | Three Years Ago | Now |
|---|---|---|
| Bill | | $x + 6$ |
| Tom | | $x$ |

If Tom is $x$ years old now, three years ago he was $x - 3$ years old. If Bill is $x + 6$ years old now, three years ago he was $x + 6 - 3 =$

$x + 3$ years old. We use this information to fill in the remaining squares in the table.

|  | **Three Years Ago** | **Now** |
|---|---|---|
| Bill | $x + 3$ | $x + 6$ |
| Tom | $x - 3$ | $x$ |

**Step 3: *Reread and write an equation.***

Reading the problem again, we see that three years ago Bill's age was four times Tom's age. Writing this as an equation, we have Bill's age three years ago = 4 (Tom's age three years ago):

$$x + 3 = 4(x - 3)$$

**Step 4: *Solve the equation.***

$$x + 3 = 4(x - 3)$$

$$x + 3 = 4x - 12$$

$$x + (-x) + 3 = 4x + (-x) - 12$$

$$3 = 3x - 12$$

$$3 + 12 = 3x - 12 + 12$$

$$15 = 3x$$

$$x = 5$$

**Step 5: *Write the answer.***

Tom is 5 years old. Bill is 11 years old.

**Step 6: *Reread and check.***

If Tom is 5 and Bill is 11, then Bill is 6 years older than Tom. Three years ago Tom was 2 and Bill was 8. At that time, Bill's age was four times Tom's age. As you can see, the answers check with the original problem.

## GEOMETRY PROBLEM

To understand Example 4 completely, you need to recall from Chapter 1 that the perimeter of a rectangle is the sum of the lengths of the sides. The formula for the perimeter is $P = 2l + 2w$.

**E X A M P L E   4**     The length of a rectangle is 5 inches more than twice the width. The perimeter is 34 inches. Find the length and width.

**Solution**   When working problems that involve geometric figures, a sketch of the figure helps organize and visualize the problem.

*Step 1:*  **Read and list.**
  *Known items:*    The figure is a rectangle. The length is 5 inches more than twice the width. The perimeter is 34 inches.
  *Unknown items:*    The length and the width

*Step 2:*  **Assign a variable and translate information.**
  Since the length is given in terms of the width (the length is 5 more than twice the width), we let $x =$ the width of the rectangle. The length is 5 more than twice the width, so it must be $2x + 5$. The diagram below is a visual description of the relationships we have listed so far.

*Step 3:*  **Reread and write an equation.**
  The equation that describes the situation is

  Twice the length $+$ twice the width is the perimeter

$$2(2x + 5)    +    2x    =    34$$

*Step 4:*  **Solve the equation.**

$$2(2x + 5) + 2x = 34$$

| | |
|---|---|
| $4x + 10 + 2x = 34$ | Distributive property |
| $6x + 10 = 34$ | Add $4x$ and $2x$. |
| $6x = 24$ | Add $-10$ to each side. |
| $x = 4$ | Divide each side by 6. |

*Step 5:*  **Write the answer.**
  The width $x$ is 4 inches. The length is $2x + 5 = 2(4) + 5 = 13$ inches.

*Step 6:*  **Reread and check.**
  If the length is 13 and the width is 4, then the perimeter must be $2(13) + 2(4) = 26 + 8 = 34$, which checks with the original problem.

## COIN PROBLEM

**EXAMPLE  5**    Jennifer has $2.45 in dimes and nickels. If she has 8 more dimes than nickels, how many of each coin does she have?

**Solution**

**Step 1: *Read and list.***
*Known items:*   The type of coins, the total value of the coins, and that there are 8 more dimes than nickels
*Unknown items:*   The number of nickels and the number of dimes

**Step 2: *Assign a variable and translate information.***
If we let $x$ = the number of nickels, then $x + 8$ = the number of dimes. Since the value of each nickel is 5 cents, the amount of money in nickels is $5x$. Similarly, since each dime is worth 10 cents, the amount of money in dimes is $10(x + 8)$. Here is a table that summarizes the information we have so far:

|  | Nickels | Dimes |
|---|---|---|
| Number | $x$ | $x + 8$ |
| Value (in cents) | $5x$ | $10(x + 8)$ |

**Step 3: *Reread and write an equation.***
Since the total value of all the coins is 245 cents, the equation that describes this situation is

| Amount of money in nickels | + | Amount of money in dimes | = | Total amount of money |
|---|---|---|---|---|
| $5x$ | + | $10(x + 8)$ | = | 245 |

**Step 4: *Solve the equation.***
To solve the equation, we apply the distributive property first.

$$5x + 10x + 80 = 245 \qquad \text{Distributive property}$$

$$15x + 80 = 245 \qquad \text{Add } 5x \text{ and } 10x.$$

$$15x = 165 \qquad \text{Add } -80 \text{ to each side.}$$

$$x = 11 \qquad \text{Divide each side by 15.}$$

**Step 5: *Write the answer.***
The number of nickels is $x = 11$.

**Step 6: *Reread and check.***
The number of dimes is $x + 8 = 11 + 8 = 19$.
To check our results

$$
\begin{array}{ll}
11 \text{ nickels are worth } 5(11) = & 55 \text{ cents} \\
19 \text{ dimes are worth } 10(19) = & 190 \text{ cents} \\
\hline
\text{The total value is} & 245 \text{ cents} = \$2.45
\end{array}
$$

When you begin working the problems in the problem set that follows, there are a couple of things to remember. The first is that you may have to read the problems a number of times before you begin to see how to solve them. The second thing to remember is that word problems are not always solved correctly the first time you try them. Sometimes it takes a couple of attempts and some wrong answers before you can set up and solve these problems correctly.

# Getting Ready for Class

*After reading through the preceding section, respond in your own words and in complete sentences.*

**A.** What is the first step in the Blueprint for Problem Solving?

**B.** What is the last thing you do when solving an application problem?

**C.** What good does it do you to solve application problems even when they don't have much to do with real life?

**D.** Write an application problem whose solution depends on solving the equation $2x + 3 = 7$.

# PROBLEM SET 2.6

Solve the following word problems. Follow the steps given in the Blueprint for Problem Solving.

## Number Problems

1. The sum of a number and 5 is 13. Find the number.

2. The difference of 10 and a number is $-8$. Find the number.

3. The sum of twice a number and 4 is 14. Find the number.

4. The difference of 4 times a number and 8 is 16. Find the number.

5. 5 times the sum of a number and 7 is 30. Find the number.

6. 5 times the difference of twice a number and 6 is $-20$. Find the number.

7. One number is 2 more than another. Their sum is 8. Find both numbers.

8. One number is 3 less than another. Their sum is 15. Find the numbers.

9. One number is 4 less than 3 times another. If their sum is increased by 5, the result is 25. Find the numbers.

10. One number is 5 more than twice another. If their

sum is decreased by 10, the result is 22. Find the numbers.

## Age Problems

11. Fred is 4 years older than Barney. Five years ago the sum of their ages was 48. How old are they now? (Begin by filling in the table.)

| | Five Years Ago | Now |
|---|---|---|
| Fred | X-1 | Y+Y |
| Barney | X-S | x |

12. Tim is 5 years older than JoAnn. Six years from now the sum of their ages will be 79. How old are they now?

| | Now | Six Years From Now |
|---|---|---|
| Tim | | |
| JoAnn | x | |

13. Jack is twice as old as Lacy. In 3 years the sum of their ages will be 54. How old are they now?

**14.** John is 4 times as old as Martha. Five years ago the sum of their ages was 50. How old are they now?

**15.** Pat is 20 years older than his son Patrick. In 2 years Pat will be twice as old as Patrick. How old are they now?

**16.** Diane is 23 years older than her daughter Amy. In 6 years Diane will be twice as old as Amy. How old are they now?

## Geometry Problems

**17.** The length of a rectangle is 5 inches more than the width. The perimeter is 34 inches. Find the length and width.

$x$

$x + 5$

**18.** The width of a rectangle is 3 feet less than the length. The perimeter is 10 feet. Find the length and width.

**19.** The perimeter of a square is 48 meters. Find the length of one side.

**20.** One side of a triangle is twice the shortest side. The third side is 3 feet more than the shortest side. The perimeter is 19 feet. Find all three sides.

**21.** The length of a rectangle is 3 inches less than twice the width. The perimeter is 54 inches. Find the length and width.

**22.** The length of a rectangle is 4 feet less than 3 times the width. The sum of the length and width is 14 more than the width. Find the width.

## Coin Problems

**23.** Sue has $2.10 in dimes and nickels. If she has 9 more dimes than nickels, how many of each coin does she have? (Completing the table may help you get started.)

|  | Nickels | Dimes |
|---|---|---|
| Number | $x$ | $x + 9$ |
| Value (in cents) | $5x$ | $10x + 90$ |

**24.** Mike has $1.55 in dimes and nickels. If he has 7 more nickels than dimes, how many of each coin does he have?

|  | Nickels | Dimes |
|---|---|---|
| Number |  | $x$ |
| Value (in cents) |  |  |

**25.** Suppose you have $9.00 in dimes and quarters. How many of each coin do you have if you have twice as many quarters as dimes?

**26.** A collection of dimes and quarters has a total value of $2.20. If there are 3 times as many dimes as quarters, how many of each coin is in the collection?

**27.** Katie has a collection of nickels, dimes, and quarters with a total value of $4.35. There are 3 more dimes than nickels and 5 more quarters than nickels. How many of each coin is in her collection? (*Hint:* Let $x$ = the number of nickels.)

**28.** Mary Jo has $3.90 worth of nickels, dimes, and quarters. The number of nickels is 3 more than the number of dimes. The number of quarters is 7 more than the number of dimes. How many of each coin does she have? (*Hint:* Let $x$ = the number of dimes.)

## Review Problems

The following problems review material we covered in Sections 1.1 and 1.2.

Write an equivalent statement in words. [1.1]

**29.** $4 < 10$      **30.** $4 \leq 10$

**31.** $9 \geq -5$      **32.** $x - 2 > 4$

Place the symbol $<$ or the symbol $>$ between the quantities in each expression. [1.2]

**33.** 12    20      **34.** $-12$    20

**35.** $-8$    $-6$      **36.** $-10$    $-20$

Simplify. [1.2]

**37.** $|8 - 3| - |5 - 2|$      **38.** $|9 - 2| - |10 - 8|$

**39.** $15 - |9 - 3(7 - 5)|$      **40.** $10 - |7 - 2(5 - 3)|$

# More Applications

Now that you have worked through a number of application problems using our blueprint, you have probably noticed that step 3, in which we write an equation that describes the situation, is the key step. Anyone with experience solving application problems will tell you that there will be times when your first attempt at step 3 results in the wrong equation. Remember, mistakes are part of the process of learning to do things correctly. Many times the correct equation will become obvious after you have written an equation that is partially wrong. In any case, it is better to write an equation that is partially wrong and be actively involved with the problem than to write nothing at all. Application problems, like other problems in algebra, are not always solved correctly the first time.

In this section we continue our work with application problems. There are two main differences between the problems in this section and the problems you worked with in the previous section. First of all, not all of the problems in the problem set that follows this section fall into specific categories. Second, many of the problems in this section and problem set are more realistic in nature.

**EXAMPLE 1**   The total price for one of the cameras I used to produce the videotapes that accompany this book was $3,200. The total price is the retail price of the camera plus the sales tax. At that time, the sales tax rate in California was 7.5%. What was the retail price of the camera?

**Solution**   We solve this problem using our six-step Blueprint for Problem Solving:

Step 1: *Read and list.*
*Known items:*   The total bill is $3,200. The sales tax rate is 7.5%, which is 0.075 in decimal form.
*Unknown item:*   The retail price of the camera

Step 2: *Assign a variable and translate information.*
If we let $x =$ the retail price of the camera, then to calculate the sales tax, we multiply the retail price $x$ by the sales tax rate:

$$\text{Sales tax} = (\text{retail price})(\text{sales tax rate})$$

$$= 0.075x$$

Step 3: *Reread and write an equation.*

$$\text{Retail price} + \text{sales tax} = \text{total price}$$
$$x \quad + \quad 0.075x \quad = \quad 3,200$$

Step 4: *Solve the equation.*

$$x + 0.075x = 3,200$$
$$1.075x = 3,200$$
$$x = \frac{3,200}{1.075}$$
$$x = 2,976.74 \qquad \text{to the nearest hundredth}$$

*Step 5:* **Write the answer.**
The retail price of the camera was $2,976.74. The sales tax is
$3,200 − $2,976.74 = $223.26.

*Step 6:* **Reread and check.**
The retail price of the camera was is $2,976.74. The tax on this is
0.075(2,976.74) = $223.26. Adding the retail price and the sales tax
we have the total bill, $3,200.00.

**EXAMPLE 2**    Suppose you invest a certain amount of money in an account
that earns 8% in annual interest. At the same time, you invest $2,000 more than that
in an account that pays 9% in annual interest. If the total interest from both accounts at the end of the year is $690, how much is invested in each account?

**Solution**

*Step 1:* **Read and list.**
*Known items:*    The interest rates, the total interest earned, and how
much more is invested at 9%
*Unknown items:*    The amounts invested in each account

*Step 2:* **Assign a variable and translate information.**
Let $x$ = the amount of money invested at 8%. From this,
$x + 2,000$ = the amount of money invested at 9%. The interest earned
on $x$ dollars invested at 8% is $0.08x$. The interest earned on $x + 2,000$
dollars invested at 9% is $0.09(x + 2,000)$.

Here is a table that summarizes this information:

|  | **Dollars Invested at 8%** | **Dollars Invested at 9%** |
|---|---|---|
| Number of | $x$ | $x + 2,000$ |
| Interest on | $0.08x$ | $0.09(x + 2,000)$ |

*Step 3:* **Reread and write an equation.**
Since the total amount of interest earned from both accounts is $690,
the equation that describes the situation is

| Interest earned at 8% | + | interest earned at 9% | = | total interest earned |
|---|---|---|---|---|
| $0.08x$ | + | $0.09(x + 2,000)$ | = | 690 |

*Step 4:* **Solve the equation.**

$$0.08x + 0.09(x + 2,000) = 690$$

$$0.08x + 0.09x + 180 = 690 \qquad \text{Distributive property}$$

$$0.17x + 180 = 690 \qquad \text{Add } 0.08x \text{ and } 0.09x.$$

$$0.17x = 510 \qquad \text{Add } -180 \text{ to each side.}$$

$$x = 3,000 \qquad \text{Divide each side by 0.17.}$$

*Step 5: **Write the answer.***
The amount of money invested at 8% is $3,000, while the amount of money invested at 9% is $x + 2{,}000 = 3{,}000 + 2{,}000 = \$5{,}000$.

*Step 6: **Reread and check.***

The interest at 8% is 8% of 3,000 = 0.08(3,000) = $240
The interest at 9% is 9% of 5,000 = 0.09(5,000) = $450

The total interest is                                                    $690

---

## FACTS FROM

# Geometry

## Labeling Triangles and the Sum of the Angles in a Triangle

One way to label the important parts of a triangle is to label the vertices with capital letters and the sides with small letters, as shown in Figure 1.

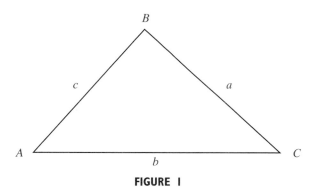

**FIGURE 1**

In Figure 1, notice that side $a$ is opposite vertex $A$, side $b$ is opposite vertex $B$, and side $c$ is opposite vertex $C$. Also, since each vertex is the vertex of one of the angles of the triangle, we refer to the three interior angles as $A$, $B$, and $C$.

In any triangle, the sum of the interior angles is 180°. For the triangle shown in Figure 1, the relationship is written

$$A + B + C = 180°$$

---

**EXAMPLE 3**    The angles in a triangle are such that one angle is twice the smallest angle, while the third angle is three times as large as the smallest angle. Find the measure of all three angles.

**Solution**

*Step 1: **Read and list.***
*Known items:*    The sum of all three angles is 180°, one angle is twice the smallest angle, the largest angle is three times the smallest angle.
*Unknown items:*    The measure of each angle

***Step 2: Assign a variable and translate information.***

Let $x$ be the smallest angle; then $2x$ will be the measure of another angle, and $3x$ will be the measure of the largest angle.

***Step 3: Reread and write an equation.***

When working with geometric objects, drawing a generic diagram will sometimes help us visualize what it is that we are asked to find. In Figure 2, we draw a triangle with angles $A$, $B$, and $C$.

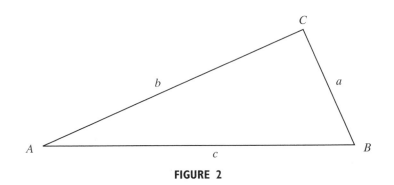

**FIGURE 2**

We can let the value of $A = x$, the value of $B = 2x$, and the value of $C = 3x$. We know that the sum of angles $A$, $B$, and $C$ will be 180°, so our equation becomes

$$x + 2x + 3x = 180°$$

***Step 4: Solve the equation.***

$$x + 2x + 3x = 180°$$
$$6x = 180°$$
$$x = 30°$$

***Step 5: Write the answer.***

The smallest angle $A$ measures 30°.
Angle $B$ measures $2x$, or $2(30°) = 60°$.
Angle $C$ measures $3x$, or $3(30°) = 90°$.

***Step 6: Reread and check.***

The angles must add to 180°:

$$A + B + C = 180°$$
$$30° + 60° + 90° = 180°$$
$$180° = 180° \qquad \text{Our answers check.}$$

**Getting Ready for Class**

*After reading through the preceding section, respond in your own words and in complete sentences.*

**A.** How do we label triangles?

**B.** What rule is always true about the three angles in a triangle?

**C.** Write an application problem whose solution depends on solving the equation $x + 0.075x = 500$.

**D.** Write an application problem whose solution depends on solving the equation $0.05x + 0.06(x + 200) = 67$.

## PROBLEM SET 2.7

### Interest Problems

**1.** Suppose you invest money in two accounts. One of the accounts pays 8% annual interest, while the other pays 9% annual interest. If you have $2,000 more invested at 9% than you have invested at 8%, how much do you have invested in each account if the total amount of interest you earn in a year is $860? (Begin by completing the following table.)

| | Dollars Invested at 8% | Dollars Invested at 9% |
|---|---|---|
| Number of | $x$ | |
| Interest on | | |

**2.** Suppose you invest a certain amount of money in an account that pays 11% interest annually and $4,000 more than that in an account that pays 12% annually. How much money do you have in each account if the total interest for a year is $940?

| | Dollars Invested at 11% | Dollars Invested at 12% |
|---|---|---|
| Number of | $x$ | |
| Interest on | | |

**3.** Tyler has two savings accounts that his grandparents opened for him. The two accounts pay 10% and 12% in annual interest; there is $500 more in the account that pays 12% than there is in the other account. If the total interest for a year is $214, how much money does he have in each account?

**4.** Travis has a savings account that his parents opened for him. It pays 6% annual interest. His uncle also opened an account for him, but it pays 8% annual interest. If there is $800 more in the account that pays 6% and the total interest from both accounts is $104, how much money is in each of the accounts?

**5.** A stockbroker has money in three accounts. The interest rates on the three accounts are 8%, 9%, and 10%. If she has twice as much money invested at 9% as she has invested at 8%, and three times as much at 10% as she has at 8%, and the total interest for the year is $280, how much is invested at each rate? (*Hint:* Let $x =$ the amount invested at 8%.)

**6.** An accountant has money in three accounts that pay 9%, 10%, and 11% in annual interest. He has twice as much invested at 9% as he does at 10% and three times as much invested at 11% as he does at 10%. If the total interest from the three accounts is $610 for the year, how much is invested at each rate? (*Hint:* Let $x =$ the amount invested at 10%.)

### Geometry Problems

**7.** Two angles in a triangle are equal, and their sum is equal to the third angle in the triangle. What are the measures of each of the three interior angles?

**8.** One angle in a triangle measures twice the smallest angle, while the largest angle is six times the smallest angle. Find the measures of all three angles.

9. The smallest angle in a triangle is $\frac{1}{5}$ as large as the largest angle. The third angle is twice the smallest angle. Find the three angles.

10. One angle in a triangle is half the largest angle, but three times the smallest. Find all three angles.

11. A right triangle has one 37° angle. Find the other two angles.

12. In a right triangle, one of the acute angles is twice as large as the other acute angle. Find the measure of the two acute angles.

## Miscellaneous Problems

13. **Phone Bill**    The cost of a long-distance phone call is $0.41 for the first minute and $0.32 for each additional minute. If the total charge for a long-distance call is $5.21, how many minutes was the call?

14. **Phone Bill**    Danny, who is 1 year old, is playing with the telephone when he accidentally presses one of the buttons his mother has programmed to dial her friend Sue's number. Sue answers the phone and realizes Danny is on the other end. She talks to Danny, trying to get him to hang up. The cost for a call is $0.23 for the first minute and $0.14 for every minute after that. If the total charge for the call is $3.73, how long did it take Sue to convince Danny to hang up the phone?

15. **Hourly Wages**    JoAnn works in the publicity office at the state university. She is paid $12 an hour for the first 35 hours she works each week and $18 an hour for every hour after that. If she makes $492 one week, how many hours did she work?

16. **Hourly Wages**    Diane has a part-time job that pays her $6.50 an hour. During one week she works 26 hours and is paid $178.10. She realizes when she sees her check that she has been given a raise. How much per hour is that raise?

17. **Ticket Sales**    Stacey is selling tickets to the school play. The tickets are $6.00 for adults and $4.50 for children. She sells twice as many adult tickets as children's tickets and brings in a total of $115.50. How many of each kind of ticket did she sell?

18. **Piano Lessons**    Tyler is taking piano lessons. Since he doesn't practice as often as his parents would like him to, he has to pay for part of the lessons himself. His parents pay him $0.50 to do the laundry and $1.25 to mow the lawn. In one month, he does the laundry 6 more times than he

mows the lawn. If his parents pay him $13.50 that month, how many times did he mow the lawn?

19. **Arrival Time**    Jeff and Carla Cole are driving separately from San Luis Obispo, California, to the north shore of Lake Tahoe, a distance of 425 miles. Jeff leaves San Luis Obispo at 11:00 A.M. and averages 55 miles per hour on the drive. Carla leaves later, at 1:00 P.M., but averages 65 miles per hour. Which person arrives in Lake Tahoe first?

20. **Arrival Time**    Referring to the information in Problem 19, how many minutes after the first person arrives does the second person arrive?

**Cost of a Taxi Ride**    Recently, the Texas Junior College Teachers Association annual conference was held in Austin. At that time a taxi ride in Austin was $1.25 for the first $\frac{1}{5}$ of a mile and $0.25 for each additional $\frac{1}{5}$ of a mile. The charge for a taxi to wait is $12.00 per hour. Use this information for Problems 21 through 24.

21. If the distance from one of the convention hotels to the airport is 7.5 miles, how much will it cost to take a taxi from that hotel to the airport?

22. If you were to tip the driver of the taxi in Problem 21 15%, how much would it cost to take a taxi from the hotel to the airport?

23. Suppose the distance from one of the hotels to one of the western dance clubs in Austin is 12.4 miles. If the fare meter in the taxi gives the charge for that trip as $16.50, is the meter working correctly?

24. Suppose that the distance from a hotel to the airport is 8.2 miles and the ride takes 20 minutes. Is it more expensive to take a taxi to the airport or just to sit in the taxi?

**Dance Lessons**    Ike and Nancy Lara give western dance lessons at the Elks Lodge on Sunday nights. The lessons cost $3.00 for members of the lodge and $5.00 for nonmembers. Half of the money collected for the lessons is paid to Ike and Nancy. The Elks Lodge keeps the other half. One Sunday night Ike counts 36 people in the dance lesson. Use this information to work Problems 25 through 28.

25. What is the least amount of money Ike and Nancy will make?

26. What is the largest amount of money Ike and Nancy will make?

**27.** At the end of the evening, the Elks Lodge gives Ike and Nancy a check for $80 to cover half of the receipts. Can this amount be correct?

**28.** Besides the number of people in the dance lesson, what additional information does Ike need to know in order to always be sure he is being paid the correct amount?

## Applying the Concepts

**29. Starting Salaries**   *USA Today* reported in 1999 that the median starting salary of graduates of Stanford University's business school was $85,200, which was 7% more than it was the year before. What was the median starting salary of the 1998 graduates of Stanford University's business school? Round your answer to the nearest dollar.

**30. Starting Salaries**   In the same article mentioned in Problem 29, *USA Today* reported that the total compensation, salary, bonus, housing allowance, and so forth, for the 1999 graduates of Stanford's business school was $122,000, an increase of 8% over the previous year. What was the total compensation package for Stanford's business school graduates in 1998? Round your answer to the nearest dollar.

## Review Problems

The problems that follow review material we covered in Chapter 1 on number sequences.

Find the next number in each sequence.

**31.** $8, 4, 0, -4, \ldots$

**32.** $8, -16, 32, -64, \ldots$

**33.** $12, -6, 3, -\dfrac{3}{2}, \ldots$

**34.** $-12, -6, 0, 6, \ldots$

**35.** $1, -4, 9, -16, \ldots$

**36.** $-1, 8, -27, 64, \ldots$

**37.** $2, \dfrac{3}{2}, 1, \dfrac{1}{2}, \ldots$

**38.** $1, \dfrac{1}{2}, 0, -\dfrac{1}{2}, \ldots$

---

**2.8**   ## Linear Inequalities

Linear inequalities are solved by a method similar to the one used in solving linear equations. The only real differences between the methods are in the multiplication property for inequalities and in graphing the solution set.

An inequality differs from an equation only with respect to the comparison symbol between the two quantities being compared. In place of the equal sign, we use $<$ (less than), $\leq$ (less than or equal to), $>$ (greater than), or $\geq$ (greater than or equal to). The addition property for inequalities is almost identical to the addition property for equality.

---

**ADDITION PROPERTY FOR INEQUALITIES**

For any three algebraic expressions $A$, $B$, and $C$,

$$\text{if} \qquad A < B$$

$$\text{then} \qquad A + C < B + C$$

*In words:*   Adding the same quantity to both sides of an inequality will not change the solution set.

It makes no difference which inequality symbol we use to state the property. Adding the same amount to both sides always produces an inequality equivalent to the original inequality. Also, since subtraction can be thought of as addition of the opposite, this property holds for subtraction as well as addition.

**E X A M P L E  1**   Solve the inequality $x + 5 < 7$.

**Solution**   To isolate $x$, we add $-5$ to both sides of the inequality:

$$x + 5 < 7$$

$$x + 5 + (\mathbf{-5}) < 7 + (\mathbf{-5})  \qquad \text{Addition property for inequalities}$$

$$x < 2$$

We can go one step further here and graph the solution set. The solution set is all real numbers less than 2. To graph this set, we simply draw a straight line and label the center 0 (zero) for reference. Then we label the 2 on the right side of zero and extend an arrow beginning at 2 and pointing to the left. We use an open circle at 2, since it is not included in the solution set. Here is the graph.

We can go one step further here and graph the solution set. The solution set is

**E X A M P L E  2**   Solve $x - 6 \le -3$.

**Solution**   Adding 6 to each side will isolate $x$ on the left side:

$$x - 6 \le -3$$

$$x - 6 + \mathbf{6} \le -3 + \mathbf{6}  \qquad \text{Add } \mathbf{6} \text{ to both sides.}$$

$$x \le 3$$

The graph of the solution set is

Notice that the dot at the 3 is darkened because 3 is included in the solution set. We will always use open circles on the graphs of solution sets with $<$ or $>$ and closed (darkened) circles on the graphs of solution sets with $\le$ or $\ge$.

To see the idea behind the multiplication property for inequalities, we will consider three true inequality statements and explore what happens when we multiply both sides by a positive number and then what happens when we multiply by a negative number.

Consider the following three true statements:

$$3 < 5 \qquad -3 < 5 \qquad -5 < -3$$

Now multiply both sides by the positive number 4:

$$4(3) < 4(5) \qquad 4(-3) < 4(5) \qquad 4(-5) < 4(-3)$$

$$12 < 20 \qquad -12 < 20 \qquad -20 < -12$$

In each case, the inequality symbol in the result points in the same direction it did in the original inequality. We say the "sense" of the inequality doesn't change when we multiply both sides by a positive quantity.

   Notice what happens when we go through the same process but multiply both sides by $-4$ instead of 4:

$$3 < 5 \qquad\qquad -3 < 5 \qquad\qquad -5 < -3$$

$$-4(3) > -4(5) \qquad -4(-3) > -4(5) \qquad -4(-5) > -4(-3)$$

$$-12 > -20 \qquad\qquad 12 > -20 \qquad\qquad 20 > 12$$

In each case, we have to change the direction in which the inequality symbol points to keep each statement true. Multiplying both sides of an inequality by a negative quantity *always* reverses the sense of the inequality. Our results are summarized in the multiplication property for inequalities.

## MULTIPLICATION PROPERTY FOR INEQUALITIES

For any three algebraic expressions $A$, $B$, and $C$,

$$\text{if} \qquad A < B$$

$$\text{then} \qquad AC < BC \qquad \text{when } C \text{ is positive}$$

$$\text{and} \qquad AC > BC \qquad \text{when } C \text{ is negative}$$

*In words:*    Multiplying both sides of an inequality by a positive number does not change the solution set. When multiplying both sides of an inequality by a negative number, it is necessary to reverse the inequality symbol in order to produce an equivalent inequality.

We can multiply both sides of an inequality by any nonzero number we choose. If that number happens to be negative, we must also reverse the sense of the inequality.

**Note:**  Since division is defined in terms of multiplication, this property is also true for division. We can divide both sides of an inequality by any number we choose. If that number happens to be negative, we must also reverse the direction of the inequality symbol.

E X A M P L E   **3**    Solve $3a < 15$ and graph the solution.

**Solution**   We begin by multiplying each side by $\frac{1}{3}$. Since $\frac{1}{3}$ is a positive number, we do not reverse the direction of the inequality symbol:

$$3a < 15$$

$$\frac{1}{3}(3a) < \frac{1}{3}(15) \qquad \text{Multiply each side by } \frac{1}{3}.$$

$$a < 5$$

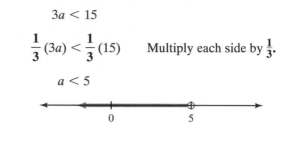

**E X A M P L E   4**   Solve $-3a \le 18$ and graph the solution.

**Solution**   We begin by multiplying both sides by $-\frac{1}{3}$. Since $-\frac{1}{3}$ is a negative number, we must reverse the direction of the inequality symbol at the same time that we multiply by $-\frac{1}{3}$.

$$-3a \le 18$$

$$-\frac{1}{3}(-3a) \ge -\frac{1}{3}(18) \qquad \begin{array}{l}\text{Multiply both sides by } -\frac{1}{3} \text{ and reverse} \\ \text{the direction of the inequality symbol.}\end{array}$$

$$a \ge -6$$

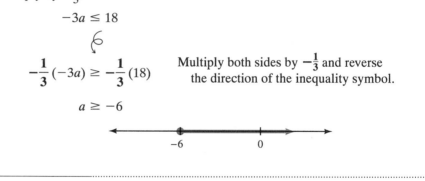

**E X A M P L E   5**   Solve $-\dfrac{x}{4} > 2$ and graph the solution.

**Solution**   To isolate $x$, we multiply each side by $-4$. Since $-4$ is a negative number, we must also reverse the direction of the inequality symbol:

$$-\frac{x}{4} > 2$$

$$-4\left(-\frac{x}{4}\right) < -4(2) \qquad \begin{array}{l}\text{Multiply each side by } -4 \text{ and reverse the} \\ \text{direction of the inequality symbol.}\end{array}$$

$$x < -8$$

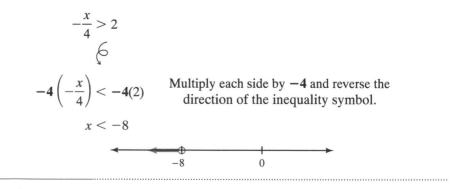

To solve more complicated inequalities, we use the following steps.

> ### STRATEGY FOR SOLVING LINEAR INEQUALITIES IN ONE VARIABLE
>
> ***Step 1a:*** Use the distributive property to separate terms, if necessary.
> **1b:** If fractions are present, consider multiplying both sides by the LCD to eliminate the fractions. If decimals are present, consider multiplying both sides by a power of 10 to clear the inequality of decimals.
> **1c:** Combine similar terms on each side of the inequality.
> ***Step 2:*** Use the addition property for inequalities to get all variable terms on one side of the inequality and all constant terms on the other side.
> ***Step 3:*** Use the multiplication property for inequalities to get $x$ by itself on one side of the inequality.
> ***Step 4:*** Graph the solution set.

**E X A M P L E   6**   Solve $2.5x - 3.48 < -4.9x + 2.07$.

**Solution**   We have two methods we can use to solve this inequality. We can simply apply our properties to the inequality the way it is currently written and work with the decimal numbers, or we can eliminate the decimals to begin with and solve the resulting inequality.

***Method 1***   Working with the decimals.

$$2.5x - 3.48 < -4.9x + 2.07 \qquad \text{Original inequality}$$

$$2.5x + \mathbf{4.9x} - 3.48 < -4.9x + \mathbf{4.9x} + 2.07 \qquad \text{Add } \mathbf{4.9x} \text{ to each side.}$$

$$7.4x - 3.48 < 2.07$$

$$7.4x - 3.48 + \mathbf{3.48} < 2.07 + \mathbf{3.48} \qquad \text{Add } \mathbf{3.48} \text{ to each side.}$$

$$7.4x < 5.55$$

$$\frac{7.4x}{\mathbf{7.4}} < \frac{5.55}{\mathbf{7.4}} \qquad \text{Divide each side by } \mathbf{7.4.}$$

$$x < 0.75$$

***Method 2***   Eliminating the decimals in the beginning.

Since the greatest number of places to the right of the decimal point in any of the numbers is 2, we can multiply each side of the inequality by 100 and we will be left with an equivalent inequality that contains only whole numbers.

$$2.5x - 3.48 < -4.9x + 2.07 \qquad \text{Original inequality}$$

$$\mathbf{100}(2.5x - 3.48) < \mathbf{100}(-4.9x + 2.07) \qquad \text{Multiply each side by 100.}$$

$$\mathbf{100}(2.5x) - \mathbf{100}(3.48) < \mathbf{100}(-4.9x) + \mathbf{100}(2.07) \qquad \text{Distributive property}$$

$$250x - 348 < -490x + 207 \qquad \text{Multiplication}$$

$$740x - 348 < 207 \qquad \text{Add } 490x \text{ to each side.}$$

$$740x < 555 \qquad \text{Add 348 to each side.}$$

$$\frac{740x}{\mathbf{740}} < \frac{555}{\mathbf{740}} \qquad \text{Divide each side by } \mathbf{740.}$$

$$x < 0.75$$

The solution by either method is $x < 0.75$. Here is the graph:

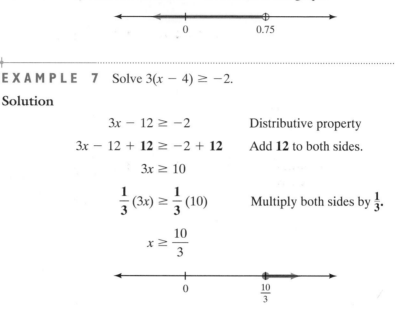

**EXAMPLE  7**    Solve $3(x - 4) \geq -2$.

**Solution**

$$3x - 12 \geq -2 \qquad \text{Distributive property}$$

$$3x - 12 + \mathbf{12} \geq -2 + \mathbf{12} \qquad \text{Add } \mathbf{12} \text{ to both sides.}$$

$$3x \geq 10$$

$$\frac{\mathbf{1}}{\mathbf{3}}(3x) \geq \frac{\mathbf{1}}{\mathbf{3}}(10) \qquad \text{Multiply both sides by } \tfrac{1}{3}.$$

$$x \geq \frac{10}{3}$$

**EXAMPLE  8**    Solve and graph $2(1 - 3x) + 4 < 4x - 14$.

**Solution**

$$2 - 6x + 4 < 4x - 14 \qquad \text{Distributive property}$$

$$-6x + 6 < 4x - 14 \qquad \text{Simplify.}$$

$$-6x + 6 + (\mathbf{-6}) < 4x - 14 + (\mathbf{-6}) \qquad \text{Add } \mathbf{-6} \text{ to both sides.}$$

$$-6x < 4x - 20$$

$$-6x + (\mathbf{-4x}) < 4x + (\mathbf{-4x}) - 20 \qquad \text{Add } \mathbf{-4x} \text{ to both sides.}$$

$$-10x < -20$$

$$\left(-\frac{1}{10}\right)(-10x) > \left(-\frac{1}{10}\right)(-20) \qquad \text{Multiply by } -\tfrac{1}{10} \text{ and reverse the sense of the inequality.}$$

$$x > 2$$

**E X A M P L E   9**   Solve $2x - 3y < 6$ for $y$.

**Solution**   We can solve this formula for $y$ by first adding $-2x$ to each side and then multiplying each side by $-\frac{1}{3}$. When we multiply by $-\frac{1}{3}$ we must reverse the direction of the inequality symbol. Since this is a formula, we will not graph the solution.

$$2x - 3y < 6 \qquad\qquad \text{Original formula}$$

$$2x + (\boldsymbol{-2x}) - 3y < (\boldsymbol{-2x}) + 6 \qquad\qquad \text{Add } \boldsymbol{-2x} \text{ to each side.}$$

$$-3y < -2x + 6$$

$$-\frac{1}{3}(-3y) > -\frac{1}{3}(-2x + 6) \qquad\qquad \text{Multiply each side by } -\frac{1}{3}.$$

$$y > \frac{2}{3}x - 2 \qquad\qquad \text{Distributive property}$$

When working application problems that involve inequalities, the phrases "at least" and "at most" translate as follows:

| In Words | In Symbols |
|---|---|
| $x$ is at least 30. | $x \geq 30$ |
| $x$ is at most 20. | $x \leq 20$ |

## APPLYING THE CONCEPTS

Our next example involves consecutive integers. When we ask for consecutive integers, we mean integers that are next to each other on the number line, like 5 and 6, or 13 and 14, or $-4$ and $-3$. In the dictionary, consecutive is defined as following one another in uninterrupted order. If we ask for consecutive *odd* integers, then we mean odd integers that follow one another on the number line. For example, 3 and 5, 11 and 13, and $-9$ and $-7$ are consecutive odd integers. As you can see, to get from one odd integer to the next consecutive odd integer we add 2.

We can modify our Blueprint for Problem Solving to solve application problems whose solutions depend on writing and then solving inequalities.

**E X A M P L E   1 0**   The sum of two consecutive odd integers is at most 28. What are the possibilities for the first of the two integers?

**Solution**   When we use the phrase "their sum is at most 28," we mean that their sum is less than or equal to 28.

> *Step 1: Read and list.*
>   *Known items:*   Two consecutive odd integers. Their sum is less than or equal to 28.
>   *Unknown items:*   The numbers in question

**Step 2:** *Assign a variable and translate information.*
If we let $x$ = the first of the two consecutive integers, then $x + 2$ is the next consecutive one.

**Step 3:** *Reread and write an inequality.*
Their sum is at most 28.

$$x + (x + 2) \leq 28$$

**Step 4:** *Solve the inequality.*

$$2x + 2 \leq 28 \qquad \text{Simplify the left side.}$$

$$2x \leq 26 \qquad \text{Add } -2 \text{ to each side.}$$

$$x \leq 13 \qquad \text{Multiply each side by } \tfrac{1}{2}.$$

**Step 5:** *Write the answer.*
The first of the two integers must be an odd integer that is less than or equal to 13. The second of the two integers will be 2 more than whatever the first one is.

**Step 6:** *Reread and check.*
Suppose the first integer is 13. The next consecutive odd integer is 15. The sum of 15 and 13 is 28. If the first odd integer is less than 13, the sum of it and the next consecutive odd integer will be less than 28.

## Getting Ready for Class

*After reading through the preceding section, respond in your own words and in complete sentences.*

**A.** State the addition property for inequalities.

**B.** How is the multiplication property for inequalities different from the multiplication property of equality?

**C.** When do we reverse the direction of an inequality symbol?

**D.** Under what conditions do we not change the direction of the inequality symbol when we multiply both sides of an inequality by a number?

## PROBLEM SET 2.8

Solve the following inequalities using the addition property of inequalities. Graph each solution set.

**1.** $x - 5 < 7$

**2.** $x + 3 < -5$

**3.** $a - 4 \leq 8$

**4.** $a + 3 \leq 10$

**5.** $x - 4.3 > 8.7$

**6.** $x - 2.6 > 10.4$

**7.** $y + 6 \geq 10$

**8.** $y + 3 \geq 12$

**9.** $2 < x - 7$

**10.** $3 < x + 8$

Solve the following inequalities using the multiplication property of inequalities. If you multiply both sides by a negative number, be sure to reverse the direction of the inequality symbol. Graph the solution set.

**11.** $3x < 6$

**12.** $2x < 14$

**13.** $5a \le 25$

**14.** $4a \le 16$

**15.** $\dfrac{x}{3} > 5$

**16.** $\dfrac{x}{7} > 1$

**17.** $-2x > 6$

**18.** $-3x \ge 9$

**19.** $-3x \ge -18$

**20.** $-8x \ge -24$

**21.** $-\dfrac{x}{5} \le 10$

**22.** $-\dfrac{x}{9} \ge -1$

**23.** $-\dfrac{2}{3}y > 4$

**24.** $-\dfrac{3}{4}y > 6$

Solve the following inequalities. Graph the solution set in each case.

**25.** $2x - 3 < 9$

**26.** $3x - 4 < 17$

**27.** $-\dfrac{1}{5}y - \dfrac{1}{3} \le \dfrac{2}{3}$

**28.** $-\dfrac{1}{6}y - \dfrac{1}{2} \le \dfrac{2}{3}$

**29.** $-7.2x + 1.8 > -19.8$

**30.** $-7.8x - 1.3 > 22.1$

**31.** $\dfrac{2}{3}x - 5 \le 7$

**32.** $\dfrac{3}{4}x - 8 \le 1$

**33.** $-\dfrac{2}{5}a - 3 > 5$

**34.** $-\dfrac{4}{5}a - 2 > 10$

**35.** $5 - \dfrac{3}{5}y > -10$

**36.** $4 - \dfrac{5}{6}y > -11$

**37.** $0.3(a + 1) \le 1.2$

**38.** $0.4(a - 2) \le 0.4$

**39.** $2(5 - 2x) \le -20$

**40.** $7(8 - 2x) > 28$

**41.** $3x - 5 > 8x$

**42.** $8x - 4 > 6x$

**43.** $\dfrac{1}{3}y - \dfrac{1}{2} \le \dfrac{5}{6}y + \dfrac{1}{2}$

**44.** $\dfrac{7}{6}y + \dfrac{4}{3} \le \dfrac{11}{6}y - \dfrac{7}{6}$

**45.** $-2.8x + 8.4 < -14x - 2.8$

**46.** $-7.2x - 2.4 < -2.4x + 12$

**47.** $3(m - 2) - 4 \ge 7m + 14$

**48.** $2(3m - 1) + 5 \ge 8m - 7$

**49.** $3 - 4(x - 2) \le -5x + 6$

**50.** $8 - 6(x - 3) \le -4x + 12$

Solve each of the following formulas for $y$.

**51.** $3x + 2y < 6$

**52.** $-3x + 2y < 6$

**53.** $2x - 5y > 10$

**54.** $-2x - 5y > 5$

**55.** $-3x + 7y \le 21$

**56.** $-7x + 3y \le 21$

**57.** $2x - 4y \ge -4$

**58.** $4x - 2y \ge -8$

For each graph below, write an inequality whose solution is the graph.

**59.**

**60.**

**61.**

**62.**

## Applying the Concepts

**63. Consecutive Integers**   The sum of two consecutive integers is at least 583. What are the possibilities for the first of the two integers?

**64. Consecutive Integers**   The sum of two consecutive integers is at most 583. What are the possibilities for the first of the two integers?

**65. Number Problem**   The sum of twice a number and six is less than ten. Find all solutions.

**66. Number Problem**   Twice the difference of a number. and three is greater than or equal to the number increased by five. Find all solutions.

**67. Number Problem**   The product of a number and four is greater than the number minus eight. Find the solution set.

**68. Number Problem**   The quotient of a number and five is less than the sum of seven and two. Find the solution set.

**69. Geometry Problem**   The length of a rectangle is 3 times the width. If the perimeter is to be at least 48 meters, what are the possible values for the width? (If the perimeter is at least 48 meters, then it is greater than or equal to 48 meters.)

**70. Geometry Problem**   The length of a rectangle is 3 more than twice the width. If the perimeter is to be at least 51 meters, what are the possible values for the width? (If the perimeter is at least 51 meters, then it is greater than or equal to 51 meters.)

**71. Geometry Problem**   The numerical values of the three sides of a triangle are given by three consecutive even integers. If the perimeter is greater than 24 inches, what are the possibilities for the shortest side?

**72. Geometry Problem**    The numerical values of the three sides of a triangle are given by three consecutive odd integers. If the perimeter is greater than 27 inches, what are the possibilities for the shortest side?

**73. Car Heaters**    If you have ever gotten in a cold car early in the morning you know that the heater does not work until the engine warms up. This is because the heater relies on the heat coming off the engine. Write an equation using an inequality sign to express when the heater will work if the heater works only after the engine is 100°F.

**74. Exercise**    When Kate exercises, she either swims or runs. She wants to spend a minimum of 8 hours a week exercising, and she wants to swim 3 times the amount she runs. What is the minimum amount of time she must spend doing each exercise?

**75. Profit and Loss**    Movie theaters pay a certain price for the movies that you and I see. Suppose a theater pays $1,500 for each showing of a popular movie. If they charge $7.50 for each ticket they sell, then they will lose money if ticket sales are less than $1,500. On the other hand, they will make a profit if ticket sales are greater than $1,500. What is the range of tickets they can sell and still lose money? What is the range of tickets they can sell and make a profit?

**76. Stock Sales**    Suppose you purchase $x$ shares of a stock at $12 per share. After 6 months you decide to sell all your shares at $20 per share. Your broker charges you $15 for the trade. If your profit is at least $3,985, how many shares did you purchase in the first place?

## Review Problems

Problems 77–82 review material we covered in Section 1.5.

Match each numbered expression with one or more of the properties in (a)–(e).

**77.** $x + 4 = 4 + x$

**78.** $2(3x) = (2 \cdot 3)x$

**79.** $5(x - 3) = 5x - 15$

**80.** $x + (y + 4) = (x + y) + 4$

**81.** $x + (y + 4) = (x + 4) + y$

**82.** $7 \cdot 5 = 5 \cdot 7$

(a) Distributive property

(b) Commutative property of addition

(c) Associative property of addition

(d) Commutative property of multiplication

(e) Associative property of multiplication

---

## 2.9    Compound Inequalities

The instrument panel on most cars includes a temperature gauge. The one shown below indicates that the normal operating temperature for the engine is from 50°F to 270°F.

We can represent the same situation with an inequality by writing $50 \leq F \leq 270$, where $F$ is the temperature in degrees Fahrenheit. This inequality is a compound inequality. In this section we present the notation and definitions associated with compound inequalities.

The *union* of two sets $A$ and $B$ is the set of all elements that are in $A$ or in $B$. The word *or* is the key word in the definition. The *intersection* of two sets $A$ and $B$ is the set of elements contained in both $A$ and $B$. The key word in this definition is *and*. We can put the words *and* and *or* together with our methods of graphing inequalities to find the solution sets for compound inequalities.

> **DEFINITION**  A **compound inequality** is two or more inequalities connected by the word *and* or *or.*

**EXAMPLE  1**  Graph the solution set for the compound inequality

$$x < -1 \qquad \text{or} \qquad x \geq 3$$

**Solution**  Graphing each inequality separately, we have

Since the two inequalities are connected by *or,* we want to graph their union: that is, we graph all points that are on either the first graph or the second graph. Essentially, we put the two graphs together on the same number line.

**EXAMPLE  2**  Graph the solution set for the compound inequality

$$x > -2 \qquad \text{and} \qquad x < 3$$

**Solution**  Graphing each inequality separately, we have

Since the two inequalities are connected by the word *and,* we will graph their intersection, which consists of all points that are common to both graphs; that is, we graph the region where the two graphs overlap.

**EXAMPLE 3** Solve and graph the solution set for

$$2x - 1 \geq 3 \quad \text{and} \quad -3x > -12$$

**Solution** Solving the two inequalities separately, we have

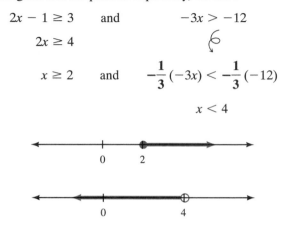

$$2x - 1 \geq 3 \quad \text{and} \quad -3x > -12$$

$$2x \geq 4$$

$$x \geq 2 \quad \text{and} \quad -\frac{1}{3}(-3x) < -\frac{1}{3}(-12)$$

$$x < 4$$

Since the word *and* connects the two graphs, we will graph their intersection — the points they have in common:

*Notation* Sometimes compound inequalities that use the word *and* can be written in a shorter form. For example, the compound inequality $-2 < x$ and $x < 3$ can be written as $-2 < x < 3$. The word *and* does not appear when an inequality is written in this form. It is implied. The solution set for $-2 < x$ and $x < 3$ is

It is all the numbers between $-2$ and $3$ on the number line. It seems reasonable, then, that this graph should be the graph of

$$-2 < x < 3$$

In both the graph and the inequality, $x$ is said to be between $-2$ and $3$.

**EXAMPLE 4** Solve and graph $-3 \leq 2x - 1 \leq 9$.

**Solution** To solve for $x$, we must add 1 to the center expression and then divide the result by 2. Whatever we do to the center expression, we must also do to the two expressions on the ends. In this way we can be sure we are producing equivalent inequalities. The solution set will not be affected.

$$-3 \leq 2x - 1 \leq 9$$

$$-2 \leq \quad 2x \quad \leq 10 \qquad \text{Add 1 to each expression.}$$

$$-1 \leq \quad x \quad \leq 5 \qquad \text{Multiply each expression by } \tfrac{1}{2}.$$

## Getting Ready for Class

*After reading through the preceding section, respond in your own words and in complete sentences.*

**A.** What is a compound inequality?

**B.** Explain the shorthand notation that can be used to write two inequalities connected by the word *and*.

**C.** Write two inequalities connected by the word *and* that together are equivalent to $-1 < x < 2$.

**D.** Explain in words how you would graph the compound inequality

$$x < 2 \text{ or } x > -3$$

# PROBLEM SET 2.9

Graph the following compound inequalities.

**1.** $x < -1$ or $x > 5$      **2.** $x \leq -2$ or $x \geq -1$

**3.** $x < -3$ or $x \geq 0$      **4.** $x < 5$ and $x > 1$

**5.** $x \leq 6$ and $x > -1$      **6.** $x \leq 7$ and $x > 0$

**7.** $x > 2$ and $x < 4$      **8.** $x < 2$ or $x > 4$

**9.** $x \geq -2$ and $x \leq 4$      **10.** $x \leq 2$ or $x \geq 4$

**11.** $x < 5$ and $x > -1$      **12.** $x > 5$ or $x < -1$

**13.** $-1 < x < 3$      **14.** $-1 \leq x \leq 3$

**15.** $-3 < x \leq -2$      **16.** $-5 \leq x \leq 0$

Solve the following. Graph the solution set in each case.

**17.** $3x - 1 < 5$ or $5x - 5 > 10$

**18.** $x + 1 < -3$ or $x - 2 > 6$

**19.** $x - 2 > -5$ and $x + 7 < 13$

**20.** $3x + 2 \leq 11$ and $2x + 2 \geq 0$

**21.** $11x < 22$ or $12x > 36$

**22.** $-5x < 25$ and $-2x \geq -12$

**23.** $3x - 5 < 10$ and $2x + 1 > -5$

**24.** $5x + 8 < -7$ or $3x - 8 > 10$

**25.** $2x - 3 < 8$ and $3x + 1 > -10$

**26.** $11x - 8 > 3$ or $12x + 7 < -5$

**27.** $2x - 1 < 3$ and $3x - 2 > 1$

**28.** $3x + 9 < 7$ or $2x - 7 > 11$

Solve and graph each of the following.

**29.** $-1 \leq x - 5 \leq 2$      **30.** $0 \leq x + 2 \leq 3$

**31.** $-4 \leq 2x \leq 6$      **32.** $-5 < 5x < 10$

**33.** $-3 < 2x + 1 < 5$      **34.** $-7 \leq 2x - 3 \leq 7$

**35.** $0 \leq 3x + 2 \leq 7$      **36.** $2 \leq 5x - 3 \leq 12$

**37.** $-7 < 2x + 3 < 11$    **38.** $-5 < 6x - 2 < 8$

**39.** $-1 \leq 4x + 5 \leq 9$    **40.** $-8 \leq 7x - 1 \leq 13$

For each graph below, write an inequality whose solution is the graph.

**41.**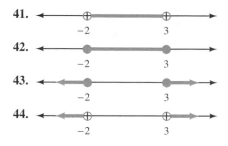

**42.**

**43.**

**44.**

## Applying the Concepts

**45. Triangle Inequality**    For any triangle, the triangle inequality states that the sum of any two sides must be greater than the third side. Given the following triangle *RST*

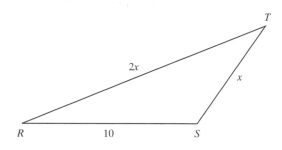

(a) Write three inequalities, based on the triangle inequality, that must be true.

(b) Use your results from part (a) to fill in the blanks in the following inequality:

$$\underline{\quad} < x < \underline{\quad}$$

**46. Civil War Cost**    After the American Civil War was fully begun, its cost for the North was at least $30,000,000 per month and occasionally as great as $90,000,000 per month.

(a) Write an inequality for the total cost to the North during the 4 years of the Civil War.

(b) One textbook gave the total cost of the Civil War for the North as $2,845,907,626.56. Is this

number a solution to the inequality you wrote in part (a)?

**47. Engine Temperature**    The engine in a car gives off a lot of heat due to the combustion in the cylinders. The water used to cool the engine keeps the temperature within the range $50 \leq F \leq 270$ where *F* is in degrees Fahrenheit. Graph this inequality on the number line.

**48. Engine Temperature**    To find the engine temperature range from Problem 47 in degrees Celsius, we use the fact that $F = \dfrac{9}{5} C + 32$ to rewrite the inequality as

$$50 \leq \frac{9}{5} C + 32 \leq 270$$

Solve this inequality and graph the solution set.

**49. Number Problem**    The sum of a number and 5 is between 10 and 20. Find the number.

**50. Number Problem**    The difference of a number and 2 is between 6 and 14. Find the number.

**51. Number Problem**    The difference of twice a number and 3 is between 5 and 7. Find the number.

**52. Number Problem**    The sum of twice a number and 5 is between 7 and 13. Find the number.

**53. Perimeter**    The length of a rectangle is 4 inches longer than the width. If the perimeter is between 20 inches and 30 inches, find all possible values for the width.

**54. Perimeter**    The length of a rectangle is 6 feet longer than the width. If the perimeter is between 24 feet and 36 feet, find all possible values for the width.

## Review Problems

The problems that follow review the material we covered in Chapter 1.

Simplify each expression.

**55.** $-|-5|$

**56.** $\left(-\dfrac{2}{3}\right)^3$

**57.** $-3 - 4(-2)$

**58.** $2^4 + 3^3 \div 9 - 4^2$

**59.** $5|3 - 8| - 6|2 - 5|$

**60.** $7 - 3(2 - 6)$

**61.** $5 - 2[-3(5 - 7) - 8]$

**62.** $\dfrac{5 + 3(7 - 2)}{2(-3) - 4}$

**63.** Find the difference of $-3$ and $-9$.

**64.** If you add $-4$ to the product of $-3$ and 5, what number results?

**65.** Apply the distributive property to $\dfrac{1}{2}(4x - 6)$.

**66.** Use the associative property to simplify $-6\left(\dfrac{1}{3}x\right)$.

For the set $\{-3, -\frac{4}{5}, 0, \frac{5}{8}, 2, \sqrt{5}\}$, which numbers are

**67.** Integers?

**68.** Rational numbers?

# CHAPTER 2 SUMMARY

**Examples**

**1.** The terms $2x$, $5x$, and $-7x$ are all similar since their variable parts are the same.

**Similar Terms [2.1]**

A *term* is a number or a number and one or more variables multiplied together. *Similar terms* are terms with the same variable part.

**2.** Simplify $3x + 4x$.

$$3x + 4x = (3 + 4)x$$
$$= 7x$$

**Simplifying Expressions [2.1]**

In this chapter we simplified expressions that contained variables by using the distributive property to combine similar terms.

**3.** The solution set for the equation $x + 2 = 5$ is $\{3\}$ because when $x$ is 3 the equation is $3 + 2 = 5$, or $5 = 5$.

**Solution Set [2.2]**

The *solution set* for an equation (or inequality) is all the numbers that, when used in place of the variable, make the equation (or inequality) a true statement.

**4.** The equations $a - 4 = 3$ and $a - 2 = 5$ are equivalent since both have solution set $\{7\}$.

**Equivalent Equations [2.2]**

Two equations are called *equivalent* if they have the same solution set.

**5.** Solve $x - 5 = 12$.

$$x - 5 + \mathbf{5} = 12 + \mathbf{5}$$
$$x + 0 = 17$$
$$x = 17$$

**Addition Property of Equality [2.2]**

When the same quantity is added to both sides of an equation, the solution set for the equation is unchanged. Adding the same amount to both sides of an equation produces an equivalent equation.

**6.** Solve $3x = 18$.

$$\frac{1}{3}(3x) = \frac{1}{3}(18)$$
$$x = 6$$

**Multiplication Property of Equality [2.3]**

If both sides of an equation are multiplied by the same nonzero number, the solution set is unchanged. Multiplying both sides of an equation by a nonzero quantity produces an equivalent equation.

**7.** Solve $2(x + 3) = 10$.

$$2x + 6 = 10$$
$$2x + 6 + (\mathbf{-6}) = 10 + (\mathbf{-6})$$
$$2x = 4$$
$$\frac{1}{2}(2x) = \frac{1}{2}(4)$$
$$x = 2$$

**Strategy for Solving Linear Equations in One Variable [2.4]**

*Step 1a:* Use the distributive property to separate terms, if necessary.

*1b:* If fractions are present, consider multiplying both sides by the LCD to eliminate the fractions. If decimals are present, consider multiplying both sides by a power of 10 to clear the equation of decimals.

*1c:* Combine similar terms on each side of the equation.

*Step 2:* Use the addition property of equality to get all variable terms on one side of the equation and all constant terms on the other side. A variable term is a term that contains the variable (for example, $5x$). A constant term is a term that does not contain the variable (the number 3, for example).

*Step 3:* Use the multiplication property of equality to get $x$ (that is, $1x$) by itself on one side of the equation.

*Step 4:* Check your solution in the original equation to be sure that you have not made a mistake in the solution process.

**8.** Solving $P = 2l + 2w$ for $l$, we have

$$P - 2w = 2l$$

$$\frac{P - 2w}{2} = l$$

### Formulas [2.5]

A *formula* is an equation with more than one variable. To solve a formula for one of its variables, we use the addition and multiplication properties of equality to move everything except the variable in question to one side of the equal sign so that the variable in question is alone on the other side.

### *Blueprint for Problem Solving [2.6, 2.7]*

*Step 1:* **Read** the problem, and then mentally **list** the items that are known and the items that are unknown.

*Step 2:* **Assign a variable** to one of the unknown items. (In most cases this will amount to letting $x =$ the item that is asked for in the problem.) Then **translate** the other **information** in the problem to expressions involving the variable.

*Step 3:* **Reread** the problem, and then **write an equation,** using the items and variables listed in steps 1 and 2, that describes the situation.

*Step 4:* **Solve the equation** found in step 3.

*Step 5:* **Write** your **answer** using a complete sentence.

*Step 6:* **Reread** the problem, and **check** your solution with the original words in the problem.

**9.** Solve $x + 5 < 7$.

$$x + 5 + (-5) < 7 + (-5)$$

$$x < 2$$

### Addition Property for Inequalities [2.8]

Adding the same quantity to both sides of an inequality produces an equivalent inequality, one with the same solution set.

**10.** Solve $-3a \le 18$.

$$-\frac{1}{3}(-3a) \ge -\frac{1}{3}(18)$$

$$a \ge -6$$

### Multiplication Property for Inequalities [2.8]

Multiplying both sides of an inequality by a positive number never changes the solution set. If both sides are multiplied by a negative number, the sense of the inequality must be reversed to produce an equivalent inequality.

**11.** Solve $3(x - 4) \ge -2$.

### Strategy for Solving Linear Inequalities in One Variable [2.8]

*Step 1a:* Use the distributive property to separate terms, if necessary.

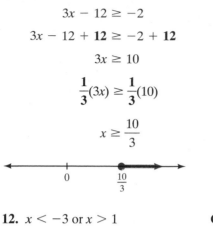

$$3x - 12 \geq -2$$

$$3x - 12 + \mathbf{12} \geq -2 + \mathbf{12}$$

$$3x \geq 10$$

$$\frac{1}{3}(3x) \geq \frac{1}{3}(10)$$

$$x \geq \frac{10}{3}$$

**1b:** If fractions are present, consider multiplying both sides by the LCD to eliminate the fractions. If decimals are present, consider multiplying both sides by a power of 10 to clear the inequality of decimals.

**1c:** Combine similar terms on each side of the inequality.

**Step 2:** Use the addition property for inequalities to get all variable terms on one side of the inequality and all constant terms on the other side.

**Step 3:** Use the multiplication property for inequalities to get $x$ by itself on one side of the inequality.

**Step 4:** Graph the solution set.

**12.** $x < -3$ or $x > 1$

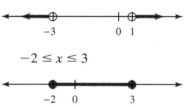

$$-2 \leq x \leq 3$$

## Compound Inequalities [2.9]

Two inequalities connected by the word *and* or *or* form a compound inequality. If the connecting word is *or*, we graph all points that are on either graph. If the connecting word is *and*, we graph only those points that are common to both graphs. The inequality $-2 \leq x \leq 3$ is equivalent to the compound inequality $-2 \leq x$ and $x \leq 3$.

### COMMON MISTAKES

**1.** Trying to subtract away coefficients (the number in front of variables) when solving equations. For example:

$$4x = 12$$

$$4x - 4 = 12 - 4$$

$$x = 8 \leftarrow \text{Mistake}$$

It is not incorrect to add $(-4)$ to both sides; it's just that $4x - 4$ is not equal to $x$. Both sides should be multiplied by $\frac{1}{4}$ to solve for $x$.

**2.** Forgetting to reverse the direction of the inequality symbol when multiplying both sides of an inequality by a negative number. For instance:

$$-3x < 12$$

$$-\frac{1}{3}(-3x) < -\frac{1}{3}(12) \leftarrow \text{Mistake}$$

$$x < -4$$

It is not incorrect to multiply both sides by $-\frac{1}{3}$. But if we do, we must also reverse the sense of the inequality.

# CHAPTER 2    REVIEW

The numbers in brackets refer to the sections of the text in which similar problems can be found.

Simplify each expression as much as possible. [2.1]

**1.** $5x - 8x$
**2.** $6x - 3 - 8x$
**3.** $-a + 2 + 5a - 9$
**4.** $5(2a - 1) - 4(3a - 2)$
**5.** $6 - 2(3y + 1) - 4$
**6.** $4 - 2(3x - 1) - 5$

Find the value of each expression when $x$ is 3. [2.1]

**7.** $7x - 2$
**8.** $-4x - 5 + 2x$
**9.** $-x - 2x - 3x$

Find the value of each expression when $x$ is $-2$. [2.1]

**10.** $5x - 3$
**11.** $-3x + 2$
**12.** $7 - x - 3$

Solve each equation. [2.2, 2.3]

**13.** $x + 2 = -6$
**14.** $x - \dfrac{1}{2} = \dfrac{4}{7}$

**15.** $10 - 3y + 4y = 12$
**16.** $-3 - 4 = -y - 2 + 2y$
**17.** $2x = -10$
**18.** $3x = 0$

**19.** $\dfrac{x}{3} = 4$
**20.** $-\dfrac{x}{4} = 2$

**21.** $3a - 2 = 5a$
**22.** $\dfrac{7}{10} a = \dfrac{1}{5} a + \dfrac{1}{2}$

**23.** $3x + 2 = 5x - 8$
**24.** $6x - 3 = 2x + 7$
**25.** $0.7x - 0.1 = 0.5x - 0.1$
**26.** $0.2x - 0.3 = 0.8x - 0.3$

Solve each equation. Be sure to simplify each side first. [2.4]

**27.** $2(x - 5) = 10$
**28.** $12 = 2(5x - 4)$

**29.** $\dfrac{1}{2}(3t - 2) + \dfrac{1}{2} = \dfrac{5}{2}$

**30.** $\dfrac{3}{5}(5x - 10) = \dfrac{2}{3}(9x + 3)$

**31.** $2(3x + 7) = 4(5x - 1) + 18$
**32.** $7 - 3(y + 4) = 10$

Use the formula $4x - 5y = 20$ to find $y$ if [2.5]

**33.** $x$ is 5
**34.** $x$ is 0
**35.** $x$ is $-5$
**36.** $x$ is 10

Solve each of the following formulas for the indicated variable. [2.5]

**37.** $2x - 5y = 10$ for $y$
**38.** $5x - 2y = 10$ for $y$
**39.** $V = \pi r^2 h$ for $h$
**40.** $P = 2l + 2w$ for $w$
**41.** What number is 86% of 240? [2.5]
**42.** What percent of 2,000 is 180? [2.5]

Solve each of the following word problems. In each case, be sure to show the equation that describes the situation. [2.6, 2.7]

**43. Number Problem**   The sum of twice a number and 6 is 28. Find the number.

**44. Geometry**   The length of a rectangle is 5 times as long as the width. If the perimeter is 60 meters, find the length and the width.

**45. Investing**   A man invests a certain amount of money in an account that pays 9% annual interest. He invests $300 more than that in an account that pays 10% annual interest. If his total interest after a year is $125, how much does he have invested in each account?

**46. Coin Problem**   A collection of 15 coins is worth $1.00. If the coins are dimes and nickels, how many of each coin are there?

Solve each inequality. [2.8]

**47.** $-2x < 4$
**48.** $-5x > -10$

**49.** $-\dfrac{a}{2} \le -3$
**50.** $-\dfrac{a}{3} > 5$

Solve each inequality, and graph the solution. [2.8]

**51.** $-4x + 5 > 37$
**52.** $2x + 10 < 5x - 11$
**53.** $2(3t + 1) + 6 \ge 5(2t + 4)$

Graph the solution to each of the following compound inequalities. [2.9]

**54.** $x < -2$ or $x > 5$
**55.** $-5x \ge 25$ or $2x - 3 \ge 9$
**56.** $-1 < 2x + 1 < 9$

# CHAPTER 2    TEST

Simplify each of the following expressions. [2.1]

1. $3x + 2 - 7x + 3$        2. $4a - 5 - a + 1$
3. $7 - 3(y + 5) - 4$
4. $8(2x + 1) - 5(x - 4)$
5. Find the value of $2x - 3 - 7x$ when $x = -5$. [2.1]
6. Find the value of $x^2 + 2xy + y^2$ when $x = 2$ and $y = 3$. [2.1]

Solve the following equations. [2.2, 2.3, 2.4]

7. $2x - 5 = 7$        8. $2y + 4 = 5y$
9. $\dfrac{1}{2}x - \dfrac{1}{10} = \dfrac{1}{5}x + \dfrac{1}{2}$
10. $\dfrac{2}{5}(5x - 10) = -5$
11. $-5(2x + 1) - 6 = 19$
12. $0.04x + 0.06(100 - x) = 4.6$
13. $2(t - 4) + 3(t + 5) = 2t - 2$
14. $2x - 4(5x + 1) = 3x + 17$
15. What number is 15% of 38? [2.5]
16. 240 is 12% of what number? [2.5]
17. If $2x - 3y = 12$, find $x$ when $y = -2$. [2.5]
18. The formula for the volume of a cone is $V = \frac{1}{3}\pi r^2 h$. Find $h$ if $V = 88$ cubic inches, $\pi = \frac{22}{7}$, and $r = 3$ inches. [2.5]

19. Solve $2x + 5y = 20$ for $y$. [2.5]
20. Solve $h = x + vt + 16t^2$ for $v$. [2.5]

Solve each word problem. [2.6, 2.7]

21. **Age Problem**  Dave is twice as old as Rick. Ten years ago the sum of their ages was 40. How old are they now?

22. **Geometry**  A rectangle is twice as long as it is wide. The perimeter is 60 inches. What are the length and width?

23. **Coin Problem**  A man has a collection of dimes and quarters with a total value of $3.50. If he has 7 more dimes than quarters, how many of each coin does he have?

24. **Investing**  A woman has money in two accounts. One account pays 7% annual interest, while the other pays 9% annual interest. If she has $600 more invested at 9% than she does at 7% and her total interest for a year is $182, how much does she have in each account?

Solve each inequality, and graph the solution. [2.8]

25. $2x + 3 < 5$        26. $-5a > 20$
27. $0.4 - 0.2x \geq 1$        28. $4 - 5(m + 1) \leq 9$

Solve each inequality, and graph the solution. [2.9]

29. $3 - 4x \geq -5$ or $2x \geq 10$
30. $-7 < 2x - 1 < 9$

# CHAPTERS 1–2 CUMULATIVE REVIEW

Simplify.

**1.** $6 + 3(6 + 2)$

**2.** $-11 + 17 + (-13)$

**3.** $7 - 9 - 12$

**4.** $8 - 4 \cdot 5$

**5.** $\dfrac{1}{5}(10x)$

**6.** $-6(5 - 11) - 4(13 - 6)$

**7.** $\left(-\dfrac{2}{3}\right)^3$

**8.** $-8(9x)$

**9.** $-\dfrac{3}{4} \div \dfrac{15}{16}$

**10.** $\dfrac{2}{5} \cdot \dfrac{4}{7}$

**11.** $\dfrac{-4(-6)}{-9}$

**12.** $\dfrac{2(-9) + 3(6)}{-7 - 8}$

**13.** $\dfrac{(5 - 3)^2}{5^2 - 3^2}$

**14.** $\dfrac{5}{9} + \dfrac{1}{3}$

**15.** $\dfrac{4}{21} - \dfrac{9}{35}$

**16.** $3(2x - 1) + 5(x + 2)$

Solve each equation.

**17.** $7x = 6x + 4$

**18.** $x + 12 = -4$

**19.** $-\dfrac{3}{5}x = 30$

**20.** $3x - 4 = 11$

**21.** $5x - 7 = x - 1$

**22.** $4(3x - 8) + 5(2x + 7) = 25$

**23.** $15 - 3(2t + 4) = 1$

**24.** $3(2a - 7) - 4(a - 3) = 15$

**25.** $\dfrac{1}{3}(x - 6) = \dfrac{1}{4}(x + 8)$

**26.** Solve $P = a + b + c$ for $c$.

**27.** Solve $3x + 4y = 12$ for $y$.

Solve each inequality, and graph the solution.

**28.** $3(x - 4) \le 6$

**29.** $-5x + 9 < -6$

**30.** $x - 1 < 1$ or $2x + 1 > 7$

**31.** $-2 < x + 1 < 5$

**32.** Translate into symbols: The difference of $x$ and 5 is 12.

**33.** Give the opposite, reciprocal, and absolute value of $-\dfrac{2}{3}$.

**34.** Find the perimeter and area:

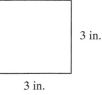

3 in.

3 in.

**35.** Find the next number in the sequence 4, 1, $-2$, $-5$, . . . .

**36.** State the property or properties that justify the following: $5 + (-5) = 0$

**37.** Use the distributive property: $\dfrac{1}{4}(8x - 4)$

**38.** List all the rational numbers from the set $\{-5, -3, -1.7, 2.3, \frac{12}{7}, \pi\}$

**39.** Reduce to lowest terms: $\dfrac{234}{312}$

**40.** Evaluate $x^2 - 8x - 9$ when $x = -2$.

**41.** Evaluate $a^2 - 2ab + b^2$ when $a = 3$, $b = -2$.

**42.** What is 30% of 50?

**43. Number Problem** Twice a number increased by 7 is 31. Find the number.

**44. Gambling** A gambler won \$5, then lost \$8, won \$1, and finally lost \$7. How much did she win or lose?

**45. Geometry** A right triangle has one 42° angle. Find the other two angles.

**46. Geometry** The length of a rectangle is 4 inches more than the width. The perimeter is 28. Find the length and width.

**47. Geometry** Two angles are complementary. If one is 25°, find the other one.

**48. Coin Problem** Arbi has \$2.20 in dimes and quarters. If he has 8 more dimes than quarters, how many coins of each does he have?

**49. Investing** I have invested money in two accounts. One account pays 5% annual interest, and the other pays 6%. I have \$200 more in the 6% account than I have in the 5% account. If the total amount of interest was \$56, how much do I have in each account?

**50. Hourly Pay** Carol tutors in the math lab. She gets paid \$8 per hour for the first 15 hours and \$10 per hour for each hour after that. She made \$150 one week. How many hours did she work?

# Linear Equations and Inequalities in Two Variables

*(© EyeWire, Inc.)*

## INTRODUCTION

When light comes into contact with a surface that does not transmit light, then all the light that contacts the surface is either reflected off the surface or absorbed into the surface. If we let $R$ represent the percentage of light reflected and $A$ the percentage of light absorbed, then the relationship between these two variables can be written as

$$R + A = 100$$

which is a linear equation in two variables. Table 1 and the following graph show the same relationship as that described by the equation. The table is a numerical description; the graph is a visual description.

| TABLE 1  Reflected and Absorbed Light | |
|---|---|
| **Percent Reflected** | **Percent Absorbed** |
| 0 | 100 |
| 20 | 80 |
| 40 | 60 |
| 60 | 40 |
| 80 | 20 |
| 100 | 0 |

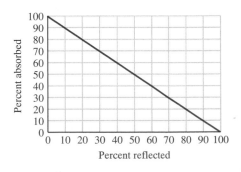

In this chapter we learn how to build tables and draw graphs from linear equations in two variables.

## STUDY SKILLS

1. **Getting Ready To Take an Exam**   Try to arrange your daily study habits so that you have very little studying to do the night before your next exam. The next two goals will help you achieve goal 1.

2. **Review With the Exam in Mind**   Each day you should review material that will be covered on the next exam. Your review should consist of working problems. Preferably, the problems you work should be problems from your list of difficult problems.

3. **Continue to List Difficult Problems**   This study skill was started in the previous chapter. You should continue to list and rework the problems that give you the most difficulty. It is this list that you will use to study for the next exam. Your goal is to go into the next exam knowing that you can successfully work any problem from your list of hard problems.

4. **Pay Attention to Instructions**   Taking a test is different from doing homework. When you take a test, the problems will be mixed up. When you do your homework, you usually work a number of similar problems. I sometimes have students who do very well on their homework, but become confused when they see the same problems on a test because they have not paid attention to the instructions on their homework. For example, suppose you see the equation $y = 3x - 2$ on your next test. By itself, the equation is simply a statement. There isn't anything to do unless the equation is accompanied by instructions. Each of the following is a valid instruction with respect to the equation $y = 3x - 2$, and the result of applying the instructions will be different in each case:

Find $x$ when $y$ is 10.        (Section 2.5)

Solve for $x$.                        (Section 2.5)

Graph the equation.            (Section 3.3)

Find the intercepts.            (Section 3.4)

There are many things to do with the equation $y = 3x - 2$. If you train yourself to pay attention to the instructions that accompany a problem as you work through the assigned problems, you will not find yourself confused about what to do with a problem when you see it on a test.

In Chapter 1, we showed the relationship between the table of values for the speed of a race car and the corresponding bar chart. Table 1 and Figure 1 from the introduction are reproduced here for reference. In Figure 1, the horizontal line that shows the elapsed time in seconds is called the *horizontal axis,* and the vertical line that shows the speed in miles per hour is called the *vertical axis.*

| TABLE 1 | Speed of a Race Car |
|---|---|
| Time (seconds) | Speed (miles per hour) |
| 0 | 0 |
| 1 | 72.7 |
| 2 | 129.9 |
| 3 | 162.8 |
| 4 | 192.2 |
| 5 | 212.4 |
| 6 | 228.1 |

**FIGURE 1**

The data in Table 1 are called *paired data* because the information is organized so that each number in the first column is paired with a specific number in the second column. Each pair of numbers is associated with one of the solid bars in Figure 1. For example, the third bar in the bar chart is associated with the pair of numbers 3 seconds and 162.8 miles per hour. The first number, 3 seconds, is associated with the horizontal axis, and the second number, 162.8 miles per hour, is associated with the vertical axis.

## SCATTER DIAGRAMS AND LINE GRAPHS

The information in Table 1 can be visualized with a *scatter diagram* and *line graph* as well. Figure 2 is a scatter diagram of the information in Table 1. We use dots instead of the bars shown in Figure 1 to show the speed of the race car at each second during the race. Figure 3 is called a *line graph.* It is constructed by taking the dots in Figure 2 and connecting each one to the next with a straight line. Notice that we have labeled the axes in these two figures a little differently than we did with the histogram (or bar chart), by making the axes intersect at the number 0.

The number sequences we have worked with in the past can also be written as paired data by associating each number in the sequence with its position in the sequence. For instance, in the sequence of odd numbers

$$1, 3, 5, 7, 9, \ldots$$

the number 7 is the fourth number in the sequence. Its position is 4, and its value is

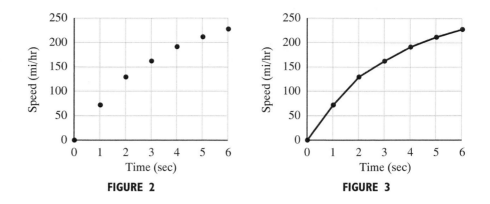

FIGURE 2          FIGURE 3

7. Here is the sequence of odd numbers written so that the position of each term is noted:

Position     1, 2, 3, 4, 5, . . .
Value        1, 3, 5, 7, 9, . . .

**E X A M P L E  1**    Tables 2 and 3 give the first five terms of the sequence of odd numbers and the sequence of squares as paired data. In each case construct a scatter diagram.

| TABLE 2  Odd Numbers | |
|---|---|
| **Position** | **Value** |
| 1 | 1 |
| 2 | 3 |
| 3 | 5 |
| 4 | 7 |
| 5 | 9 |

| TABLE 3  Squares | |
|---|---|
| **Position** | **Value** |
| 1 | 1 |
| 2 | 4 |
| 3 | 9 |
| 4 | 16 |
| 5 | 25 |

**Solution**    The two scatter diagrams are based on the data from Tables 2 and 3 shown here. Notice how the dots in Figure 4 seem to line up in a straight line, while

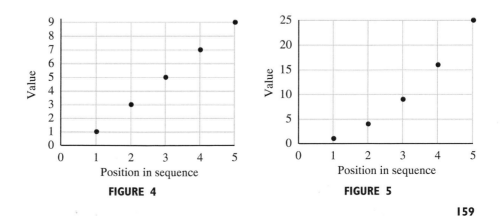

FIGURE 4          FIGURE 5

the dots in Figure 5 give the impression of a curve. We say the points in Figure 4 suggest a *linear* relationship between the two sets of data, while the points in Figure 5 suggest a *nonlinear* relationship.

As you know, each dot in Figures 4 and 5 corresponds to a pair of numbers, one of which is associated with the horizontal axis and the other with the vertical axis. Paired data play a very important role in the equations we will solve in the next section. To prepare ourselves for those equations, we need to expand the concept of paired data to include negative numbers. At the same time, we want to standardize the position of the axes in the diagrams that we use to visualize paired data.

> **DEFINITION** A pair of numbers enclosed in parentheses and separated by a comma, such as $(-2, 1)$, is called an **ordered pair** of numbers. The first number in the pair is called the ***x*-coordinate** of the ordered pair; the second number is called the ***y*-coordinate**. For the ordered pair $(-2, 1)$, the *x*-coordinate is $-2$, and the *y*-coordinate is 1.

Ordered pairs of numbers are important in the study of mathematics because they give us a way to visualize solutions to equations. To see the visual component of ordered pairs, we need the diagram shown in Figure 6. It is called the *rectangular coordinate system*.

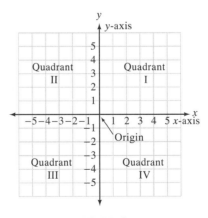

**FIGURE 6**

The rectangular coordinate system is built from two number lines oriented perpendicular to each other. The horizontal number line is exactly the same as our real number line and is called the *x*-axis. The vertical number line is also the same as our real number line, with the positive direction up and the negative direction down. It is called the *y*-axis. The point where the two axes intersect is called the origin. As you can see from Figure 6, the axes divide the plane into four quadrants, which are numbered I through IV in a counterclockwise direction.

## GRAPHING ORDERED PAIRS

To graph the ordered pair $(a, b)$, we start at the origin and move $a$ units forward or back (forward if $a$ is positive and back if $a$ is negative). Then we move $b$ units up or down (up if $b$ is positive, down if $b$ is negative). The point where we end up is the graph of the ordered pair $(a, b)$.

**EXAMPLE 2**   Graph the ordered pairs $(3, 4)$, $(3, -4)$, $(-3, 4)$, and $(-3, -4)$.

**Solution**

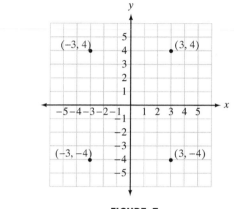

**FIGURE 7**

We can see in Figure 7 that when we graph ordered pairs, the $x$-coordinate corresponds to movement parallel to the $x$-axis (horizontal) and the $y$-coordinate corresponds to movement parallel to the $y$-axis (vertical).

**EXAMPLE 3**   Graph the ordered pairs $(-1, 3)$, $(2, 5)$, $(0, 0)$, $(0, -3)$, and $(4, 0)$.

**Solution**   See Figure 8.

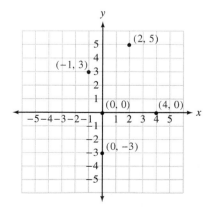

**FIGURE 8**

**Note:**  If we do not label the axes of a coordinate system, we assume that each square is 1 unit long and 1 unit wide.

 ## Getting Ready for Class

*After reading through the preceding section, respond in your own words and in complete sentences.*

**A.** What is an ordered pair of numbers?

**B.** Explain in words how you would graph the ordered pair (3, 4).

**C.** How do you construct a rectangular coordinate system?

**D.** Where is the origin on a rectangular coordinate system?

## PROBLEM SET 3.1

Graph the following ordered pairs.

| | | |
|---|---|---|
| **1.** (3, 2) | **2.** (3, −2) | **3.** (−3, 2) |
| **4.** (−3, −2) | **5.** (5, 1) | **6.** (5, −1) |
| **7.** (1, 5) | **8.** (1, −5) | **9.** (−1, 5) |
| **10.** (−1, −5) | **11.** $(2, \frac{1}{2})$ | **12.** $(3, \frac{3}{2})$ |
| **13.** $(-4, -\frac{5}{2})$ | **14.** $(-5, -\frac{3}{2})$ | **15.** (3, 0) |
| **16.** (−2, 0) | **17.** (0, 5) | **18.** (0, 0) |

**19–28.** Give the coordinates of each numbered point in the figure.

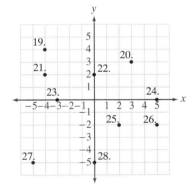

Graph the points (4, 3) and (−4, −1), and draw a straight line that passes through both of them. Then answer the following questions.

**29.** Does the point (2, 2) lie on the line?

**30.** Does the point (−2, 0) lie on the line?

**31.** Does the point (0, −2) lie on the line?

**32.** Does the point (−6, 2) lie on the line?

Graph the points (−2, 4) and (2, −4), and draw a straight line that passes through both of them. Then answer the following questions.

**33.** Does the point (0, 0) lie on the line?

**34.** Does the point (−1, 2) lie on the line? *y*

**35.** Does the point (2, −1) lie on the line? *u*

**36.** Does the point (1, −2) lie on the line?

Draw a straight line that passes through the points (3, 4) and (3, −4). Then answer the following questions.

**37.** Is the point (3, 0) on this line?

**38.** Is the point (0, 3) on this line?

**39.** Is there any point on this line with an *x*-coordinate other than 3?

**40.** If you extended the line, would it pass through a point with a *y*-coordinate of 10?

Draw a straight line that passes through the points (3, 4) and (−3, 4). Then answer the following questions.

**41.** Is the point (4, 0) on this line?

**42.** Is the point (0, 4) on this line?

**43.** Is there any point on this line with a *y*-coordinate other than 4?

**44.** If you extended the line, would it pass through a point with an *x*-coordinate of 10?

**45.** Draw a straight line that passes through the points (−2, −3) and (4, −3). What do all the points on this line have in common?

**46.** Draw a straight line that passes through the points (−2, −3) and (−2, 4). What do all the points on this line have in common?

**47.** For the points on a rectangular coordinate system, where will you find all the ordered pairs of the form (0, *y*)?

**48.** For the points on a rectangular coordinate system, where will you find all the ordered pairs of the form (*x*, 0)?

## Applying the Concepts

**49. Hourly Wages** Jane is considering a job at Marcy's department store. The job pays $8.00 per hour. The following line graph shows how much Jane will make for working from 0 to 40 hours in a week. List three ordered pairs that lie on the line graph.

**50. Hourly Wages** Jane is also considering a job at Gigi's Boutique. This job pays $6.00 per hour plus $50 per week in commission. The following line

graph shows how much Jane will make for working from 0 to 40 hours in a week at Gigi's. List three ordered pairs that lie on the line graph.

**51. Garbage Production** The table and bar chart from Problem 89 in Problem Set 1.4 are shown here. Each gives the annual production of garbage in the United States for some specific years.

The Growing Garbage Problem

| Year | Garbage (millions of tons) |
|------|-----------------------------|
| 1960 | 88 |
| 1970 | 121 |
| 1980 | 152 |
| 1990 | 205 |
| 1997 | 217 |

Use the information from the table and bar chart to construct a line graph using the template below.

**52. Grass Height** The table and bar chart from Problem 90 in Problem Set 1.4 are shown here. Each gives the growth of a certain species of grass over time.

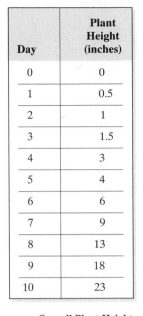

| Day | Plant Height (inches) |
|-----|-----------------------|
| 0   | 0   |
| 1   | 0.5 |
| 2   | 1   |
| 3   | 1.5 |
| 4   | 3   |
| 5   | 4   |
| 6   | 6   |
| 7   | 9   |
| 8   | 13  |
| 9   | 18  |
| 10  | 23  |

**Overall Plant Height**

Use the information from the table and chart to construct a line graph using the following template.

**Overall Plant Height**

**53. Wireless Phone Costs**  The table and bar chart from Problem 91 in Problem Set 1.4 are shown here. Each shows the projected cost of wireless phone use. Use the information from the table and chart to construct a line graph.

| Year | Cents per Minute |
|------|------------------|
| 1998 | 33 |
| 1999 | 28 |
| 2000 | 25 |
| 2001 | 23 |
| 2002 | 22 |
| 2003 | 20 |

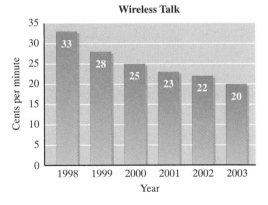

**Wireless Talk**

**54. Digital Camera Sales**  The table and bar chart from Problem 78 in Problem Set 1.3 are shown here. Each shows the sales of digital cameras from 1996 to 1999. Use the information from the table and chart to construct a line graph.

| Year | Sales (in millions) |
|------|---------------------|
| 1996 | $386 |
| 1997 | $518 |
| 1998 | $573 |
| 1999 | $819 |

**Digital Camera Sales**

**55. Unemployment Rates** The following table shows adjusted unemployment rates in the United States for the years 1983 to 1993. Construct a scatter diagram of this information.

| Unemployment Rates | |
| --- | --- |
| **Year** | **Unemployment Rate (%)** |
| 1983 | 10.2 |
| 1984 | 7.8 |
| 1985 | 7.4 |
| 1986 | 7.0 |
| 1987 | 6.3 |
| 1988 | 5.4 |
| 1989 | 5.2 |
| 1990 | 5.4 |
| 1991 | 6.5 |
| 1992 | 7.3 |
| 1993 | 7.0 |

**56. Tourist Attraction Attendance** The annual number of visitors to some of the most popular tourist attractions in the mid-1990s is shown in the following table. Construct a line graph for the information in this table.

| Visitors to Tourist Attractions | |
| --- | --- |
| **Attraction** | **Number of Visitors** |
| Walt Disney World | 30,000,000 |
| Disneyland | 12,000,000 |
| Great Smoky Mountains | 9,000,000 |
| Universal Studios, Orlando | 6,000,000 |
| Branson, Missouri | 5,000,000 |
| Grand Canyon | 3,900,000 |
| Sea World, Orlando | 3,850,000 |
| Sea World, San Diego | 3,820,000 |
| Knott's Berry Farm | 3,800,000 |

## Review Problems

The problems that follow review the material we covered on number sequences in Chapter 1.

Find the next number in each sequence.

**57.** 3, 10, 17, 24, . . .    **58.** 3, −4, −11, −18, . . .

**59.** 3, 1, $\frac{1}{3}$, $\frac{1}{9}$, . . .    **60.** 3, 21, 147, 1,029, . . .

**61.** 7, 4, 1, −2, . . .    **62.** 7, 10, 13, 16, . . .

**63.** 7, 21, 63, 189, . . .    **64.** 7, 1, $\frac{1}{7}$, $\frac{1}{49}$, . . .

**65.** 5, 6, 8, 11, . . .    **66.** 10, 11, 13, 16, . . .

**3.2    Solutions to Linear Equations in Two Variables**

In this section we will begin to investigate equations in two variables. As you will see, equations in two variables have pairs of numbers for solutions. Since we know how to use paired data to construct tables, histograms, and other charts, we can take our work with paired data further by using equations in two variables to construct tables of paired data. Let's begin this section by reviewing the relationship between equations in one variable and their solutions.

If we solve the equation $3x - 2 = 10$, the solution is $x = 4$. If we graph this solution, we simply draw the real number line and place a dot at the point whose coordinate is 4. The relationship between linear equations in one variable, their solutions, and the graphs of those solutions looks like this:

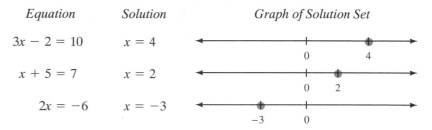

| Equation | Solution | Graph of Solution Set |
|----------|----------|----------------------|

When the equation has one variable, the solution is a single number whose graph is a point on a line.

Now, consider the equation $2x + y = 3$. The first thing we notice is that there are two variables instead of one. Therefore, a solution to the equation $2x + y = 3$ will be not a single number but a pair of numbers, one for $x$ and one for $y$, that makes the equation a true statement. One pair of numbers that works is $x = 2, y = -1$, because when we substitute them for $x$ and $y$ in the equation, we get a true statement. That is,

$$2(2) + (-1) \stackrel{?}{=} 3$$

$$4 - 1 = 3$$

$$3 = 3 \qquad \text{A true statement}$$

The pair of numbers $x = 2, y = -1$ is written as $(2, -1)$. As you know from Section 3.1, $(2, -1)$ is called an *ordered pair* because it is a pair of numbers written in a specific order. The first number is always associated with the variable $x$, and the second number is always associated with the variable $y$. We call the first number in the ordered pair the *x-coordinate* (or $x$-component) and the second number the *y-coordinate* (or $y$-component) of the ordered pair.

Let's look back to the equation $2x + y = 3$. The ordered pair $(2, -1)$ is not the only solution. Another solution is $(0, 3)$, because when we substitute 0 for $x$ and 3 for $y$ we get

$$2(0) + 3 \stackrel{?}{=} 3$$

$$0 + 3 = 3$$

$$3 = 3 \qquad \text{A true statement}$$

Still another solution is the ordered pair $(5, -7)$, because

$$2(5) + (-7) \stackrel{?}{=} 3$$

$$10 - 7 = 3$$

$$3 = 3 \qquad \text{A true statement}$$

As a matter of fact, for any number we want to use for $x$, there is another number we can use for $y$ that will make the equation a true statement. An infinite number of ordered pairs satisfy (are solutions to) the equation $2x + y = 3$; we have listed just a few of them.

**EXAMPLE 1**   Given the equation $2x + 3y = 6$, complete the following ordered pairs so that they will be solutions to the equation: $(0, \ )$, $( \ , 1)$, $(3, \ )$.

**Solution**   To complete the ordered pair $(0, \ )$, we substitute 0 for $x$ in the equation and then solve for $y$:

$$2(0) + 3y = 6$$

$$3y = 6$$

$$y = 2$$

The ordered pair is $(0, 2)$.

To complete the ordered pair $( \ , 1)$, we substitute 1 for $y$ in the equation and solve for $x$:

$$2x + 3(1) = 6$$

$$2x + 3 = 6$$

$$2x = 3$$

$$x = \frac{3}{2}$$

The ordered pair is $(\frac{3}{2}, 1)$.

To complete the ordered pair $(3, \ )$, we substitute 3 for $x$ in the equation and solve for $y$:

$$2(3) + 3y = 6$$

$$6 + 3y = 6$$

$$3y = 0$$

$$y = 0$$

The ordered pair is $(3, 0)$.

Notice in each case that once we have used a number in place of one of the variables, the equation becomes a linear equation in one variable. We then use the method explained in Chapter 2 to solve for that variable.

**EXAMPLE 2**   Complete the following table for the equation $2x - 5y = 20$.

| $x$ | $y$ |
|---|---|
| 0 |  |
|  | 2 |
|  | 0 |
| −5 |  |

**Solution**   Filling in the table is equivalent to completing the following ordered pairs: (0,  ), (  , 2), (  , 0), (−5,  ). So we proceed as in Example 1.

When $x = 0$, we have          When $y = 2$, we have

$$2(0) - 5y = 20 \qquad\qquad 2x - 5(2) = 20$$

$$0 - 5y = 20 \qquad\qquad 2x - 10 = 20$$

$$-5y = 20 \qquad\qquad 2x = 30$$

$$y = -4 \qquad\qquad x = 15$$

When $y = 0$, we have          When $x = -5$, we have

$$2x - 5(0) = 20 \qquad\qquad 2(-5) - 5y = 20$$

$$2x - 0 = 20 \qquad\qquad -10 - 5y = 20$$

$$2x = 20 \qquad\qquad -5y = 30$$

$$x = 10 \qquad\qquad y = -6$$

The completed table looks like this:

| $x$ | $y$ |
|---|---|
| 0 | −4 |
| 15 | 2 |
| 10 | 0 |
| −5 | −6 |

which is equivalent to the ordered pairs $(0, -4)$, $(15, 2)$, $(10, 0)$, and $(-5, -6)$.

**EXAMPLE 3**   Complete the following table for the equation $y = 2x - 1$.

| $x$ | $y$ |
|---|---|
| 0 |  |
| 5 |  |
|  | 7 |
|  | 3 |

**Solution**   When $x = 0$, we have        When $x = 5$, we have

$$y = 2(0) - 1 \qquad\qquad y = 2(5) - 1$$

$$y = 0 - 1 \qquad\qquad y = 10 - 1$$

$$y = -1 \qquad\qquad\qquad y = 9$$

When $y = 7$, we have        When $y = 3$, we have

$$7 = 2x - 1 \qquad\qquad 3 = 2x - 1$$

$$8 = 2x \qquad\qquad\qquad 4 = 2x$$

$$4 = x \qquad\qquad\qquad 2 = x$$

The completed table is

| $x$ | $y$ |
|-----|-----|
| 0 | $-1$ |
| 5 | 9 |
| 4 | 7 |
| 2 | 3 |

which means the ordered pairs $(0, -1)$, $(5, 9)$, $(4, 7)$, and $(2, 3)$ are among the solutions to the equation $y = 2x - 1$.

**E X A M P L E   4**   Which of the ordered pairs $(2, 3)$, $(1, 5)$, and $(-2, -4)$ are solutions to the equation $y = 3x + 2$?

**Solution**   If an ordered pair is a solution to the equation, then it must satisfy the equation. That is, when the coordinates are used in place of the variables in the equation, the equation becomes a true statement.

Try $(2, 3)$ in $y = 3x + 2$:

$$3 \stackrel{?}{=} 3(2) + 2$$

$$3 = 6 + 2$$

$$3 = 8 \qquad\qquad \text{A false statement}$$

Try $(1, 5)$ in $y = 3x + 2$:

$$5 \stackrel{?}{=} 3(1) + 2$$

$$5 = 3 + 2$$

$$5 = 5 \qquad\qquad \text{A true statement}$$

Try $(-2, -4)$ in $y = 3x + 2$:

$$-4 \stackrel{?}{=} 3(-2) + 2$$

$$-4 = -6 + 2$$

$$-4 = -4 \qquad\qquad \text{A true statement}$$

The ordered pairs (1, 5) and (−2, −4) are solutions to the equation $y = 3x + 2$, and (2, 3) is not.

**EXAMPLE 5**     One of the rates listed in the GTE Mobilnet rate card in 1993 called for a flat rate of $15 per month plus $1.50 for each minute you used the phone. Write an equation in two variables that will let you calculate the monthly charge for talking $x$ minutes. Then make a table that shows the cost for talking 10, 15, 20, 25, or 30 minutes in one month. Use the data in the table to construct a histogram.

**Solution**     If we let $x$ = the number of minutes of phone use, then the charge for using the phone $x$ minutes in one month will be $1.5x + 15$. If $y$ is the charge for talking $x$ minutes, then

$$y = 1.5x + 15$$

Now, GTE Mobilnet charges $1.50 for each minute or fraction of a minute, so the preceding equation will work only when $x$ is a nonnegative integer. If we substitute $x$ = 10, 15, 20, 25, and 30 into the equation for $x$, then the equation gives us the corresponding costs as shown in Table 1. A histogram of the data in Table 1 is shown in Figure 1.

**TABLE 1   Monthly Phone Costs**

| Time (in minutes) | Cost (in dollars) |
|---|---|
| 10 | $30.00 |
| 15 | $37.50 |
| 20 | $45.00 |
| 25 | $52.50 |
| 30 | $60.00 |

**FIGURE 1**

![Getting Ready icon]

# Getting Ready for Class

*After reading through the preceding section, respond in your own words and in complete sentences.*

**A.** How can you tell if an ordered pair is a solution to an equation?

**B.** How would you find a solution to $y = 3x - 5$?

**C.** Why is (3, 2) not a solution to $y = 3x - 5$?

**D.** How many solutions are there to an equation that contains two variables?

# PROBLEM SET 3.2

For each equation, complete the given ordered pairs.

**1.** $2x + y = 6$ $\quad$ $(0, \ ), (3, \ ), ( \ , -6)$
**2.** $3x - y = 5$ $\quad$ $(0, \ ), (1, \ ), ( \ , 5)$
**3.** $3x + 4y = 12$ $\quad$ $(0, \ ), ( \ , 0), (-4, \ )$
**4.** $5x - 5y = 20$ $\quad$ $(0, \ ), ( \ , -2), (1, \ )$
**5.** $y = 4x - 3$ $\quad$ $(1, \ ), ( \ , 0), (5, \ )$
**6.** $y = 3x - 5$ $\quad$ $( \ , 13), (0, \ ), (-2, \ )$
**7.** $y = 7x - 1$ $\quad$ $(2, \ ), ( \ , 6), (0, \ )$
**8.** $y = 8x + 2$ $\quad$ $(3, \ ), ( \ , 0), ( \ , -6)$
**9.** $x = -5$ $\quad$ $( \ , 4), ( \ , -3), ( \ , 0)$
**10.** $y = 2$ $\quad$ $(5, \ ), (-8, \ ), (\frac{1}{2}, \ )$

For each of the following equations, complete the given table.

**11.** $y = 3x$ $\qquad$ **12.** $y = -2x$

| $x$ | $y$ |
|---|---|
| 1 | |
| $-3$ | |
| | 12 |
| | 18 |

| $x$ | $y$ |
|---|---|
| $-4$ | |
| 0 | |
| | 10 |
| | 12 |

**13.** $y = 4x$ $\qquad$ **14.** $y = -5x$

| $x$ | $y$ |
|---|---|
| 0 | |
| | $-2$ |
| $-3$ | |
| | 12 |

| $x$ | $y$ |
|---|---|
| 3 | |
| | 0 |
| $-2$ | |
| | $-20$ |

**15.** $x + y = 5$ $\qquad$ **16.** $x - y = 8$

| $x$ | $y$ |
|---|---|
| 2 | |
| 3 | |
| | 0 |
| | $-4$ |

| $x$ | $y$ |
|---|---|
| 0 | |
| 4 | |
| | $-3$ |
| | $-2$ |

**17.** $2x - y = 4$ $\qquad$ **18.** $3x - y = 9$

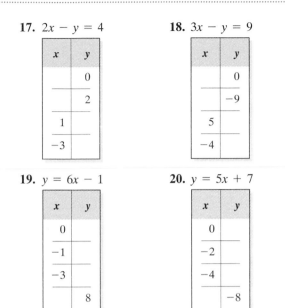

| $x$ | $y$ |
|---|---|
| | 0 |
| | 2 |
| 1 | |
| $-3$ | |

| $x$ | $y$ |
|---|---|
| | 0 |
| | $-9$ |
| 5 | |
| $-4$ | |

**19.** $y = 6x - 1$ $\qquad$ **20.** $y = 5x + 7$

| $x$ | $y$ |
|---|---|
| 0 | |
| $-1$ | |
| $-3$ | |
| | 8 |

| $x$ | $y$ |
|---|---|
| 0 | |
| $-2$ | |
| $-4$ | |
| | $-8$ |

For the following equations, tell which of the given ordered pairs are solutions.

**21.** $2x - 5y = 10$ $\quad$ $(2, 3), (0, -2), (\frac{5}{2}, 1)$
**22.** $3x + 7y = 21$ $\quad$ $(0, 3), (7, 0), (1, 2)$
**23.** $y = 7x - 2$ $\quad$ $(1, 5), (0, -2), (-2, -16)$
**24.** $y = 8x - 3$ $\quad$ $(0, 3), (5, 16), (1, 5)$
**25.** $y = 6x$ $\quad$ $(1, 6), (-2, -12), (0, 0)$
**26.** $y = -4x$ $\quad$ $(0, 0), (2, 4), (-3, 12)$
**27.** $x + y = 0$ $\quad$ $(1, 1), (2, -2), (3, 3)$
**28.** $x - y = 1$ $\quad$ $(0, 1), (0, -1), (1, 2)$
**29.** $x = 3$ $\quad$ $(3, 0), (3, -3), (5, 3)$
**30.** $y = -4$ $\quad$ $(3, -4), (-4, 4), (0, -4)$

## Applying the Concepts

**31. Perimeter** If the perimeter of a rectangle is 30 inches, then the relationship between the length $l$ and the width $w$ is given by the equation

$$2l + 2w = 30$$

What is the length when the width is 3 inches?

**32. Perimeter** The relationship between the perimeter $P$ of a square and the length of its side $s$ is given by the formula $P = 4s$. If each side of a square is 5 inches, what is the perimeter? If the perimeter of a square is 28 inches, how long is a side?

33. **Phone Rates** MCIA, a long-distance phone company, charges $3.00 a month plus $0.10 per minute for long-distance calls. If you talk for $x$ minutes during a month, then your total monthly cost $y$ will be $y = 3 + 0.10x$. Use this equation to fill in the following table; then use the results to create a line graph on the template.

| Minutes | Cost |
|---------|------|
| 0       |      |
| 10      |      |
| 20      |      |
| 30      |      |
| 40      |      |
| 50      |      |
| 60      |      |
| 70      |      |
| 80      |      |
| 90      |      |
| 100     |      |

34. **Phone Rates** One Bell, a long-distance phone company, charges $5.00 a month plus $0.05 per minute for long distance calls. If you talk for $x$ minutes during a month, then your total monthly cost $y$ will be $y = 5 + 0.05x$. Use this equation to fill in the following table; then use the results to create a line graph on the template.

| Minutes | Cost |
|---------|------|
| 0       |      |
| 10      |      |
| 20      |      |
| 30      |      |
| 40      |      |
| 50      |      |
| 60      |      |
| 70      |      |
| 80      |      |
| 90      |      |
| 100     |      |

35. **Internet Access** Billy is looking for an Internet service provider to get online. Computer Service

| Hours | Cost |
|-------|------|
| 0     |      |
| 1     |      |
| 2     |      |
| 3     |      |
| 4     |      |
| 5     |      |
| 6     |      |
| 7     |      |
| 8     |      |
| 9     |      |
| 10    |      |

offers an Internet connection for $10 a month plus $3 for every hour you are connected to the Internet. If you are on the Internet for $x$ hours, then your total cost $y$ will be $y = 10 + 3x$. Use this equation to fill in the preceding table; then use the results to create a line graph on the template.

**36. Internet Access**   Joan wants to connect to the Internet. ICM World offers their Internet connection

for $18 a month plus $1 for every hour you are connected. If you are on the Internet for $x$ hours in 1 month, then your monthly cost $y$ will be $y = 18 + x$. Use this equation to fill in the preceding table; then use the results to create a line graph on the template.

**37. Cost of Trash Service**   San Luis Garbage charges $13.00 per month for their basic service, plus $1.50 per bag for each trash bag left by the curb on the day the trash is picked up. Write an equation in two variables that will let you calculate the total monthly charge $y$ for picking up $x$ trash bags. Then use this equation to make a table that gives the cost for picking up 5, 7, 9, 11, and 13 trash bags in one month. Complete the table, and then use the information in the table to construct a bar chart.

| Bags Collected | Cost (in dollars) |
|:---:|:---:|
| 5 | |
| 7 | |
| 9 | |
| 11 | |
| 13 | |

**38. Cost of Trash Service**   Five Cities Garbage charges $21.85 per month for their basic service, plus $1.00 per bag for each trash bag left by the curb on the day the trash is picked up. Write an equation in two variables that will let you calculate the total monthly charge $y$ for picking up $x$ trash bags. Then use this equation to make a table that gives the cost for picking up 5, 7, 9, 11, and 13 trash bags in one month. Complete the table, and then use the information in the table to construct a bar chart.

| Bags Collected | Cost (in dollars) |
|:---:|:---:|
| 5 | |
| 7 | |
| 9 | |
| 11 | |
| 13 | |

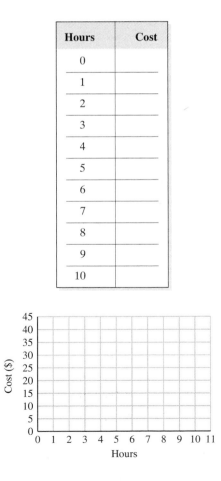

**39. Cost of Bottled Water**   A water bottling company charges $7.00 per month for their water dispenser and $1.10 for each gallon of water delivered. Write an equation in two variables that will let you calculate the total monthly charge $y$ for drinking $x$ gallons of water in a month. Then use this equation to make a table that gives the cost for drinking 15, 20, 25, 30, and 35 gallons of water in 1 month. Use the information in the table to construct a bar chart.

**40. Car Rental Cost**   A car rental company charges $21.99 per day and $0.20 per mile to rent one of their cars. Write an equation in two variables that will let you calculate the total charge to rent one of these cars for a week and drive it $x$ miles.

**41.** Find $y$ when $x$ is 4 in the formula $3x + 2y = 6$.

**42.** Find $y$ when $x$ is 0 in the formula $3x + 2y = 6$.

**43.** Find $y$ when $x$ is 0 in $y = -\dfrac{1}{3}x + 2$.

**44.** Find $y$ when $x$ is 3 in $y = -\dfrac{1}{3}x + 2$.

**45.** Find $y$ when $x$ is 2 in $y = \dfrac{3}{2}x - 3$.

**46.** Find $y$ when $x$ is 4 in $y = \dfrac{3}{2}x - 3$.

**47.** Solve $5x + y = 4$ for $y$.

**48.** Solve $-3x + y = 5$ for $y$.

**49.** Solve $3x - 2y = 6$ for $y$.

**50.** Solve $2x - 3y = 6$ for $y$.

## Review Problems

The following problems review material we covered in Section 2.5.

## 3.3   Graphing Linear Equations in Two Variables

At the end of the previous section we used a histogram to obtain a visual picture of *some* of the solutions to the equations $y = 1.5x + 15$. In this section we will use the rectangular coordinate system introduced in Section 3.1 to obtain a visual picture of *all* solutions to a linear equation in two variables. The process we use to obtain a visual picture of all solutions to an equation is called *graphing*. The picture itself is called the *graph* of the equation.

**E X A M P L E  I**   Graph the solution set for $x + y = 5$.

**Solution**   We know from the previous section that an infinite number of ordered pairs are solutions to the equation $x + y = 5$. We can't possibly list them all. What we can do is list a few of them and see if there is any pattern to their graphs.

Some ordered pairs that are solutions to $x + y = 5$ are (0, 5), (2, 3), (3, 2), (5, 0). The graph of each is shown in Figure 1.

Now, by passing a straight line through these points we can graph the solution set for the equation $x + y = 5$. Linear equations in two variables always have graphs that are straight lines. The graph of the solution set for $x + y = 5$ is shown in Figure 2.

Every ordered pair that satisfies $x + y = 5$ has its graph on the line, and any point on the line has coordinates that satisfy the equation. So, there is a one-to-one correspondence between points on the line and solutions to the equation.

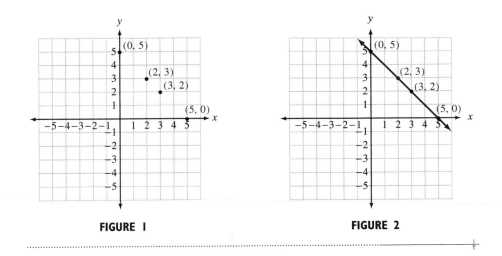

**FIGURE 1**                                                          **FIGURE 2**

Our ability to graph an equation as we have done in Example 1 is due to the invention of the rectangular coordinate system. The French philosopher René Descartes (1595–1650) is the person usually credited with the invention of the rectangular coordinate system. As a philosopher, Descartes is responsible for the statement "I think, therefore I am." Until Descartes invented his coordinate system in 1637, algebra and geometry were treated as separate subjects. The rectangular coordinate system allows us to connect algebra and geometry by associating geometric shapes with algebraic equations.

Here is the precise definition for a linear equation in two variables.

> **DEFINITION**   Any equation that can be put in the form $ax + by = c$, where $a$, $b$, and $c$ are real numbers and $a$ and $b$ are not both 0, is called a **linear equation in two variables**. The graph of any equation of this form is a straight line (that is why these equations are called "linear"). The form $ax + by = c$ is called **standard form**.

To graph a linear equation in two variables, we simply graph its solution set. That is, we draw a line through all the points whose coordinates satisfy the equation. Here are the steps to follow.

### TO GRAPH A STRAIGHT LINE

*Step 1:* Find any three ordered pairs that satisfy the equation. This can be done by using a convenient number for one variable and solving for the other variable.

*Step 2:* Graph the three ordered pairs found in step 1. Actually, we need only two points to graph a straight line. The third point serves as a check. If all three points do not line up, there is a mistake in our work.

*Step 3:* Draw a straight line through the three points graphed in step 2.

**EXAMPLE 2**    Graph the equation $y = 3x - 1$.

**Solution**    Since $y = 3x - 1$ can be put in the form $ax + by = c$, it is a linear equation in two variables. Hence, the graph of its solution set is a straight line. We can find some specific solutions by substituting numbers for $x$ and then solving for the corresponding values of $y$. We are free to choose any numbers for $x$, so let's use 0, 2, and $-1$.

Let $x = 0$:    $y = 3(0) - 1$

$y = 0 - 1$

$y = -1$

The ordered pair $(0, -1)$ is one solution.

Let $x = 2$:    $y = 3(2) - 1$

$y = 6 - 1$

$y = 5$

The ordered pair $(2, 5)$ is a second solution.

Let $x = -1$:    $y = 3(-1) - 1$

$y = -3 - 1$

$y = -4$

The ordered pair $(-1, -4)$ is a third solution.

In table form

| $x$ | $y$ |
|-----|-----|
| 0 | $-1$ |
| 2 | 5 |
| $-1$ | $-4$ |

Next, we graph the ordered pairs $(0, -1)$, $(2, 5)$, $(-1, -4)$ and draw a straight line through them.

The line we have drawn in Figure 3 is the graph of $y = 3x - 1$.

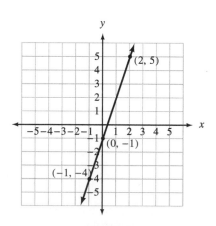

**FIGURE 3**

Example 2 again illustrates the connection between algebra and geometry that we mentioned previously. Descartes' rectangular coordinate system allows us to associate the equation $y = 3x - 1$ (an algebraic concept) with a specific straight line (a geometric concept). The study of the relationship between equations in algebra and their associated geometric figures is called *analytic geometry*. The rectangular coordinate system is often referred to as the *Cartesian coordinate system* in honor of Descartes.

**EXAMPLE 3**  Graph the equation $y = -\frac{1}{3}x + 2$.

**Solution**  We need to find three ordered pairs that satisfy the equation. To do so, we can let $x$ equal any numbers we choose and find corresponding values of $y$. Since every value of $x$ we substitute into the equation is going to be multiplied by $-\frac{1}{3}$, let's use numbers for $x$ that are divisible by 3, like $-3$, 0, and 3. That way, when we multiply them by $-\frac{1}{3}$, the result will be an integer.

Let $x = -3$:     $y = -\dfrac{1}{3}(-3) + 2$

$$y = 1 + 2$$

$$y = 3$$

The ordered pair $(-3, 3)$ is one solution.

Let $x = 0$:     $y = -\dfrac{1}{3}(0) + 2$

$$y = 0 + 2$$

$$y = 2$$

The ordered pair $(0, 2)$ is a second solution.

Let $x = 3$:     $y = -\dfrac{1}{3}(3) + 2$

$$y = -1 + 2$$

$$y = 1$$

The ordered pair $(3, 1)$ is a third solution.

In table form

| $x$ | $y$ |
|-----|-----|
| $-3$ | 3 |
| 0 | 2 |
| 3 | 1 |

Graphing the ordered pairs $(-3, 3)$, $(0, 2)$, and $(3, 1)$ and drawing a straight line through their graphs, we have the graph of the equation $y = -\frac{1}{3}x + 2$, as shown in Figure 4.

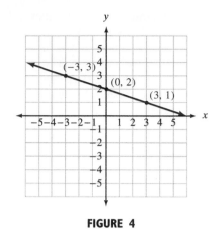

**FIGURE 4**

**Note:**    In Example 3 the values of $x$ we used, $-3$, 0, and 3, are referred to as convenient values of $x$ because they are easier to work with than some other numbers. For instance, if we let $x = 2$ in our original equation, we would have to add $-\frac{2}{3}$ and 2 to find the corresponding value of $y$. Not only would the arithmetic be more difficult, the ordered pair we obtained would have a fraction for its $y$-coordinate, making it more difficult to graph accurately.

**E X A M P L E    4**    Graph the lines $x = 2$ and $y = -3$ on the same coordinate system.

**Solution**    The line $x = 2$ is the set of all points whose $x$-coordinate is 2. When the variable $y$ does not appear in the equation, it means that $y$ can be any number. The points $(2, 0)$, $(2, 3)$, $(2, -4)$, and $(2, 2)$ therefore all satisfy the equation $x = 2$, simply because their $x$-coordinates are 2. The graph of $x = 2$, along with the points we named, is given in Figure 5.

**FIGURE 5**

The line $y = -3$ is the set of all points whose $y$-coordinate is $-3$. The points $(0, -3)$, $(4, -3)$, and $(-4, -3)$ all satisfy the equation $y = -3$, because their $y$-coordinates are all $-3$. The graph of the line $y = -3$ is also shown in Figure 5.

**EXAMPLE  5**    Graph the solution set for $3x - 2y = 6$.

**Solution**    It will be easier to find convenient values of $x$ to use in the equation if we first solve the equation for $y$. To do so, we add $-3x$ to each side, and then we multiply each side by $-\frac{1}{2}$.

$$3x - 2y = 6 \qquad\qquad \text{Original equation}$$

$$-2y = -3x + 6 \qquad\qquad \text{Add } -3x \text{ to each side.}$$

$$-\frac{1}{2}(-2y) = -\frac{1}{2}(-3x + 6) \qquad \text{Multiply each side by } -\frac{1}{2}.$$

$$y = \frac{3}{2}x - 3 \qquad\qquad \text{Simplify each side.}$$

Now, since each value of $x$ will be multiplied by $\frac{3}{2}$, it will be to our advantage to choose values of $x$ that are divisible by 2. That way, we will obtain values of $y$ that do not contain fractions. This time, let's use 0, 2, and 4 for $x$.

When $x = 0$: $\qquad y = \frac{3}{2}(0) - 3$

$$y = 0 - 3$$

$$y = -3 \qquad\qquad (0, -3) \text{ is one solution.}$$

When $x = 2$: $\qquad y = \frac{3}{2}(2) - 3$

$$y = 3 - 3$$

$$y = 0 \qquad\qquad (2, 0) \text{ is a second solution.}$$

When $x = 4$: $\qquad y = \frac{3}{2}(4) - 3$

$$y = 6 - 3$$

$$y = 3 \qquad\qquad (4, 3) \text{ is a third solution.}$$

Graphing the ordered pairs $(0, -3)$, $(2, 0)$, and $(4, 3)$ and drawing a line through them, we have the graph shown in Figure 6.

**Note:**    After reading through Example 5, many students ask why we didn't use $-2$ for $x$ when we were finding ordered pairs that were solutions to the original equation. The answer is,

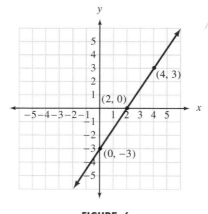

**FIGURE 6**

we could have. If we were to let $x = -2$, the corresponding value of $y$ would have been $-6$. As you can see by looking at the graph in Figure 6, the ordered pair $(-2, -6)$ is on the graph.

## Getting Ready for Class

*After reading through the preceding section, respond in your own words and in complete sentences.*

**A.** Explain how you would go about graphing the line $x + y = 5$.

**B.** When graphing straight lines, why is it a good idea to find three points, when every straight line is determined by only two points?

**C.** What kind of equations have vertical lines for graphs?

**D.** What kind of equations have horizontal lines for graphs?

## PROBLEM SET 3.3

For the following equations, complete the given ordered pairs, and use the results to graph the solution set for the equation.

**1.** $x + y = 4$     $(0, \quad), (2, \quad), (\quad, 0)$

**2.** $x - y = 3$     $(0, \quad), (2, \quad), (\quad, 0)$

**3.** $x + y = 3$     $(0, \quad), (2, \quad), (\quad, -1)$

**4.** $x - y = 4$     $(1, \quad), (-1, \quad), (\quad, 0)$

**5.** $y = 2x$     $(0, \quad), (-2, \quad), (2, \quad)$

**6.** $y = \dfrac{1}{2}x$     $(0, \quad), (-2, \quad), (2, \quad)$

**7.** $y = \dfrac{1}{3}x$     $(-3, \quad), (0, \quad), (3, \quad)$

**8.** $y = 3x$     $(-2, \quad), (0, \quad), (2, \quad)$

**9.** $y = 2x + 1$     $(0, \quad), (-1, \quad), (1, \quad)$

**10.** $y = -2x + 1$     $(0, \quad), (-1, \quad), (1, \quad)$

**11.** $y = 4$     $(0, \quad), (-1, \quad), (2, \quad)$

**12.** $x = 3$     $(\quad, -2), (\quad, 0), (\quad, 5)$

**13.** $y = \dfrac{1}{2}x + 3$     $(-2, \quad), (0, \quad), (2, \quad)$

**14.** $y = \dfrac{1}{2}x - 3$      $(-2, \quad), (0, \quad), (2, \quad)$

**15.** $y = -\dfrac{2}{3}x + 1$      $(-3, \quad), (0, \quad), (3, \quad)$

**16.** $y = -\dfrac{2}{3}x - 1$      $(-3, \quad), (0, \quad), (3, \quad)$

Solve each equation for $y$. Then, complete the given ordered pairs, and use them to draw the graph.

**17.** $2x + y = 3$      $(-1, \quad), (0, \quad), (1, \quad)$
**18.** $3x + y = 2$      $(-1, \quad), (0, \quad), (1, \quad)$
**19.** $3x + 2y = 6$      $(0, \quad), (2, \quad), (4, \quad)$
**20.** $2x + 3y = 6$      $(0, \quad), (3, \quad), (6, \quad)$
**21.** $-x + 2y = 6$      $(-2, \quad), (0, \quad), (2, \quad)$
**22.** $-x + 3y = 6$      $(-3, \quad), (0, \quad), (3, \quad)$

Find three solutions to each of the following equations, and then graph the solution set.

**23.** $y = -\dfrac{1}{2}x$          **24.** $y = -2x$

**25.** $y = 3x - 1$          **26.** $y = -3x - 1$

**27.** $-2x + y = 1$          **28.** $-3x + y = 1$

**29.** $3x + 4y = 8$          **30.** $3x - 4y = 8$

**31.** $x = -2$          **32.** $y = 3$

**33.** $y = 2$          **34.** $x = -3$

**35. Perimeter**  If the perimeter of a rectangle is 10 inches, then the equation that describes the relationship between the length $l$ and width $w$ is

$$2l + 2w = 10$$

Graph this equation using a coordinate system in which the horizontal axis is labeled $l$ and the vertical axis is labeled $w$.

**36. Perimeter**  The perimeter of a rectangle is 6 inches. Graph the equation that describes the relationship between the length $l$ and width $w$.

**37.** The solutions to the equation $y = x$ are ordered pairs of the form $(x, x)$. The $x$- and $y$-coordinates are equal. Graph the line $y = x$.

**38.** The solutions to a certain equation are ordered pairs of the form $(x, 2x)$. Find the equation and its graph.

**39.** Use the graph below to complete the table.

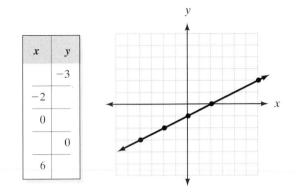

| $x$ | $y$ |
|---|---|
| | $-3$ |
| $-2$ | |
| $0$ | |
| | $0$ |
| $6$ | |

**40.** Use the graph below to complete the table. (*Hint:* Some parts have two answers.)

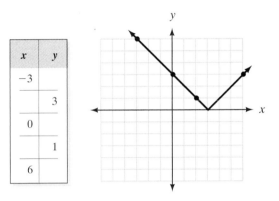

| $x$ | $y$ |
|---|---|
| $-3$ | |
| | $3$ |
| $0$ | |
| | $1$ |
| $6$ | |

**41.** The $y$-coordinate of each ordered pair that satisfies the equation $y = |x|$ will never be negative. Each value of $y$ is the absolute value of the corresponding value of $x$. Fill in the following ordered pairs so that they are solutions to the equation $y = |x|$. Then graph them and connect their graphs in a way that makes the most sense to you.

$(-3, \quad)(-2, \quad)(-1, \quad)(0, \quad)(1, \quad)(2, \quad)(3, \quad)$

**42.** Fill in the following ordered pairs so that they are solutions to the equation $y = |x - 2|$. Graph each point, and then connect the graphs in a way that makes the most sense to you.

$(-1, \quad)(0, \quad)(1, \quad)(2, \quad)(3, \quad)(4, \quad)(5, \quad)$

**43.** Graph the lines $y = x + 1$ and $y = x - 3$ on the same coordinate system. Can you tell from looking at these first two graphs where the graph of $y = x + 3$ would be?

**44.** Graph the lines $y = 2x + 2$ and $y = 2x - 1$ on the same coordinate system. Use the similarities between these two graphs to graph the line $y = 2x - 4$.

## Applying the Concepts

**45. Airlines**   Ocean Wide Airlines is leaving LAX going to New York. The airplane is on the ground at the airport. The plane takes off and maintains a constant climb rate. The airplane leaves from the airport, (0, 0) on the coordinate system, and reaches its cruising altitude of 35,000 feet (approximately 6.6 miles) when it is 90 miles from the airport. The equation for the plane's altitude $y$, when it is $x$ miles away from the airport, is $y = \dfrac{175}{2,376}x$.

Use the equation to fill in the following table, rounding your answers to the nearest tenth of a mile. Use the results to graph the equation on the template. (The graph will appear in the first quadrant only because $x$ is greater than or equal to 0.)

| Distance From Airport (miles) | Height (miles) |
|---|---|
| 0 | |
| 15 | |
| 30 | |
| 45 | |
| 60 | |
| 75 | |
| 90 | |

**46. Airlines**   Ocean Wide Airlines is flying into LAX from Hawaii. The airplane has an altitude of 30,000

feet (approximately 5.7 miles) and is 65 miles from the airport when it begins its descent to the airport. The equation for the airplane's flight path during its descent is given by $y = -\dfrac{57}{650}x$.

Use the equation to fill in the following table, rounding your answers to the nearest tenth of a mile. Then use your results to graph the equation on the template. (The graph will appear in the second quadrant only because we are giving its distance from the airport in negative numbers.)

| Distance From Airport (miles) | Height (miles) |
|---|---|
| −65 | |
| −55 | |
| −45 | |
| −35 | |
| −25 | |
| −15 | |
| −5 | |
| 0 | |

**47. Taxes**   Suppose that 21% of your monthly pay is withheld for federal income taxes and another 8% is withheld for Social Security, state income tax, and other miscellaneous items. If $G$ is your monthly pay before any money is deducted (your gross pay), the amount of money that you take home each month is given by $T = 0.71G$. Graph this equation on a coordinate system in which the horizontal axis is the $G$-axis and the vertical axis is the $T$-axis. Start your graph at $G = 0$ and end it at $G = 2,500$.

**48. Temperature and Altitude**  As we mentioned previously, there is a relationship between air temperature and altitude. The higher above the Earth we are, the lower the temperature. If the temperature on the ground is 56°F, then the temperature $T$ (in degrees Fahrenheit), at an altitude of $A$ feet, is $T = -0.0035A + 56$. Graph this equation on a coordinate system in which the horizontal axis is the $A$-axis and the vertical axis is the $T$-axis. Start your graph at $A = 0$ and end it at $A = 40{,}000$.

## Review Problems

The following problems review material we covered in Sections 2.3 and 2.4.

Solve each equation.

**49.** $3(x - 2) = 9$

**50.** $-4(x - 3) = -16$

**51.** $2(3x - 1) + 4 = -10$

**52.** $-5(2x + 3) - 10 = 15$

**53.** $6 - 2(4x - 7) = -4$

**54.** $5 - 3(2 - 3x) = 8$

**55.** $\dfrac{1}{2}x + 4 = \dfrac{2}{3}x + 5$

**56.** $\dfrac{1}{4}x - 3 = \dfrac{3}{2}x + 7$

## 3.4  More on Graphing: Intercepts

In this section we continue our work with graphing lines by finding the points where a line crosses the axes of our coordinate system. To do so, we use the fact that any point on the $x$-axis has a $y$-coordinate of 0, and any point on the $y$-axis has an $x$-coordinate of 0. We begin with the following definition.

> **DEFINITION**  The *x-intercept* of a straight line is the $x$-coordinate of the point where the graph crosses the $x$-axis. The **y-intercept** is defined similarly. It is the $y$-coordinate of the point where the graph crosses the $y$-axis.

If the $x$-intercept is $a$, then the point $(a, 0)$ lies on the graph. (This is true because any point on the $x$-axis has a $y$-coordinate of 0.)

If the $y$-intercept is $b$, then the point $(0, b)$ lies on the graph. (This is true because any point on the $y$-axis has an $x$-coordinate of 0.)

Graphically, the relationship is shown in Figure 1.

**FIGURE 1**

**EXAMPLE 1**     Find the $x$- and $y$-intercepts for $3x - 2y = 6$, and then use them to draw the graph.

**Solution**     To find where the graph crosses the $x$-axis, we let $y = 0$. (The $y$-coordinate of any point on the $x$-axis is 0.)

*x-intercept:*

When $\qquad\qquad\qquad\qquad\ y = 0$

the equation $\qquad\quad 3x - 2y = 6$

becomes $\qquad\qquad 3x - 2(0) = 6$

$\qquad\qquad\qquad\qquad 3x - 0 = 6$

$\qquad\qquad\qquad\qquad\qquad x = 2 \qquad$ Multiply each side by $\frac{1}{3}$.

The graph crosses the $x$-axis at $(2, 0)$, which means the $x$-intercept is 2.

*y-intercept:*

When $\qquad\qquad\qquad\qquad\ x = 0$

the equation $\qquad\quad 3x - 2y = 6$

becomes $\qquad\qquad 3(0) - 2y = 6$

$\qquad\qquad\qquad\qquad 0 - 2y = 6$

$\qquad\qquad\qquad\qquad\quad -2y = 6$

$\qquad\qquad\qquad\qquad\qquad y = -3 \qquad$ Multiply each side by $-\frac{1}{2}$.

The graph crosses the $y$-axis at $(0, -3)$, which means the $y$-intercept is $-3$.

Plotting the $x$- and $y$-intercepts and then drawing a line through them, we have the graph of $3x - 2y = 6$, as shown in Figure 2.

**FIGURE 2**

**E X A M P L E   2**   Graph $-x + 2y = 4$ by finding the intercepts and using them to draw the graph.

**Solution**   Again, we find the x-intercept by letting $y = 0$ in the equation and solving for x. Similarly, we find the y-intercept by letting $x = 0$ and solving for y.

*x-intercept:*

$$\text{When} \qquad\qquad\qquad y = 0$$

$$\text{the equation} \qquad -x + 2y = 4$$

$$\text{becomes} \qquad -x + 2(0) = 4$$

$$-x + 0 = 4$$

$$-x = 4$$

$$x = -4 \qquad \text{Multiply each side by } -1.$$

The x-intercept is $-4$, indicating that the point $(-4, 0)$ is on the graph of $-x + 2y = 4$.

*y-intercept:*

$$\text{When} \qquad\qquad\qquad x = 0$$

$$\text{the equation} \qquad -x + 2y = 4$$

$$\text{becomes} \qquad -0 + 2y = 4$$

$$2y = 4$$

$$y = 2 \qquad \text{Multiply each side by } \tfrac{1}{2}.$$

The y-intercept is 2, indicating that the point $(0, 2)$ is on the graph of $-x + 2y = 4$.

   Plotting the intercepts and drawing a line through them, we have the graph of $-x + 2y = 4$, as shown in Figure 3.

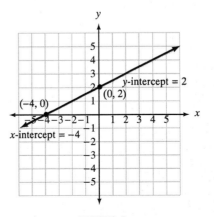

**FIGURE 3**

Graphing a line by finding the intercepts, as we have done in Examples 1 and 2, is an easy method of graphing if the equation has the form $ax + by = c$ and both the numbers $a$ and $b$ divide the number $c$ evenly.

In our next example we use the intercepts to graph a line in which $y$ is given in terms of $x$.

**EXAMPLE 3**    Use the intercepts for $y = -\frac{1}{3}x + 2$ to draw its graph.

**Solution**    We graphed this line previously in Example 3 of Section 3.3 by substituting three different values of $x$ into the equation and solving for $y$. This time we will graph the line by finding the intercepts.

***x-intercept:***

When $\qquad\qquad y = 0$

the equation $\qquad y = -\dfrac{1}{3}x + 2$

becomes $\qquad 0 = -\dfrac{1}{3}x + 2$

$\qquad\qquad -2 = -\dfrac{1}{3}x \qquad\qquad$ Add $-2$ to each side.

$\qquad\qquad 6 = x \qquad\qquad\qquad$ Multiply each side by $-3$.

The $x$-intercept is 6, which means the graph passes through the point $(6, 0)$.

***y-intercept:***

When $\qquad\qquad x = 0$

the equation $\qquad y = -\dfrac{1}{3}x + 2$

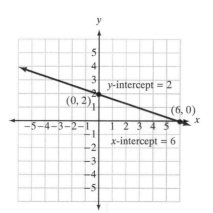

**FIGURE 4**

becomes $\qquad y = -\dfrac{1}{3}(0) + 2$

$$y = 2$$

The $y$-intercept is 2, which means the graph passes through the point $(0, 2)$.

The graph of $y = -\dfrac{1}{3}x + 2$ is shown in Figure 4. Compare this graph, and the method used to obtain it, with Example 3 in Section 3.3.

## Getting Ready for Class

*After reading through the preceding section, respond in your own words and in complete sentences.*

**A.** What is the $x$-intercept for a graph?

**B.** What is the $y$-intercept for a graph?

**C.** How do we find the $y$-intercept for a line from the equation?

**D.** How do we graph a line using its intercepts?

## PROBLEM SET 3.4

Find the $x$- and $y$-intercepts for the following equations. Then use the intercepts to graph each equation.

**1.** $2x + y = 4$

**2.** $2x + y = 2$

**3.** $-x + y = 3$

**4.** $-x + y = 4$

**5.** $-x + 2y = 2$

**6.** $-x + 2y = 4$

**7.** $5x + 2y = 10$

**8.** $2x + 5y = 10$

**9.** $4x - 2y = 8$

**10.** $2x - 4y = 8$

**11.** $-4x + 5y = 20$

**12.** $-5x + 4y = 20$

**13.** $y = 2x - 6$

**14.** $y = 2x + 6$

**15.** $y = 2x + 2$

**16.** $y = -2x + 2$

**17.** $y = 2x - 1$

**18.** $y = -2x - 1$

**19.** $y = \dfrac{1}{2}x + 3$

**20.** $y = \dfrac{1}{2}x - 3$

**21.** $y = -\dfrac{1}{3}x - 2$

**22.** $y = -\dfrac{1}{3}x + 2$

For each of the following lines the $x$-intercept and the $y$-intercept are both 0, which means the graph of each will go through the origin, $(0, 0)$. Graph each line by finding a point on each, other than the origin, and then drawing a line through that point and the origin.

**23.** $y = -2x$

**24.** $y = \dfrac{1}{2}x$

**25.** $y = 2x$

**26.** $y = -\dfrac{1}{2}x$

**27.** $y = \dfrac{1}{3}x$

**28.** $y = 3x$

**29.** $y = -\dfrac{1}{3}x$

**30.** $y = -3x$

**31.** $y = \dfrac{2}{3}x$

**32.** $y = \dfrac{3}{2}x$

**33.** Graph the line that passes through the point $(-4, 4)$ and has an $x$-intercept of $-2$. What is the $y$-intercept of this line?

**34.** Graph the line that passes through the point $(-3, 4)$ and has a $y$-intercept of 3. What is the $x$-intercept of this line?

**35.** A line passes through the point (1, 4) and has a *y*-intercept of 3. Graph the line and name its *x*-intercept.

**36.** A line passes through the point (3, 4) and has an *x*-intercept of 1. Graph the line and name its *y*-intercept.

**37.** Graph the line that passes through the points (−2, 5) and (5, −2). What are the *x*- and *y*-intercepts for this line?

**38.** Graph the line that passes through the points (5, 3) and (−3, −5). What are the *x*- and *y*-intercepts for this line?

**39.** Use the graph below to complete the following table.

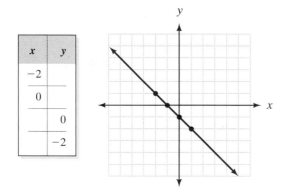

| *x* | *y* |
|-----|-----|
| −2  |     |
| 0   |     |
|     | 0   |
|     | −2  |

**40.** Use the graph below to complete the following table.

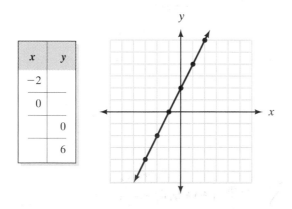

| *x* | *y* |
|-----|-----|
| −2  |     |
| 0   |     |
|     | 0   |
|     | 6   |

**41.** The vertical line *x* = 3 has only one intercept. Graph *x* = 3, and name its intercept. (Remember, ordered pairs (*x*, *y*) that are solutions to the equa-

tion *x* = 3 are ordered pairs with an *x*-coordinate of 3 and any *y*-coordinate.)

**42.** Graph the vertical line *x* = −2. Then name its intercept.

**43.** The horizontal line *y* = 4 has only one intercept. Graph *y* = 4, and name its intercept. (Ordered pairs (*x*, *y*) that are solutions to the equation *y* = 4 are ordered pairs with a *y*-coordinate of 4 and any *x*-coordinate.)

**44.** Graph the horizontal line *y* = −3. Then name its intercept.

## Applying the Concepts

**45. Complementary Angles**   The following diagram shows sunlight hitting the ground. Angle α (alpha) is called the angle of inclination, and angle θ (theta) is called the angle of incidence. As the sun moves across the sky, the values of these angles change. Assume that α + θ = 90, where both α and θ are in degrees measure. Graph this equation on a coordinate system where the horizontal axis is the α-axis and the vertical axis is the θ-axis. Find the intercepts first, and limit your graph to the first quadrant only.

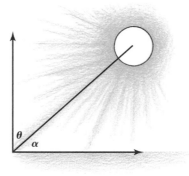

**46. Light**   When light comes into contact with an impenetrable object, such as a thick piece of wood or metal, it is reflected or absorbed, but not transmitted, as shown in the following diagram. If we let *R* represent the percentage of light reflected and *A* the percentage of light absorbed by a surface, then the relationship between *R* and *A* is *R* + *A* = 100. Graph this equation on a coordinate system where the horizontal axis is the *A*-axis and the vertical axis is the *R*-axis. Find the intercepts first, and limit your graph to the first quadrant.

## Review Problems

The following problems review material we covered in Section 2.8.

Solve each inequality.

**47.** $-3x \geq 12$

**48.** $-2x > 10$

**49.** $-\dfrac{x}{3} \leq -1$

**50.** $-\dfrac{x}{5} < -2$

**51.** $-4x + 1 < 17$

**52.** $-3x + 2 \leq -7$

---

| **3.5** | **The Slope of a Line** |
| --- | --- |

In defining the slope of a straight line, we are looking for a number to associate with a straight line that does two things. First of all, we want the slope of a line to measure the "steepness" of the line. That is, in comparing two lines, the slope of the steeper line should have the larger numerical value. Second, we want a line that *rises* going from left to right to have a *positive* slope. We want a line that *falls* going from left to right to have a *negative* slope. (A line that neither rises nor falls going from left to right must, therefore, have 0 slope.) These are illustrated in Figure 1.

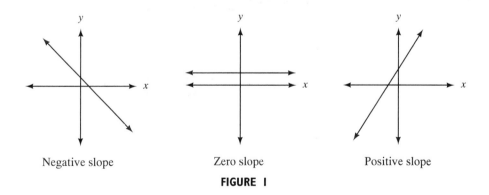

Negative slope          Zero slope          Positive slope

**FIGURE 1**

Suppose we know the coordinates of two points on a line. Since we are trying to develop a general formula for the slope of a line, we will use general points — call the two points $P_1\,(x_1, y_1)$ and $P_2\,(x_2, y_2)$. They represent the coordinates of any two different points on our line. We define the *slope* of our line to be the ratio of the vertical change to the horizontal change as we move from point $(x_1, y_1)$ to point $(x_2, y_2)$ on the line. (See Figure 2.)

**Note:**   The 2 in $x_2$ is called a *subscript*. It is notation that allows us to distinguish between the variables $x_1$ and $x_2$, while still showing that they are both *x*-coordinates.

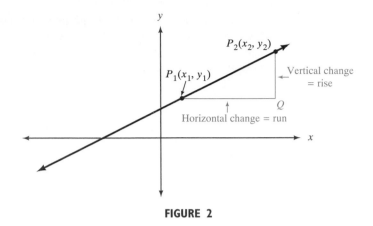

**FIGURE 2**

We call the vertical change the *rise* in the graph and the horizontal change the *run* in the graph. The slope, then, is

$$\text{Slope} = \frac{\text{vertical change}}{\text{horizontal change}} = \frac{\text{rise}}{\text{run}}$$

We would like to have a numerical value to associate with the rise in the graph and a numerical value to associate with the run in the graph. A quick study of Figure 2 shows that the coordinates of point $Q$ must be $(x_2, y_1)$, since $Q$ is directly below point $P_2$ and right across from point $P_1$. We can draw our diagram again in the manner shown in Figure 3. It is apparent from this graph that the rise can be expressed as $(y_2 - y_1)$ and the run as $(x_2 - x_1)$. We usually denote the slope of a line by the letter $m$. The complete definition of slope follows along with a diagram (Figure 3) that illustrates the definition.

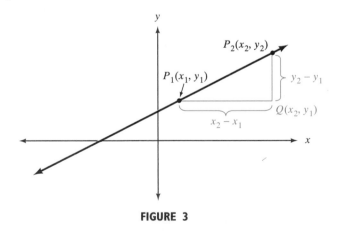

**FIGURE 3**

**DEFINITION**    If points $(x_1, y_1)$ and $(x_2, y_2)$ are any two different points, then the **slope** of the line on which they lie is

$$\text{Slope} = m = \frac{\text{rise}}{\text{run}} = \frac{y_2 - y_1}{x_2 - x_1}$$

This definition of the slope of a line does just what we want it to do. If the line rises going from left to right, the slope will be positive. If the line falls from left to right, the slope will be negative. Also, the steeper the line, the larger the numerical value the slope will have.

**EXAMPLE  1**    Find the slope of the line between the points $(1, 2)$ and $(3, 5)$.

**Solution**    We can let

$$(x_1, y_1) = (1, 2)$$

and

$$(x_2, y_2) = (3, 5)$$

then

$$m = \frac{y_2 - y_1}{x_2 - x_1} = \frac{5 - 2}{3 - 1} = \frac{3}{2}$$

The slope is $\frac{3}{2}$. For every vertical change of 3 units, there will be a corresponding horizontal change of 2 units. (See Figure 4.)

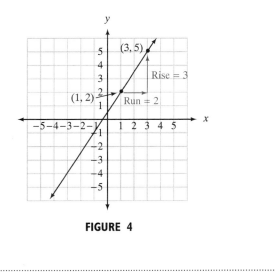

**FIGURE 4**

**EXAMPLE  2**    Find the slope of the line through $(-2, 1)$ and $(5, -4)$.

**Solution**    It makes no difference which ordered pair we call $(x_1, y_1)$ and which we call $(x_2, y_2)$.

$$\text{Slope} = m = \frac{y_2 - y_1}{x_2 - x_1} = \frac{-4 - 1}{5 - (-2)} = \frac{-5}{7}$$

The slope is $-\frac{5}{7}$. Every vertical change of $-5$ units (down 5 units) is accompanied by a horizontal change of 7 units (to the right 7 units). (See Figure 5.)

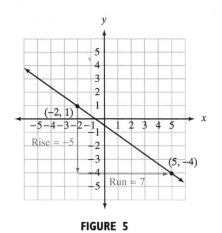

**FIGURE 5**

**E X A M P L E   3**     Graph the line with slope $\frac{3}{2}$ and $y$-intercept 1.

**Solution**     Since the $y$-intercept is 1, we know that one point on the line is (0, 1). So, we begin by plotting the point (0, 1), as shown in Figure 6.

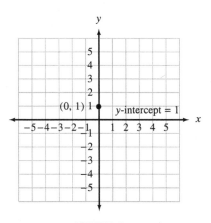

**FIGURE 6**

Many lines pass through the point shown in Figure 6, but only one of those lines has a slope of $\frac{3}{2}$. The slope, $\frac{3}{2}$, can be thought of as the rise in the graph divided by the run in the graph. Therefore, if we start at the point (0, 1) and move 3 units up (that's a rise of 3) and then 2 units to the right (a run of 2), we will be at another point on the graph. Figure 7 shows that the point we reach by doing so is the point (2, 4).

$$\text{Slope} = m = \frac{\text{rise}}{\text{run}} = \frac{3}{2}$$

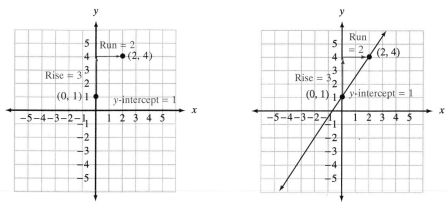

**FIGURE 7**                              **FIGURE 8**

To graph the line with slope $\frac{3}{2}$ and $y$-intercept 1, we simply draw a line through the two points in Figure 7 to obtain the graph shown in Figure 8.

# Getting Ready for Class

*After reading through the preceding section, respond in your own words and in complete sentences.*

**A.** What is the slope of a line?

**B.** Would you rather climb a hill with a slope of 1 or a slope of 3? Explain why.

**C.** Describe how to obtain the slope of a line if you know the coordinates of two points on the line.

**D.** Describe how you would graph a line from its slope and $y$-intercept.

# PROBLEM SET 3.5

Find the slope of the line through the following pairs of points. Then plot each pair of points, draw a line through them, and indicate the rise and run in the graph in the manner shown in Examples 1 and 2.

**1.** $(2, 1), (4, 4)$                **2.** $(3, 1), (5, 4)$

**3.** $(1, 4), (5, 2)$                **4.** $(1, 3), (5, 2)$

**5.** $(1, -3), (4, 2)$              **6.** $(2, -3), (5, 2)$

**7.** $(-3, -2), (1, 3)$            **8.** $(-3, -1), (1, 4)$

**9.** $(-3, 2), (3, -2)$            **10.** $(-3, 3), (3, -1)$

**11.** $(2, -5), (3, -2)$           **12.** $(2, -4), (3, -1)$

In each of the following problems, graph the line with the given slope and $y$-intercept $b$.

**13.** $m = \dfrac{2}{3}, b = 1$                **14.** $m = \dfrac{3}{4}, b = -2$

**15.** $m = \dfrac{3}{2}, b = -3$               **16.** $m = \dfrac{4}{3}, b = 2$

**17.** $m = -\dfrac{4}{3}, b = 5$              **18.** $m = -\dfrac{3}{5}, b = 4$

**19.** $m = 2, b = 1$                         **20.** $m = -2, b = 4$

**21.** $m = 3, b = -1$                        **22.** $m = 3, b = -2$

Find the slope and $y$-intercept for each line.

**23.**

**24.**

**25.**

**26.**

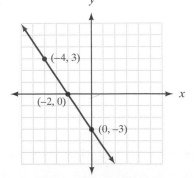

**27.** Graph the line that has an $x$-intercept of 3 and a $y$-intercept of $-2$. What is the slope of this line?

**28.** Graph the line that has an $x$-intercept of 2 and a $y$-intercept of $-3$. What is the slope of this line?

**29.** Graph the line with $x$-intercept 4 and $y$-intercept 2. What is the slope of this line?

**30.** Graph the line with $x$-intercept $-4$ and $y$-intercept $-2$. What is the slope of this line?

**31.** Graph the line $y = 2x - 3$, then name the slope and $y$-intercept by looking at the graph.

**32.** Graph the $y = -2x + 3$, then name the slope and $y$-intercept by looking at the graph.

**33.** Graph the line $y = \frac{1}{2}x + 1$, then name the slope and $y$-intercept by looking at the graph.

**34.** Graph the line $y = -\frac{1}{2}x - 2$, then name the slope and $y$-intercept by looking at the graph.

**35.** Find $y$ if the line through $(4, 2)$ and $(6, y)$ has a slope of 2.

**36.** Find $y$ if the line through $(1, y)$ and $(7, 3)$ has a slope of 6.

## Applying the Concepts

**37. Garbage Production**  The table and completed line graph from Problem 51 in Problem Set 3.1 are shown here. Each gives the annual production of

| Year | Garbage (million of tons) |
|------|---------------------------|
| 1960 | 88 |
| 1970 | 121 |
| 1980 | 152 |
| 1990 | 205 |
| 1997 | 217 |

garbage in the United States for some specific years. Find the slope of each of the four line segments, $A$, $B$, $C$, and $D$.

**38. Grass Height** The table and completed line graph from Problem 52 in Problem Set 3.1 are shown here. Each gives the growth of a certain species over time. Find the slopes of the line segments labeled $A$, $B$, and $C$.

| Day | Plant Height |
|-----|--------------|
| 0 | 0 |
| 1 | 0.5 |
| 2 | 1 |
| 3 | 1.5 |
| 4 | 3 |
| 5 | 4 |
| 6 | 6 |
| 7 | 9 |
| 8 | 13 |
| 9 | 18 |
| 10 | 23 |

**Overall Plant Height**

**39. Wireless Phone Costs** The table from Problem 53 in Problem Set 3.1 and a line graph are shown here. Each shows the projected cost of wireless phone use. Find the slope of each of the three line segments, $A$, $B$, and $C$.

| Year | Cents per Minute |
|------|------------------|
| 1998 | 33 |
| 1999 | 28 |
| 2000 | 25 |
| 2001 | 23 |
| 2002 | 22 |
| 2003 | 20 |

**Wireless Talk**

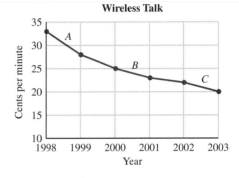

**40. Digital Camera Sales** The table from Problem 54 in Problem Set 3.1 and a line graph are shown below. Each shows the sales of digital cameras from 1996 to 1999. Find the slopes of line segments $A$, $B$, and $C$.

| Year | Sales (in millions) |
|------|---------------------|
| 1996 | $386 |
| 1997 | $518 |
| 1998 | $573 |
| 1999 | $819 |

**Digital Camera Sales**

## Review Problems

The following problems review material we covered in Section 2.1.

Find the value of each expression when $x$ is $-3$.

**41.** $2x - 9$

**42.** $-4x + 3$

**43.** $9 - 6x$

**44.** $7 - 5x$

**45.** $4(3x + 2) + 1$

**46.** $3(6x + 5) + 10$

**47.** $2x^2 + 3x + 4$

**48.** $4x^2 + 3x + 2$

---

## 3.6   Finding the Equation of a Line

To this point in the chapter, most of the problems we have worked have used the equation of a line to find different types of information about the line. For instance, given the equation of a line, we can find points on the line, the graph of the line, the intercepts, and the slope of the line. In this section we reverse things somewhat and move in the other direction; we will use information about a line, such as its slope and $y$-intercept, to find the equation of a line.

There are three main types of problems to solve in this section.

**1.** Find the equation of a line from the slope and $y$-intercept.

**2.** Find the equation of a line given one point on the line and the slope of the line.

**3.** Find the equation of a line given two points on the line.

Examples 1 and 2 illustrate the first type of problem. Example 5 solves the second type of problem. The third type of problem is solved in Example 6.

### THE SLOPE-INTERCEPT FORM OF AN EQUATION OF A STRAIGHT LINE

**E X A M P L E   I**   Find the equation of the line with slope $\frac{3}{2}$ and $y$-intercept 1.

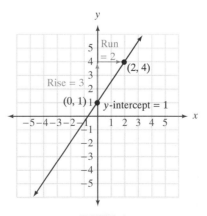

**FIGURE I**

**Solution**    We graphed the line with slope $\frac{3}{2}$ and $y$-intercept 1 in the last example of the previous section. Figure 1 shows that graph.

What we want to do now is find the equation of the line shown in Figure 1. To do so, we take any other point $(x, y)$ on the line and apply our slope formula to that point and the point $(0, 1)$. We set that result equal to $\frac{3}{2}$, since $\frac{3}{2}$ is the slope of our line. The work is as follows, with a diagram of the situation following.

$$\frac{y - 1}{x - 0} = \frac{3}{2} \qquad \text{Slope} = \frac{\text{vertical change}}{\text{horizontal change}}$$

$$\frac{y - 1}{x} = \frac{3}{2} \qquad x - 0 = x$$

$$y - 1 = \frac{3}{2}x \qquad \text{Multiply each side by } x.$$

$$y = \frac{3}{2}x + 1 \qquad \text{Add 1 to each side.}$$

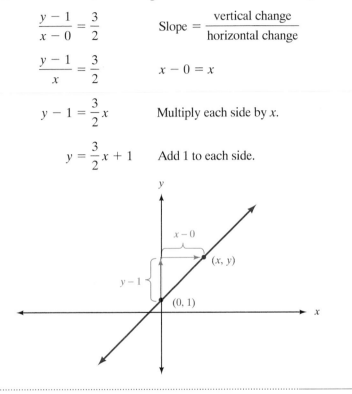

What is interesting and useful about the equation we have just found is that the number in front of $x$ is the slope of the line and the constant term is the $y$-intercept. It is no coincidence that it turned out this way. Whenever an equation has the form $y = mx + b$, the graph is always a straight line with slope $m$ and $y$-intercept $b$. To see that this is true in general, suppose we want the equation of a line with slope $m$ and $y$-intercept $b$. Since the $y$-intercept is $b$, then the point $(0, b)$ is on the line. If $(x, y)$ is any other point on the line, then we apply our slope formula to get

$$\frac{y - b}{x - 0} = m \qquad \text{Slope} = \frac{\text{vertical change}}{\text{horizontal change}}$$

$$\frac{y - b}{x} = m \qquad x - 0 = x$$

$$y - b = mx \qquad \text{Multiply each side by } x.$$

$$y = mx + b \qquad \text{Add } b \text{ to each side.}$$

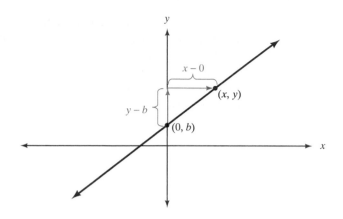

Here is a summary of what we have just found.

## SLOPE-INTERCEPT FORM OF THE EQUATION OF A LINE

The equation of the line with slope $m$ and $y$-intercept $b$ is always given by

$$y = mx + b$$

**E X A M P L E   2**     Find the equation of the line with slope $-\frac{4}{3}$ and $y$-intercept 5. Then, graph the line.

**Solution**     Substituting $m = -\frac{4}{3}$ and $b = 5$ into the equation $y = mx + b$, we have

$$y = -\frac{4}{3}x + 5$$

Finding the equation from the slope and $y$-intercept is just that easy. If the slope is $m$ and the $y$-intercept is $b$, then the equation is always $y = mx + b$.

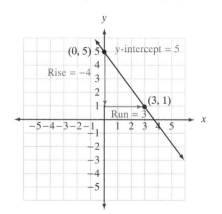

**FIGURE 2**

Since the *y*-intercept is 5, the graph goes through the point (0, 5). To find a second point on the graph, we start at (0, 5) and move 4 units down (that's a rise of −4) and 3 units to the right (a run of 3). The point we reach is (3, 1). Drawing a line that passes through (0, 5) and (3, 1), we have the graph of our equation. (Note that we could also let the rise = 4 and the run = −3 and obtain the same graph.) The graph is shown in Figure 2.

**E X A M P L E   3**   Find the slope and *y*-intercept for $-2x + y = -4$. Then, use them to draw the graph.

**Solution**   To identify the slope and *y*-intercept from the equation, the equation must be in the form $y = mx + b$ (slope-intercept form). To write our equation in this form, we must solve the equation for *y*. To do so, we simply add $2x$ to each side of the equation.

$$-2x + y = -4 \qquad \text{Original equation}$$

$$y = 2x - 4 \qquad \text{Add } 2x \text{ to each side.}$$

The equation is now in slope-intercept form, so the slope must be 2 and the *y*-intercept −4. The graph, therefore, crosses the *y*-axis at (0, −4). Since the slope is 2, we can let the rise = 2 and the run = 1 and find a second point on the graph. The graph is shown in Figure 3.

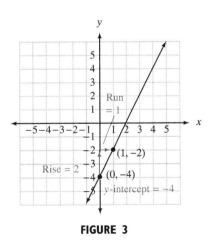

**FIGURE 3**

**E X A M P L E   4**   Find the slope and *y*-intercept for $3x - 2y = 6$.

**Solution**   To find the slope and *y*-intercept from the equation, we must write the equation in the form $y = mx + b$. This means we must solve the equation $3x - 2y = 6$ for *y*.

$$3x - 2y = 6 \qquad\qquad \text{Original equation}$$

$$-2y = -3x + 6 \qquad \text{Add } -3x \text{ to each side.}$$

$$-\frac{1}{2}(-2y) = -\frac{1}{2}(-3x + 6) \qquad \text{Multiply each side by } -\frac{1}{2}.$$

$$y = \frac{3}{2}x - 3 \qquad \text{Simplify each side.}$$

Now that the equation is written in slope-intercept form, we can identify the slope as $\frac{3}{2}$ and the $y$-intercept as $-3$. The graph is shown in Figure 4.

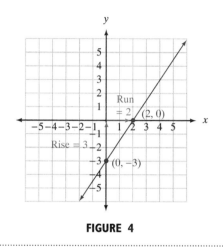

**FIGURE 4**

## THE POINT-SLOPE FORM OF AN EQUATION OF A STRAIGHT LINE

A second useful form of the equation of a straight line is the point-slope form.

Let line $l$ contain the point $(x_1, y_1)$ and have slope $m$. If $(x, y)$ is any other point on $l$, then by the definition of slope we have

$$\frac{y - y_1}{x - x_1} = m$$

Multiplying both sides by $(x - x_1)$ gives us

$$(x - x_1) \cdot \frac{y - y_1}{x - x_1} = m(x - x_1)$$

$$y - y_1 = m(x - x_1)$$

This last equation is known as the *point-slope form* of the equation of a straight line.

## POINT-SLOPE FORM OF THE EQUATION OF A LINE

The equation of the line through $(x_1, y_1)$ with slope $m$ is given by

$$y - y_1 = m(x - x_1)$$

This form is used to find the equation of a line, either given one point on the line and the slope, or two points on the line.

**E X A M P L E   5**    Find the equation of the line with slope $-2$ that contains the point $(-4, 3)$. Write the answer in slope-intercept form.

**Solution**

Using        $(x_1, y_1) = (-4, 3)$   and   $m = -2$

in           $y - y_1 = m(x - x_1)$        Point-slope form

gives us      $y - 3 = -2(x + 4)$        Note: $x - (-4) = x + 4$

$y - 3 = -2x - 8$        Multiply out right side.

$y = -2x - 5$        Add 3 to each side.

Figure 5 is the graph of the line that contains $(-4, 3)$ and has a slope of $-2$. Notice that the $y$-intercept on the graph matches that of the equation we found.

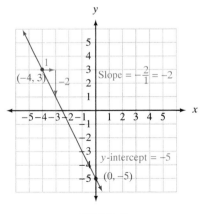

**FIGURE 5**

**E X A M P L E   6**    Find the equation of the line that passes through the points $(-3, 3)$ and $(3, -1)$.

**Solution**  We begin by finding the slope of the line:

$$m = \frac{3 - (-1)}{-3 - 3} = \frac{4}{-6} = -\frac{2}{3}$$

Using $(x_1, y_1) = (3, -1)$ and $m = -\frac{2}{3}$ in $y - y_1 = m(x - x_1)$ yields

$$y + 1 = -\frac{2}{3}(x - 3)$$

$$y + 1 = -\frac{2}{3}x + 2 \qquad \text{Multiply out right side.}$$

$$y = -\frac{2}{3}x + 1 \qquad \text{Add } -1 \text{ to each side.}$$

Figure 6 shows the graph of the line that passes through the points $(-3, 3)$ and $(3, -1)$. As you can see, the slope and $y$-intercept are $-\frac{2}{3}$ and 1, respectively.

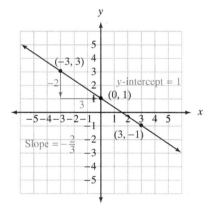

**FIGURE 6**

**Note:**  In Example 6 we could have used the point $(-3, 3)$ instead of $(3, -1)$ and obtained the same equation. That is, using $(x_1, y_1) = (-3, 3)$ and $m = -\frac{2}{3}$ in $y - y_1 = m(x - x_1)$ gives us

$$y - 3 = -\frac{2}{3}(x + 3)$$

$$y - 3 = -\frac{2}{3}x - 2$$

$$y = -\frac{2}{3}x + 1$$

which is the same result we obtained using $(3, -1)$.

### METHODS OF GRAPHING LINES

1. Substitute convenient values of $x$ into the equation, and find the corresponding values of $y$. We used this method first for equations like $y = 2x - 3$. To use this method for equations that look like $2x - 3y = 6$, we first solve them for $y$.
2. Find the $x$- and $y$-intercepts. This method works best for equations of the form $3x + 2y = 6$ where the numbers in front of $x$ and $y$ divide the constant term evenly.
3. Find the slope and $y$-intercept. This method works best when the equation has the form $y = mx + b$ and $b$ is an integer.

## Getting Ready for Class

*After reading through the preceding section, respond in your own words and in complete sentences.*

**A.** What are $m$ and $b$ in the equation $y = mx + b$?

**B.** How would you find the slope and $y$-intercept for the line $3x - 2y = 6$?

**C.** What is the point-slope form of the equation of a line?

**D.** How would you find the equation of a line from two points on the line?

## PROBLEM SET 3.6

In each of the following problems, give the equation of the line with the given slope and $y$-intercept.

**1.** $m = \dfrac{2}{3}, b = 1$

**2.** $m = \dfrac{3}{4}, b = -2$

**3.** $m = \dfrac{3}{2}, b = -1$

**4.** $m = \dfrac{4}{3}, b = 2$

**5.** $m = -\dfrac{2}{5}, b = 3$

**6.** $m = -\dfrac{3}{5}, b = 4$

**7.** $m = 2, b = -4$

**8.** $m = -2, b = 4$

**9.** $m = -3, b = 2$

**10.** $m = 3, b = -2$

Find the slope and $y$-intercept for each of the following equations by writing them in the form $y = mx + b$. Then, graph each equation.

**11.** $-2x + y = 4$

**12.** $-2x + y = 2$

**13.** $3x + y = 3$

**14.** $3x + y = 6$

**15.** $3x + 2y = 6$

**16.** $2x + 3y = 6$

**17.** $4x - 5y = 20$

**18.** $2x - 5y = 10$

**19.** $-2x - 5y = 10$

**20.** $-4x + 5y = 20$

For each of the following problems, the slope and one point on a line are given. In each case use the point-slope form to find the equation of that line. (Write your answers in slope-intercept form.)

**21.** $(-2, -5), m = 2$

**22.** $(-1, -5), m = 2$

**23.** $(-4, 1), m = -\dfrac{1}{2}$

**24.** $(-2, 1), m = -\dfrac{1}{2}$

**25.** $(2, -3), m = \dfrac{3}{2}$

**26.** $(3, -4), m = \dfrac{4}{3}$

**27.** $(-1, 4), m = -3$

**28.** $(-2, 5), m = -3$

**29.** $(2, 4), m = 1$

**30.** $(4, 2), m = -1$

Find the equation of the line that passes through each pair of points. Write your answers in slope-intercept form.

**31.** $(-2, -4), (1, -1)$　**32.** $(2, 4), (-3, -1)$

**33.** $(-1, -5), (2, 1)$　**34.** $(-1, 6), (1, 2)$

**35.** $(-3, -2), (3, 6)$　**36.** $(-3, 6), (3, -2)$

**37.** $(-3, -1), (3, -5)$　**38.** $(-3, -5), (3, 1)$

**39.** $(-2, 1), (1, -2)$　**40.** $(1, -2), (2, 1)$

Find the slope and $y$-intercept for each line. Then write the equation of each line in slope-intercept form.

**41.**

**42.**

**43.**

**44.**

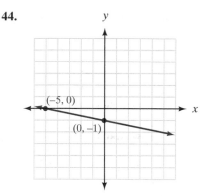

**45.** Find the equation of the line with $x$-intercept 3 and $y$-intercept 2.

**46.** Find the equation of the line with $x$-intercept 2 and $y$-intercept 3.

**47.** Find the equation of the line with $x$-intercept $-2$ and $y$-intercept $-5$.

**48.** Find the equation of the line with $x$-intercept $-3$ and $y$-intercept $-5$.

**49.** The equation of the vertical line that passes through the points $(3, -2)$ and $(3, 4)$ is either $x = 3$ or $y = 3$. Which one is it?

**50.** Find the equation of the vertical line that passes through the points $(2, 5)$ and $(2, -3)$.

**51.** The equation of the horizontal line that passes through the points $(2, 3)$ and $(-1, 3)$ is either $x = 3$ or $y = 3$. Which one is it?

**52.** Find the equation of the horizontal line that passes through the points $(-4, -2)$ and $(3, -2)$.

## Applying the Concepts

**53. Video Production** Pat has a small business duplicating videotapes. If he does all the work himself, he has a profit of $300 a day. When he hires an employee to help out, his profit decreases depending on how much he pays the employee. The graph of the relationship between his profit and employee costs is shown in the following figure.

(a) When is profit is zero, what is the employee cost?

(b) What is the maximum profit?

(c) Use the intercepts to calculate the slope of the line.

(d) Find the equation of this line.

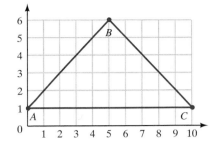

## Review Problems

The problems that follow review material we covered in Sections 2.5, 3.3, and 3.4.

Translate each of the following percent problems into an equation, and then solve the equation. [2.5]

**55.** What number is 25% of 300?

**56.** 30 is what percent of 120?

**57.** 60 is 15% of what number?

**58.** 75 is 30% of what number?

Graph each of the following lines. [3.3, 3.4]

**59.** $x + y = 4$    **60.** $x - y = -2$    **61.** $y = 2x - 3$

**62.** $y = 2x + 3$    **63.** $y = 2x$    **64.** $y = -2x$

**54. Geometry**   The triangle in the following figure has vertices $A$, $B$, and $C$ at $(0, 1)$, $(5, 6)$, and $(10, 1)$.

(a) Find the slope of each of the three sides of the triangle.

(b) Find the equation of the line through $A$ and $B$.

(c) Find the equation of the line through $B$ and $C$.

(d) Find the equation of the line through $A$ and $C$.

## 3.7    Linear Inequalities in Two Variables

A linear inequality in two variables is any expression that can be put in the form

$$ax + by < c$$

where $a$, $b$, and $c$ are real numbers ($a$ and $b$ not both 0). The inequality symbol can be any one of the following four: $<, \leq, >, \geq$.

Some examples of linear inequalities are

$$2x + 3y < 6 \qquad y \geq 2x + 1 \qquad x - y \leq 0$$

Although not all of these inequalities have the form $ax + by < c$, each one can be put in that form.

The solution set for a linear inequality is a section of the coordinate plane. The boundary for the section is found by replacing the inequality symbol with an equal sign and graphing the resulting equation. The boundary is included in the solution set (and represented with a solid line) if the inequality symbol used originally is

$\leq$ or $\geq$. The boundary is not included (and is represented with a broken line) if the original symbol is $<$ or $>$.

Let's look at some examples.

**EXAMPLE 1**   Graph the solution set for $x + y \leq 4$.

**Solution**   The boundary for the graph is the graph of $x + y = 4$. The boundary is included in the solution set because the inequality symbol is $\leq$.

The graph of the boundary is shown in Figure 1.

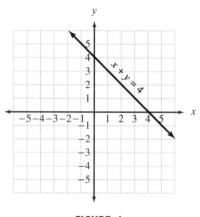

**FIGURE 1**

The boundary separates the coordinate plane into two sections, or regions: the region above the boundary and the region below the boundary. The solution set for $x + y \leq 4$ is one of these two regions along with the boundary. To find the correct region, we simply choose any convenient point that is *not* on the boundary. We then substitute the coordinates of the point into the original inequality $x + y \leq 4$. If the point we choose satisfies the inequality, then it is a member of the solution set, and we can assume that all points on the same side of the boundary as the chosen point are also in the solution set. If the coordinates of our point do not satisfy the original inequality, then the solution set lies on the other side of the boundary.

In this example a convenient point not on the boundary is the origin. Substituting $(0, 0)$ into $x + y \leq 4$ gives us

$$0 + 0 \overset{?}{\leq} 4$$

$$0 \leq 4 \qquad \text{A true statement}$$

Since the origin is a solution to the inequality $x + y \leq 4$ and the origin is below the boundary, all other points below the boundary are also solutions.

The graph of $x + y \leq 4$ is shown in Figure 2.

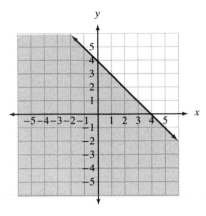

**FIGURE 2**

The region above the boundary is described by the inequality $x + y > 4$.

Here is a list of steps to follow when graphing the solution set for linear inequalities in two variables.

## TO GRAPH THE SOLUTION SET FOR LINEAR INEQUALITIES IN TWO VARIABLES

*Step 1:* Replace the inequality symbol with an equal sign. The resulting equation represents the boundary for the solution set.

*Step 2:* Graph the boundary found in step 1 using a *solid line* if the boundary is included in the solution set (that is, if the original inequality symbol was either ≤ or ≥). Use a *broken line* to graph the boundary if it is *not* included in the solution set. (It is not included if the original inequality was either < or >.)

*Step 3:* Choose any convenient point not on the boundary and substitute the co-ordinates into the *original* inequality. If the resulting statement is *true*, the graph lies on the *same* side of the boundary as the chosen point. If the resulting statement is *false*, the solution set lies on the *opposite* side of the boundary.

**E X A M P L E   2**    Graph the solution set for $y < 2x - 3$.

**Solution**    The boundary is the graph of $y = 2x - 3$. The boundary is not included because the original inequality symbol is <. We therefore use a broken line to represent the boundary, as shown in Figure 3.

A convenient test point is again the origin. Using $(0, 0)$ in $y < 2x - 3$, we have

$$0 \overset{?}{<} 2(0) - 3$$

$$0 < -3 \qquad \text{A false statement}$$

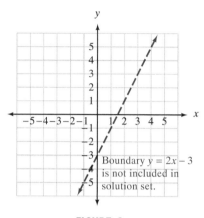

**FIGURE 3**

Since our test point gives us a false statement and it lies above the boundary, the solution set must lie on the other side of the boundary, as shown in Figure 4.

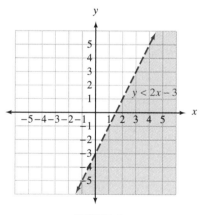

**FIGURE 4**

**E X A M P L E  3**    Graph the inequality $2x + 3y \le 6$.

**Solution**    We begin by graphing the boundary $2x + 3y = 6$. The boundary is included in the solution because the inequality symbol is $\le$.

If we use (0, 0) as our test point, we see that it yields a true statement when its coordinates are substituted into $2x + 3y \leq 6$. The graph, therefore, lies below the boundary, as shown in Figure 5.

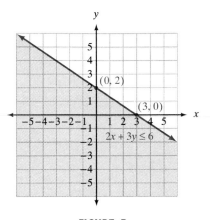

**FIGURE 5**

The ordered pair (0, 0) is a solution to $2x + 3y \leq 6$; all points on the same side of the boundary as (0, 0) must also be solutions to the inequality $2x + 3y \leq 6$.

**E X A M P L E　4**　Graph the solution set for $x \leq 5$.

**Solution**　The boundary is $x = 5$, which is a vertical line. All points to the left have $x$-coordinates less than 5, and all points to the right have $x$-coordinates greater than 5, as shown in Figure 6.

**FIGURE 6**

# Getting Ready for Class

*After reading through the preceding section, respond in your own words and in complete sentences.*

**A.** When graphing a linear inequality in two variables, how do you find the equation of the boundary line?

**B.** What is the significance of a broken line in the graph of an inequality?

**C.** When graphing a linear inequality in two variables, how do you know which side of the boundary line to shade?

**D.** Describe the set of ordered pairs that are solutions to $x + y < 6$.

## PROBLEM SET 3.7

Graph the following linear inequalities.

**1.** $2x - 3y < 6$
**2.** $3x + 2y \geq 6$
**3.** $x - 2y \leq 4$
**4.** $2x + y > 4$
**5.** $x - y \leq 2$
**6.** $x - y \leq 1$
**7.** $3x - 4y \geq 12$
**8.** $4x + 3y < 12$
**9.** $5x - y \leq 5$
**10.** $4x + y > 4$
**11.** $2x + 6y \leq 12$
**12.** $x - 5y > 5$
**13.** $x \geq 1$
**14.** $x < 5$
**15.** $x \geq -3$
**16.** $y \leq -4$
**17.** $y < 2$
**18.** $3x - y > 1$
**19.** $2x + y > 3$
**20.** $5x + 2y < 2$
**21.** $y \leq 3x - 1$
**22.** $y \geq 3x + 2$
**23.** $y \geq x - 5$
**24.** $y > x + 3$
**25.** $y \leq -\dfrac{1}{2}x + 2$
**26.** $y < \dfrac{1}{3}x + 3$
**27.** $y < -x + 4$
**28.** $y \geq -x - 3$

### Review Problems

The following problems review material we covered in Chapter 2.

**29.** Simplify the expression $7 - 3(2x - 4) - 8$.

**30.** Find the value of $x^2 - 2xy + y^2$ when $x = 3$ and $y = -4$.

Solve each equation.

**31.** $-\dfrac{3}{2}x = 12$
**32.** $2x - 4 = 5x + 2$

**33.** $8 - 2(x + 7) = 2$

**34.** $3(2x - 5) - (2x - 4) = 6 - (4x + 5)$

**35.** Solve the formula $P = 2l + 2w$ for $w$.

Solve each inequality, and graph the solution.

**36.** $-4x < 20$
**37.** $3 - 2x > 5$

**38.** $3 - 4(x - 2) \geq -5x + 6$

**39.** Solve the formula $3x - 2y \leq 12$ for $y$.

**40.** What number is 12% of 2,000?

**41. Geometry** The length of a rectangle is 5 inches more than 3 times the width. If the perimeter is 26 inches, find the length and width.

# CHAPTER 3    SUMMARY

**Examples**

**1.** The equation $3x + 2y = 6$ is an example of a linear equation in two variables.

### Linear Equations in Two Variables [3.1, 3.2]

A linear equation in two variables is any equation that can be put in the form $ax + by = c$. The graph of every linear equation is a straight line.

**2.** The graph of $y = -\frac{2}{3}x - 1$ is shown below.

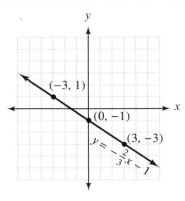

### Strategy for Graphing a Straight Line [3.3]

*Step 1:* Find any three ordered pairs that satisfy the equation. This can be done by using a convenient number for one variable and solving for the other variable.

*Step 2:* Graph the three ordered pairs found in step 1. Actually, we need only two points to graph a straight line. The third point serves as a check. If all three points do not line up, there is a mistake in our work.

*Step 3:* Draw a straight line through the three points graphed in step 2.

**3.** To find the $x$-intercept for $3x + 2y = 6$, we let $y = 0$ and get

$$3x = 6$$

$$x = 2$$

In this case the $x$-intercept is 2, and the graph crosses the $x$-axis at $(2, 0)$.

### Intercepts [3.4]

The $x$-intercept of an equation is the $x$-coordinate of the point where the graph crosses the $x$-axis. The $y$-intercept is the $y$-coordinate of the point where the graph crosses the $y$-axis. We find the $y$-intercept by substituting $x = 0$ into the equation and solving for $y$. The $x$-intercept is found by letting $y = 0$ and solving for $x$.

**4.** The slope of the line through $(3, -5)$ and $(-2, 1)$ is

$$m = \frac{-5 - 1}{3 - (-2)} = \frac{-6}{5} = -\frac{6}{5}$$

### Slope of a Line [3.5]

The *slope* of the line containing the points $(x_1, y_1)$ and $(x_2, y_2)$ is given by

$$\text{Slope} = m = \frac{y_2 - y_1}{x_2 - x_1} = \frac{\text{rise}}{\text{run}}$$

**5.** The equation of the line through $(1, 2)$ with a slope of 3 is

$$y - 2 = 3(x - 1)$$

$$y - 2 = 3x - 3$$

$$y = 3x - 1$$

### Point-Slope Form of a Straight Line [3.6]

If a line has a slope of $m$ and contains the point $(x_1, y_1)$, the equation can be written as

$$y - y_1 = m(x - x_1)$$

**6.** The equation of the line with a slope of 2 and a $y$-intercept 5 is

$$y = 2x + 5$$

**Slope-Intercept Form of a Straight Line [3.6]**

The equation of the line with a slope of $m$ and a $y$-intercept $b$ is

$$y = mx + b$$

**7.** Graph $x - y \geq 3$.

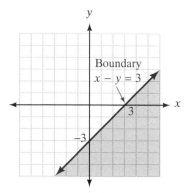

**To Graph a Linear Inequality in Two Variables [3.7]**

**Step 1:** Replace the inequality symbol with an equal sign. The resulting equation represents the boundary for the solution set.

**Step 2:** Graph the boundary found in step 1, using a *solid line* if the original inequality symbol was either $\leq$ or $\geq$. Use a *broken line* otherwise.

**Step 3:** Choose any convenient point not on the boundary and substitute the coordinates into the *original* inequality. If the resulting statement is *true*, the graph lies on the *same* side of the boundary as the chosen point. If the resulting statement is *false*, the solution set lies on the *opposite* side of the boundary.

---

# CHAPTER 3 REVIEW

The numbers in brackets refer to the sections of the text in which similar problems can be found.

For each equation, complete the given ordered pairs. [3.2]

**1.** $3x + y = 6$  $(4, \ ), (0, \ ), ( \ , 3), ( \ , 0)$

**2.** $2x - 5y = 20$  $(5, \ ), (0, \ ), ( \ , 2), ( \ , 0)$

**3.** $y = 2x - 6$  $(4, \ ), ( \ , -2), ( \ , 3)$

**4.** $y = 5x + 3$  $(2, \ ), ( \ , 0), ( \ , -3)$

**5.** $y = -3$  $(2, \ ), (-1, \ ), (-3, \ )$

**6.** $x = 6$  $( \ , 5), ( \ , 0), ( \ , -1)$

For the following equations, tell which of the given ordered pairs are solutions. [3.2]

**7.** $3x - 4y = 12$  $\left(-2, \dfrac{9}{2}\right), (0, 3), \left(2, -\dfrac{3}{2}\right)$

**8.** $y = 3x + 7$  $\left(-\dfrac{8}{3}, -1\right), \left(\dfrac{7}{3}, 0\right), (-3, -2)$

Graph the following ordered pairs. [3.1]

**9.** $(4, 2)$    **10.** $(-3, 1)$    **11.** $(0, 5)$

**12.** $(-2, -3)$    **13.** $(-3, 0)$    **14.** $\left(5, -\dfrac{3}{2}\right)$

For the following equations, complete the given ordered pairs, and use the results to graph the solution set for the equations. [3.3]

**15.** $x + y = -2$         **16.** $y = 3x$
  $( \ , 0), (0, \ ), (1, \ )$         $(-1, \ ), (1, \ ), ( \ , 0)$

**17.** $y = 2x - 1$
  $(1, \ ), (0, \ ), ( \ , -3)$

Graph the following equations. [3.3]

**18.** $3x - y = 3$   **19.** $y = -\dfrac{1}{3}x$   **20.** $y = 2x + 1$

**21.** $x = 5$         **22.** $y = -3$

Find the $x$- and $y$-intercepts for each equation. [3.4]

**23.** $3x - y = 6$      **24.** $2x - 6y = 24$

**25.** $y = x - 3$      **26.** $y = 3x - 6$

Find the slope of the line through the given pair of points. [3.5]

**27.** $(2, 3), (3, 5)$      **28.** $(-2, 3), (6, -5)$

**29.** $(-1, -4), (-3, -8)$

**30.** Find $x$ if the line through $(3, 3)$ and $(x, 9)$ has slope 2.

Find the equation of the line that contains the given point and has the given slope. Write answers in slope-intercept form. [3.6]

**31.** $(-1, 4); m = -2$      **32.** $(4, 3); m = \dfrac{1}{2}$

**33.** $(3, -2); m = -\dfrac{3}{4}$

Find the equation of the line with the given slope and $y$-intercept. [3.6]

**34.** $m = 3, b = 2$      **35.** $m = -1, b = 6$

**36.** $m = -\dfrac{1}{3}, b = \dfrac{3}{4}$

For each of the following equations, determine the slope and $y$-intercept. [3.6]

**37.** $y = 4x - 1$      **38.** $2x + y = -5$

**39.** $6x + 3y = 9$      **40.** $5x + 2y = 8$

Graph the following linear inequalities. [3.7]

**41.** $x - y < 3$      **42.** $x \geq -3$

**43.** $y \leq -2x + 3$

---

**C H A P T E R   3   TEST**

**1.** Fill in the following ordered pairs for the equation $2x - 5y = 10$. [3.2]

     $(0, \ )$    $( \ , 0)$    $(10, \ )$    $( \ , -3)$

**2.** Which of the following ordered pairs are solutions to $y = 4x - 3$? [3.2]

     $(2, 5)$    $(0, -3)$    $(3, 0)$    $(-2, 11)$

Graph each line. [3.3]

**3.** $y = 3x - 2$      **4.** $x = -2$

Find the $x$- and $y$-intercepts. [3.4]

**5.** $3x - 5y = 15$      **6.** $y = \dfrac{3}{2}x + 1$

Find the slope of the line through each pair of points. [3.5]

**7.** $(2, -3), (4, -7)$      **8.** $(-3, 5), (2, -8)$

**9.** $(a, b), (c, d)$      **10.** $(5x, 3), (2x, 4)$

**11.** Find the equation of the line through $(-2, 5)$ with a slope of 3. [3.6]

**12.** Find the equation of the line with a slope of 4 and $y$-intercept 8. [3.6]

**13.** Line $l$ passes through the points $(3, 1)$ and $(-2, 4)$. Find the equation of line $l$. [3.6]

**14.** A straight line has an $x$-intercept 1 and contains the point $(3, 4)$. Find its equation. [3.6]

Graph each linear inequality in two variables. [3.7]

**15.** $y < -x + 4$      **16.** $3x - 4y \geq 12$

# CHAPTERS 1–3 CUMULATIVE REVIEW

Simplify.

**1.** $7 - 2 \cdot 6$

**2.** $2 + 5(-4 + 8)$

**3.** $4 \cdot 6 + 12 \div 4 - 3^2$

**4.** $5[6 + (-12)] - 10(4 - 11)$

**5.** $(4 - 9)(-3 - 8)$

**6.** $\dfrac{18 + (-34)}{7 - 7}$

**7.** $\dfrac{11}{60} - \dfrac{13}{84}$

**8.** $5(8x)$

**9.** $5a + 3 - 4a - 6$

Solve each equation.

**10.** $5x + 6 - 4x - 3 = 7$

**11.** $4x - 5 = 3$

**12.** $-4x + 7 = 5x - 11$

**13.** $6(t + 5) - 4 = 2$

**14.** $3 = \dfrac{1}{4}(5x + 2)$

**15.** $0.05x + 0.07(200 - x) = 11$

Solve each inequality, and graph the solution.

**16.** $-4x > 28$

**17.** $5 - 7x \geq 19$

Graph on a rectangular coordinate system.

**18.** $y = 2x + 1$

**19.** $x + 2y = 3$

**20.** $y = -\dfrac{3}{2}x$

**21.** $x - 2y \leq 4$

**22.** $y \geq 3$

**23.** Graph the points $(-3, -2)$ and $(2, 3)$, and draw a line that passes through both of them. Does $(4, 7)$ lie on the line? Does $(1, 2)$?

**24.** Graph the line through $(-2, -2)$ with $x$-intercept $-1$. What is the $y$-intercept?

**25.** Find the $x$- and $y$-intercepts, and graph $2x + 5y = 10$.

**26.** Graph the line with slope $\dfrac{2}{3}$ and $y$-intercept $-1$.

**27.** Find the slope of the line through $(2, 3)$, $(5, 7)$.

**28.** Find the equation of the line with slope 3 and $y$-intercept 1.

**29.** Find the slope of the line $2x + 3y = 6$.

**30.** Find the equation of the line with slope $\dfrac{1}{2}$ that passes through $(-4, 2)$. Write your answer in slope-intercept form.

**31.** Find the equation of the line through $(2, 3)$ and $(4, 7)$. Write your answer in slope-intercept form.

**32.** Find the equation of the line given with $x$-intercept 3, $y$-intercept $-4$. Write your answer in slope-intercept form.

**33.** Complete the table for $3x - 2y = 6$.

| $x$ | $y$ |
|-----|-----|
|     | 3   |
| 0   |     |

**34.** Which ordered pairs are solutions to $2x + 5y = 10$: $(0, 2)$, $(-5, 0)$, or $(4, \frac{2}{5})$?

**35.** Translate into symbols: The sum of $x$ and 7 is 4.

**36.** Write the expression, and then simplify: the difference of 15 and twice 11.

**37.** Find the next number in the sequence: $\dfrac{1}{4}, \dfrac{1}{9}, \dfrac{1}{16}, \dfrac{1}{25}, \cdots$

**38.** Give the opposite, reciprocal, and absolute value of 2.

**39.** Five subtracted from the product of 6 and $-2$ is what number?

**40.** List all the integers from the set $\{-3, -0.2, 0, \frac{3}{4}, \sqrt{7}, 5\}$.

**41.** Reduce to lowest terms: $\dfrac{75}{135}$

**42.** Find the next term in the sequence: $\dfrac{1}{5}, 1, \dfrac{9}{5}, \cdots$

**43.** Evaluate $x^2 + 6x - 7$ when $x = 2$.

**44.** Use the formula $2x - 3y = 7$ to find $y$ when $x = -1$.

**45.** What percent of 36 is 27?

**46. Number Problem** The sum of a number and 9 is 23. Find the number.

**47. Geometry Problem** The length of a rectangle is 5 centimeters more than twice the width. The perimeter is 44 centimeters. Find the width and length.

**48. Cost of a Letter** The cost of mailing a letter was 32¢ for the first ounce and 29¢ for each additional ounce. If the cost of mailing a letter was 90¢, how many ounces did the letter weigh?

**49. Investing** Barbara had money in two accounts. She had $900 more in an account paying 8% annual interest than she had in an account paying 6% annual interest. If she earned $240 in interest for the year, how much did she have invested in each account?

**50. Geometry** Two angles are supplementary. If one angle is 20° more than the other, find the angles.
(a) 35°, 55°     (b) 80°, 60°
(c) 90°, 110°    (d) 80°, 100°

# Systems of Linear Equations

**FAST** and **RELIABLE**
**Internet Service**
*Low Monthly Fees!!!*
*Low Hourly Charges!*
*Program includes an e-mail account,
browser, and e-mail software

## INTRODUCTION

Two companies offer Internet access to their customers. Company A charges $10 a month plus $3 for every hour of Internet connection. Company B charges $18 a month plus $1 for every hour of Internet connection. To compare the monthly charges of the two companies we form what is called a system of equations. Here is that system.

$$y = 3x + 10$$
$$y = x + 18$$

The top equation gives us information on company A; the bottom equation gives information on company B. Tables 1 and 2 and the graphs in Figure 1 give us additional information about this system of equations.

| TABLE 1 Company A | |
|---|---|
| **Hours** | **Cost** |
| 0 | $10 |
| 1 | $13 |
| 2 | $16 |
| 3 | $19 |
| 4 | $22 |
| 5 | $25 |
| 6 | $28 |
| 7 | $31 |
| 8 | $34 |
| 9 | $37 |
| 10 | $40 |

| TABLE 2 Company B | |
|---|---|
| **Hours** | **Cost** |
| 0 | $18 |
| 1 | $19 |
| 2 | $20 |
| 3 | $21 |
| 4 | $22 |
| 5 | $23 |
| 6 | $24 |
| 7 | $25 |
| 8 | $26 |
| 9 | $27 |
| 10 | $28 |

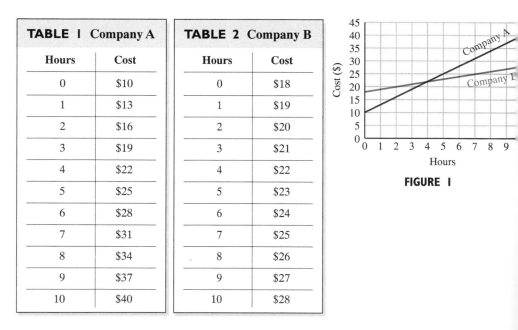

**FIGURE 1**

As you can see from looking at the tables and at the graphs in Figure 1, the monthly charges for the two companies will be equal if Internet use is exactly 4 hours. In this chapter we work with systems of linear equations.

## STUDY SKILLS

The study skills for this chapter are about attitude. They are points of view that point toward success.

1. **Be Focused, Not Distracted**   I have students who begin their assignments by asking themselves, "Why am I taking this class?" Or, "When am I ever going to use this stuff?" If you are asking yourself similar questions, you may be distracting yourself away from doing the things that will produce the results you want in this course. Don't dwell on questions and evaluations of the class that can be used as excuses for not doing well. If you want to succeed in this course, focus your energy and efforts toward success, rather than distracting yourself away from your goals.

2. **Be Resilient**   Don't let setbacks keep you from your goals. You want to put yourself on the road to becoming a person who can succeed in this class or any class in college. Failing a test or quiz, or having a difficult time on some topics, is normal. No one goes through college without some setbacks. Don't let a temporary disappointment keep you from succeeding in this course. A low grade on a test or quiz is simply a signal that some reevaluation of your study habits needs to take place.

3. **Intend to Succeed**   I always have a few students who simply go through the motions of studying without intending to master the material. It is more important to them to look like they are studying than to actually study. You need to study with the intention of being successful in the course. Intend to master the material, no matter what it takes.

Two linear equations considered at the same time make up what is called a *system* of linear equations. Both equations contain two variables and, of course, have graphs that are straight lines. The following are systems of linear equations:

$$x + y = 3 \qquad y = 2x + 1 \qquad 2x - y = 1$$
$$3x + 4y = 2 \qquad y = 3x + 2 \qquad 3x - 2y = 6$$

The solution set for a *system* of linear equations is all ordered pairs that are solutions to both equations. Since each linear equation has a graph that is a straight line, we can expect the intersection of the graphs to be a point whose coordinates are solutions to the system. That is, if we graph both equations on the same coordinate system, we can read the coordinates of the point of intersection and have the solution to our system. Here is an example.

**EXAMPLE 1** Solve the following system by graphing.

$$x + y = 4$$
$$x - y = -2$$

**Solution** On the same set of coordinate axes we graph each equation separately. Figure 1 shows both graphs, without showing the work necessary to get them. We can see from the graphs that they intersect at the point (1, 3). The point (1, 3) must therefore be the solution to our system, since it is the only ordered pair whose graph lies on both lines. Its coordinates satisfy both equations.

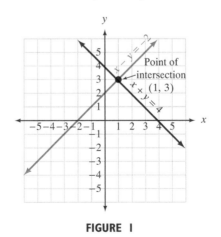

**FIGURE 1**

We can check our results by substituting the coordinates $x = 1$, $y = 3$ into both equations to see if they work.

| When | $x = 1$ | When | $x = 1$ |
| --- | --- | --- | --- |
| and | $y = 3$ | and | $y = 3$ |

| the equation | $x + y = 4$ | the equation | $x - y = -2$ |
|---|---|---|---|
| becomes | $1 + 3 \overset{?}{=} 4$ | becomes | $1 - 3 \overset{?}{=} -2$ |
| or | $4 = 4$ | or | $-2 = -2$ |

The point (1, 3) satisfies both equations.

Here are some steps to follow in solving linear systems by graphing.

### STRATEGY FOR SOLVING A LINEAR SYSTEM BY GRAPHING

*Step 1:* Graph the first equation by the methods described in Section 3.2 or 3.3.
*Step 2:* Graph the second equation on the same set of axes used for the first equation.
*Step 3:* Read the coordinates of the point of intersection of the two graphs. The ordered pair is the solution to the system.
*Step 4:* Check the solution in both equations, if necessary.

**EXAMPLE 2**   Solve the following system by graphing.

$$x + 2y = 8$$
$$2x - 3y = 2$$

**Solution**   Graphing each equation on the same coordinate system, we have the lines shown in Figure 2.

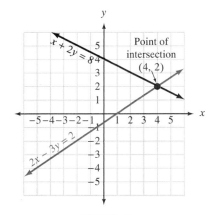

**FIGURE 2**

From Figure 2, we can see the solution for our system is (4, 2). We check this solution as follows.

| When | $x = 4$ | When | $x = 4$ |
|---|---|---|---|
| and | $y = 2$ | and | $y = 2$ |

the equation     $x + 2y = 8$     the equation     $2x - 3y = 2$

becomes     $4 + 2(2) \overset{?}{=} 8$     becomes     $2(4) - 3(2) \overset{?}{=} 2$

$$4 + 4 = 8$$     $$8 - 6 = 2$$

$$8 = 8$$     $$2 = 2$$

The point (4, 2) satisfies both equations and therefore must be the solution to our system.

**EXAMPLE 3**   Solve this system by graphing.

$$y = 2x - 3$$
$$x = 3$$

**Solution**   Graphing both equations on the same set of axes, we have Figure 3.

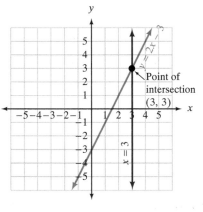

Point of intersection (3, 3)

**FIGURE 3**

The solution to the system is the point (3, 3).

**EXAMPLE 4**   Solve by graphing.

$$y = x - 2$$
$$y = x + 1$$

**Solution**   Graphing both equations produces the lines shown in Figure 4. We can see in Figure 4 that the lines are parallel and therefore do not intersect. Our system has no ordered pair as a solution, since there is no ordered pair that satisfies both equations. We say the solution set is the empty set and write $\varnothing$.

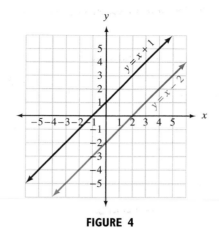

**FIGURE 4**

Example 4 is one example of two special cases associated with linear systems. The other special case happens when the two graphs coincide. Here is an example.

**E X A M P L E    5**    Graph the system.

$$2x + y = 4$$

$$4x + 2y = 8$$

**Solution**    Both graphs are shown in Figure 5. The two graphs coincide. The reason becomes apparent when we multiply both sides of the first equation by 2:

$$2x + y = 4$$

$$\mathbf{2}(2x + y) = \mathbf{2}(4) \qquad \text{Multiply both sides by } \mathbf{2}.$$

$$4x + 2y = 8$$

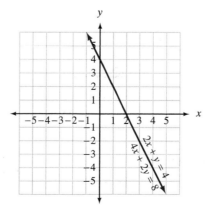

**FIGURE 5**

The equations have the same solution set. Any ordered pair that is a solution to one is a solution to the system. The system has an infinite number of solutions. (Any point on the line is a solution to the system.)

----

The two special cases illustrated in the previous two examples do not happen often. Usually, a system has a single ordered pair as a solution. Solving a system of linear equations by graphing is useful only when the ordered pair in the solution set has integers for coordinates. Two other solution methods work well in all cases. We will develop the other two methods in the next two sections.

**Note:** We sometimes use special vocabulary to describe the special cases shown in Examples 4 and 5. When a system of equations has no solution because the lines are parallel (as in Example 4), we say the system is *inconsistent*. When the lines coincide (as in Example 5), we say the system is *dependent*.

## Getting Ready for Class

*After reading through the preceding section, respond in your own words and in complete sentences.*

**A.** What is a system of two linear equations in two variables?

**B.** What is a solution to a system of linear equations?

**C.** How do we solve a system of linear equations by graphing?

**D.** Under what conditions will a system of linear equations not have a solution?

## PROBLEM SET 4.1

Solve the following systems of linear equations by graphing.

**1.** $x + y = 3$
    $x - y = 1$

**2.** $x + y = 2$
    $x - y = 4$

**3.** $x + y = 1$
    $-x + y = 3$

**4.** $x + y = 1$
    $x - y = -5$

**5.** $x + y = 8$
    $-x + y = 2$

**6.** $x + y = 6$
    $-x + y = -2$

**7.** $3x - 2y = 6$
    $x - y = 1$

**8.** $5x - 2y = 10$
    $x - y = -1$

**9.** $6x - 2y = 12$
    $3x + y = -6$

**10.** $4x - 2y = 8$
    $2x + y = -4$

**11.** $4x + y = 4$
    $3x - y = 3$

**12.** $5x - y = 10$
    $2x + y = 4$

**13.** $x + 2y = 0$
    $2x - y = 0$

**14.** $3x + y = 0$
    $5x - y = 0$

**15.** $3x - 5y = 15$
    $-2x + y = 4$

**16.** $2x - 4y = 8$
    $2x - y = -1$

**17.** $y = 2x + 1$
    $y = -2x - 3$

**18.** $y = 3x - 4$
    $y = -2x + 1$

**19.** $x + 3y = 3$
    $y = x + 5$

**20.** $2x + y = -2$
    $y = x + 4$

**21.** $x + y = 2$
    $x = -3$

**22.** $x + y = 6$
    $y = 2$

**23.** $x = -4$
    $y = 6$

**24.** $x = 5$
    $y = -1$

**25.** $x + y = 4$
    $2x + 2y = -6$

**26.** $x - y = 3$
    $2x - 2y = 6$

**27.** $4x - 2y = 8$
$\quad\ 2x - \ y = 4$

**28.** $3x - 6y = 6$
$\qquad x - 2y = 4$

**29.** As you have probably guessed by now, it can be difficult to solve a system of equations by graphing if the solution to the system contains a fraction. The solution to the following system is $(\frac{1}{2}, 1)$. Solve the system by graphing.

$$y = -2x + 2$$
$$y = \ \ 4x - 1$$

**30.** The solution to the following system is $(\frac{1}{3}, -2)$. Solve the system by graphing.

$$y = \ \ 3x - 3$$
$$y = -3x - 1$$

**31.** A second difficulty can arise in solving a system of equations by graphing if one or both of the equations is difficult to graph. The solution to the following system is $(2, 1)$. Solve the system by graphing.

$$3x - 8y = -2$$
$$x - \ y = 1$$

**32.** The solution to the following system is $(-3, 2)$. Solve the system by graphing.

$$2x + 5y = 4$$
$$x - \ y = -5$$

## Applying the Concepts

**33. Job Comparison**   Jane is deciding between two sales positions. She can work for Marcy's and receive $8.00 per hour or for Gigi's, where she earns $6.00 per hour but also receives a $50 commission per week. The two lines in the following figure represent the money Jane will make for working at each of the jobs.

(a) From the figure, how many hours would Jane have to work in order to earn the same amount at each of the positions?

(b) If Jane expects to work less than 20 hours a week, which job should she choose?

(c) If Jane expects to work more than 30 hours a week, which job should she choose?

**34. Truck Rental**   You need to rent a moving truck for two days. Rider Moving Trucks charges $50 per day and $0.50 per mile. UMove Trucks charges $45 per day and $0.75 per mile. The following figure represents the cost of renting each of the trucks.

(a) From the figure, after how many miles would the trucks cost the same?

(b) Which company will give you a better deal if you drive less than 30 miles?

(c) Which company will give you a better deal if you drive more than 60 miles?

**35. Internet Access**   Patrice has seen prices from two Internet companies, ICM World and Computer Service. She wants to make a decision about which one is cheaper. ICM World is $18 a month plus $1 for every hour on the Internet. Computer Service is $10 a month plus $3 for every hour on the Internet. The following graph represents the cost of the two Internet providers.

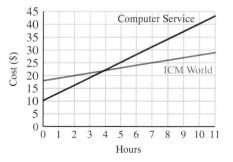

(a) From the graph, at how many hours would the cost of the two Internet providers be the same?
(b) If Patrice decides that she would be on the Internet for less than 3 hours a month, which plan is cheaper?
(c) If Patrice decides that she would be on the Internet for more than 6 hours a month, which plan is cheaper?

**36. Phone Rates** MaBell has a monthly charge of $5.00 and a rate of $0.05 per minute for long-distance calls. MCIA charges $3.00 per month and $0.10 per minute. The two lines in the following figure represent the total monthly cost for the two phone companies.

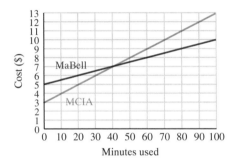

(a) From the figure, how many minutes would have to be used in order for the two plans to cost the same amount?
(b) If less than 30 minutes are used a month, which company has a better deal?
(c) If more than 60 minutes are used a month, which company has a better deal?

## Review Problems

The following problems review material we covered in Section 3.5.

Find the slope of the line through the following pairs of points.

**37.** $(-3, -5), (3, 1)$      **38.** $(-1, -5), (2, 1)$

**39.** $(0, 3), (5, 0)$      **40.** $(0, 3), (-5, 0)$

In each of the following problems, graph the line with slope $m$ and $y$-intercept $b$.

**41.** $m = \dfrac{1}{2}, b = -3$      **42.** $m = -\dfrac{1}{2}, b = 3$

---

### 4.2    The Elimination Method

The addition property states that if equal quantities are added to both sides of an equation, the solution set is unchanged. In the past we have used this property to help solve equations in one variable. We will now use it to solve systems of linear equations. Here is another way to state the addition property of equality.

Let $A, B, C,$ and $D$ represent algebraic expressions.

$$\begin{array}{ll} \text{If} & A = B \\ \text{and} & \underline{C = D} \\ \text{then} & A + C = B + D \end{array}$$

Since $C$ and $D$ are equal (that is, they represent the same number), what we have done is to add the same amount to both sides of the equation $A = B$. Let's see how we can use this form of the addition property of equality to solve a system of linear equations.

**EXAMPLE 1** Solve the following system.

$$x + y = 4$$
$$x - y = 2$$

**Solution**  The system is written in the form of the addition property of equality as written in this section. It looks like this:

$$A = B$$
$$C = D$$

where $A$ is $x + y$, $B$ is 4, $C$ is $x - y$, and $D$ is 2.

We use the addition property of equality to add the left sides together and the right sides together.

$$\begin{aligned} x + y &= 4 \\ \underline{x - y} &= \underline{2} \\ 2x + 0 &= 6 \end{aligned}$$

We now solve the resulting equation for $x$.

$$2x + 0 = 6$$
$$2x = 6$$
$$x = 3$$

The value we get for $x$ is the value of the $x$-coordinate of the point of intersection of the two lines $x + y = 4$ and $x - y = 2$. To find the $y$-coordinate, we simply substitute $x = 3$ into either of the two original equations and get

$$3 + y = 4$$
$$y = 1$$

The solution to our system is the ordered pair (3, 1). It satisfies both equations.

| When | $x = 3$ | When | $x = 3$ |
|---|---|---|---|
| and | $y = 1$ | and | $y = 1$ |
| the equation | $x + y = 4$ | the equation | $x - y = 2$ |
| becomes | $3 + 1 \overset{?}{=} 4$ | becomes | $3 - 1 \overset{?}{=} 2$ |
| or | $4 = 4$ | or | $2 = 2$ |

Figure 1 is visual evidence that the solution to our system is (3, 1).

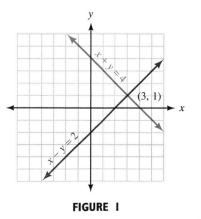

**FIGURE  1**

**Note:**  The graphs shown for the first three examples are not part of the solution shown in each example. The graphs are there simply to show you that the results we obtain by the elimination method are consistent with the results we would obtain by graphing.

The most important part of this method of solving linear systems is eliminating one of the variables when we add the left and right sides together. In our first example, the equations were written so that the $y$ variable was eliminated when we added the left and right sides together. If the equations are not set up this way to begin with, we have to work on one or both of them separately before we can add them together to eliminate one variable.

**E X A M P L E   2**   Solve the following system.

$$x + 2y = \phantom{-}4$$
$$x - \phantom{2}y = -5$$

**Solution**   Notice that if we were to add the equations together as they are, the resulting equation would have terms in both $x$ and $y$. Let's eliminate the variable $x$ by multiplying both sides of the second equation by $-1$ before we add the equations together. (As you will see, we can choose to eliminate either the $x$ or the $y$ variable.) Multiplying both sides of the second equation by $-1$ will not change its solution, so we do not need to be concerned that we have altered the system.

$$x + 2y = \phantom{-}4 \xrightarrow{\text{No change}} x + 2y = 4 \qquad \text{Add left and right sides to get}$$
$$x - \phantom{2}y = -5 \xrightarrow[\text{Multiply by } -1]{} \underline{-x + \phantom{2}y = 5}$$
$$0 + 3y = 9$$
$$3y = 9$$
$$y = 3 \quad \begin{cases} y\text{-coordinate of the} \\ \text{point of intersection} \end{cases}$$

Substituting $y = 3$ into either of the two original equations, we get $x = -2$.

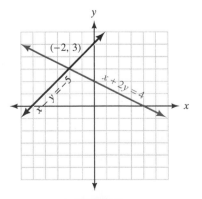

**FIGURE 2**

The solution to the system is $(-2, 3)$. It satisfies both equations. Figure 2 shows the solution to the system as the point where the two lines cross.

**EXAMPLE 3** Solve the following system.

$$2x - y = 6$$
$$x + 3y = 3$$

**Solution** Let's eliminate the $y$ variable from the two equations. We can do this by multiplying the first equation by 3 and leaving the second equation unchanged.

$$2x - y = 6 \xrightarrow{\text{3 times both sides}} 6x - 3y = 18$$
$$x + 3y = 3 \xrightarrow[\text{No change}]{} x + 3y = 3$$

The important thing about our system now is that the coefficients (the numbers in front) of the $y$ variables are opposites. When we add the terms on each side of the equal sign, then the terms in $y$ will add to zero and be eliminated.

$$
\begin{array}{r}
6x - 3y = 18 \\
x + 3y = 3 \\
\hline
7x \quad\quad = 21
\end{array}
$$
Add corresponding terms.

This gives us $x = 3$. Using this value of $x$ in the second equation of our original system, we have

$$3 + 3y = 3$$
$$3y = 0$$
$$y = 0$$

We could substitute $x = 3$ into any of the equations with both $x$ and $y$ variables and also get $y = 0$. The solution to our system is the ordered pair $(3, 0)$. Figure 3 is a picture of the system of equations showing the solution $(3, 0)$.

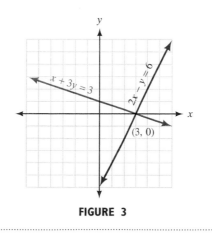

**FIGURE 3**

**EXAMPLE 4**   Solve the system.

$$2x + 3y = -1$$
$$3x + 5y = -2$$

**Solution**   Let's eliminate $x$ from the two equations. If we multiply the first equation by 3 and the second by $-2$, the coefficients of $x$ will be 6 and $-6$, respectively. The $x$ terms in the two equations will then add to zero.

$$2x + 3y = -1 \xrightarrow{\text{Multiply by 3}} 6x + 9y = -3$$
$$3x + 5y = -2 \xrightarrow[\text{Multiply by }-2]{} -6x - 10y = 4$$

We now add the left and right sides of our new system together.

$$6x + 9y = -3$$
$$\underline{-6x - 10y = 4}$$
$$-y = 1$$
$$y = -1$$

Substituting $y = -1$ into the first equation in our original system, we have

$$2x + 3(-1) = -1$$
$$2x - 3 = -1$$
$$2x = 2$$
$$x = 1$$

The solution to our system is $(1, -1)$. It is the only ordered pair that satisfies both equations.

**EXAMPLE 5**   Solve the system.

$$3x + 5y = -7$$
$$5x + 4y = 10$$

**Solution**   Let's eliminate $y$ by multiplying the first equation by $-4$ and the second equation by 5.

$$3x + 5y = -7 \xrightarrow{\text{Multiply by }-4} -12x - 20y = 28$$
$$5x + 4y = 10 \xrightarrow[\text{Multiply by 5}]{} \underline{25x + 20y = 50}$$
$$13x = 78$$
$$x = 6$$

Substitute $x = 6$ into either equation in our original system, and the result will be $y = -5$. The solution is therefore $(6, -5)$.

**E X A M P L E   6**   Solve the system.

$$\frac{1}{2}x - \frac{1}{3}y = 2$$

$$\frac{1}{4}x + \frac{2}{3}y = 6$$

**Solution**   Although we could solve this system without clearing the equations of fractions, there is probably less chance for error if we have only integer coefficients to work with. So let's begin by multiplying both sides of the top equation by 6 and both sides of the bottom equation by 12, to clear each equation of fractions.

$$\frac{1}{2}x - \frac{1}{3}y = 2 \xrightarrow{\text{Multiply by 6}} 3x - 2y = 12$$

$$\frac{1}{4}x + \frac{2}{3}y = 6 \xrightarrow{\text{Multiply by 12}} 3x + 8y = 72$$

Now we can eliminate $x$ by multiplying the top equation by $-1$ and leaving the bottom equation unchanged.

$$3x - 2y = 12 \xrightarrow{\text{Multiply by } -1} -3x + 2y = -12$$

$$3x + 8y = 72 \xrightarrow{\text{No change}} \underline{\phantom{-}3x + 8y = \phantom{0}72}$$

$$10y = \phantom{0}60$$

$$y = \phantom{0}6$$

We can substitute $y = 6$ into any equation that contains both $x$ and $y$. Let's use $3x - 2y = 12$.

$$3x - 2(6) = 12$$

$$3x - 12 = 12$$

$$3x = 24$$

$$x = \phantom{0}8$$

The solution to the system is $(8, 6)$.

Our next two examples will show what happens when we apply the elimination method to a system of equations consisting of parallel lines and to a system in which the lines coincide.

**E X A M P L E   7**   Solve the system.

$$2x - \phantom{0}y = \phantom{0}2$$

$$4x - 2y = 12$$

**Solution**   Let us choose to eliminate $y$ from the system. We can do this by multiplying the first equation by $-2$ and leaving the second equation unchanged.

$$2x - y = 2 \xrightarrow{\text{Multiply by } -2} -4x + 2y = -4$$

$$4x - 2y = 12 \xrightarrow[\text{No change}]{} 4x - 2y = 12$$

If we add both sides of the resulting system, we have

$$-4x + 2y = -4$$
$$\underline{4x - 2y = 12}$$
$$0 + 0 = 8$$

or                     $0 = 8$        A false statement

Both variables have been eliminated, and we end up with the false statement $0 = 8$. We have tried to solve a system that consists of two parallel lines. There is no solution, and that is the reason we end up with a false statement. Figure 4 is a visual representation of the situation and is conclusive evidence that there is no solution to our system.

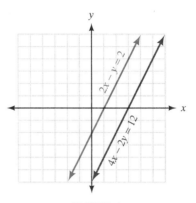

**FIGURE 4**

**EXAMPLE 8**   Solve the system.

$$4x - 3y = 2$$
$$8x - 6y = 4$$

**Solution**   Multiplying the top equation by $-2$ and adding, we can eliminate the variable $x$.

$$4x - 3y = 2 \xrightarrow{\text{Multiply by } -2} -8x + 6y = -4$$

$$8x - 6y = 4 \xrightarrow[\text{No change}]{} \underline{8x - 6y = 4}$$

$$0 = 0$$

Both variables have been eliminated, and the resulting statement $0 = 0$ is true. In this case the lines coincide because the equations are equivalent. The solution set consists of all ordered pairs that satisfy either equation.

The preceding two examples illustrate the two special cases in which the graphs of the equations in the system either coincide or are parallel.

Here is a summary of our results from these two examples:

| | | |
|---|---|---|
| Both variables are eliminated, and the resulting statement is false. | $\leftrightarrow$ | The lines are parallel, and there is no solution to the system. |
| Both variables are eliminated, and the resulting statement is true. | $\leftrightarrow$ | The lines coincide, and there are an infinite number of solutions to the system. |

The main idea in solving a system of linear equations by the elimination method is to use the multiplication property of equality on one or both of the original equations, if necessary, to make the coefficients of either variable opposites. The following box shows some steps to follow when solving a system of linear equations by the elimination method.

### STRATEGY FOR SOLVING A SYSTEM OF LINEAR EQUATIONS BY THE ELIMINATION METHOD

*Step 1:* Decide which variable to eliminate. (In some cases one variable will be easier to eliminate than the other. With some practice you will notice which one it is.)

*Step 2:* Use the multiplication property of equality on each equation separately to make the coefficients of the variable that is to be eliminated opposites.

*Step 3:* Add the respective left and right sides of the system together.

*Step 4:* Solve for the remaining variable.

*Step 5:* Substitute the value of the variable from step 4 into an equation containing both variables and solve for the other variable.

*Step 6:* Check your solution in both equations, if necessary.

## Getting Ready for Class

*After reading through the preceding section, respond in your own words and in complete sentences.*

**A.** How is the addition property of equality used in the elimination method of solving a system of linear equations?

**B.** What happens when we use the elimination method to solve a system of linear equations consisting of two parallel lines?

**C.** What does it mean when we solve a system of linear equations by the elimination method and we end up with the statement $0 = 8$?

**D.** What is the first step in solving a system of linear equations that contains fractions?

## PROBLEM SET 4.2

Solve the following systems of linear equations by elimination.

**1.** $x + y = 3$
$x - y = 1$

**2.** $x + y = -2$
$x - y = 6$

**3.** $x + y = 10$
$-x + y = 4$

**4.** $x - y = 1$
$-x - y = -7$

**5.** $x - y = 7$
$-x - y = 3$

**6.** $x - y = 4$
$2x + y = 8$

**7.** $x + y = -1$
$3x - y = -3$

**8.** $2x - y = -2$
$-2x - y = 2$

**9.** $3x + 2y = 1$
$-3x - 2y = -1$

**10.** $-2x - 4y = 1$
$2x + 4y = -1$

Solve each of the following systems by eliminating the $y$ variable.

**11.** $3x - y = 4$
$2x + 2y = 24$

**12.** $2x + y = 3$
$3x + 2y = 1$

**13.** $5x - 3y = -2$
$10x - y = 1$

**14.** $4x - y = -1$
$2x + 4y = 13$

**15.** $11x - 4y = 11$
$5x + y = 5$

**16.** $3x - y = 7$
$10x - 5y = 25$

Solve each of the following systems by eliminating the $x$ variable.

**17.** $3x - 5y = 7$
$-x + y = -1$

**18.** $4x + 2y = 32$
$x + y = -2$

**19.** $-x - 8y = -1$
$-2x + 4y = 13$

**20.** $-x + 10y = 1$
$-5x + 15y = -9$

**21.** $-3x - y = 7$
$6x + 7y = 11$

**22.** $-5x + 2y = -6$
$10x + 7y = 34$

Solve each of the following systems of linear equations by the elimination method.

**23.** $6x - y = -8$
$2x + y = -16$

**24.** $5x - 3y = -3$
$3x + 3y = -21$

**25.** $x + 3y = 9$
$2x - y = 4$

**26.** $x + 2y = 0$
$2x - y = 0$

**27.** $x - 6y = 3$
$4x + 3y = 21$

**28.** $8x + y = -1$
$4x - 5y = 16$

**29.** $2x + 9y = 2$
$5x + 3y = -8$

**30.** $5x + 2y = 11$
$7x + 8y = 7$

**31.** $\dfrac{1}{3}x + \dfrac{1}{4}y = \dfrac{7}{6}$
$\dfrac{3}{2}x - \dfrac{1}{3}y = \dfrac{7}{3}$

**32.** $\dfrac{7}{12}x - \dfrac{1}{2}y = \dfrac{1}{6}$
$\dfrac{2}{5}x - \dfrac{1}{3}y = \dfrac{11}{15}$

**33.** $3x + 2y = -1$
$6x + 4y = 0$

**34.** $8x - 2y = 2$
$4x - y = 2$

**35.** $11x + 6y = 17$
$5x - 4y = 1$

**36.** $3x - 8y = 7$
$10x - 5y = 45$

**37.** $\dfrac{1}{2}x + \dfrac{1}{6}y = \dfrac{1}{3}$
$-x - \dfrac{1}{3}y = -\dfrac{1}{6}$

**38.** $-\dfrac{1}{3}x - \dfrac{1}{2}y = -\dfrac{2}{3}$
$-\dfrac{2}{3}x - y = -\dfrac{4}{3}$

**39.** For some systems of equations it is necessary to apply the addition property of equality to each equation to line up the $x$-variables and $y$-variables before trying to eliminate a variable. Solve the following system by first writing each equation so that the variable terms with $x$ in them come first, the variable terms with $y$ in them come second, and the constant terms are on the right side of each equation.

$$4x - 5y = 17 - 2x$$

$$5y = 3x + 4$$

**40.** Solve the following system by first writing each equation so that the variable terms with $x$ in them come first, the variable terms with $y$ in them come second, and the constant terms are on the right side of each equation.

$$3x - 6y = -20 + 7x$$

$$4x = 3y - 34$$

**41.** Multiply both sides of the second equation in the following system by 100, and then solve as usual.

$$x + \phantom{0}y = 22$$

$$0.05x + 0.10y = 1.70$$

**42.** Multiply both sides of the second equation in the following system by 100, and then solve as usual.

$$x + \phantom{0}y = 15,000$$

$$0.06x + 0.07y = 980$$

## Review Problems

The problems that follow review material we covered in Section 3.6.

Find the slope and *y*-intercept for each of the following equations by first writing them in the form $y = mx + b$.

**43.** $-2x + 4y = 8$          **44.** $5x - 3y = 15$

For each of the following problems, the slope and one point on a line are given. Find the equation.

**45.** $(-2, -6), m = 3$          **46.** $(4, 2), m = \dfrac{1}{2}$

Find the equation of the line that passes through each pair of points.

**47.** $(-3, -5), (3, 1)$          **48.** $(-1, -5), (2, 1)$

---

### 4.3     The Substitution Method

There is a third method of solving systems of equations. It is the substitution method, and, like the elimination method, it can be used on any system of linear equations. Some systems, however, lend themselves more to the substitution method than others do.

**E X A M P L E   I**   Solve the following system.

$$x + y = 2$$

$$y = 2x - 1$$

**Solution**   If we were to solve this system by the methods used in the previous section, we would have to rearrange the terms of the second equation so that similar terms would be in the same column. There is no need to do this, however, since the second equation tells us that *y* is $2x - 1$. We can replace the *y* variable in the first equation with the expression $2x - 1$ from the second equation. That is, we *substitute* $2x - 1$ from the second equation for *y* in the first equation. Here is what it looks like:

$$x + (2x - 1) = 2$$

The equation we end up with contains only the variable *x*. The *y*-variable has been eliminated by substitution.

Solving the resulting equation, we have

$$x + (2x - 1) = 2$$
$$3x - 1 = 2$$
$$3x = 3$$
$$x = 1$$

This is the $x$-coordinate of the solution to our system. To find the $y$-coordinate, we substitute $x = 1$ into the second equation of our system. (We could substitute $x = 1$ into the first equation also and have the same result.)

$$y = 2(1) - 1$$
$$y = 2 - 1$$
$$y = 1$$

   The solution to our system is the ordered pair $(1, 1)$. It satisfies both of the original equations. Figure 1 provides visual evidence that the substitution method yields the correct solution.

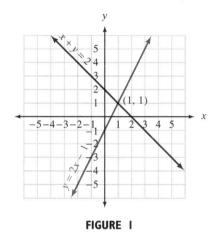

**FIGURE I**

**EXAMPLE 2**   Solve the following system by the substitution method.

$$2x - 3y = 12$$
$$y = 2x - 8$$

**Solution**   Again, the second equation says $y$ is $2x - 8$. Since we are looking for the ordered pair that satisfies both equations, the $y$ in the first equation must also be $2x - 8$. Substituting $2x - 8$ from the second equation for $y$ in the first equation, we have

$$2x - 3(2x - 8) = 12$$

This equation can still be read as $2x - 3y = 12$, because $2x - 8$ is the same as $y$. Solving the equation, we have

$$2x - 3(2x - 8) = 12$$
$$2x - 6x + 24 = 12$$
$$-4x + 24 = 12$$
$$-4x = -12$$
$$x = 3$$

To find the $y$-coordinate of our solution, we substitute $x = 3$ into the second equation in the original system.

$$\text{When} \qquad x = 3$$
$$\text{the equation} \qquad y = 2x - 8$$
$$\text{becomes} \qquad y = 2(3) - 8$$
$$y = 6 - 8 = -2$$

The solution to our system is $(3, -2)$.

**E X A M P L E   3**    Solve the following system by solving the first equation for $x$ and then using the substitution method:

$$x - 3y = -1$$
$$2x - 3y = 4$$

**Solution**    We solve the first equation for $x$ by adding $3y$ to both sides to get

$$x = 3y - 1$$

Using this value of $x$ in the second equation, we have

$$2(3y - 1) - 3y = 4$$
$$6y - 2 - 3y = 4$$
$$3y - 2 = 4$$
$$3y = 6$$
$$y = 2$$

Next, we find $x$.

$$\text{When} \qquad y = 2$$
$$\text{the equation} \qquad x = 3y - 1$$
$$\text{becomes} \qquad x = 3(2) - 1$$
$$x = 6 - 1$$
$$x = 5$$

The solution to our system is $(5, 2)$.

Here are the steps to use in solving a system of equations by the substitution method.

---

### STRATEGY FOR SOLVING A SYSTEM OF EQUATIONS BY THE SUBSTITUTION METHOD

*Step 1:* Solve either one of the equations for $x$ or $y$. (This step is not necessary if one of the equations is already in the correct form, as in Examples 1 and 2.)

*Step 2:* Substitute the expression for the variable obtained in step 1 into the other equation and solve it.

*Step 3:* Substitute the solution from step 2 into any equation in the system that contains both variables and solve it.

*Step 4:* Check your results, if necessary.

---

**EXAMPLE 4**    Solve by substitution.

$$-2x + 4y = 14$$
$$-3x + y = 6$$

**Solution**    We can solve either equation for either variable. If we look at the system closely, it becomes apparent that solving the second equation for $y$ is the easiest way to go. If we add $3x$ to both sides of the second equation, we have

$$y = 3x + 6$$

Substituting the expression $3x + 6$ back into the first equation in place of $y$ yields the following result.

$$-2x + 4(3x + 6) = 14$$
$$-2x + 12x + 24 = 14$$
$$10x + 24 = 14$$
$$10x = -10$$
$$x = -1$$

Substituting $x = -1$ into the equation $y = 3x + 6$ leaves us with

$$y = 3(-1) + 6$$
$$y = -3 + 6$$
$$y = 3$$

The solution to our system is $(-1, 3)$.

---

**EXAMPLE  5**   Solve by substitution.

$$4x + 2y = 8$$

$$y = -2x + 4$$

**Solution**   Substituting the expression $-2x + 4$ for $y$ from the second equation into the first equation, we have

$$4x + 2(-2x + 4) = 8$$

$$4x - 4x + 8 = 8$$

$$8 = 8 \qquad \text{A true statement}$$

Both variables have been eliminated, and we are left with a true statement. Recall from the last section that a true statement in this situation tells us the lines coincide. That is, the equations $4x + 2y = 8$ and $y = -2x + 4$ have exactly the same graph. Any point on that graph has coordinates that satisfy both equations and is a solution to the system.

**EXAMPLE  6**   The following table shows two contract rates charged by GTE Wireless for cellular phone use. At how many minutes will the two rates cost the same amount?

|        | **Flat Rate** | **Plus** | **Per Minute Charge** |
|--------|---------------|----------|-----------------------|
| Plan 1 | $15           |          | $1.50                 |
| Plan 2 | $24.95        |          | $0.75                 |

**Solution**   If we let $y = $ the monthly charge for $x$ minutes of phone use, then the equations for each plan are

$$\text{Plan 1:} \quad y = 1.5x + 15$$
$$\text{Plan 2:} \quad y = 0.75x + 24.95$$

We can solve this system by substitution by replacing the variable $y$ in Plan 2 with the expression $1.5x + 15$ from Plan 1. If we do so, we have

$$1.5x + 15 = 0.75x + 24.95$$

$$0.75x + 15 = 24.95$$

$$0.75x = 9.95$$

$$x = 13.27 \qquad \text{to the nearest hundredth}$$

The monthly bill is based on the number of minutes you use the phone, with any fraction of a minute moving you up to the next minute. If you talk for a total of 13 minutes, you are billed for 13 minutes. If you talk for 13 minutes, 10 seconds, you are billed for 14 minutes. The number of minutes on your bill will always be a whole number. So, to calculate the cost for talking 13.27 minutes, we would replace $x$ with 14 and find $y$. Let's compare the two plans at $x = 13$ minutes and at $x = 14$ minutes.

$$\text{Plan 1:} \quad y = 1.5x + 15$$

$$\text{When } x = 13, y = \$34.50$$

$$\text{When } x = 14, y = \$36.00$$

$$\text{Plan 2:} \quad y = 0.75x + 24.95$$

$$\text{When } x = 13, y = \$34.70$$

$$\text{When } x = 14, y = \$35.45$$

The two plans will never give the same cost for talking $x$ minutes. If you talk 13 or fewer minutes, Plan 1 will cost less. If you talk for more than 13 minutes, you will be billed for 14 minutes, and Plan 2 will cost less than Plan 1.

## Getting Ready for Class

*After reading through the preceding section, respond in your own words and in complete sentences.*

**A.** What is the first step in solving a system of linear equations by substitution?

**B.** When would substitution be more efficient than the elimination method in solving two linear equations?

**C.** What does it mean when we solve a system of linear equations by the substitution method and we end up with the statement $8 = 8$?

**D.** How would you begin solving the following system using the substitution method?

$$x + y = 2$$
$$y = 2x - 1$$

## PROBLEM SET 4.3

Solve the following systems by substitution. Substitute the expression in the second equation into the first equation and solve.

**1.** $x + y = 11$
$\quad y = 2x - 1$

**2.** $x - y = -3$
$\quad y = 3x + 5$

**3.** $x + y = 20$
$\quad y = 5x + 2$

**4.** $3x - y = -1$
$\quad x = 2y - 7$

**5.** $-2x + y = -1$
$\quad y = -4x + 8$

**6.** $4x - y = 5$
$\quad y = -4x + 1$

**7.** $3x - 2y = -2$
$\quad x = -y + 6$

**8.** $2x - 3y = 17$
$\quad x = -y + 6$

**9.** $5x - 4y = -16$
$\quad y = 4$

**10.** $6x + 2y = 18$
$\quad x = 3$

**11.** $5x + 4y = 7$
$\quad y = -3x$

**12.** $10x + 2y = -6$
$\quad y = -5x$

Solve the following systems by solving one of the equations for $x$ or $y$ and then using the substitution method.

**13.** $x + 3y = 4$
  $x - 2y = -1$

**14.** $x - y = 5$
  $x + 2y = -1$

**15.** $2x + y = 1$
  $x - 5y = 17$

**16.** $2x - 2y = 2$
  $x - 3y = -7$

**17.** $3x + 5y = -3$
  $x - 5y = -5$

**18.** $2x - 4y = -4$
  $x + 2y = 8$

**19.** $5x + 3y = 0$
  $x - 3y = -18$

**20.** $x - 3y = -5$
  $x - 2y = 0$

**21.** $-3x - 9y = 7$
  $x + 3y = 12$

**22.** $2x + 6y = -18$
  $x + 3y = -9$

Solve the following systems using the substitution method.

**23.** $5x - 8y = 7$
  $y = 2x - 5$

**24.** $3x + 4y = 10$
  $y = 8x - 15$

**25.** $7x - 6y = -1$
  $x = 2y - 1$

**26.** $4x + 2y = 3$
  $x = 4y - 3$

**27.** $-3x + 2y = 6$
  $y = 3x$

**28.** $-2x - y = -3$
  $y = -3x$

**29.** $5x - 6y = -4$
  $x = y$

**30.** $2x - 4y = 0$
  $y = x$

**31.** $3x + 3y = 9$
  $y = 2x - 12$

**32.** $7x + 6y = -9$
  $y = -2x + 1$

**33.** $7x - 11y = 16$
  $y = 10$

**34.** $9x - 7y = -14$
  $x = 7$

**35.** $-4x + 4y = -8$
  $y = x - 2$

**36.** $-4x + 2y = -10$
  $y = 2x - 5$

Solve each system by substitution. You can eliminate the decimals if you like, but you don't have to. The solution will be the same in either case.

**37.** $0.05x + 0.10y = 1.70$
  $y = 22 - x$

**38.** $0.20x + 0.50y = 3.60$
  $y = 12 - x$

## Applying the Concepts

**39. Gas Mileage**   Daniel is trying to decide whether to buy a car or a truck. The truck he is considering will cost him $150 a month in loan payments, and it gets 20 miles per gallon in gas mileage. The car will cost $180 a month in loan payments, but it

gets 35 miles per gallon in gas mileage. Daniel estimates that he will pay $1.40 per gallon for gas. This means that the monthly cost to drive the truck $x$ miles will be $y = \frac{1.40}{20}x + 150$. The total monthly cost to drive the car $x$ miles will be $y = \frac{1.40}{35}x + 180$. The following figure shows the graph of each equation.

(a) At how many miles do the car and the truck cost the same to operate?
(b) If Daniel drives more than 1,200 miles, which will be cheaper?
(c) If Daniel drives fewer than 800 miles, which will be cheaper?
(d) Why do the graphs appear in the first quadrant only?

**40. Video Production**   Pat runs a small company that duplicates videotapes. The daily cost and daily revenue for a company duplicating videos are shown in the following figure. The daily cost for duplicating $x$ videos is $y = \frac{6}{5}x + 20$; the daily revenue (the amount of money he brings in each day) for duplicating $x$ videos is $y = 1.7x$. The graphs of the two lines are shown in the following figure.

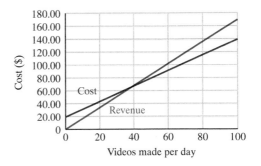

(a) Pat will "break even" when his cost and his revenue are equal. How many videos does he need to duplicate to break even?

(b) Pat will incur a loss when his revenue is less than his cost. If he duplicates 30 videos in one day, will he incur a loss?

(c) Pat will make a profit when his revenue is larger than his costs. For what values of $x$ will Pat make a profit?

(d) Why does the graph appear in the first quadrant only?

**41. Cost of Bottled Water**   The table shows the rates charged by two water-bottling companies for their monthly service, which includes rent on the water dispenser and the per gallon charge for water. If the water comes only in 5-gallon bottles, how many gallons must be used in a month for the two companies to charge the same amount?

|           | Flat Rate | Plus | Per Gallon Charge |
|-----------|-----------|------|-------------------|
| Company 1 | $7.00     |      | $1.10             |
| Company 2 | $5.00     |      | $1.15             |

**42. Cost of Renting a Car**   The daily rates for renting a compact car from two rental companies are shown in the table. How many miles would a car have to be driven in a *week* for the two companies to charge the same amount?

|           | Daily Rate | Plus | Per Mile Charge |
|-----------|------------|------|-----------------|
| Company 1 | $21.99     |      | $0.20           |
| Company 2 | $19.99     |      | $0.22           |

## Review Problems

The problems that follow review material we covered in Sections 2.5, 2.6, and 2.7. Reviewing these problems will help you understand the next section.

**43. Geometry**   A rectangle is 3 times as long as it is wide. If the perimeter is 24 meters, find the length and width.

**44. Geometry**   The perimeter of a square is 48 inches. Find the length of its side.

**45. Coin Problem**   A collection of coins consists of nickels and dimes and is worth $2.10. If there are 3 more dimes than nickels, how many of each coin are in the collection?

**46. Coin Problem**   Mary has $5.75 in nickels and quarters. If she has 5 more quarters than nickels, how many of each coin does she have?

**47.** What number is 8% of 6,000?

**48.** 540 is 9% of what number?

**49. Investing**   A man invests twice as much money at 10% annual interest as he does at 8% annual interest. If his total interest for the year is $224, how much does he have invested at each rate?

**50. Investing**   A woman invests 3 times as much money at 8% annual interest as she does at 6% annual interest. If the total interest for the year is $66, how much does she have invested at each rate?

---

## 4.4   Applications

I have often heard students remark about the word problems in beginning algebra: "What does this have to do with real life?" Most of the word problems we will encounter don't have much to do with "real life." We are actually just practicing. Ultimately, all problems requiring the use of algebra are word problems. That is, they are stated in words first, then translated to symbols. The problem then is solved by some system of mathematics, like algebra. Most real applications involve calculus or higher levels of mathematics. So, if the problems we solve are upsetting or frustrating to you, then you are probably taking them too seriously.

The word problems in this section have two unknown quantities. We will write two equations in two variables (each of which represents one of the unknown quantities), which of course is a system of equations. We then solve the system by one of the methods developed in the previous sections of this chapter. Here are the steps to follow in solving these word problems.

> *Blueprint for Problem Solving*
> *Using a System of Equations*
>
> *Step 1:* ***Read*** the problem, and then mentally ***list*** the items that are known and the items that are unknown.
>
> *Step 2:* ***Assign variables*** to each of the unknown items. That is, let $x =$ one of the unknown items and $y =$ the other unknown item. Then ***translate*** the other ***information*** in the problem to expressions involving the two variables.
>
> *Step 3:* ***Reread*** the problem, and then ***write a system of equations,*** using the items and variables listed in steps 1 and 2, that describes the situation.
>
> *Step 4:* ***Solve the system*** found in step 3.
>
> *Step 5:* ***Write*** your ***answers*** using complete sentences.
>
> *Step 6:* ***Reread*** the problem, and ***check*** your solution with the original words in the problem.

Remember, the more problems you work, the more problems you will be able to work. If you have trouble getting started on the problem set, come back to the examples and work through them yourself. The examples are similar to the problems found in the problem set.

## NUMBER PROBLEM

**EXAMPLE 1**   One number is 2 more than 5 times another number. Their sum is 20. Find the two numbers.

**Solution**   Applying the steps in our blueprint, we have the following:

*Step 1:* We know that the two numbers have a sum of 20 and that one of them is 2 more than 5 times the other. We don't know what the numbers themselves are.

*Step 2:* Let $x$ represent one of the numbers and $y$ represent the other. "One number is 2 more than 5 times another" translates to

$$y = 5x + 2$$

"Their sum is 20" translates to

$$x + y = 20$$

*Step 3:* The system that describes the situation must be

$$x + y = 20$$
$$y = 5x + 2$$

***Step 4:*** We can solve this system by substituting the expression $5x + 2$ in the second equation for $y$ in the first equation:

$$x + 5x + 2 = 20$$

$$6x + 2 = 20$$

$$6x = 18$$

$$x = 3$$

Using $x = 3$ in either of the first two equations and then solving for $y$, we get $y = 17$.

***Step 5:*** So 17 and 3 are the numbers we are looking for.

***Step 6:*** The number 17 is 2 more than 5 times 3, and the sum of 17 and 3 is 20.

## INTEREST PROBLEM

**EXAMPLE 2**    Mr. Hicks had $15,000 to invest. He invested part at 6% and the rest at 7%. If he earned $980 in interest, how much did he invest at each rate?

**Solution**    Remember, step 1 is done mentally.

***Step 1:*** We do not know the specific amounts invested in the two accounts. We do know that their sum is $15,000 and that the interest rates on the two accounts are 6% and 7%.

***Step 2:*** Let $x$ = the amount invested at 6% and $y$ = the amount invested at 7%. Since Mr. Hicks invested a total of $15,000, we have

$$x + y = 15{,}000$$

The interest he earns comes from 6% of the amount invested at 6% and 7% of the amount invested at 7%. To find 6% of $x$, we multiply $x$ by 0.06, which gives us $0.06x$. To find 7% of $y$, we multiply 0.07 times $y$ and get $0.07y$.

$$\underset{\text{at 6%}}{\text{Interest}} + \underset{\text{at 7%}}{\text{interest}} = \underset{\text{interest}}{\text{total}}$$

$$0.06x \quad + \quad 0.07y \quad = \quad 980$$

***Step 3:*** The system is

$$x + \quad y = 15{,}000$$

$$0.06x + 0.07y = 980$$

***Step 4:*** We multiply the first equation by $-6$ and the second by 100 to eliminate $x$:

$$x + \quad y = 15{,}000 \xrightarrow{\text{Multiply by } -6} -6x - 6y = -90{,}000$$

$$0.06x + 0.07y = 980 \xrightarrow[\text{Multiply by 100}]{} \underline{\phantom{-}6x + 7y = \phantom{-}98{,}000}$$

$$y = \phantom{-}8{,}000$$

Substituting $y = 8,000$ into the first equation and solving for $x$, we get $x = 7,000$.

*Step 5:* He invested $7,000 at 6% and $8,000 at 7%.

*Step 6:* Checking our solutions in the original problem, we have: The sum of $7,000 and $8,000 is $15,000, the total amount he invested. To complete our check, we find the total interest earned from the two accounts:

The interest on $7,000 at 6% is 0.06(7,000) = 420

The interest on $8,000 at 7% is 0.07(8,000) = 560

The total interest is                                                    $980

## COIN PROBLEM

**EXAMPLE   3**   John has $1.70, all in dimes and nickels. He has a total of 22 coins. How many of each kind does he have?

**Solution**

*Step 1:* We know that John has 22 coins that are dimes and nickels. We know that a dime is worth 10 cents and a nickel is worth 5 cents. We do not know the specific number of dimes and nickels he has.

*Step 2:* Let $x =$ the number of nickels and $y =$ the number of dimes. The total number of coins is 22, so

$$x + y = 22$$

The total amount of money he has is $1.70, which comes from nickels and dimes:

| Amount of money in nickels | + | amount of money in dimes | = | total amount of money |
|---|---|---|---|---|
| 0.05x | + | 0.10y | = | 1.70 |

*Step 3:* The system that represents the situation is

$$x + \quad y = \quad 22 \qquad \text{The number of coins}$$

$$0.05x + 0.10y = 1.70 \qquad \text{The value of the coins}$$

*Step 4:* We multiply the first equation by $-5$ and the second by 100 to eliminate the variable $x$:

$$x + \quad y = \quad 22 \xrightarrow{\text{Multiply by } -5} -5x - \quad 5y = -110$$

$$0.05x + 0.10y = 1.70 \xrightarrow{\text{Multiply by 100}} \underline{\quad 5x + 10y = \quad 170}$$

$$5y = \quad 60$$

$$y = \quad 12$$

Substituting $y = 12$ into our first equation, we get $x = 10$.

*Step 5:* John has 12 dimes and 10 nickels.

*Step 6:* Twelve dimes and 10 nickels total 22 coins.

$$12 \text{ dimes are worth } 12(0.10) = 1.20$$

$$\underline{10 \text{ nickels are worth } 10(0.05) = 0.50}$$

$$\text{The total value is} \qquad \$1.70$$

## MIXTURE PROBLEM

**E X A M P L E   4**   How much 20% alcohol solution and 50% alcohol solution must be mixed to get 12 gallons of 30% alcohol solution?

**Solution**   To solve this problem we must first understand that a 20% alcohol solution is 20% alcohol and 80% water.

*Step 1:* We know there are two solutions that together must total 12 gallons. We know 20% of one of the solutions is alcohol and the rest is water, while the other solution is 50% alcohol and 50% water. We do not know how many gallons of each individual solution we need.

*Step 2:* Let $x =$ the number of gallons of 20% alcohol solution needed and $y =$ the number of gallons of 50% alcohol solution needed. Since the total number of gallons we will end up with is 12, and this 12 gallons must come from the two solutions we are mixing, our first equation is

$$x + y = 12$$

To obtain our second equation, we look at the amount of alcohol in our two original solutions and our final solution. The amount of alcohol in the $x$ gallons of 20% solution is $0.20x$, while the amount of alcohol in $y$ gallons of 50% solution is $0.50y$. The amount of alcohol in the 12 gallons of 30% solution is $0.30(12)$. Since the amount of alcohol we start with must equal the amount of alcohol we end up with, our second equation is

$$0.20x + 0.50y = 0.30(12)$$

The information we have so far can also be summarized with a table. Sometimes it is easier to see where the equations come from by looking at a table like the one that follows.

|  | 20% Solution | 50% Solution | Final Solution |
|---|---|---|---|
| Number of gallons | $x$ | $y$ | 12 |
| Gallons of alcohol | $0.20x$ | $0.50y$ | $0.30(12)$ |

***Step 3:*** Our system of equations is

$$x + \quad y = 12$$

$$0.20x + 0.50y = 0.30(12)$$

***Step 4:*** We can solve this system by substitution. Solving the first equation for $y$ and substituting the result into the second equation, we have

$$0.20x + 0.50(12 - x) = 0.30(12)$$

Multiplying each side by 10 gives us an equivalent equation that is a little easier to work with.

$$2x + 5(12 - x) = 3(12)$$

$$2x + 60 - 5x = 36$$

$$-3x + 60 = 36$$

$$-3x = -24$$

$$x = 8$$

If $x$ is 8, then $y$ must be 4, because $x + y = 12$.

***Step 5:*** It takes 8 gallons of 20% alcohol solution and 4 gallons of 50% alcohol solution to produce 12 gallons of 30% alcohol solution.

***Step 6:*** Try it and see.

---

 # Getting Ready for Class

*After reading through the preceding section, respond in your own words and in complete sentences.*

**A.** If you were to apply the Blueprint for Problem Solving from Section 2.6 to the examples in this section, what would be the first step?

**B.** If you were to apply the Blueprint for Problem Solving from Section 2.6 to the examples in this section, what would be the last step?

**C.** When working application problems involving boats moving in rivers, how does the current of the river affect the speed of the boat?

**D.** Write an application problem for which the solution depends on solving a system of equations.

# PROBLEM SET 4.4

Solve the following word problems. Be sure to show the equations used.

## Number Problems

1. Two numbers have a sum of 25. One number is 5 more than the other. Find the numbers.

2. The difference of two numbers is 6. Their sum is 30. Find the two numbers.

3. The sum of two numbers is 15. One number is 4 times the other. Find the numbers.

4. The difference of two positive numbers is 28. One number is 3 times the other. Find the two numbers.

5. Two positive numbers have a difference of 5. The larger number is 1 more than twice the smaller. Find the two numbers.

6. One number is 2 more than 3 times another. Their sum is 26. Find the two numbers.

7. One number is 5 more than 4 times another. Their sum is 35. Find the two numbers.

8. The difference of two positive numbers is 8. The larger is twice the smaller decreased by 7. Find the two numbers.

## Interest Problems

9. Mr. Wilson invested money in two accounts. His total investment was $20,000. If one account pays 6% in interest and the other pays 8% in interest, how much did he have in each account if he earned a total of $1,380 in interest in 1 year?

10. A total of $11,000 was invested. Part of the $11,000 was invested at 4%, and the rest was invested at 7%. If the investments earn $680 per year, how much was invested at each rate?

11. A woman invested 4 times as much at 5% as she did at 6%. The total amount of interest she earns in 1 year from both accounts is $520. How much did she invest at each rate?

12. Ms. Hagan invested twice as much money in an account that pays 7% interest as she did in an account that pays 6% in interest. Her total investment pays her $1,000 a year in interest. How much did she invest at each rate?

## Coin Problems

13. Ron has 14 coins with a total value of $2.30. The coins are nickels and quarters. How many of each coin does he have?

14. Diane has $0.95 in dimes and nickels. She has a total of 11 coins. How many of each kind does she have?

15. Suppose Tom has 21 coins totaling $3.45. If he has only dimes and quarters, how many of each type does he have?

16. A coin collector has 31 dimes and nickels with a total face value of $2.40. (They are actually worth a lot more.) How many of each coin does she have?

## Mixture Problems

17. How many liters of 50% alcohol solution and 20% alcohol solution must be mixed to obtain 18 liters of 30% alcohol solution?

|  | 50% Solution | 20% Solution | Final Solution |
|---|---|---|---|
| Number of liters | $x$ | $y$ | |
| Liters of alcohol | | | |

18. How many liters of 10% alcohol solution and 5% alcohol solution must be mixed to obtain 40 liters of 8% alcohol solution?

|  | 10% Solution | 5% Solution | Final Solution |
|---|---|---|---|
| Number of liters | | | |
| Liters of alcohol | | | |

19. A mixture of 8% disinfectant solution is to be made from 10% and 7% disinfectant solutions. How much of each solution should be used if 30 gallons of 8% solution are needed?

**20.** How much 50% antifreeze solution and 40% antifreeze solution should be combined to give 50 gallons of 46% antifreeze solution?

## Miscellaneous Problems

**21.** For a Saturday matinee, adult tickets cost $5.50, while children under 12 pay only $4.00. If 70 tickets are sold for a total of $310, how many of the tickets were adult tickets and how many were sold to children under 12?

**22.** The Bishop's Peak 4-H club is having its annual fund-raising dinner. Adults pay $15 apiece, and children pay $10 apiece. If the number of adult tickets sold is twice the number of children's tickets sold and the total income for the dinner is $1,600, how many of each kind of ticket did the 4-H club sell?

**23.** A farmer has 96 feet of fence with which to make a corral. If he arranges it into a rectangle that is twice as long as it is wide, what are the dimensions?

**24.** If a 22-inch rope is to be cut into two pieces so that one piece is 3 inches longer than twice the other, how long is each piece?

**25.** A gambler finishes a session of blackjack with $5 chips and $25 chips. If he has 45 chips in all, with a total value of $465, how many of each kind of chip does the gambler have?

**26.** Tyler has been saving his winning lottery tickets. He has 23 tickets that are worth a total of $175. If each ticket is worth either $5 or $10, how many of each does he have?

**27.** Mary Jo spends $2,550 to buy stock in two companies. She pays $11 a share to one of the companies and $20 a share to the other. If she ends up with a total of 150 shares, how many shares did she buy at $11 a share, and how many did she buy at $20 a share?

**28.** Kelly sells 62 shares of stock she owns for a total of $433. If the stock was in two different companies, one selling at $6.50 a share and the other at $7.25 a share, how many of each did she sell?

## Review Problems

The following problems review material we covered in Chapter 3.

**29.** Fill in each ordered pair so that it is a solution to $y = \frac{1}{2}x + 3$. $(-2, \quad)$ $(0, \quad)$ $(2, \quad)$.

**30.** Graph the line $y = \frac{1}{2}x + 3$.

**31.** Graph the line $x = -2$.

**32.** Graph $3x - 2y = 6$.

**33.** Find the slope of the line through $(2, 5)$ and $(0, 1)$.

**34.** Find the slope and $y$-intercept for the line $2x - 5y = 10$.

**35.** Find the equation of the line through $(-2, 1)$ with slope $\frac{1}{2}$.

**36.** Write the equation of the line with slope $-2$ and $y$-intercept $\frac{3}{2}$.

**37.** Find the equation of the line through $(2, 5)$ and $(0, 1)$.

**38.** Graph the solution set for $2x - y < 4$.

# CHAPTER 4    SUMMARY

**Examples**

**1.** The solution to the system

$$x + 2y = 4$$
$$x - y = 1$$

is the ordered pair (2, 1). It is the only ordered pair that satisfies both equations.

## Definitions [4.1]

**1.** A *system of linear equations,* as the term is used in this book, is two linear equations that each contain the same two variables.

**2.** The *solution set* for a system of equations is the set of all ordered pairs that satisfy *both* equations. The solution set to a system of linear equations will contain the following:

> **Case I**    One ordered pair when the graphs of the two equations intersect at only one point (this is the most common situation)
>
> **Case II**    No ordered pairs when the graphs of the two equations are parallel lines
>
> **Case III**    An infinite number of ordered pairs when the graphs of the two equations coincide (are the same line)

**2.** Solving the system in Example 1 by graphing looks like

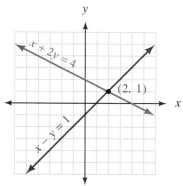

## Strategy for Solving a System by Graphing [4.1]

> **Step 1:** Graph the first equation.
> **Step 2:** Graph the second equation on the same set of axes.
> **Step 3:** The solution to the system consists of the coordinates of the point where the graphs cross each other (the coordinates of the point of intersection).
> **Step 4:** Check the solution to see that it satisfies *both* equations, if necessary.

**3.** We can eliminate the *y* variable from the system in Example 1 by multiplying both sides of the second equation by 2 and adding the result to the first equation.

$$x + 2y = 4 \xrightarrow{\text{No change}} x + 2y = 4$$
$$x - y = 1 \xrightarrow{\text{Multiply by 2}} \underline{2x - 2y = 2}$$
$$3x \quad\quad = 6$$
$$x = 2$$

Substituting $x = 2$ into either of the original two equations gives $y = 1$. The solution is (2, 1).

## Strategy for Solving a System by the Elimination Method [4.2]

> **Step 1:** Look the system over to decide which variable will be easier to eliminate.
> **Step 2:** Use the multiplication property of equality on each equation separately to ensure that the coefficients of the variable to be eliminated are opposites.
> **Step 3:** Add the left and right sides of the system produced in step 2, and solve the resulting equation.
> **Step 4:** Substitute the solution from step 3 back into any equation with both *x* and *y* variables, and solve.
> **Step 5:** Check your solution in both equations, if necessary.

**4.** We can apply the substitution method to the system in Example 1 by first solving the second equation for $x$ to get $x = y + 1$.

Substituting this expression for $x$ into the first equation, we have

$$y + 1 + 2y = 4$$
$$3y + 1 = 4$$
$$3y = 3$$
$$y = 1$$

Using $y = 1$ in either of the original equations gives $x = 2$.

**Strategy for Solving a System by the Substitution Method [4.3]**

*Step 1:* Solve either of the equations for one of the variables (this step is not necessary if one of the equations has the correct form already).

*Step 2:* Substitute the results of step 1 into the other equation, and solve.

*Step 3:* Substitute the results of step 2 into an equation with both $x$ and $y$ variables, and solve. (The equation produced in step 1 is usually a good one to use.)

*Step 4:* Check your solution, if necessary.

**Special Cases [4.1, 4.2, 4.3]**

In some cases, using the elimination or substitution method eliminates both variables. The situation is interpreted as follows.

**1.** If the resulting statement is *false,* then the lines are parallel, and there is no solution to the system.

**2.** If the resulting statement is *true,* then the equations represent the same line (the lines coincide). In this case any ordered pair that satisfies either equation is a solution to the system.

### COMMON MISTAKES

The most common mistake encountered in solving linear systems is the failure to complete the problem. Here is an example.

$$x + y = 8$$
$$\underline{x - y = 4}$$
$$2x \quad\quad = 12$$
$$x = 6$$

This is only half the solution. To find the other half, we must substitute the 6 back into one of the original equations and then solve for $y$.

Remember, solutions to systems of linear equations always consist of ordered pairs. We need an $x$-coordinate and a $y$-coordinate; $x = 6$ can never be a solution to a system of linear equations.

# CHAPTER 4  REVIEW

Solve the following systems by graphing. [4.1]

**1.** $x + y = 2$
  $x - y = 6$

**2.** $x + y = -1$
  $-x + y = 5$

**3.** $2x - 3y = 12$
  $-2x + y = -8$

**4.** $4x - 2y = 8$
  $3x + y = 6$

**5.** $y = 2x - 3$
  $y = -2x + 5$

**6.** $y = -x - 3$
  $y = 3x + 1$

Solve the following systems by the elimination method. [4.2]

**7.** $x - y = 4$
  $x + y = -2$

**8.** $-x - y = -3$
  $2x + y = 1$

**9.** $5x - 3y = 2$
  $-10x + 6y = -4$

**10.** $2x + 3y = -2$
  $3x - 2y = 10$

**11.** $-3x + 4y = 1$
  $-4x + y = -3$

**12.** $-4x - 2y = 3$
  $2x + y = 1$

**13.** $-2x + 5y = -11$
  $7x - 3y = -5$

**14.** $-2x + 5y = -15$
  $3x - 4y = 19$

Solve the following systems by substitution. [4.3]

**15.** $x + y = 5$
  $y = -3x + 1$

**16.** $x - y = -2$
  $y = -2x - 10$

**17.** $4x - 3y = -16$
  $y = 3x + 7$

**18.** $5x + 2y = -2$
  $y = -8x + 10$

**19.** $x - 4y = 2$
  $-3x + 12y = -8$

**20.** $4x - 2y = 8$
  $3x + y = -19$

**21.** $10x - 5y = 20$
  $x + 6y = -11$

**22.** $3x - y = 2$
  $-6x + 2y = -4$

Solve the following word problems. Be sure to show the equations used. [4.4]

**23. Number Problem**  The sum of two numbers is 18. If twice the smaller number is 6 more than the larger, find the two numbers.

**24. Number Problem**  The difference of two positive numbers is 16. One number is 3 times the other. Find the two numbers.

**25. Investing**  A total of $12,000 was invested. Part of the $12,000 was invested at 4%, and the rest was invested at 5%. If the interest for one year is $560, how much was invested at each rate?

**26. Investing**  A total of $14,000 was invested. Part of the $14,000 was invested at 6%, and the rest was invested at 8%. If the interest for one year is $1,060, how much was invested at each rate?

**27. Coin Problem**  Barbara has $1.35 in dimes and nickels. She has a total of 17 coins. How many of each does she have?

**28. Coin Problem**  Tom has $2.40 in dimes and quarters. He has a total of 15 coins. How many of each does he have?

**29. Mixture Problem**  How many liters of 20% alcohol solution and 10% alcohol solution must be mixed to obtain 50 liters of a 12% alcohol solution?

**30. Mixture Problem**  How many liters of 25% alcohol solution and 15% alcohol solution must be mixed to obtain 40 liters of a 20% alcohol solution?

# CHAPTER 4  TEST

Solve each system by graphing. [4.1]

**1.** $x + 2y = 5$
  $y = 2x$

**2.** $x - y = 5$
  $x = 3$

Solve each system by the elimination method. [4.2]

**3.** $x - y = 1$
  $2x + y = -10$

**4.** $2x + y = 7$
  $3x + y = 12$

**5.** $7x + 8y = -2$
$3x - 2y = 10$

**6.** $6x - 10y = 6$
$9x - 15y = 9$

Solve each system by the substitution method. [4.3]

**7.** $3x + 2y = 20$
$y = 2x + 3$

**8.** $3x - 6y = -6$
$x = y + 1$

**9.** $7x - 2y = -4$
$-3x + y = 3$

**10.** $2x - 3y = -7$
$x + 3y = -8$

Solve the following word problems. In each case, be sure to show the system of equations that describes the situation. [4.4]

**11. Number Problem** The sum of two numbers is 12. Their difference is 2. Find the numbers.

**12. Number Problem** The sum of two numbers is 15. One number is 6 more than twice the other. Find the two numbers.

**13. Investing** Dr. Stork has $10,000 to invest. He would like to earn $980 per year in interest. How much should he invest at 9% if the rest is to be invested at 11%?

**14. Coin Problem** Diane has 12 coins that total $1.60. If the coins are all nickels and quarters, how many of each type does she have?

# CHAPTERS 1 – 4 CUMULATIVE REVIEW

Simplify.

**1.** $3 \cdot 4 + 5$

**2.** $4 \cdot 3^2 + 4(6 - 3)$

**3.** $7[8 + (-5)] + 3(-7 + 12)$

**4.** $2(x - 5) + 8$

**5.** $8 - 6(5 - 9)$

**6.** $\dfrac{5(4 - 12) - 8(14 - 3)}{3 - 5 - 6}$

**7.** $\dfrac{2}{3} + \dfrac{3}{4} - \dfrac{1}{6}$

**8.** $7 - 5(2a - 3) + 7$

Solve each equation.

**9.** $-5 - 6 = -y - 3 + 2y$

**10.** $-2x = 0$

**11.** $3(x - 4) = 9$

**12.** $8 - 2(y + 4) = 12$

Solve each inequality, and graph the solution.

**13.** $0.3x + 0.7 \le -2$

**14.** $5x + 10 \le 7x - 14$

Graph on a rectangular coordinate system.

**15.** $y = -2x + 1$

**16.** $y = x$

**17.** $y = -\dfrac{2}{3}x$

**18.** Graph the line with slope $\dfrac{3}{4}$ and $y$-intercept 1.

**19.** Solve the following system by graphing.

$$2x + 3y = 3$$
$$4x + 6y = -4$$

**20.** Solve the following system by graphing.

$$3x - y = 3$$
$$2x + 3y = 2$$

Solve each system.

**21.** $x + y = 7$
$2x + 2y = 14$

**22.** $2x + y = -1$
$2x - 3y = 11$

**23.** $2x + 3y = 13$
$x - y = -1$

**24.** $x + y = 13$
$0.05x + 0.10y = 1$

**25.** $2x + 5y = 33$
$x - 3y = 0$

**26.** $3x + 4y = 8$
$x - y = 5$

**27.** $3x - 7y = 12$
$2x + y = 8$

**28.** $5x + 6y = 9$
$x - 2y = 5$

**29.** $2x - 3y = 7$
$y = 5x + 2$

**30.** $3x - 6y = 9$
$x = 2y + 3$

**31.** State the property or properties that justify the following: $a + 2 = 2 + a$

**32.** Find the next number in the geometric sequence: 2, 4, 8, 16, . . .

**33.** What is the quotient of $-30$ and 6?

**34.** Factor into primes: 180

**35.** What percent of 82 is 20.5?

**36.** What is 62% of 25?

Find the value of each expression when $x$ is 3.

**37.** $-3x + 7 + 5x$

**38.** $-x - 4x - 2x$

**39.** Find the value of the expression $4x - 5$ when $x$ is $-2$.

**40.** Given $4x - 5y = 12$, complete the ordered pair $(-2, \ )$.

**41.** Write the expression, and then simplify: The difference of 5 and $-8$.

**42.** Complete the table: $y = \dfrac{3}{5}x - 2$

| $x$ | $y$ |
|---|---|
| 2 | |
| | $-1$ |

**43.** Find the $x$- and $y$-intercepts: $3x - 4y = 12$

**44.** Find the slope of the line $x = 3y - 6$.

**45.** Find the slope of the line through $(-1, 1)$, $(-5, -4)$.

**46.** Find the slope of the line through $(6, \frac{1}{2})$, $(-3, \frac{5}{4})$.

**47.** Find the equation of the line with slope $\frac{2}{3}$ and $y$-intercept 3.

**48.** Find the equation of the line with slope 2 if it passes through (4, 1).

**49.** Find the equation of the line through (4, 3) and (6, 6).

**50. Coin Problem**   Joy has 15 coins with a value of $1.10. The coins are nickels and dimes. How many of each does she have?

# Exponents and Polynomials

## INTRODUCTION

If you were given a penny on the first day of September, and then each day after that you were given twice the amount of money you received the day before, how much money would you receive on September 30? To begin, Table 1 and Figure 1 show the amount of money you would receive on each of the first 10 days of the month. As you can see, on the tenth day of the month you would receive $5.12.

| TABLE 1 | Money That Doubles Each Day |
|---------|------------------------------|
| **Day** | **Money (in cents)** |
| 1 | $1 = 2^0$ |
| 2 | $2 = 2^1$ |
| 3 | $4 = 2^2$ |
| 4 | $8 = 2^3$ |
| 5 | $16 = 2^4$ |
| 6 | $32 = 2^5$ |
| 7 | $64 = 2^6$ |
| 8 | $128 = 2^7$ |
| 9 | $256 = 2^8$ |
| 10 | $512 = 2^9$ |

**FIGURE 1**

To find the amount of money on day 30, we could continue to double the amount on each of the next 20 days. Or, we could notice the pattern of exponents in the second column of the table and reason that the amount of money on day 30 would be $2^{29}$ cents, which is a very large number. In fact, $2^{29}$ cents is $5,368,709.12 — a little less than 5.4 million dollars. When you are finished with this chapter, you will have a good working knowledge of exponents.

## STUDY SKILLS

The study skills for this chapter concern the way you approach new situations in mathematics. The first study skill applies to your natural instincts for what does and doesn't work in mathematics. The second study skill gives you a way of testing your instincts.

1. **Don't Let Your Intuition Fool You**  As you become more experienced and more successful in mathematics, you will be able to trust your mathematical intuition. For now, though, it can get in the way of success. For example, if you ask a beginning algebra student to "subtract 3 from $-5$" many will answer $-2$ or 2. Both answers are incorrect, even though they may seem intuitively true.

2. **Test Properties About Which You Are Unsure**  From time to time you will be in a situation in which you would like to apply a property or rule, but you are not sure it is true. You can always test a property or statement by substituting numbers for variables. For instance, I always have students who rewrite $(x + 3)^2$ as $x^2 + 9$, thinking that the two expressions are equivalent. The fact that the two expressions are not equivalent becomes obvious when we substitute 10 for $x$ in each one.

When $x = 10$, the expression $(x + 3)^2$ is

$$(10 + 3)^2 = 13^2 = 169$$

When $x = 10$, the expression $x^2 + 9$ is

$$10^2 + 9 = 100 + 9 = 109$$

Along the same lines, there may come a time when you are wondering if $\sqrt{x^2 + 25}$ is the same as $x + 5$. If you try $x = 10$ in each expression, you will find out quickly that the two expressions are not the same.

When $x = 10$, the expression $\sqrt{x^2 + 25}$ is

$$\sqrt{10^2 + 25} = \sqrt{125} \approx 11.2$$

When $x = 10$, the expression $x + 5$ is

$$10 + 5 = 15$$

When you test the equivalence of expressions by substituting numbers for the variable, make it easy on yourself by choosing numbers that are easy to work with, such as 10. Don't try to verify the equivalence of expressions by substituting 0, 1, or 2 for the variable, as these numbers make expressions seem equivalent when in fact they are not.

It is not unusual, nor is it wrong, to try occasionally to apply a property that doesn't exist. If you have any doubt about generalizations you are making, test them by replacing variables with numbers and simplifying.

# Multiplication With Exponents

Recall that an *exponent* is a number written just above and to the right of another number, which is called the *base*. In the expression $5^2$, for example, the exponent is 2 and the base is 5. The expression $5^2$ is read "5 to the second power" or "5 squared." The meaning of the expression is

$$5^2 = 5 \cdot 5 = 25$$

In the expression $5^3$, the exponent is 3 and the base is 5. The expression $5^3$ is read "5 to the third power" or "5 cubed." The meaning of the expression is

$$5^3 = 5 \cdot 5 \cdot 5 = 125$$

Here are some further examples.

---

**EXAMPLES**

**1.** $4^3 = 4 \cdot 4 \cdot 4 = 16 \cdot 4 = 64$          Exponent 3, base 4

**2.** $-3^4 = -3 \cdot 3 \cdot 3 \cdot 3 = -81$          Exponent 4, base 3

**3.** $(-2)^5 = (-2)(-2)(-2)(-2)(-2) = -32$      Exponent 5, base $-2$

**4.** $\left(-\dfrac{3}{4}\right)^2 = \left(-\dfrac{3}{4}\right)\left(-\dfrac{3}{4}\right) = \dfrac{9}{16}$      Exponent 2, base $-\dfrac{3}{4}$

---

QUESTION: In what way are $(-5)^2$ and $-5^2$ different?
ANSWER: In the first case, the base is $-5$. In the second case, the base is 5. The answer to the first is 25. The answer to the second is $-25$. Can you tell why? Would there be a difference in the answers if the exponent in each case were changed to 3?

We can simplify our work with exponents by developing some properties of exponents. We want to list the things we know are true about exponents and then use these properties to simplify expressions that contain exponents.

The first property of exponents applies to products with the same base. We can use the definition of exponents, as indicating repeated multiplication, to simplify expressions such as $7^4 \cdot 7^2$.

$$7^4 \cdot 7^2 = (7 \cdot 7 \cdot 7 \cdot 7)(7 \cdot 7)$$

$$= (7 \cdot 7 \cdot 7 \cdot 7 \cdot 7 \cdot 7)$$

$$= 7^6 \qquad \textit{Notice}: 4 + 2 = 6$$

As you can see, multiplication with the same base resulted in addition of exponents. We can summarize this result with the following property.

### PROPERTY 1 FOR EXPONENTS

If $a$ is any real number and $r$ and $s$ are integers, then

$$a^r \cdot a^s = a^{r+s}$$

*In words:*  To multiply two expressions with the same base, add exponents and use the common base.

Here are some examples using Property 1.

**EXAMPLES**  Use Property 1 to simplify the following expressions. Leave your answers in terms of exponents.

**5.** $5^3 \cdot 5^6 = 5^{3+6} = 5^9$

**6.** $x^7 \cdot x^8 = x^{7+8} = x^{15}$

**7.** $3^4 \cdot 3^8 \cdot 3^5 = 3^{4+8+5} = 3^{17}$

**Note:**  In Examples 5, 6, and 7, notice that in each case the base in the original problem is the same as the base that appears in the answer and that the base is written only once in the answer. A very common mistake that people make when they first begin to use Property 1 is to write a 2 in front of the base in the answer. For example, people making this mistake would get $2x^{15}$ or $(2x)^{15}$ as the result in Example 6. To avoid this mistake, you must be sure you understand the meaning of Property 1 exactly as it is written.

Another common type of expression involving exponents is one in which an expression containing an exponent is raised to another power. The expression $(5^3)^2$ is an example:

$$\begin{aligned} (5^3)^2 &= (5^3)(5^3) \\ &= 5^{3+3} \\ &= 5^6 \qquad \text{\textit{Notice}: } 3 \cdot 2 = 6. \end{aligned}$$

This result offers justification for the second property of exponents.

### PROPERTY 2 FOR EXPONENTS

If $a$ is any real number and $r$ and $s$ are integers, then

$$(a^r)^s = a^{r \cdot s}$$

*In words:*  A power raised to another power is the base raised to the product of the powers.

**EXAMPLES**  Simplify the following expressions:

**8.** $(4^5)^6 = 4^{5 \cdot 6} = 4^{30}$

**9.** $(x^3)^5 = x^{3 \cdot 5} = x^{15}$

The third property of exponents applies to expressions in which the product of two or more numbers or variables is raised to a power. Let's look at how the expression $(2x)^3$ can be simplified:

$$(2x)^3 = (2x)(2x)(2x)$$
$$= (2 \cdot 2 \cdot 2)(x \cdot x \cdot x)$$
$$= 2^3 \cdot x^3 \qquad \textit{Notice: The exponent 3 distributes}$$
$$= 8x^3 \qquad \qquad \text{over the product } 2x.$$

We can generalize this result into a third property of exponents.

### PROPERTY 3 FOR EXPONENTS

If $a$ and $b$ are any two real numbers and $r$ is an integer, then

$$(ab)^r = a^r b^r$$

*In words:*   The power of a product is the product of the powers.

Here are some examples using Property 3 to simplify expressions.

**EXAMPLES**   Simplify the following expressions:

**10.** $(5x)^2 = 5^2 \cdot x^2$                                          Property 3
         $= 25x^2$

**11.** $(2xy)^3 = 2^3 \cdot x^3 \cdot y^3$                            Property 3
           $= 8x^3 y^3$

**12.** $(3x^2)^3 = 3^3 (x^2)^3$                             Property 3
           $= 27x^6$                                            Property 2

**13.** $\left(-\dfrac{1}{4} x^2 y^3\right)^2 = \left(-\dfrac{1}{4}\right)^2 (x^2)^2 (y^3)^2$         Property 3

                    $= \dfrac{1}{16} x^4 y^6$                         Property 2

**14.** $(x^4)^3 (x^2)^5 = x^{12} \cdot x^{10}$                        Property 2
              $= x^{22}$                                              Property 1

**15.** $(2y)^3 (3y^2) = 2^3 y^3 (3y^2)$                  Property 3
             $= 8 \cdot 3(y^3 \cdot y^2)$             Commutative and associative
                                                properties

             $= 24y^5$                                         Property 1

**16.** $(2x^2 y^5)^3 (3x^4 y)^2 = 2^3 (x^2)^3 (y^5)^3 \cdot 3^2 (x^4)^2 y^2$    Property 3
                     $= 8x^6 y^{15} \cdot 9x^8 y^2$                Property 2
                     $= (8 \cdot 9)(x^6 x^8)(y^{15} y^2)$      Commutative and associative
                                                    properties

                     $= 72x^{14} y^{17}$                      Property 1

## FACTS FROM
# Geometry

### Volume of a Rectangular Solid

It is easy to see why the phrase "five squared" is associated with the expression $5^2$. Simply find the area of the square shown in Figure 1 with a side of 5.

**FIGURE 1**

5

5 | Area = $5^2$

To see why the phrase "five cubed" is associated with the expression $5^3$, we have to find the *volume* of a cube for which all three dimensions are 5 units long. The volume of a cube is a measure of the space occupied by the cube. To calculate the volume of the cube shown in Figure 2, we multiply the three dimensions together to get $5 \cdot 5 \cdot 5 = 5^3$.

**FIGURE 2**

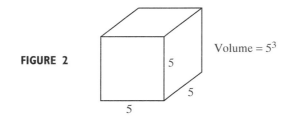

Volume = $5^3$

5

5

5

The cube shown in Figure 2 is a special case of a general category of three-dimensional geometric figures called *rectangular solids*. Rectangular solids have rectangles for sides, and all connecting sides meet at right angles. The three dimensions are length, width, and height. To find the volume of a rectangular solid, we find the product of the three dimensions, as shown in Figure 3.

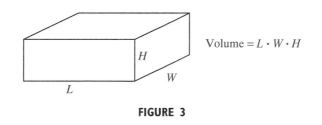

Volume = $L \cdot W \cdot H$

H

W

L

**FIGURE 3**

**Note:** If we include units with the dimensions of the diagrams, then the units for the area will be square units and the units for volume will be cubic units. More specifically, if a square has a side 5 inches long, then its area will be

$$A = (5 \text{ inches})^2 = 25 \text{ inches}^2$$

where the unit inches$^2$ stands for square inches.

If a cube has a side 5 inches long, then its volume will be

$$V = (5 \text{ inches})^3 = 125 \text{ inches}^3$$

where the unit inches$^3$ stands for cubic inches.

If a rectangular solid has a length of 5 inches, a width of 4 inches, and a height of 3 inches, then its volume is

$$V = (5 \text{ inches})(4 \text{ inches})(3 \text{ inches})$$

$$= 60 \text{ inches}^3$$

## SCIENTIFIC NOTATION

Many branches of science require working with very large numbers. In astronomy, for example, distances commonly are given in light-years. A light-year is the distance light travels in a year. It is approximately

$$5{,}880{,}000{,}000{,}000 \text{ miles}$$

This number is difficult to use in calculations because of the number of zeros it contains. Scientific notation provides a way of writing very large numbers in a more manageable form.

---

**DEFINITION**   A number is in **scientific notation** when it is written as the product of a number between 1 and 10 and an integer power of 10. A number written in scientific notation has the form

$$n \times 10^r$$

where $1 \le n < 10$ and $r =$ an integer.

---

**EXAMPLE  17**   Write 376,000 in scientific notation.

**Solution**   We must rewrite 376,000 as the product of a number between 1 and 10 and a power of 10. To do so, we move the decimal point 5 places to the left so that it appears between the 3 and the 7. Then we multiply this number by $10^5$. The number that results has the same value as our original number and is written in scientific notation:

$$376{,}000 = 3.76 \times 10^5$$

Moved 5 places.

Decimal point originally here.

Keeps track of the 5 places we moved the decimal point.

**EXAMPLE 18** Write $4.52 \times 10^3$ in expanded form.

**Solution** Since $10^3$ is 1,000, we can think of this as simply a multiplication problem. That is,

$$4.52 \times 10^3 = 4.52 \times 1,000 = 4,520$$

On the other hand, we can think of the exponent 3 as indicating the number of places we need to move the decimal point in order to write our number in expanded form. Since our exponent is positive 3, we move the decimal point three places to the right:

$$4.52 \times 10^3 = 4,520$$

# Getting Ready for Class

*After reading through the preceding section, respond in your own words and in complete sentences.*

**A.** Explain the difference between $-5^2$ and $(-5)^2$.

**B.** How do you multiply two expressions containing exponents when they each have the same base?

**C.** What is Property 2 for exponents?

**D.** When is a number written in scientific notation?

# PROBLEM SET 5.1

Name the base and exponent in each of the following expressions. Then use the definition of exponents as repeated multiplication to simplify.

**1.** $4^2$

**2.** $6^2$

**3.** $(0.3)^2$

**4.** $(0.03)^2$

**5.** $4^3$

**6.** $10^3$

**7.** $(-5)^2$

**8.** $-5^2$

**9.** $-2^3$

**10.** $(-2)^3$

**11.** $3^4$

**12.** $(-3)^4$

**13.** $\left(\dfrac{2}{3}\right)^2$

**14.** $\left(\dfrac{2}{3}\right)^3$

**15.** $\left(\dfrac{1}{2}\right)^4$

**16.** $\left(\dfrac{4}{5}\right)^2$

**17.** (a) Complete the following table.

| Number $x$ | Square $x^2$ |
|---|---|
| 1 | |
| 2 | |
| 3 | |
| 4 | |
| 5 | |
| 6 | |
| 7 | |

(b) Using the results of part (a), fill in the blank in the following statement: For numbers larger than 1, the square of the number is _____ than the number.

**18.** (a) Complete the following table.

| Number $x$ | Square $x^2$ |
|---|---|
| $\frac{1}{2}$ | |
| $\frac{1}{3}$ | |
| $\frac{1}{4}$ | |
| $\frac{1}{5}$ | |
| $\frac{1}{6}$ | |
| $\frac{1}{7}$ | |
| $\frac{1}{8}$ | |

(b) Using the results of part (a), fill in the blank in the following statement: For numbers between 0 and 1, the square of the number is _____ than the number.

Use Property 1 for exponents to simplify each expression. Leave all answers in terms of exponents.

**19.** $x^4 \cdot x^5$          **20.** $x^7 \cdot x^3$

**21.** $y^{10} \cdot y^{20}$          **22.** $y^{30} \cdot y^{30}$

**23.** $2^5 \cdot 2^4 \cdot 2^3$          **24.** $4^2 \cdot 4^3 \cdot 4^4$

**25.** $x^4 \cdot x^6 \cdot x^8 \cdot x^{10}$          **26.** $x^{20} \cdot x^{18} \cdot x^{16} \cdot x^{14}$

Use Property 2 for exponents to write each of the following problems with a single exponent. (Assume all variables are positive numbers.)

**27.** $(x^2)^5$          **28.** $(x^5)^2$

**29.** $(5^4)^3$          **30.** $(5^3)^4$

**31.** $(y^3)^3$          **32.** $(y^2)^2$

**33.** $(2^5)^{10}$          **34.** $(10^5)^2$

**35.** $(a^3)^x$          **36.** $(a^5)^x$

**37.** $(b^x)^y$          **38.** $(b^r)^s$

Use Property 3 for exponents to simplify each of the following expressions.

**39.** $(4x)^2$          **40.** $(2x)^4$

**41.** $(2y)^5$          **42.** $(5y)^2$

**43.** $(-3x)^4$          **44.** $(-3x)^3$

**45.** $(0.5ab)^2$          **46.** $(0.4ab)^2$

**47.** $(4xyz)^3$          **48.** $(5xyz)^3$

Simplify the following expressions by using the properties of exponents.

**49.** $(2x^4)^3$          **50.** $(3x^5)^2$

**51.** $(4a^3)^2$          **52.** $(5a^2)^2$

**53.** $(x^2)^3(x^4)^2$          **54.** $(x^5)^2(x^3)^5$

**55.** $(a^3)^1(a^2)^4$          **56.** $(a^4)^1(a^1)^3$

**57.** $(2x)^3(2x)^4$          **58.** $(3x)^2(3x)^3$

**59.** $(3x^2)^3(2x)^4$          **60.** $(3x)^3(2x^3)^2$

**61.** $(4x^2y^3)^2$          **62.** $(9x^3y^5)^2$

**63.** $\left(\dfrac{2}{3} a^4b^5\right)^3$          **64.** $\left(\dfrac{3}{4} ab^7\right)^3$

**65.** Complete the following table, and then use the template to construct a line graph of the information in the table.

| Number $x$ | Square $x^2$ |
|---|---|
| $-3$ | |
| $-2$ | |
| $-1$ | |
| $0$ | |
| $1$ | |
| $2$ | |
| $3$ | |

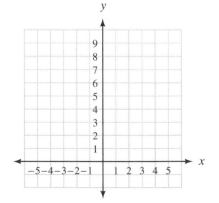

**66.** Complete the table, and then use the template to construct a line graph of the information in the table.

| Number $x$ | Cube $x^3$ |
|---|---|
| $-3$ | |
| $-2$ | |
| $-1$ | |
| $0$ | |
| $1$ | |
| $2$ | |
| $3$ | |

**67.** Complete the table. When you are finished, notice how the points in this table could be used to refine the line graph you created in Problem 65.

| Number $x$ | Square $x^2$ |
|---|---|
| $-2.5$ | |
| $-1.5$ | |
| $-0.5$ | |
| $0$ | |
| $0.5$ | |
| $1.5$ | |
| $2.5$ | |

**68.** Complete the following table. When you are finished, notice that this table contains exactly the same entries as the table from Problem 67. This table uses fractions, while the table from Problem 67 uses decimals.

| Number $x$ | Square $x^2$ |
|---|---|
| $-\frac{5}{2}$ | |
| $-\frac{3}{2}$ | |
| $-\frac{1}{2}$ | |
| $0$ | |
| $\frac{1}{2}$ | |
| $\frac{3}{2}$ | |
| $\frac{5}{2}$ | |

Write each number in scientific notation.

**69.** 43,200

**70.** 432,000

**71.** 570

**72.** 5,700

**73.** 238,000

**74.** 2,380,000

Write each number in expanded form.

**75.** $2.49 \times 10^3$

**76.** $2.49 \times 10^4$

**77.** $3.52 \times 10^2$

**78.** $3.52 \times 10^5$

**79.** $2.8 \times 10^4$

**80.** $2.8 \times 10^3$

## Applying the Concepts

**81. Volume of a Cube**   Find the volume of a cube if each side is 3 inches long.

**82. Volume of a Cube**   Find the volume of a cube if each side is 3 feet long.

**83. Volume of a Cube**   A bottle of perfume is packaged in a box that is in the shape of a cube (see next page). Find the volume of the box if each side is 2.5 inches long. Round to the nearest tenth.

2.5 in.

2.5 in.

2.5 in.

**84. Volume of a Cube** A television set is packaged in a box that is in the shape of a cube. Find the volume of the box if each side is 18 inches long.

18 in.

18 in.

18 in.

**85. Volume of a Box** A rented videotape is in a plastic container that has the shape of a rectangular solid. Find the volume of the container if the length is 8 inches, the width is 4.5 inches, and the height is 1 inch.

8 in.

4.5 in.    1 in.

**86. Volume of Your Book** Your textbook is in the shape of a rectangular solid. Find the volume in cubic inches.

**87. Volume of a Box** If a box has a volume of 42 cubic feet, is it possible for you to fit inside the box? Explain your answer.

**88. Volume of a Box** A box has a volume of 45 cubic inches. Will a can of soup fit inside the box? Explain your answer.

**89. Age in Seconds** If you are 21 years old, you have been alive for over 650,000,000 seconds. Write this last number in scientific notation.

**90. Distance Around the Earth** The distance around the Earth at the equator is more than 130,000,000 feet. Write this number in scientific notation.

**91. Lifetime Earnings** If you earn at least $12 an hour and work full-time for 30 years, you will make at least $7.4 \times 10^5$ dollars. Write this last number in expanded form.

**92. Heart Beats per Year** If your pulse is 72, then in 1 year your heart will beat at least $3.78 \times 10^7$ times. Write this last number in expanded form.

**93. Investing** If you put $1,000 into a savings account every year from the time you are 25 years old until you are 55 years old, you will have more than $1.8 \times 10^5$ dollars in the account when you reach 55 years of age (assuming 10% annual interest). Write $1.8 \times 10^5$ in expanded form.

**94. Investing** If you put $20 into a savings account every month from the time you are 20 years old until you are 30 years old, you will have over $3.27 \times 10^3$ dollars in the account when you reach 30 years of age (assuming 6% annual interest compounded monthly). Write $3.27 \times 10^3$ in expanded form.

**Displacement** The displacement, in cubic inches, of a car engine is given by the formula

$$d = \pi \cdot s \cdot c \cdot \left(\frac{1}{2} \cdot b\right)^2$$

where $s$ is the stroke and $b$ is the bore, as shown in the figure, and $c$ is the number of cylinders.

Calculate the engine displacement for each of the following cars. Use 3.14 to approximate $\pi$.

**95. Ferrari Modena**  8 cylinders, 3.35 inches of bore, 3.11 inches of stroke

**96. Audi A8**  8 cylinders, 3.32 inches of bore, 3.66 inches of stroke

**97. Mitsubishi Eclipse**  6 cylinders, 3.59 inches of bore, 2.99 inches of stroke

**98. Porsche 911 GT3**  6 cylinders, 3.94 inches of bore, 3.01 inches of stroke

**99. Fractals**  Fractals are infinitely complex shapes of repeating patterns, which can be used to model the world around us. An example of a fractal is the Koch snowflake curve, which we construct as follows: Start with a straight line segment divided into three equal sections. This is Stage 0.

  Stage 0

Next, cut out the middle section, and replace it with two pieces, each of which is the same length as the two outside sections. This becomes Stage 1.

  Stage 1

To construct Stage 2, we remove the middle third of each of the four line segments in Stage 1 and re-

place them with two segments equal in length to the remaining segments. Here is Stage 2:

  Stage 2

(a) If the original line (Stage 0) was 1 foot long, how long is the line in Stage 1?
(b) How long is the line in Stage 2?
(c) Construct Stage 3. How long is the line created in Stage 3?
(d) If you were to continue this process until you reached Stage 10, how long would the line in Stage 10 be?

**100. Fractals**  Here is another fractal construction, similar to the Koch snowflake curve constructed in the previous problem. We start with a straight line segment divided into three equal sections. This is Stage 0.

———————|————|—————  Stage 0

Next, cut out the middle section, and replace it with three pieces, each of which is the same length as the two outside sections. Arrange the three new pieces so that they form three sides of a square. This becomes Stage 1.

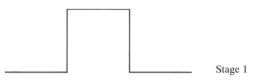  Stage 1

To construct Stage 2, we remove the middle third of each of the five line segments in Stage 1 and replace them with three segments equal in length to the remaining segments. Arrange each new set of segments so that they form three sides of a square. Here is Stage 2:

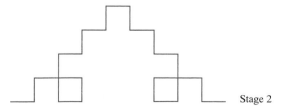  Stage 2

(a) If the original line (Stage 0) was 1 foot long, how long is the line in Stage 1?
(b) How long is the line in Stage 2?
(c) Construct Stage 3. How long is the line created in Stage 3?
(d) If you were to continue this process until you reached Stage 10, how long would the line in Stage 10 be?

## Review Problems

The following problems review material on subtraction with negative numbers that we covered in Section 1.4.

Reviewing these problems will help you understand the material in the next section.

Subtract.

**101.** $4 - 7$             **102.** $-4 - 7$

**103.** $4 - (-7)$        **104.** $-4 - (-7)$

**105.** $15 - 20$        **106.** $15 - (-20)$

**107.** $-15 - (-20)$    **108.** $-15 - 20$

---

## 5.2    Division With Exponents

In Section 5.1 we found that multiplication with the same base results in addition of exponents, that is, $a^r \cdot a^s = a^{r+s}$. Since division is the inverse of multiplication, we can expect division with the same base to result in subtraction of exponents.

To develop the properties for exponents under division, we again apply the definition of exponents:

$$\frac{x^5}{x^3} = \frac{x \cdot x \cdot x \cdot x \cdot x}{x \cdot x \cdot x}$$

$$= \frac{x \cdot x \cdot x}{x \cdot x \cdot x}(x \cdot x)$$

$$= 1(x \cdot x)$$

$$= x^2 \quad \textit{Notice: } 5 - 3 = 2$$

$$\frac{2^4}{2^7} = \frac{2 \cdot 2 \cdot 2 \cdot 2}{2 \cdot 2 \cdot 2 \cdot 2 \cdot 2 \cdot 2 \cdot 2}$$

$$= \frac{2 \cdot 2 \cdot 2 \cdot 2}{2 \cdot 2 \cdot 2 \cdot 2} \cdot \frac{1}{2 \cdot 2 \cdot 2}$$

$$= \frac{1}{2 \cdot 2 \cdot 2}$$

$$= \frac{1}{2^3} \quad \textit{Notice: } 7 - 4 = 3$$

In both cases division with the same base resulted in subtraction of the smaller exponent from the larger. The problem is deciding whether the answer is a fraction or not. The problem is resolved quite easily by the following definition.

---

**DEFINITION**    If $r$ is a positive integer, then $a^{-r} = \dfrac{1}{a^r} = \left(\dfrac{1}{a}\right)^r$    $(a \neq 0)$

---

The following examples illustrate how we use this definition to simplify expressions that contain negative exponents.

**E X A M P L E S**   Write each expression with a positive exponent and then simplify:

**1.** $2^{-3} = \dfrac{1}{2^3} = \dfrac{1}{8}$

*Notice:* Negative exponents do not indicate negative numbers. They indicate reciprocals.

**2.** $5^{-2} = \dfrac{1}{5^2} = \dfrac{1}{25}$

**3.** $3x^{-6} = 3 \cdot \dfrac{1}{x^6} = \dfrac{3}{x^6}$

Now let us look back to our original problem and try to work it again with the help of a negative exponent. We know that $2^4/2^7 = 1/2^3$. Let us decide now that with division of the same base, we will always subtract the exponent in the denominator from the exponent in the numerator and see if this conflicts with what we know is true.

$$\dfrac{2^4}{2^7} = 2^{4-7} \qquad \text{Subtracting the bottom exponent from the top exponent}$$

$$= 2^{-3} \qquad \text{Subtraction}$$

$$= \dfrac{1}{2^3} \qquad \text{Definition of negative exponents}$$

Subtracting the exponent in the denominator from the exponent in the numerator and then using the definition of negative exponents gives us the same result we obtained previously. We can now continue the list of properties of exponents we started in Section 5.1.

### PROPERTY 4 FOR EXPONENTS

If $a$ is any real number and $r$ and $s$ are integers, then

$$\dfrac{a^r}{a^s} = a^{r-s} \qquad (a \neq 0)$$

*In words:*   To divide with the same base, subtract the exponent in the denominator from the exponent in the numerator and raise the base to the exponent that results.

The following examples show how we use Property 4 and the definition for negative exponents to simplify expressions involving division.

**E X A M P L E S**   Simplify the following expressions:

**4.** $\dfrac{x^9}{x^6} = x^{9-6} = x^3$

**5.** $\dfrac{x^4}{x^{10}} = x^{4-10} = x^{-6} = \dfrac{1}{x^6}$

**6.** $\dfrac{2^{15}}{2^{20}} = 2^{15-20} = 2^{-5} = \dfrac{1}{2^5} = \dfrac{1}{32}$

Our final property of exponents is similar to Property 3 from Section 5.1, but it involves division instead of multiplication. After we have stated the property, we will give a proof of it. The proof shows why this property is true.

### PROPERTY 5 FOR EXPONENTS

If $a$ and $b$ are any two real numbers ($b \neq 0$) and $r$ is an integer, then

$$\left(\frac{a}{b}\right)^r = \frac{a^r}{b^r}$$

*In words:*  A quotient raised to a power is the quotient of the powers.

*Proof*

$$\left(\frac{a}{b}\right)^r = \left(a \cdot \frac{1}{b}\right)^r \qquad \text{By the definition of division}$$

$$= a^r \cdot \left(\frac{1}{b}\right)^r \qquad \text{By Property 3}$$

$$= a^r \cdot b^{-r} \qquad \text{By the definition of negative exponents}$$

$$= a^r \cdot \frac{1}{b^r} \qquad \text{By the definition of negative exponents}$$

$$= \frac{a^r}{b^r} \qquad \text{By the definition of division}$$

**EXAMPLES**  Simplify the following expressions:

**7.** $\left(\dfrac{x}{2}\right)^3 = \dfrac{x^3}{2^3} = \dfrac{x^3}{8}$

**8.** $\left(\dfrac{5}{y}\right)^2 = \dfrac{5^2}{y^2} = \dfrac{25}{y^2}$

**9.** $\left(\dfrac{2}{3}\right)^4 = \dfrac{2^4}{3^4} = \dfrac{16}{81}$

## ZERO AND ONE AS EXPONENTS

We have two special exponents left to deal with before our rules for exponents are complete: 0 and 1. To obtain an expression for $x^1$, we will solve a problem two different ways:

$$\left. \begin{array}{l} \dfrac{x^3}{x^2} = \dfrac{x \cdot x \cdot x}{x \cdot x} = x \\[12pt] \dfrac{x^3}{x^2} = x^{3-2} = x^1 \end{array} \right\} \quad \text{Hence, } x^1 = x$$

Stated generally, this rule says that $a^1 = a$. This seems reasonable, and we will use it, since it is consistent with our property of division using the same base.

We use the same procedure to obtain an expression for $x^0$:

$$\left. \begin{array}{l} \dfrac{5^2}{5^2} = \dfrac{25}{25} = 1 \\[12pt] \dfrac{5^2}{5^2} = 5^{2-2} = 5^0 \end{array} \right\} \quad \text{Hence, } 5^0 = 1$$

It seems, therefore, that the best definition of $x^0$ is 1 for all $x$ except $x = 0$. In the case of $x = 0$, we have $0^0$, which we will not define. This definition will probably seem awkward at first. Most people would like to define $x^0$ as 0 when they first encounter it. Remember, the zero in this expression is an exponent, so $x^0$ does not mean to multiply by zero. Thus, we can make the general statement that $a^0 = 1$ for all real numbers except $a = 0$.

Here are some examples involving the exponents 0 and 1.

**EXAMPLES**   Simplify the following expressions:

**10.** $8^0 = 1$

**11.** $8^1 = 8$

**12.** $4^0 + 4^1 = 1 + 4 = 5$

**13.** $(2x^2y)^0 = 1$

Here is a summary of the definitions and properties of exponents we have developed so far. For each definition or property in the list, $a$ and $b$ are real numbers, and $r$ and $s$ are integers.

| **DEFINITIONS** | **PROPERTIES** |
|---|---|
| $a^{-r} = \dfrac{1}{a^r} = \left(\dfrac{1}{a}\right)^r \quad (a \neq 0)$ | **1.** $a^r \cdot a^s = a^{r+s}$ |
| | **2.** $(a^r)^s = a^{rs}$ |
| $a^1 = a$ | **3.** $(ab)^r = a^r b^r$ |
| $a^0 = 1 \quad (a \neq 0)$ | **4.** $\dfrac{a^r}{a^s} = a^{r-s} \quad (a \neq 0)$ |
| | **5.** $\left(\dfrac{a}{b}\right)^r = \dfrac{a^r}{b^r} \quad (b \neq 0)$ |

Here are some additional examples. These examples use a combination of the preceding properties and definitions.

---

**EXAMPLES** Simplify each expression. Write all answers with positive exponents only:

**14.** $\dfrac{(5x^3)^2}{x^4} = \dfrac{25x^6}{x^4}$     Properties 2 and 3

$\phantom{\dfrac{(5x^3)^2}{x^4}} = 25x^2$     Property 4

**15.** $\dfrac{x^{-8}}{(x^2)^3} = \dfrac{x^{-8}}{x^6}$     Property 2

$\phantom{\dfrac{x^{-8}}{(x^2)^3}} = x^{-8-6}$     Property 4

$\phantom{\dfrac{x^{-8}}{(x^2)^3}} = x^{-14}$     Subtraction

$\phantom{\dfrac{x^{-8}}{(x^2)^3}} = \dfrac{1}{x^{14}}$     Definition of negative exponents

**16.** $\left(\dfrac{y^5}{y^3}\right)^2 = \dfrac{(y^5)^2}{(y^3)^2}$     Property 5

$\phantom{\left(\dfrac{y^5}{y^3}\right)^2} = \dfrac{y^{10}}{y^6}$     Property 2

$\phantom{\left(\dfrac{y^5}{y^3}\right)^2} = y^4$     Property 4

Notice in Example 16 that we could have simplified inside the parentheses first and then raised the result to the second power:

$$\left(\frac{y^5}{y^3}\right)^2 = (y^2)^2 = y^4$$

**17.** $(3x^5)^{-2} = \dfrac{1}{(3x^5)^2}$     Definition of negative exponents

$\phantom{(3x^5)^{-2}} = \dfrac{1}{9x^{10}}$     Properties 2 and 3

**18.** $x^{-8} \cdot x^5 = x^{-8+5}$     Property 1

$\phantom{x^{-8} \cdot x^5} = x^{-3}$     Addition

$\phantom{x^{-8} \cdot x^5} = \dfrac{1}{x^3}$     Definition of negative exponents

**19.** $\dfrac{(a^3)^2 a^{-4}}{(a^{-4})^3} = \dfrac{a^6 a^{-4}}{a^{-12}}$     Property 2

$\phantom{\dfrac{(a^3)^2 a^{-4}}{(a^{-4})^3}} = \dfrac{a^2}{a^{-12}}$     Property 1

$\phantom{\dfrac{(a^3)^2 a^{-4}}{(a^{-4})^3}} = a^{14}$     Property 4

---

In the next two examples we use division to compare the area and volume of geometric figures.

**E X A M P L E   2 0**   Suppose you have two squares, one of which is larger than the other. If a side of the larger square is 3 times as long as a side of the smaller square, how many of the smaller squares will it take to cover up the larger square?

**Solution**   If we let $x$ represent the length of a side of the smaller square, then the length of a side of the larger square is $3x$. The area of each square, along with a diagram of the situation, is given in Figure 1.

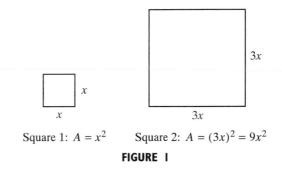

Square 1: $A = x^2$        Square 2: $A = (3x)^2 = 9x^2$

**FIGURE  1**

To find out how many smaller squares it will take to cover up the larger square, we divide the area of the larger square by the area of the smaller square.

$$\frac{\text{Area of square 2}}{\text{Area of square 1}} = \frac{9x^2}{x^2} = 9$$

It will take 9 of the smaller squares to cover the larger square.

**E X A M P L E   2 1**   Suppose you have two boxes, each of which is a cube. If the length of a side in the second box is 3 times as long as the length of a side of the first box, how many of the smaller boxes will fit inside the larger box?

**Solution**   If we let $x$ represent the length of a side of the smaller box, then the length of a side of the larger box is $3x$. The volume of each box, along with a diagram of the situation, is given in Figure 2.

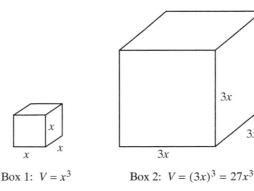

Box 1: $V = x^3$         Box 2: $V = (3x)^3 = 27x^3$

**FIGURE  2**

To find out how many smaller boxes will fit inside the larger box, we divide the volume of the larger box by the volume of the smaller box.

$$\frac{\text{Volume of box 2}}{\text{Volume of box 1}} = \frac{27x^3}{x^3} = 27$$

We can fit 27 of the smaller boxes inside the larger box.

## MORE ON SCIENTIFIC NOTATION

Now that we have completed our list of definitions and properties of exponents, we can expand the work we did previously with scientific notation.

Recall that a number is in scientific notation when it is written in the form

$$n \times 10^r$$

where $1 \leq n < 10$ and $r$ is an integer.

Since negative exponents give us reciprocals, we can use negative exponents to write very small numbers in scientific notation. For example, the number 0.00057, when written in scientific notation, is equivalent to $5.7 \times 10^{-4}$. Here's why:

$$5.7 \times 10^{-4} = 5.7 \times \frac{1}{10^4} = 5.7 \times \frac{1}{10,000} = \frac{5.7}{10,000} = 0.00057$$

Example 22 lists some other numbers in both scientific notation and expanded form.

**EXAMPLE 22**

| Number Written the Long Way | | Number Written Again in Scientific Notation |
|---|---|---|
| 376,000 | = | $3.76 \times 10^5$ |
| 49,500 | = | $4.95 \times 10^4$ |
| 3,200 | = | $3.2 \times 10^3$ |
| 591 | = | $5.91 \times 10^2$ |
| 46 | = | $4.6 \times 10^1$ |
| 8 | = | $8 \times 10^0$ |
| 0.47 | = | $4.7 \times 10^{-1}$ |
| 0.093 | = | $9.3 \times 10^{-2}$ |
| 0.00688 | = | $6.88 \times 10^{-3}$ |
| 0.0002 | = | $2 \times 10^{-4}$ |
| 0.000098 | = | $9.8 \times 10^{-5}$ |

Notice that in each case, when the number is written in scientific notation, the decimal point in the first number is placed so that the number is between 1 and 10.

The exponent on 10 in the second number keeps track of the number of places we moved the decimal point in the original number to get a number between 1 and 10:

$$376{,}000 = 3.76 \times 10^5$$

Moved 5 places.

Decimal point was originally here.

Keeps track of the 5 places we moved the decimal point.

$$0.00688 = 6.88 \times 10^{-3}$$

Moved 3 places.

Keeps track of the 3 places we moved the decimal point.

 # Getting Ready for Class

*After reading through the preceding section, respond in your own words and in complete sentences.*

**A.** How do you divide two expressions containing exponents when they each have the same base?

**B.** Explain the difference between $3^2$ and $3^{-2}$.

**C.** If a positive base is raised to a negative exponent, can the result be a negative number?

**D.** Explain what happens when we use 0 as an exponent.

## PROBLEM SET 5.2

Write each of the following with positive exponents, and then simplify, when possible.

**1.** $3^{-2}$  **2.** $3^{-3}$  **3.** $6^{-2}$

**4.** $2^{-6}$  **5.** $8^{-2}$  **6.** $3^{-4}$

**7.** $5^{-3}$  **8.** $9^{-2}$  **9.** $2x^{-3}$

**10.** $5x^{-1}$  **11.** $(2x)^{-3}$  **12.** $(5x)^{-1}$

**13.** $(5y)^{-2}$  **14.** $5y^{-2}$  **15.** $10^{-2}$

**16.** $10^{-3}$

**17.** Complete the following table.

| Number $x$ | Square $x^2$ | Power of 2 $2^x$ |
|---|---|---|
| $-3$ | | |
| $-2$ | | |
| $-1$ | | |
| $0$ | | |
| $1$ | | |
| $2$ | | |
| $3$ | | |

**18.** Complete the following table.

| Number $x$ | Cube $x^3$ | Power of 3 $3^x$ |
|---|---|---|
| $-3$ | | |
| $-2$ | | |
| $-1$ | | |
| $0$ | | |
| $1$ | | |
| $2$ | | |
| $3$ | | |

Use Property 4 to simplify each of the following expressions. Write all answers that contain exponents with positive exponents only.

**19.** $\dfrac{5^1}{5^3}$    **20.** $\dfrac{7^6}{7^8}$    **21.** $\dfrac{x^{10}}{x^4}$

**22.** $\dfrac{x^4}{x^{10}}$    **23.** $\dfrac{4^3}{4^0}$    **24.** $\dfrac{4^0}{4^3}$

**25.** $\dfrac{(2x)^7}{(2x)^4}$    **26.** $\dfrac{(2x)^4}{(2x)^7}$    **27.** $\dfrac{6^{11}}{6}$

**28.** $\dfrac{8^7}{8}$    **29.** $\dfrac{6}{6^{11}}$    **30.** $\dfrac{8}{8^7}$

**31.** $\dfrac{2^{-5}}{2^3}$    **32.** $\dfrac{2^{-5}}{2^{-3}}$    **33.** $\dfrac{2^5}{2^{-3}}$

**34.** $\dfrac{2^{-3}}{2^{-5}}$    **35.** $\dfrac{(3x)^{-5}}{(3x)^{-8}}$    **36.** $\dfrac{(2x)^{-10}}{(2x)^{-15}}$

Simplify the following expressions. Any answers that contain exponents should contain positive exponents only.

**37.** $(3xy)^4$    **38.** $(4xy)^3$    **39.** $10^0$

**40.** $10^1$    **41.** $(2a^2b)^1$    **42.** $(2a^2b)^0$

**43.** $(7y^3)^{-2}$    **44.** $(5y^4)^{-2}$    **45.** $x^{-3}x^{-5}$

**46.** $x^{-6}\cdot x^8$    **47.** $y^7\cdot y^{-10}$    **48.** $y^{-4}\cdot y^{-6}$

**49.** $\dfrac{(x^2)^3}{x^4}$    **50.** $\dfrac{(x^5)^3}{x^{10}}$    **51.** $\dfrac{(a^4)^3}{(a^3)^2}$

**52.** $\dfrac{(a^5)^3}{(a^5)^2}$    **53.** $\dfrac{y^7}{(y^2)^8}$    **54.** $\dfrac{y^2}{(y^3)^4}$

**55.** $\left(\dfrac{y^7}{y^2}\right)^8$    **56.** $\left(\dfrac{y^2}{y^3}\right)^4$    **57.** $\dfrac{(x^{-2})^3}{x^{-5}}$

**58.** $\dfrac{(x^2)^{-3}}{x^{-5}}$    **59.** $\left(\dfrac{x^{-2}}{x^{-5}}\right)^3$    **60.** $\left(\dfrac{x^2}{x^{-5}}\right)^{-3}$

**61.** $\dfrac{(a^3)^2(a^4)^5}{(a^5)^2}$    **62.** $\dfrac{(a^4)^8(a^2)^5}{(a^3)^4}$    **63.** $\dfrac{(a^{-2})^3(a^4)^2}{(a^{-3})^{-2}}$

**64.** $\dfrac{(a^{-5})^{-3}(a^7)^{-1}}{(a^{-3})^5}$

**65.** Complete the following table, and then use the template to construct a line graph of the information in the table.

| Number $x$ | Power of 2 $2^x$ |
|---|---|
| $-3$ | |
| $-2$ | |
| $-1$ | |
| $0$ | |
| $1$ | |
| $2$ | |
| $3$ | |

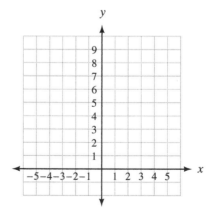

**66.** Complete the following table, and then use the template to construct a line graph of the information in the table.

| Number $x$ | Power of 3 $3^x$ |
|:---:|:---:|
| $-3$ | |
| $-2$ | |
| $-1$ | |
| $0$ | |
| $1$ | |
| $2$ | |
| $3$ | |

| Expanded Form | Scientific Notation $n \times 10^r$ |
|:---:|:---:|
| 0.000357 | $3.57 \times 10^{-4}$ |
| 0.00357 | |
| 0.0357 | |
| 0.357 | |
| 3.57 | |
| 35.7 | |
| 357 | |
| 3,570 | |
| 35,700 | |

**74.** Complete the following table.

| Expanded Form | Scientific Notation $n \times 10^r$ |
|:---:|:---:|
| 0.000123 | $1.23 \times 10^{-4}$ |
| | $1.23 \times 10^{-3}$ |
| | $1.23 \times 10^{-2}$ |
| | $1.23 \times 10^{-1}$ |
| | $1.23 \times 10^{0}$ |
| | $1.23 \times 10^{1}$ |
| | $1.23 \times 10^{2}$ |
| | $1.23 \times 10^{3}$ |
| | $1.23 \times 10^{4}$ |

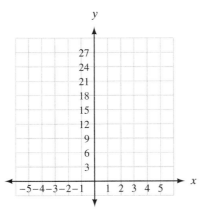

Write each of the following numbers in scientific notation.

**67.** 0.0048

**68.** 0.000048

**69.** 25

**70.** 35

**71.** 0.000009

**72.** 0.0009

**73.** Complete the following table.

Write each of the following numbers in expanded form.

**75.** $4.23 \times 10^{-3}$     **76.** $4.23 \times 10^{3}$

**77.** $8 \times 10^{-5}$     **78.** $8 \times 10^{5}$

**79.** $4.2 \times 10^{0}$     **80.** $4.2 \times 10^{1}$

## Applying the Concepts

### Scientific Notation Problems

**81.** Some home computers can do a calculation in $2 \times 10^{-3}$ second. Write this number in expanded form.

**82.** Some of the cells in the human body have a radius of $3 \times 10^{-5}$ inch. Write this number in expanded form.

**83.** One of the smallest ants in the world has a length of only 0.006 inch. Write this number in scientific notation.

**84.** Some cameras used in scientific research can take 1 picture every 0.000000167 second. Write this number in scientific notation.

**85.** The number $25 \times 10^3$ is not in scientific notation because 25 is larger than 10. Write $25 \times 10^3$ in scientific notation.

**86.** The number $0.25 \times 10^3$ is not in scientific notation because 0.25 is less than 1. Write $0.25 \times 10^3$ in scientific notation.

**87.** The number $23.5 \times 10^4$ is not in scientific notation because 23.5 is not between 1 and 10. Rewrite $23.5 \times 10^4$ in scientific notation.

**88.** The number $375 \times 10^3$ is not in scientific notation because 375 is not between 1 and 10. Rewrite $375 \times 10^3$ in scientific notation.

**89.** The number $0.82 \times 10^{-3}$ is not in scientific notation because 0.82 is not between 1 and 10. Rewrite $0.82 \times 10^{-3}$ in scientific notation.

**90.** The number $0.93 \times 10^{-2}$ is not in scientific notation because 0.93 is not between 1 and 10. Rewrite $0.93 \times 10^{-2}$ in scientific notation.

**Comparing Areas** Suppose you have two squares, one of which is larger than the other. Suppose further that the side of the larger square is twice as long as the side of the smaller square.

**91.** If the length of the side of the smaller square is 10 inches, give the area of each square. Then find the number of smaller squares it will take to cover the larger square.

**92.** How many smaller squares will it take to cover the larger square if the length of the side of the smaller square is 1 foot?

**93.** If the length of the side of the smaller square is $x$, find the area of each square. Then find the number of smaller squares it will take to cover the larger square.

**94.** Suppose the length of the side of the larger square is 1 foot. How many smaller squares will it take to cover the larger square?

**Comparing Volumes** Suppose you have two boxes, each of which is a cube. Suppose further that the length of a side of the second box is twice as long as the length of a side of the first box.

**95.** If the length of a side of the first box is 6 inches, give the volume of each box. Then find the number of smaller boxes that will fit inside the larger box.

**96.** How many smaller boxes can be placed inside the larger box if the length of a side of the second box is 1 foot?

**97.** If the length of a side of the first box is $x$, find the volume of each box. Then find the number of smaller boxes that will fit inside the larger box.

**98.** Suppose the length of a side of the larger box is 12 inches. How many smaller boxes will fit inside the larger box?

## Review Problems

The following problems review material we covered in Section 2.1. They will help you understand some of the material in the next section.

Simplify the following expressions.

**99.** $4x + 3x$

**100.** $9x + 7x$

**101.** $5a - 3a$

**102.** $10a - 2a$

**103.** $4y + 5y + y$

**104.** $6y - y + 2y$

---

**5.3**   **Operations With Monomials**

We have developed all the tools necessary to perform the four basic operations on the simplest of polynomials: monomials.

> **DEFINITION**  A **monomial** is a one-term expression that is either a constant (number) or the product of a constant and one or more variables raised to whole-number exponents.

The following are examples of monomials:

$$-3 \qquad 15x \qquad -23x^2y \qquad 49x^4y^2z^4 \qquad \frac{3}{4}a^2b^3$$

The numerical part of each monomial is called the *numerical coefficient,* or just the *coefficient.* Monomials are also called *terms.*

## MULTIPLICATION AND DIVISION OF MONOMIALS

There are two basic steps involved in the multiplication of monomials. First, we rewrite the products using the commutative and associative properties. Then, we simplify by multiplying coefficients and adding exponents of like bases.

**EXAMPLES**  Multiply:

**1.** $(-3x^2)(4x^3) = (-3 \cdot 4)(x^2 \cdot x^3)$    Commutative and associative properties

$$= -12x^5 \qquad \text{Multiply coefficients; add exponents.}$$

**2.** $\left(\dfrac{4}{5}x^5 \cdot y^2\right)(10x^3 \cdot y) = \left(\dfrac{4}{5} \cdot 10\right)(x^5 \cdot x^3)(y^2 \cdot y)$    Commutative and associative properties

$$= 8x^8y^3 \qquad \text{Multiply coefficients; add exponents.}$$

You can see that in each case the work was the same — multiply coefficients and add exponents of the same base. We can expect division of monomials to proceed in a similar way. Since our properties are consistent, division of monomials will result in division of coefficients and subtraction of exponents of like bases.

**EXAMPLES**  Divide:

**3.** $\dfrac{15x^3}{3x^2} = \dfrac{15}{3} \cdot \dfrac{x^3}{x^2}$    Write as separate fractions.

$$= 5x \qquad \text{Divide coefficients; subtract exponents.}$$

**4.** $\dfrac{39x^2y^3}{3xy^5} = \dfrac{39}{3} \cdot \dfrac{x^2}{x} \cdot \dfrac{y^3}{y^5}$    Write as separate fractions.

$$= 13x \cdot \dfrac{1}{y^2} \qquad \text{Divide coefficients; subtract exponents.}$$

$$= \dfrac{13x}{y^2} \qquad \text{Write answer as a single fraction.}$$

In Example 4, the expression $y^3/y^5$ simplifies to $1/y^2$ because of Property 4 for exponents and the definition of negative exponents. If we were to show all the work in this simplification process, it would look like this:

$$\frac{y^3}{y^5} = y^{3-5} \qquad \text{Property 4 for exponents}$$

$$= y^{-2} \qquad \text{Subtraction}$$

$$= \frac{1}{y^2} \qquad \text{Definition of negative exponents}$$

The point of this explanation is this: Even though we may not show all the steps when simplifying an expression involving exponents, the result we obtain can still be justified using the properties of exponents. We have not introduced any new properties in Example 4; we have just not shown the details of each simplification.

**EXAMPLE 5**  Divide:

$$\frac{25a^5b^3}{50a^2b^7} = \frac{25}{50} \cdot \frac{a^5}{a^2} \cdot \frac{b^3}{b^7} \qquad \text{Write as separate fractions.}$$

$$= \frac{1}{2} \cdot a^3 \cdot \frac{1}{b^4} \qquad \text{Divide coefficients; subtract exponents.}$$

$$= \frac{a^3}{2b^4} \qquad \text{Write answer as a single fraction.}$$

Notice in Example 5 that dividing 25 by 50 results in $\frac{1}{2}$. This is the same result we would obtain if we reduced the fraction $\frac{25}{50}$ to lowest terms, and there is no harm in thinking of it that way. Also, notice that the expression $b^3/b^7$ simplifies to $1/b^4$ by Property 4 for exponents and the definition of negative exponents, even though we have not shown the steps involved in doing so.

## MULTIPLICATION AND DIVISION OF NUMBERS WRITTEN IN SCIENTIFIC NOTATION

We multiply and divide numbers written in scientific notation using the same steps we used to multiply and divide monomials.

**EXAMPLE 6**  Multiply $(4 \times 10^7)(2 \times 10^{-4})$.

**Solution**  Since multiplication is commutative and associative, we can rearrange the order of these numbers and group them as follows:

$$(4 \times 10^7)(2 \times 10^{-4}) = (4 \times 2)(10^7 \times 10^{-4})$$

$$= 8 \times 10^3$$

Notice that we add exponents, $7 + (-4) = 3$, when we multiply with the same base.

**EXAMPLE  7**   Divide $\dfrac{9.6 \times 10^{12}}{3 \times 10^4}$.

**Solution**   We group the numbers between 1 and 10 separately from the powers of 10 and proceed as we did in Example 6:

$$\frac{9.6 \times 10^{12}}{3 \times 10^4} = \frac{9.6}{3} \times \frac{10^{12}}{10^4}$$

$$= 3.2 \times 10^8$$

Notice that the procedure we used in both of these examples is very similar to multiplication and division of monomials, for which we multiplied or divided coefficients and added or subtracted exponents.

## ADDITION AND SUBTRACTION OF MONOMIALS

Addition and subtraction of monomials will be almost identical, since subtraction is defined as addition of the opposite. With multiplication and division of monomials, the key was rearranging the numbers and variables using the commutative and associative properties. With addition, the key is application of the distributive property. We sometimes use the phrase *combine monomials* to describe addition and subtraction of monomials.

> **DEFINITION**   Two terms (monomials) with the same variable part (same variables raised to the same powers) are called **similar** (or *like*) **terms.**

You can add only similar terms. This is because the distributive property (which is the key to addition of monomials) cannot be applied to terms that are not similar.

**EXAMPLES**   Combine the following monomials.

**8.** $-3x^2 + 15x^2 = (-3 + 15)x^2$     Distributive property

    $\phantom{-3x^2 + 15x^2} = 12x^2$     Add coefficients.

**9.** $9x^2y - 20x^2y = (9 - 20)x^2y$     Distributive property

    $\phantom{9x^2y - 20x^2y} = -11x^2y$     Add coefficients.

**10.** $5x^2 + 8y^2$     In this case we cannot apply the distributive property, so we cannot add the monomials.

The next examples show how we simplify expressions containing monomials when more than one operation is involved.

**EXAMPLE 11**    Simplify $\dfrac{(6x^4y)(3x^7y^5)}{9x^5y^2}$.

**Solution**    We begin by multiplying the two monomials in the numerator:

$$\frac{(6x^4y)(3x^7y^5)}{9x^5y^2} = \frac{18x^{11}y^6}{9x^5y^2} \qquad \text{Simplify numerator.}$$

$$= 2x^6y^4 \qquad \text{Divide.}$$

**EXAMPLE 12**    Simplify $\dfrac{(6.8 \times 10^5)(3.9 \times 10^{-7})}{7.8 \times 10^{-4}}$.

**Solution**    We group the numbers between 1 and 10 separately from the powers of 10:

$$\frac{(6.8)(3.9)}{7.8} \times \frac{(10^5)(10^{-7})}{10^{-4}} = 3.4 \times 10^{5+(-7)-(-4)}$$

$$= 3.4 \times 10^2$$

**EXAMPLE 13**    Simplify $\dfrac{14x^5}{2x^2} + \dfrac{15x^8}{3x^5}$.

**Solution**    Simplifying each expression separately and then combining similar terms gives

$$\frac{14x^5}{2x^2} + \frac{15x^8}{3x^5} = 7x^3 + 5x^3 \qquad \text{Divide.}$$

$$= 12x^3 \qquad \text{Add.}$$

**EXAMPLE 14**    A rectangular solid is twice as long as it is wide and half as high as it is wide. Write an expression for the volume.

**Solution**    We begin by making a diagram of the object (Figure 1) with the dimensions labeled as given in the problem.

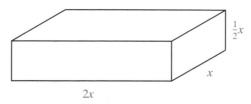

**FIGURE 1**

The volume is the product of the three dimensions:

$$V = 2x \cdot x \cdot \frac{1}{2}x = x^3$$

The box has the same volume as a cube with side $x$, as shown in Figure 2.

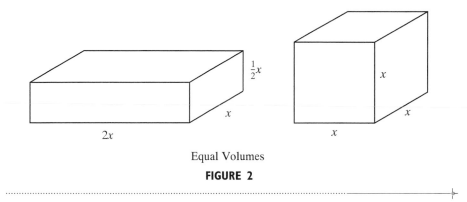

Equal Volumes

**FIGURE 2**

## MORE ABOUT GRAPHING

Up to this point, the graphing we have done in this chapter has produced line graphs. On the next graph we draw, we connect the points with a smooth curve instead of line segments. This will give us an accurate picture of the graph of the equation $y = x^2$.

**EXAMPLE  15**    Graph the equation $y = x^2$.

**Solution**   We start by finding some ordered pairs that are solutions to the equation. We can choose any convenient numbers for $x$ and then use the equation $y = x^2$ to find the corresponding values for $y$. Let's use the values $-3, -2, -1, 0, 1, 2$, and $3$ for $x$ and find corresponding values for $y$. Here is how the table looks when we let $x$ have these values:

| $x$ | $y = x^2$ | $y$ |
|-----|-----------|-----|
| $-3$ | $y = (-3)^2 = 9$ | 9 |
| $-2$ | $y = (-2)^2 = 4$ | 4 |
| $-1$ | $y = (-1)^2 = 1$ | 1 |
| $0$ | $y = 0^2 = 0$ | 0 |
| $1$ | $y = 1^2 = 1$ | 1 |
| $2$ | $y = 2^2 = 4$ | 4 |
| $3$ | $y = 3^2 = 9$ | 9 |

The table gives us the solutions $(-3, 9), (-2, 4), (-1, 1), (0, 0), (1, 1), (2, 4)$, and $(3, 9)$ for the equation $y = x^2$. We plot each of the points on a rectangular coordinate system and draw a smooth curve between them, as shown in Figure 3.

The graph is called a **parabola.** All equations of the form $y = ax^2$ ($a \neq 0$) produce parabolas when graphed. The point (0, 0) is called the **vertex** of the parabola in Figure 3.

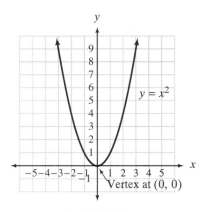

FIGURE 3

![getting ready icon] **Getting Ready for Class**

*After reading through the preceding section, respond in your own words and in complete sentences.*

**A.** What is a monomial?

**B.** Describe how you would multiply $3x^2$ and $5x^2$.

**C.** Describe how you would add $3x^2$ and $5x^2$.

**D.** Describe how you would multiply two numbers written in scientific notation.

## PROBLEM SET 5.3

Multiply.

**1.** $(3x^4)(4x^3)$    **2.** $(6x^5)(-2x^2)$

**3.** $(-2y^4)(8y^7)$    **4.** $(5y^{10})(2y^5)$

**5.** $(8x)(4x)$    **6.** $(7x)(5x)$

**7.** $(10a^3)(10a)(2a^2)$    **8.** $(5a^4)(10a)(10a^4)$

**9.** $(6ab^2)(-4a^2b)$    **10.** $(-5a^3b)(4ab^4)$

**11.** $(4x^2y)(3x^3y^3)(2xy^4)$

**12.** $(5x^6)(-10xy^4)(-2x^2y^6)$

Divide. Write all answers with positive exponents only.

**13.** $\dfrac{15x^3}{5x^2}$    **14.** $\dfrac{25x^5}{5x^4}$

**15.** $\dfrac{18y^9}{3y^{12}}$    **16.** $\dfrac{24y^4}{-8y^7}$

**17.** $\dfrac{32a^3}{64a^4}$    **18.** $\dfrac{25a^5}{75a^6}$

**19.** $\dfrac{21a^2b^3}{-7ab^5}$    **20.** $\dfrac{32a^5b^6}{8ab^5}$

**21.** $\dfrac{3x^3y^2z}{27xy^2z^3}$    **22.** $\dfrac{5x^5y^4z}{30x^3yz^2}$

**23.** Fill in the table.

| $a$ | $b$ | $ab$ | $\dfrac{a}{b}$ | $\dfrac{b}{a}$ |
|---|---|---|---|---|
| 10 | $5x$ | | | |
| $20x^3$ | $6x^2$ | | | |
| $25x^5$ | $5x^4$ | | | |
| $3x^{-2}$ | $3x^2$ | | | |
| $-2y^4$ | $8y^7$ | | | |

**24.** Fill in the table.

| $a$ | $b$ | $ab$ | $\dfrac{a}{b}$ | $\dfrac{b}{a}$ |
|---|---|---|---|---|
| $10y$ | $2y^2$ | | | |
| $10y^2$ | $2y$ | | | |
| $5y^3$ | 15 | | | |
| 5 | $15y^3$ | | | |
| $4y^{-3}$ | $4y^3$ | | | |

Find each product. Write all answers in scientific notation.

**25.** $(3 \times 10^3)(2 \times 10^5)$    **26.** $(4 \times 10^8)(1 \times 10^6)$

**27.** $(3.5 \times 10^4)(5 \times 10^{-6})$

**28.** $(7.1 \times 10^5)(2 \times 10^{-8})$

**29.** $(5.5 \times 10^{-3})(2.2 \times 10^{-4})$

**30.** $(3.4 \times 10^{-2})(4.5 \times 10^{-6})$

Find each quotient. Write all answers in scientific notation.

**31.** $\dfrac{8.4 \times 10^5}{2 \times 10^2}$    **32.** $\dfrac{9.6 \times 10^{20}}{3 \times 10^6}$

**33.** $\dfrac{6 \times 10^8}{2 \times 10^{-2}}$    **34.** $\dfrac{8 \times 10^{12}}{4 \times 10^{-3}}$

**35.** $\dfrac{2.5 \times 10^{-6}}{5 \times 10^{-4}}$    **36.** $\dfrac{4.5 \times 10^{-8}}{9 \times 10^{-4}}$

Combine by adding or subtracting as indicated.

**37.** $3x^2 + 5x^2$    **38.** $4x^3 + 8x^3$

**39.** $8x^5 - 19x^5$    **40.** $75x^6 - 50x^6$

**41.** $2a + a - 3a$    **42.** $5a + a - 6a$

**43.** $10x^3 - 8x^3 + 2x^3$    **44.** $7x^5 + 8x^5 - 12x^5$

**45.** $20ab^2 - 19ab^2 + 30ab^2$

**46.** $18a^3b^2 - 20a^3b^2 + 10a^3b^2$

**47.** Fill in the table.

| $a$ | $b$ | $ab$ | $a + b$ |
|---|---|---|---|
| $5x$ | $3x$ | | |
| $4x^2$ | $2x^2$ | | |
| $3x^3$ | $6x^3$ | | |
| $2x^4$ | $-3x^4$ | | |
| $x^5$ | $7x^5$ | | |

**48.** Fill in the table.

| $a$ | $b$ | $ab$ | $a - b$ |
|---|---|---|---|
| $2y$ | $3y$ | | |
| $-2y$ | $3y$ | | |
| $4y^2$ | $5y^2$ | | |
| $y^3$ | $-3y^3$ | | |
| $5y^4$ | $7y^4$ | | |

Simplify. Write all answers with positive exponents only.

**49.** $\dfrac{(3x^2)(8x^5)}{6x^4}$    **50.** $\dfrac{(7x^3)(6x^8)}{14x^5}$

**51.** $\dfrac{(9a^2b)(2a^3b^4)}{18a^5b^7}$    **52.** $\dfrac{(21a^5b)(2a^8b^4)}{14ab}$

**53.** $\dfrac{(4x^3y^2)(9x^4y^{10})}{(3x^5y)(2x^6y)}$    **54.** $\dfrac{(5x^4y^4)(10x^3y^3)}{(25xy^5)(2xy^7)}$

Simplify each expression, and write all answers in scientific notation.

**55.** $\dfrac{(6 \times 10^8)(3 \times 10^5)}{9 \times 10^7}$

**56.** $\dfrac{(8 \times 10^4)(5 \times 10^{10})}{2 \times 10^7}$

**57.** $\dfrac{(5 \times 10^3)(4 \times 10^{-5})}{2 \times 10^{-2}}$

**58.** $\dfrac{(7 \times 10^6)(4 \times 10^{-4})}{1.4 \times 10^{-3}}$

**59.** $\dfrac{(2.8 \times 10^{-7})(3.6 \times 10^4)}{2.4 \times 10^3}$

**60.** $\dfrac{(5.4 \times 10^2)(3.5 \times 10^{-9})}{4.5 \times 10^6}$

Simplify.

**61.** $\dfrac{18x^4}{3x} + \dfrac{21x^7}{7x^4}$

**62.** $\dfrac{24x^{10}}{6x^4} + \dfrac{32x^7}{8x}$

**63.** $\dfrac{45a^6}{9a^4} - \dfrac{50a^8}{2a^6}$

**64.** $\dfrac{16a^9}{4a} - \dfrac{28a^{12}}{4a^4}$

**65.** $\dfrac{6x^7y^4}{3x^2y^2} + \dfrac{8x^5y^8}{2y^6}$

**66.** $\dfrac{40x^{10}y^{10}}{8x^2y^5} + \dfrac{10x^8y^8}{5y^3}$

Use your knowledge of the properties and definitions of exponents to find $x$ in each of the following.

**67.** $4^x \cdot 4^5 = 4^7$

**68.** $\dfrac{5^x}{5^3} = 5^4$

**69.** $(7^3)^x = 7^{12}$

**70.** $\dfrac{3^x}{3^4} = 9$

## Applying the Concepts

**71. Comparing Expressions**   The statement

$$(a + b)^2 = a^2 + b^2$$

looks similar to Property 3 for exponents. However, it is not a property of exponents because almost every time we replace $a$ and $b$ with numbers, this expression becomes a false statement. Let $a = 4$ and $b = 5$ in the expressions $(a + b)^2$ and $a^2 + b^2$, and see what each simplifies to.

**72. Comparing Expressions**   Show that the statement $(a - b)^2 = a^2 - b^2$ is not, in general, true by substituting 3 for $a$ and 5 for $b$ in each of the expressions and then simplifying each result.

**73. Comparing Expressions**   Show that the expressions $(a + b)^2$ and $a^2 + 2ab + b^2$ are equal when $a = 3$ and $b = 4$.

**74. Comparing Expressions**   Show that the expressions $(a - b)^2$ and $a^2 - 2ab + b^2$ are equal when $a = 7$ and $b = 5$.

**75. Geometry**   A rectangle is twice as long as it is wide. Write an expression for the perimeter and an expression for the area of the rectangle.

**76. Geometry**   A rectangle is 3 times as long as it is wide. Write an expression for the perimeter and an expression for the area of the rectangle.

**77. Volume**   A rectangular solid is twice as long as it is wide. If the height is 4 inches, find an expression for the volume.

**78. Volume**   A rectangular solid is 4 times as long as it is wide. If the height is 2 inches, find an expression for the volume.

**79. Volume and Scientific Notation**   A delivery truck is being filled with boxes of prepackaged food. The cargo space of the truck is 8.5 feet wide, 55 feet long, and 10.1 feet high.

(a) Find the volume of the truck, round your answer to the nearest hundred, then write the answer in scientific notation.

(b) If each food box has a volume of 0.15 cubic foot, how many boxes can fit in the delivery truck? Round your answer to the nearest thousand and write your answer in scientific notation.

**80. Pyramids and Scientific Notation**   The Great Pyramid at Giza is one of the largest and oldest man-made structures in the world. It weighs over $10^{10}$ kilograms. If each stone making up the pyramid weighs approximately 4,000 kilograms, how many stones make up the structure? Write your answer in scientific notation.

*© 1985 North Wind Pictures*

For each equation, fill in the given ordered pairs and use them to draw the graph of the equation.

**81.** $y = x^2$    $(-4,\ ), (-2,\ ), (-1,\ ), (0,\ ),$
$(1,\ ), (2,\ ), (4,\ )$

**82.** $y = x^2$    $(-2,\ ), (-1.5,\ ), (-\frac{1}{2},\ ),$
$(0,\ ), (\frac{1}{2},\ ), (1.5,\ ), (2,\ )$

**83.** $y = 2x^2$    $(-3,\ ), (-2,\ ), (-1,\ ), (0,\ ),$
$(1,\ ), (2,\ ), (3,\ )$

**84.** $y = 3x^2$    $(-3,\ ), (-2,\ ), (-1,\ ), (0,\ ),$
$(1,\ ), (2,\ ), (3,\ )$

**85.** $y = \frac{1}{2}x^2$    $(-4,\ ), (-2,\ ), (-1,\ ), (0,\ ),$
$(1,\ ), (2,\ ), (4,\ )$

**86.** $y = \frac{1}{3}x^2$    $(-3,\ ), (-1,\ ), (0,\ ), (1,\ ),$
$(3,\ )$

**87.** $y = \frac{1}{4}x^2$    $(-4,\ ), (-2,\ ), (-1,\ ), (0,\ ),$
$(1,\ ), (2,\ ), (4,\ )$

**88.** $y = 4x^2$    $(-2,\ ), (-1,\ ), (0,\ ), (1,\ ),$
$(2,\ )$

## Review Problems

The problems that follow review material we covered in Sections 2.1 and 3.3. Reviewing the problems from Section 2.1 will help you understand some of the material in the next section.

Find the value of each expression when $x = -2$. [2.1]

**89.** $-2x + 5$          **90.** $-4x - 1$

**91.** $x^2 + 5x + 6$        **92.** $x^2 - 5x + 6$

For each of the following equations complete the given ordered pairs so that each is a solution to the equation; then use the ordered pairs to graph the equation. [3.3]

**93.** $y = 2x + 2$      $(-2,\ ), (0,\ ), (2,\ )$

**94.** $y = 2x - 3$      $(-1,\ ), (0,\ ), (2,\ )$

**95.** $y = \frac{1}{3}x + 1$      $(-3,\ ), (0,\ ), (3,\ )$

**96.** $y = \frac{1}{2}x - 2$      $(-2,\ ), (0,\ ), (2,\ )$

---

### 5.4    Addition and Subtraction of Polynomials

In this section we will extend what we learned in Section 5.3 to expressions called polynomials. We begin this section with the definition of a polynomial.

> **DEFINITION**    A **polynomial** is a finite sum of monomials (terms).

*Examples*   The following are polynomials:

$$3x^2 + 2x + 1 \qquad 15x^2y + 21xy^2 - y^2 \qquad 3a - 2b + 4c - 5d$$

Polynomials can be further classified by the number of terms they contain. A polynomial with two terms is called a binomial. If it has three terms, it is a trinomial. As stated before, a monomial has only one term.

> **DEFINITION** The **degree** of a polynomial in one variable is the highest power to which the variable is raised.

*Examples*

$$3x^5 + 2x^3 + 1 \qquad \text{A trinomial of degree 5}$$

$$2x + 1 \qquad \text{A binomial of degree 1}$$

$$3x^2 + 2x + 1 \qquad \text{A trinomial of degree 2}$$

$$3x^5 \qquad \text{A monomial of degree 5}$$

$$-9 \qquad \text{A monomial of degree 0}$$

There are no new rules for adding one or more polynomials. We rely only on our previous knowledge. Here are some examples.

**EXAMPLE 1**  Add $(2x^2 - 5x + 3) + (4x^2 + 7x - 8)$.

**Solution**  We use the commutative and associative properties to group similar terms together and then apply the distributive property to add

$$(2x^2 - 5x + 3) + (4x^2 + 7x - 8)$$
$$= (2x^2 + 4x^2) + (-5x + 7x) + (3 - 8) \qquad \text{Commutative and associative properties}$$
$$= (2 + 4)x^2 + (-5 + 7)x + (3 - 8) \qquad \text{Distributive property}$$
$$= 6x^2 + 2x - 5 \qquad \text{Addition}$$

The results here indicate that to add two polynomials, we add coefficients of similar terms.

**EXAMPLE 2**  Add $x^2 + 3x + 2x + 6$.

**Solution**  The only similar terms here are the two middle terms. We combine them as usual to get

$$x^2 + 3x + 2x + 6 = x^2 + 5x + 6$$

You will recall from Chapter 1 the definition of subtraction: $a - b = a + (-b)$. To subtract one expression from another, we simply add its opposite.

The letters $a$ and $b$ in the definition can each represent polynomials. The opposite of a polynomial is the opposite of each of its terms. When you subtract one polynomial from another, you subtract each of its terms.

**E X A M P L E   3**   Subtract $(3x^2 + x + 4) - (x^2 + 2x + 3)$.

**Solution**   To subtract $x^2 + 2x + 3$, we change the sign of each of its terms and add. If you are having trouble remembering why we do this, remember that we can think of $-(x^2 + 2x + 3)$ as $-1(x^2 + 2x + 3)$. If we distribute the $-1$ across $x^2 + 2x + 3$, we get $-x^2 - 2x - 3$:

$(3x^2 + x + 4) - (x^2 + 2x + 3)$

$\qquad = 3x^2 + x + 4 - x^2 - 2x - 3$   We take the opposite of each
$\qquad\qquad\qquad\qquad\qquad\qquad\qquad\quad$ term in the second polynomial.

$\qquad = (3x^2 - x^2) + (x - 2x) + (4 - 3)$

$\qquad = 2x^2 - x + 1$

**E X A M P L E   4**   Subtract $-4x^2 + 5x - 7$ from $x^2 - x - 1$.

**Solution**   The polynomial $x^2 - x - 1$ comes first, then the subtraction sign, and finally the polynomial $-4x^2 + 5x - 7$ in parentheses.

$(x^2 - x - 1) - (-4x^2 + 5x - 7)$

$\qquad = x^2 - x - 1 + 4x^2 - 5x + 7$   We take the opposite of
$\qquad\qquad\qquad\qquad\qquad\qquad\qquad\quad$ each term in the
$\qquad\qquad\qquad\qquad\qquad\qquad\qquad\quad$ second polynomial.

$\qquad = (x^2 + 4x^2) + (-x - 5x) + (-1 + 7)$

$\qquad = 5x^2 - 6x + 6$

There are two important points to remember when adding or subtracting polynomials. First, to add or subtract two polynomials, you always add or subtract *coefficients* of similar terms. Second, the exponents never increase in value when you are adding or subtracting similar terms.

The last topic we want to consider in this section is finding the value of a polynomial for a given value of the variable.

To find the value of the polynomial $3x^2 + 1$ when $x$ is 5, we replace $x$ with 5 and simplify the result:

$$\text{When} \qquad\qquad\qquad x = 5$$

$$\text{the polynomial} \qquad\quad 3x^2 + 1$$

$$\text{becomes} \qquad\quad 3(5)^2 + 1 = 3(25) + 1$$

$$= 75 + 1$$

$$= 76$$

**EXAMPLE 5** Find the value of $3x^2 - 5x + 4$ when $x = -2$.

**Solution**

When $\qquad\qquad\qquad\qquad\qquad\qquad x = -2$

the polynomial $\qquad\qquad\qquad\qquad 3x^2 - 5x + 4$

becomes $\qquad\qquad 3(-2)^2 - 5(-2) + 4 = 3(4) + 10 + 4$

$$= 12 + 10 + 4$$

$$= 26$$

 # Getting Ready for Class

*After reading through the preceding section, respond in your own words and in complete sentences.*

**A.** What are similar terms?

**B.** What is the degree of a polynomial?

**C.** Describe how you would subtract one polynomial from another.

**D.** Describe how you would find the value of the polynomial $3x^2 - 5x + 4$ when $x$ is $-2$.

## PROBLEM SET 5.4

Identify each of the following polynomials as a trinomial, binomial, or monomial, and give the degree in each case.

**1.** $2x^3 - 3x^2 + 1$
**2.** $4x^2 - 4x + 1$
**3.** $5 + 8a - 9a^3$
**4.** $6 + 12x^3 + x^4$
**5.** $2x - 1$
**6.** $4 + 7x$
**7.** $45x^2 - 1$
**8.** $3a^3 + 8$
**9.** $7a^2$
**10.** $90x$
**11.** $-4$
**12.** $56$

Perform the following additions and subtractions.

**13.** $(2x^2 + 3x + 4) + (3x^2 + 2x + 5)$
**14.** $(x^2 + 5x + 6) + (x^2 + 3x + 4)$
**15.** $(3a^2 - 4a + 1) + (2a^2 - 5a + 6)$
**16.** $(5a^2 - 2a + 7) + (4a^2 - 3a + 2)$
**17.** $x^2 + 4x + 2x + 8$   **18.** $x^2 + 5x - 3x - 15$
**19.** $6x^2 - 3x - 10x + 5$

**20.** $10x^2 + 30x - 2x - 6$
**21.** $x^2 - 3x + 3x - 9$   **22.** $x^2 - 5x + 5x - 25$
**23.** $3y^2 - 5y - 6y + 10$
**24.** $y^2 - 18y + 2y - 12$
**25.** $(6x^3 - 4x^2 + 2x) + (9x^2 - 6x + 3)$
**26.** $(5x^3 + 2x^2 + 3x) + (2x^2 + 5x + 1)$
**27.** $\left(\dfrac{2}{3}x^2 - \dfrac{1}{5}x - \dfrac{3}{4}\right) + \left(\dfrac{4}{3}x^2 - \dfrac{4}{5}x + \dfrac{7}{4}\right)$
**28.** $\left(\dfrac{3}{8}x^3 - \dfrac{5}{7}x^2 - \dfrac{2}{5}\right) + \left(\dfrac{5}{8}x^3 - \dfrac{2}{7}x^2 + \dfrac{7}{5}\right)$
**29.** $(a^2 - a - 1) - (-a^2 + a + 1)$
**30.** $(5a^2 - a - 6) - (-3a^2 - 2a + 4)$
**31.** $\left(\dfrac{5}{9}x^3 + \dfrac{1}{3}x^2 - 2x + 1\right)$

$$-\left(\dfrac{2}{3}x^3 + x^2 + \dfrac{1}{2}x - \dfrac{3}{4}\right)$$

**32.** $\left(4x^3 - \dfrac{2}{5}x^2 + \dfrac{3}{8}x - 1\right)$

$\qquad\qquad\qquad - \left(\dfrac{9}{2}x^3 + \dfrac{1}{4}x^2 - x + \dfrac{5}{6}\right)$

**33.** $(4y^2 - 3y + 2) + (5y^2 + 12y - 4)$

$\qquad\qquad\qquad\qquad - (13y^2 - 6y + 20)$

**34.** $(2y^2 - 7y - 8) - (6y^2 + 6y - 8)$

$\qquad\qquad\qquad\qquad + (4y^2 - 2y + 3)$

**35.** Subtract $10x^2 + 23x - 50$ from $11x^2 - 10x + 13$.

**36.** Subtract $2x^2 - 3x + 5$ from $4x^2 - 5x + 10$.

**37.** Subtract $3y^2 + 7y - 15$ from $11y^2 + 11y + 11$.

**38.** Subtract $15y^2 - 8y - 2$ from $3y^2 - 3y + 2$.

**39.** Add $50x^2 - 100x - 150$ to $25x^2 - 50x + 75$.

**40.** Add $7x^2 - 8x + 10$ to $-8x^2 + 2x - 12$.

**41.** Subtract $2x + 1$ from the sum of $3x - 2$ and $11x + 5$.

**42.** Subtract $3x - 5$ from the sum of $5x + 2$ and $9x - 1$.

**43.** Find the value of the polynomial $x^2 - 2x + 1$ when $x$ is 3.

**44.** Find the value of the polynomial $(x - 1)^2$ when $x$ is 3.

**45.** Find the value of the polynomial $(y - 5)^2$ when $y$ is 10.

**46.** Find the value of the polynomial $y^2 - 10y + 25$ when $y$ is 10.

**47.** Find the value of $a^2 + 4a + 4$ when $a$ is 2.

**48.** Find the value of $(a + 2)^2$ when $a$ is 2.

## Applying the Concepts

**49. Packaging**  A crystal ball with a diameter of 6 inches is being packaged for shipment. If the crys-

tal ball is placed inside a circular cylinder with radius 3 inches and height 6 inches, how much volume will need to be filled with padding? (The volume of a sphere with radius $r$ is $\dfrac{4}{3}\pi r^3$, and the volume of a right circular cylinder with radius $r$ and height $h$ is $\pi r^2 h$.)

**50. Packaging**  Suppose the circular cylinder of Problem 49 has a radius of 4 inches and a height of 7 inches. How much volume will need to be filled with padding?

## Review Problems

The following problems review material we covered in Section 5.3. Reviewing these problems will help you understand the next section.

Multiply.

**51.** $3x(-5x)$

**52.** $-3x(-7x)$

**53.** $2x(3x^2)$

**54.** $x^2(3x)$

**55.** $3x^2(2x^2)$

**56.** $4x^2(2x^2)$

---

**5.5    Multiplication With Polynomials**

We begin our discussion of multiplication of polynomials by finding the product of a monomial and a trinomial.

**EXAMPLE  1**   Multiply $3x^2(2x^2 + 4x + 5)$.

**Solution**   Applying the distributive property gives us

$$3x^2(2x^2 + 4x + 5) = 3x^2(2x^2) + 3x^2(4x) + 3x^2(5) \qquad \text{Distributive property}$$
$$= 6x^4 + 12x^3 + 15x^2 \qquad \text{Multiplication}$$

The distributive property is the key to multiplication of polynomials. We can use it to find the product of any two polynomials. There are some shortcuts we can use in certain situations, however. Let's look at an example that involves the product of two binomials.

**EXAMPLE  2**   Multiply $(3x - 5)(2x - 1)$.

**Solution**

$$(3x - 5)(2x - 1) = 3x(2x - 1) - 5(2x - 1)$$
$$= 3x(2x) + 3x(-1) + (-5)(2x) + (-5)(-1)$$
$$= 6x^2 - 3x - 10x + 5$$
$$= 6x^2 - 13x + 5$$

If we look closely at the second and third lines of work in this example, we can see that the terms in the answer come from all possible products of terms in the first binomial with terms in the second binomial. This result is generalized as follows.

**RULE**

To multiply any two polynomials, multiply each term in the first with each term in the second.

There are two ways we can put this rule to work.

## FOIL METHOD

If we look at the original problem in Example 2 and then at the answer, we see that the first term in the answer came from multiplying the first terms in each binomial:

$$3x \cdot 2x = 6x^2 \qquad \text{FIRST}$$

The middle term in the answer came from adding the products of the two outside terms with the two inside terms in each binomial:

$$3x(-1) = -3x \qquad \text{OUTSIDE}$$
$$-5(2x) = \underline{-10x} \qquad \text{INSIDE}$$
$$-13x$$

The last term in the answer came from multiplying the two last terms:

$$-5(-1) = 5 \qquad \text{LAST}$$

To summarize the FOIL method, we will multiply another two binomials.

**EXAMPLE 3**  Multiply $(2x + 3)(5x - 4)$.

**Solution**

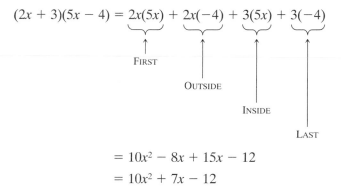

$$(2x + 3)(5x - 4) = 2x(5x) + 2x(-4) + 3(5x) + 3(-4)$$

FIRST

OUTSIDE

INSIDE

LAST

$$= 10x^2 - 8x + 15x - 12$$
$$= 10x^2 + 7x - 12$$

With practice $-8x + 15x = 7x$ can be done mentally.

## COLUMN METHOD

The FOIL method can be applied only when multiplying two binomials. To find products of polynomials with more than two terms, we use what is called the COLUMN method.

The COLUMN method of multiplying two polynomials is very similar to long multiplication with whole numbers. It is just another way of finding all possible products of terms in one polynomial with terms in another polynomial.

**EXAMPLE 4**  Multiply $(2x + 3)(3x^2 - 2x + 1)$.

**Solution**

$$
\begin{array}{r}
3x^2 - 2x + 1 \\
2x + 3 \\
\hline
6x^3 - 4x^2 + 2x \qquad \leftarrow 2x(3x^2 - 2x + 1) \\
9x^2 - 6x + 3 \leftarrow 3(3x^2 - 2x + 1) \\
\hline
6x^3 + 5x^2 - 4x + 3 \leftarrow \text{Add similar terms.}
\end{array}
$$

It will be to your advantage to become very fast and accurate at multiplying polynomials. You should be comfortable using either method. The following examples illustrate the three types of multiplication.

**EXAMPLES**   Multiply:

**5.** $4a^2(2a^2 - 3a + 5) = 4a^2(2a^2) + 4a^2(-3a) + 4a^2(5)$

$$= 8a^4 - 12a^3 + 20a^2$$

**6.** $(x - 2)(y + 3) = x(y) + x(3) + (-2)(y) + (-2)(3)$
                        F         O         I             L

$$= xy + 3x - 2y - 6$$

**7.** $(x + y)(a - b) = x(a) + x(-b) + y(a) + y(-b)$
                         F         O          I          L

$$= xa - xb + ya - yb$$

**8.** $(5x - 1)(2x + 6) = 5x(2x) + 5x(6) + (-1)(2x) + (-1)(6)$
                           F          O            I               L

$$= 10x^2 + 30x + (-2x) + (-6)$$

$$= 10x^2 + 28x - 6$$

**9.** $(3x + 2)(x^2 - 5x + 6)$

$$
\begin{array}{r}
x^2 - \phantom{0}5x + \phantom{0}6 \\
3x + \phantom{0}2 \\
\hline
3x^3 - 15x^2 + 18x \phantom{+ 12} \\
2x^2 - 10x + 12 \\
\hline
3x^3 - 13x^2 + \phantom{0}8x + 12
\end{array}
$$

**EXAMPLE  10**   The length of a rectangle is 3 units more than twice the width. Write an expression for the area of the rectangle.

**Solution**   We begin by drawing a rectangle and labeling the width $x$. Since the length is 3 more than twice the width, we label the length $2x + 3$.

Since the area $A$ of a rectangle is the product of the length and width, we write our formula for the area of this rectangle as

$$A = x(2x + 3)$$

$$A = 2x^2 + 3x \qquad \text{Multiply.}$$

## REVENUE

Suppose that a store sells $x$ items at $p$ dollars per item. The total amount of money obtained by selling the items is called the *revenue*. It can be found by multiplying the number of items sold, $x$, by the price per item, $p$. For example, if 100 items are sold for $6 each, the revenue is $100(6) = \$600$. Similarly, if 500 items are sold for $8 each, the total revenue is $500(8) = \$4,000$. If we denote the revenue with the letter $R$, then the formula that relates $R$, $x$, and $p$ is

$$\text{Revenue} = (\text{number of items sold})(\text{price of each item})$$

In symbols: $R = xp$.

**EXAMPLE 11**   A store selling diskettes for home computers knows from past experience that it can sell $x$ diskettes each day at a price of $p$ dollars per diskette, according to the equation $x = 800 - 100p$. Write a formula for the daily revenue that involves only the variables $R$ and $p$.

**Solution**   From our previous discussion we know that the revenue $R$ is given by the formula

$$R = xp$$

But, since $x = 800 - 100p$, we can substitute $800 - 100p$ for $x$ in the revenue equation to obtain

$$R = (800 - 100p)p$$

$$R = 800p - 100p^2$$

This last formula gives the revenue, $R$, in terms of the price, $p$.

## Getting Ready for Class

*After reading through the preceding section, respond in your own words and in complete sentences.*

**A.** How do we multiply two polynomials?

**B.** Describe how the distributive property is used to multiply a monomial and a polynomial.

**C.** Describe how you would use the FOIL method to multiply two binomials.

**D.** Give an example showing how the product of two binomials can be a trinomial.

# PROBLEM SET 5.5

Multiply the following by applying the distributive property.

**1.** $2x(3x + 1)$

**2.** $4x(2x - 3)$

**3.** $2x^2(3x^2 - 2x + 1)$

**4.** $5x(4x^3 - 5x^2 + x)$

**5.** $2ab(a^2 - ab + 1)$

**6.** $3a^2b(a^3 + a^2b^2 + b^3)$

**7.** $y^2(3y^2 + 9y + 12)$

**8.** $5y(2y^2 - 3y + 5)$

**9.** $4x^2y(2x^3y + 3x^2y^2 + 8y^3)$

**10.** $6xy^3(2x^2 + 5xy + 12y^2)$

Multiply the following binomials. You should do about half the problems using the FOIL method and the other half using the COLUMN method. Remember, you want to be comfortable using both methods.

**11.** $(x + 3)(x + 4)$

**12.** $(x + 2)(x + 5)$

**13.** $(x + 6)(x + 1)$

**14.** $(x + 1)(x + 4)$

**15.** $\left(x + \dfrac{1}{2}\right)\left(x + \dfrac{3}{2}\right)$

**16.** $\left(x + \dfrac{3}{5}\right)\left(x + \dfrac{2}{5}\right)$

**17.** $(a + 5)(a - 3)$

**18.** $(a - 8)(a + 2)$

**19.** $(x - a)(y + b)$

**20.** $(x + a)(y - b)$

**21.** $(x + 6)(x - 6)$

**22.** $(x + 3)(x - 3)$

**23.** $\left(y + \dfrac{5}{6}\right)\left(y - \dfrac{5}{6}\right)$

**24.** $\left(y - \dfrac{4}{7}\right)\left(y + \dfrac{4}{7}\right)$

**25.** $(2x - 3)(x - 4)$

**26.** $(3x - 5)(x - 2)$

**27.** $(a + 2)(2a - 1)$

**28.** $(a - 6)(3a + 2)$

**29.** $(2x - 5)(3x - 2)$

**30.** $(3x + 6)(2x - 1)$

**31.** $(2x + 3)(a + 4)$

**32.** $(2x - 3)(a - 4)$

**33.** $(5x - 4)(5x + 4)$

**34.** $(6x + 5)(6x - 5)$

**35.** $\left(2x - \dfrac{1}{2}\right)\left(x + \dfrac{3}{2}\right)$

**36.** $\left(4x - \dfrac{3}{2}\right)\left(x + \dfrac{1}{2}\right)$

**37.** $(1 - 2a)(3 - 4a)$

**38.** $(1 - 3a)(3 + 2a)$

For each of the following problems, fill in the area of each small rectangle and square, and then add the results together to find the indicated product.

**39.** $(x + 2)(x + 3)$

**40.** $(x + 4)(x + 5)$

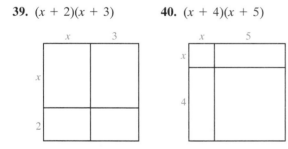

**41.** $(x + 1)(2x + 2)$

**42.** $(2x + 1)(2x + 2)$

Multiply the following.

**43.** $(a - 3)(a^2 - 3a + 2)$

**44.** $(a + 5)(a^2 + 2a + 3)$

**45.** $(x + 2)(x^2 - 2x + 4)$

**46.** $(x + 3)(x^2 - 3x + 9)$

**47.** $(2x + 1)(x^2 + 8x + 9)$

**48.** $(3x - 2)(x^2 - 7x + 8)$

**49.** $(5x^2 + 2x + 1)(x^2 - 3x + 5)$

**50.** $(2x^2 + x + 1)(x^2 - 4x + 3)$

Multiply.

**51.** $(x^2 + 3)(2x^2 - 5)$

**52.** $(4x^3 - 8)(5x^3 + 4)$

**53.** $(3a^4 + 2)(2a^2 + 5)$

**54.** $(7a^4 - 8)(4a^3 - 6)$

**55.** $(x + 3)(x + 4)(x + 5)$

**56.** $(x - 3)(x - 4)(x - 5)$

## Applying the Concepts

**57. Area** The length of a rectangle is 5 units more than twice the width. Write an expression for the area of the rectangle.

**58. Area** The length of a rectangle is 2 units more than 3 times the width. Write an expression for the area of the rectangle.

**59. Area** The width and length of a rectangle are given by two consecutive integers. Write an expression for the area of the rectangle.

**60. Area** The width and length of a rectangle are given by two consecutive even integers. Write an expression for the area of the rectangle.

**61. Revenue** A store selling typewriter ribbons knows that the number of ribbons it can sell each week, $x$, is related to the price per ribbon, $p$, by the equation $x = 1,200 - 100p$. Write an expression for the weekly revenue that involves only the variables $R$ and $p$. (*Remember:* The equation for revenue is $R = xp$.)

**62. Revenue** A store selling small portable radios knows from past experience that the number of radios it can sell each week, $x$, is related to the price per radio, $p$, by the equation $x = 1,300 - 100p$.

Write an expression for the weekly revenue that involves only the variables $R$ and $p$.

63. **Revenue**   The relationship between the number of calculators a company sells per day, $x$, and the price of each calculator, $p$, is given by the equation $x = 1,700 - 100p$. Write an expression for the daily revenue that involves only the variables $R$ and $p$.

64. **Revenue**   The relationship between the number of pencil sharpeners a company can sell each day, $x$, and the price of each sharpener, $p$, is given by the equation $x = 1,800 - 100p$. Write an expression for the daily revenue that involves only the variables $R$ and $p$.

## Review Problems

The following problems review material we covered in Chapter 4.

65. Solve this system by graphing:   $x + y = 4$
$$x - y = 2$$

Solve each system by the elimination method.

66. $3x + 2y = 1$
$2x + y = 3$

67. $2x + 3y = -1$
$3x + 5y = -2$

Solve each system by the substitution method.

68. $x + y = 20$
$y = 5x + 2$

69. $2x - 6y = 2$
$y = 3x + 1$

70. **Investing**   A total of $1,200 is invested in two accounts. One of the accounts pays 8% interest annually, and the other pays 10% interest annually. If the total amount of interest earned from both accounts for the year is $104, how much is invested in each account?

71. **Coin Problem**   Amy has $1.85 in dimes and quarters. If she has a total of 11 coins, how many of each coin does she have?

---

### 5.6   Binomial Squares and Other Special Products

In this section we will combine the results of the last section with our definition of exponents to find some special products.

**EXAMPLE 1**   Find the square of $(3x - 2)$.

**Solution**   To square $(3x - 2)$, we multiply it by itself:

$$(3x - 2)^2 = (3x - 2)(3x - 2) \qquad \text{Definition of exponents}$$
$$= 9x^2 - 6x - 6x + 4 \qquad \text{FOIL method}$$
$$= 9x^2 - 12x + 4 \qquad \text{Combine similar terms.}$$

Notice that the first and last terms in the answer are the square of the first and last terms in the original problem and that the middle term is twice the product of the two terms in the original binomial.

**EXAMPLES**

2. $(a + b)^2 = (a + b)(a + b)$
$= a^2 + 2ab + b^2$

**3.** $(a - b)^2 = (a - b)(a - b)$
$$= a^2 - 2ab + b^2$$

---

Binomial squares having the form of Examples 2 and 3 occur very frequently in algebra. It will be to your advantage to memorize the following rule for squaring a binomial.

**RULE**

The square of a binomial is the sum of the square of the first term, the square of the last term, and twice the product of the two original terms. In symbols this rule is written as follows:

$$(x + y)^2 = \underset{\text{Square of first term}}{x^2} + \underset{\text{Twice product of the two terms}}{2xy} + \underset{\text{Square of last term}}{y^2}$$

**Note:** A very common mistake when squaring binomials is to write

$$(a + b)^2 = a^2 + b^2$$

which just isn't true. The mistake becomes obvious when we substitute 2 for $a$ and 3 for $b$:

$$(2 + 3)^2 \neq 2^2 + 3^2$$

$$25 \neq 13$$

Exponents do not distribute over addition or subtraction.

---

**EXAMPLES**   Multiply using the preceding rule:

|  | | First term squared | Twice their product | Last term squared | Answer |
|---|---|---|---|---|---|
| **4.** | $(x - 5)^2 =$ | $x^2$ | $+ \ 2(x)(-5) \ +$ | $25$ | $= x^2 - 10x + 25$ |
| **5.** | $(x + 2)^2 =$ | $x^2$ | $+ \ 2(x)(2) \ +$ | $4$ | $= x^2 + 4x + 4$ |
| **6.** | $(2x - 3)^2 =$ | $4x^2$ | $+ \ 2(2x)(-3) \ +$ | $9$ | $= 4x^2 - 12x + 9$ |
| **7.** | $(5x - 4)^2 =$ | $25x^2$ | $+ \ 2(5x)(-4) \ +$ | $16$ | $= 25x^2 - 40x + 16$ |

---

Another special product that occurs frequently is $(a + b)(a - b)$. The only difference in the two binomials is the sign between the two terms. The interesting thing about this type of product is that the middle term is always zero. Here are some examples.

---

**EXAMPLES**   Multiply using the FOIL method:

**8.** $(2x - 3)(2x + 3) = 4x^2 + 6x - 6x - 9$      FOIL method
$$= 4x^2 - 9$$

**9.** $(x - 5)(x + 5) = x^2 + 5x - 5x - 25$     FOIL method
$$= x^2 - 25$$

**10.** $(3x - 1)(3x + 1) = 9x^2 + 3x - 3x - 1$     FOIL method
$$= 9x^2 - 1$$

Notice that in each case the middle term is zero and therefore doesn't appear in the answer. The answers all turn out to be the difference of two squares. Here is a rule to help you memorize the result.

**RULE**

When multiplying two binomials that differ only in the sign between their terms, subtract the square of the last term from the square of the first term.

$$(a - b)(a + b) = a^2 - b^2$$

Here are some problems that result in the difference of two squares.

**EXAMPLES**   Multiply using the preceding rule:

**11.** $(x + 3)(x - 3) = x^2 - 9$
**12.** $(a + 2)(a - 2) = a^2 - 4$
**13.** $(9a + 1)(9a - 1) = 81a^2 - 1$
**14.** $(2x - 5y)(2x + 5y) = 4x^2 - 25y^2$
**15.** $(3a - 7b)(3a + 7b) = 9a^2 - 49b^2$

Although all the problems in this section can be worked correctly using the methods in the previous section, they can be done much faster if the new rules are *memorized.* Here is a summary of the new rules:

$$(a + b)^2 = (a + b)(a + b) = a^2 + 2ab + b^2$$
$$(a - b)^2 = (a - b)(a - b) = a^2 - 2ab + b^2$$
$$(a - b)(a + b) = a^2 - b^2$$

**EXAMPLE   16**   Write an expression in symbols for the sum of the squares of three consecutive even integers. Then, simplify that expression.

**Solution**   If we let $x =$ the first of the even integers, then $x + 2$ is the next consecutive even integer and $x + 4$ is the one after that. An expression for the sum of their squares is

$x^2 + (x + 2)^2 + (x + 4)^2$     Sum of squares

$\quad = x^2 + (x^2 + 4x + 4) + (x^2 + 8x + 16)$     Expand squares.

$\quad = 3x^2 + 12x + 20$     Add similar terms.

**Getting Ready for Class**

*After reading through the preceding section, respond in your own words and in complete sentences.*

**A.** Describe how you would square the binomial $a + b$.

**B.** Explain why $(x + 3)^2$ cannot be $x^2 + 9$.

**C.** What kind of products result in the difference of two squares?

**D.** When multiplied out, how will $(x + 3)^2$ and $(x - 3)^2$ differ?

## PROBLEM SET 5.6

Perform the indicated operations.

**1.** $(x - 2)^2$
**2.** $(x + 2)^2$

**3.** $(a + 3)^2$
**4.** $(a - 3)^2$

**5.** $(x - 5)^2$
**6.** $(x - 4)^2$

**7.** $\left(a - \dfrac{1}{2}\right)^2$
**8.** $\left(a + \dfrac{1}{2}\right)^2$

**9.** $(x + 10)^2$
**10.** $(x - 10)^2$

**11.** $(a + 0.8)^2$
**12.** $(a - 0.4)^2$

**13.** $(2x - 1)^2$
**14.** $(3x + 2)^2$

**15.** $(4a + 5)^2$
**16.** $(4a - 5)^2$

**17.** $(3x - 2)^2$
**18.** $(2x - 3)^2$

**19.** $(3a + 5b)^2$
**20.** $(5a - 3b)^2$

**21.** $(4x - 5y)^2$
**22.** $(5x + 4y)^2$

**23.** $(7m + 2n)^2$
**24.** $(2m - 7n)^2$

**25.** $(6x - 10y)^2$
**26.** $(10x + 6y)^2$

**27.** $(x^2 + 5)^2$
**28.** $(x^2 + 3)^2$

**29.** $(a^2 + 1)^2$
**30.** $(a^2 - 2)^2$

**Comparing Expressions** Fill in each table.

**31.**

| $x$ | $(x + 3)^2$ | $x^2 + 9$ | $x^2 + 6x + 9$ |
|-----|-------------|-----------|----------------|
| 1 | | | |
| 2 | | | |
| 3 | | | |
| 4 | | | |

**32.**

| $x$ | $(x - 5)^2$ | $x^2 + 25$ | $x^2 - 10x + 25$ |
|-----|-------------|------------|-------------------|
| 1 | | | |
| 2 | | | |
| 3 | | | |
| 4 | | | |

**33.**

| $a$ | $b$ | $(a + b)^2$ | $a^2 + b^2$ | $a^2 + ab + b^2$ | $a^2 + 2ab + b^2$ |
|-----|-----|-------------|-------------|-------------------|--------------------|
| 1 | 1 | | | | |
| 3 | 5 | | | | |
| 3 | 4 | | | | |
| 4 | 5 | | | | |

**34.**

| $a$ | $b$ | $(a - b)^2$ | $a^2 - b^2$ | $a^2 - 2ab + b^2$ |
|-----|-----|-------------|-------------|--------------------|
| 2 | 1 | | | |
| 5 | 2 | | | |
| 2 | 5 | | | |
| 4 | 3 | | | |

Multiply.

**35.** $(a + 5)(a - 5)$
**36.** $(a - 6)(a + 6)$

**37.** $(y - 1)(y + 1)$
**38.** $(y - 2)(y + 2)$

**39.** $(9 + x)(9 - x)$
**40.** $(10 - x)(10 + x)$

**41.** $(2x + 5)(2x - 5)$
**42.** $(3x + 5)(3x - 5)$

**43.** $\left(4x + \dfrac{1}{3}\right)\left(4x - \dfrac{1}{3}\right)$
**44.** $\left(6x + \dfrac{1}{4}\right)\left(6x - \dfrac{1}{4}\right)$

**45.** $(2a + 7)(2a - 7)$    **46.** $(3a + 10)(3a - 10)$

**47.** $(6 - 7x)(6 + 7x)$    **48.** $(7 - 6x)(7 + 6x)$

**49.** $(x^2 + 3)(x^2 - 3)$    **50.** $(x^2 + 2)(x^2 - 2)$

**51.** $(a^2 + 4)(a^2 - 4)$    **52.** $(a^2 + 9)(a^2 - 9)$

**53.** $(5y^4 - 8)(5y^4 + 8)$    **54.** $(7y^5 + 6)(7y^5 - 6)$

Multiply and simplify.

**55.** $(x + 3)(x - 3) + (x - 5)(x + 5)$

**56.** $(x - 7)(x + 7) + (x - 4)(x + 4)$

**57.** $(2x + 3)^2 - (4x - 1)^2$

**58.** $(3x - 5)^2 - (2x + 3)^2$

**59.** $(a + 1)^2 - (a + 2)^2 + (a + 3)^2$

**60.** $(a - 1)^2 + (a - 2)^2 - (a - 3)^2$

**61.** $(2x + 3)^3$    **62.** $(3x - 2)^3$

**63. Shortcut**  The formula for the difference of two squares can be used as a shortcut to multiplying certain whole numbers if they have the correct form. Use the difference of two squares formula to multiply 49(51) by first writing 49 as $(50 - 1)$ and 51 as $(50 + 1)$.

**64. Shortcut**  Use the difference of two squares formula to multiply 101(99) by first writing 101 as $(100 + 1)$ and 99 as $(100 - 1)$.

**65. Comparing Expressions**  Evaluate the expression $(x + 3)^2$ and the expression $x^2 + 6x + 9$ for $x = 2$.

**66. Comparing Expressions**  Evaluate the expression $x^2 - 25$ and the expression $(x - 5)(x + 5)$ for $x = 6$.

**67. Number Problem**  Write an expression for the sum of the squares of two consecutive integers. Then, simplify that expression.

**68. Number Problem**  Write an expression for the sum of the squares of two consecutive odd integers. Then, simplify that expression.

**69. Number Problem**  Write an expression for the sum of the squares of three consecutive integers. Then, simplify that expression.

**70. Number Problem**  Write an expression for the sum of the squares of three consecutive odd integers. Then, simplify that expression.

**71. Area**  We can use the concept of area to further justify our rule for squaring a binomial. The length of each side of the square shown in the figure is $a + b$. (The longer line segment has length $a$, and the shorter line segment has length $b$.) The area of the whole square is $(a + b)^2$. On the other hand, the whole area is the sum of the areas of the two smaller squares and the two small rectangles that make it up. Write the area of the two smaller squares and the two small rectangles and then add them together to verify the formula $(a + b)^2 = a^2 + 2ab + b^2$.

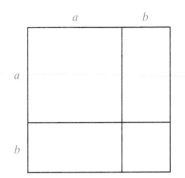

**72. Area**  The length of each side of the large square shown in the figure is $x + 5$. Therefore, its area is $(x + 5)^2$. Find the area of the two smaller squares and the two smaller rectangles that make up the large square, and then add them together to verify the formula $(x + 5)^2 = x^2 + 10x + 25$.

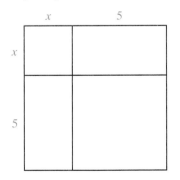

For each problem, fill in the area of each small rectangle and square, and then add the result together to find the indicated product.

**73.** $(2x + 1)^2$

**74.** $(2x + 3)^2$

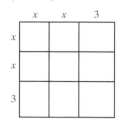

## Review Problems

The following problems review material we covered in Sections 5.3 and 4.1.

Simplify each expression (divide). [5.3]

**75.** $\dfrac{15x^2y}{3xy}$

**76.** $\dfrac{21xy^2}{3xy}$

**77.** $\dfrac{35a^6b^8}{70a^2b^{10}}$

**78.** $\dfrac{75a^2b^6}{25a^4b^3}$

Solve each system by graphing. [4.1]

**79.** $x + y = 2$
$\quad\ x - y = 4$

**80.** $x + y = 1$
$\quad\ x - y = -3$

**81.** $y = 2x + 3$
$\quad\ y = -2x - 1$

**82.** $y = 3x - 2$
$\quad\ y = -2x + 3$

---

### 5.7   Dividing a Polynomial by a Monomial

To divide a polynomial by a monomial, we will use the definition of division and apply the distributive property. Follow the steps in this example closely.

**EXAMPLE 1**   Divide $10x^3 - 15x^2$ by $5x$.

**Solution**

$$\dfrac{10x^3 - 15x^2}{5x} = (10x^3 - 15x^2)\dfrac{1}{5x}$$   Division by $5x$ is the same as multiplication by $\dfrac{1}{5x}$.

$$= 10x^3\left(\dfrac{1}{5x}\right) - 15x^2\left(\dfrac{1}{5x}\right)$$   Distribute $\dfrac{1}{5x}$ to both terms.

$$= \dfrac{10x^3}{5x} - \dfrac{15x^2}{5x}$$   Multiplication by $\dfrac{1}{5x}$ is the same as division by $5x$.

$$= 2x^2 - 3x$$   Division of monomials as done in Section 5.3

---

If we were to leave out the first steps, the problem would look like this:

$$\dfrac{10x^3 - 15x^2}{5x} = \dfrac{10x^3}{5x} - \dfrac{15x^2}{5x}$$

$$= 2x^2 - 3x$$

The problem is much shorter and clearer this way. You may leave out the first two steps from Example 1 when working problems in this section. They are part of Example 1 only to help show you why the following rule is true.

**RULE**

To divide a polynomial by a monomial, simply divide each term in the polynomial by the monomial.

Here are some further examples using our rule for division of a polynomial by a monomial.

**EXAMPLE 2**  Divide $\dfrac{3x^2 - 6}{3}$.

**Solution**  We begin by writing the 3 in the denominator under each term in the numerator. Then we simplify the result:

$$\frac{3x^2 - 6}{3} = \frac{3x^2}{3} - \frac{6}{3} \qquad \text{Divide each term in the numerator by 3.}$$

$$= x^2 - 2 \qquad \text{Simplify.}$$

**EXAMPLE 3**  Divide $\dfrac{4x^2 - 2}{2}$.

**Solution**  Dividing each term in the numerator by 2, we have

$$\frac{4x^2 - 2}{2} = \frac{4x^2}{2} - \frac{2}{2} \qquad \text{Divide each term in the numerator by 2.}$$

$$= 2x^2 - 1 \qquad \text{Simplify.}$$

**EXAMPLE 4**  Find the quotient of $27x^3 - 9x^2$ and $3x$.

**Solution**  We are again asked to divide the first polynomial by the second one:

$$\frac{27x^3 - 9x^2}{3x} = \frac{27x^3}{3x} - \frac{9x^2}{3x} \qquad \text{Divide each term by } 3x.$$

$$= 9x^2 - 3x \qquad \text{Simplify.}$$

**EXAMPLE 5**  Divide $(15x^2y - 21xy^2)$ by $(-3xy)$.

**Solution**  This is the same type of problem we have shown in the first four examples; it is just worded a little differently. Note that when we divide each term in the first polynomial by $-3xy$, the negative sign must be taken into account:

$$\frac{15x^2y - 21xy^2}{-3xy} = \frac{15x^2y}{-3xy} - \frac{21xy^2}{-3xy} \qquad \text{Divide each term by } -3xy.$$

$$= -5x - (-7y) \qquad \text{Simplify.}$$

$$= -5x + 7y \qquad \text{Simplify.}$$

**EXAMPLE 6** Divide $\dfrac{10x^3 - 5x^2 + 20x}{10x^3}$.

**Solution** Proceeding as we have in the first five examples, we distribute $10x^3$ under each term in the numerator:

$$\frac{10x^3 - 5x^2 + 20x}{10x^3} = \frac{10x^3}{10x^3} - \frac{5x^2}{10x^3} + \frac{20x}{10x^3} \qquad \text{Divide each term by } 10x^3.$$

$$= 1 - \frac{1}{2x} + \frac{2}{x^2} \qquad \text{Simplify.}$$

Note that our answer is not a polynomial.

**EXAMPLE 7** Divide $\dfrac{24x^3y^2 + 16x^2y^2 - 4x^2y^3}{8x^2y}$.

**Solution** Writing $8x^2y$ under each term in the numerator and then simplifying, we have

$$\frac{24x^3y^2 + 16x^2y^2 - 4x^2y^3}{8x^2y} = \frac{24x^3y^2}{8x^2y} + \frac{16x^2y^2}{8x^2y} - \frac{4x^2y^3}{8x^2y}$$

$$= 3xy + 2y - \frac{y^2}{2}$$

From the examples in this section, it is clear that to divide a polynomial by a monomial, we must divide each term in the polynomial by the monomial. Often, students taking algebra for the first time will make the following mistake:

$$\frac{x + 2}{2} = x + 1 \qquad \text{Mistake}$$

The mistake here is in not dividing both terms in the numerator by 2. The correct way to divide $x + 2$ by 2 looks like this:

$$\frac{x + 2}{2} = \frac{x}{2} + \frac{2}{2} = \frac{x}{2} + 1 \qquad \text{Correct}$$

# Getting Ready for Class

*After reading through the preceding section, respond in your own words and in complete sentences.*

**A.** What property of real numbers is the key to dividing a polynomial by a monomial?

**B.** Describe how you would divide a polynomial by $5x$.

**C.** What part do the properties of exponents play in dividing a polynomial by a monomial?

**D.** Explain the mistake in the problem $\dfrac{x+2}{2} = x + 1$.

# PROBLEM SET 5.7

Divide the following polynomials by $5x$.

**1.** $5x^2 - 10x$

**2.** $10x^3 - 15x$

**3.** $15x - 10x^3$

**4.** $50x^3 - 20x^2$

**5.** $25x^2y - 10xy$

**6.** $15xy^2 + 20x^2y$

**7.** $35x^5 - 30x^4 + 25x^3$

**8.** $40x^4 - 30x^3 + 20x^2$

**9.** $50x^5 - 25x^3 + 5x$

**10.** $75x^6 + 50x^3 - 25x$

Divide the following by $-2a$.

**11.** $8a^2 - 4a$

**12.** $a^3 - 6a^2$

**13.** $16a^5 + 24a^4$

**14.** $30a^6 + 20a^3$

**15.** $8ab + 10a^2$

**16.** $6a^2b - 10ab^2$

**17.** $12a^3b - 6a^2b^2 + 14ab^3$

**18.** $4ab^3 - 16a^2b^2 - 22a^3b$

**19.** $a^2 + 2ab + b^2$

**20.** $a^2b - 2ab^2 + b^3$

Perform the following divisions (find the following quotients).

**21.** $\dfrac{6x + 8y}{2}$

**22.** $\dfrac{9x - 3y}{3}$

**23.** $\dfrac{7y - 21}{-7}$

**24.** $\dfrac{14y - 12}{2}$

**25.** $\dfrac{10xy - 8x}{2x}$

**26.** $\dfrac{12xy^2 - 18x}{-6x}$

**27.** $\dfrac{x^2y - x^3y^2}{x}$

**28.** $\dfrac{x^2y - x^3y^2}{x^2}$

**29.** $\dfrac{x^2y - x^3y^2}{-x^2y}$

**30.** $\dfrac{ab + a^2b^2}{ab}$

**31.** $\dfrac{a^2b^2 - ab^2}{-ab^2}$

**32.** $\dfrac{a^2b^2c + ab^2c^2}{abc}$

**33.** $\dfrac{x^3 - 3x^2y + xy^2}{x}$

**34.** $\dfrac{x^2 - 3xy^2 + xy^3}{x}$

**35.** $\dfrac{10a^2 - 15a^2b + 25a^2b^2}{5a^2}$

**36.** $\dfrac{11a^2b^2 - 33ab}{-11ab}$

**37.** $\dfrac{26x^2y^2 - 13xy}{-13xy}$

**38.** $\dfrac{6x^2y^2 - 3xy}{6xy}$

**39.** $\dfrac{4x^2y^2 - 2xy}{4xy}$

**40.** $\dfrac{6x^2a + 12x^2b - 6x^2c}{36x^2}$

**41.** $\dfrac{5a^2x - 10ax^2 + 15a^2x^2}{20a^2x^2}$

**42.** $\dfrac{12ax - 9bx + 18cx}{6x^2}$

**43.** $\dfrac{16x^5 + 8x^2 + 12x}{12x^3}$

**44.** $\dfrac{27x^2 - 9x^3 - 18x^4}{-18x^3}$

Divide. Assume all variables represent positive numbers.

**45.** $\dfrac{9a^{5m} - 27a^{3m}}{3a^{2m}}$

**46.** $\dfrac{26a^{3m} - 39a^{5m}}{13a^{3m}}$

**47.** $\dfrac{10x^{5m} - 25x^{3m} + 35x^m}{5x^m}$

**48.** $\dfrac{18x^{2m} + 24x^{4m} - 30x^{6m}}{6x^{2m}}$

Simplify each numerator, and then divide.

**49.** $\dfrac{2x^3(3x + 2) - 3x^2(2x - 4)}{2x^2}$

**50.** $\dfrac{5x^2(6x - 3) + 6x^3(3x - 1)}{3x}$

**51.** $\dfrac{(x + 2)^2 - (x - 2)^2}{2x}$

**52.** $\dfrac{(x - 3)^2 - (x + 3)^2}{3x}$

**53.** $\dfrac{(x + 5)^2 + (x + 5)(x - 5)}{2x}$

**54.** $\dfrac{(x - 4)^2 + (x + 4)(x - 4)}{2x}$

**55. Comparing Expressions**   Evaluate the expression $\dfrac{10x + 15}{5}$ and the expression $2x + 3$ when $x = 2$.

**56. Comparing Expressions**   Evaluate the expression $\dfrac{6x^2 + 4x}{2x}$ and the expression $3x + 2$ when $x = 5$.

**57. Comparing Expressions**   Show that the expression $\dfrac{3x + 8}{2}$ is not the same as the expression $3x + 4$ by replacing $x$ with 10 in both expressions and simplifying the results.

**58. Comparing Expressions**   Show that the expression $\dfrac{x + 10}{x}$ is not equal to 10 by replacing $x$ with 5 and simplifying.

## Review Problems

The following problems review material we covered in Sections 4.2 and 4.3.

Solve each system of equations by the elimination method. [4.2]

**59.** $x + y = 6$
   $x - y = 8$

**60.** $2x + y = 5$
   $-x + y = -4$

**61.** $2x - 3y = -5$
   $x + y = 5$

**62.** $2x - 4y = 10$
   $3x - 2y = -1$

Solve each system by the substitution method. [4.3]

**63.** $x + y = 2$
   $y = 2x - 1$

**64.** $2x - 3y = 4$
   $x = 3y - 1$

**65.** $4x + 2y = 8$
   $y = -2x + 4$

**66.** $4x + 2y = 8$
   $y = -2x + 5$

---

**5.8**    Dividing a Polynomial by a Polynomial

Since long division for polynomials is very similar to long division with whole numbers, we will begin by reviewing a division problem with whole numbers. You may realize when looking at Example 1 that you don't have a very good idea why you proceed as you do with long division. What you do know is that the process always works. We are going to approach the explanations in this section in much the same manner. That is, we won't always be sure why the steps we will use are important, only that they always produce the correct result.

**EXAMPLE I**   Divide $27\overline{)3{,}962}$.

**Solution**

$$
\begin{array}{r}
1 \phantom{00} \\
27\overline{)3{,}962} \\
2\,7 \phantom{0} \\
\hline
1\,2 \phantom{0}
\end{array}
$$
← Estimate 27 into 39.
← Multiply $1 \times 27 = 27$.
← Subtract $39 - 27 = 12$.

$$
\begin{array}{r}
1 \phantom{00} \\
27\overline{)3{,}962} \\
2\,7\downarrow \phantom{} \\
\hline
1\,26
\end{array}
$$
← Bring down the 6.

These are the four basic steps in long division. Estimate, multiply, subtract, and bring down the next term. To finish the problem, we simply perform the same four steps again:

$$
\begin{array}{r}
14 \phantom{0} \\
27\overline{)3{,}962} \\
2\,7 \phantom{0} \\
1\,26 \phantom{} \\
1\,08\downarrow \\
\hline
182
\end{array}
$$
← 4 is the estimate.

← Multiply to get 108.
← Subtract to get 18, and then bring down the 2.

One more time.

$$
\begin{array}{r}
146 \\
27\overline{)3{,}962} \\
2\,7 \phantom{0} \\
\hline
1\,26 \phantom{} \\
1\,08 \phantom{} \\
\hline
182 \\
162 \\
\hline
20
\end{array}
$$
← 6 is the estimate.

← Multiply to get 162.
← Subtract to get 20.

Since there is nothing left to bring down, we have our answer.

$$
\frac{3{,}962}{27} = 146 + \frac{20}{27}, \text{ or } 146\frac{20}{27}
$$

Here is how it works with polynomials.

**E X A M P L E   2**   Divide $\dfrac{x^2 - 5x + 8}{x - 3}$.

**Solution**

$$
\begin{array}{r}
x \\
x - 3 \overline{\smash{)}\, x^2 - 5x + 8}
\end{array}
$$
← Estimate $x^2 \div x = x$.

$$
\begin{array}{r}
x \\
x - 3 \overline{\smash{)}\, x^2 - 5x + 8} \\
\underline{\phantom{-}x^2 - 3x} \\
- 2x
\end{array}
$$
← Multiply $x(x - 3) = x^2 - 3x$.
← Subtract $(x^2 - 5x) - (x^2 - 3x) = -2x$.

$$
\begin{array}{r}
x \\
x - 3 \overline{\smash{)}\, x^2 - 5x + 8} \\
\underline{\phantom{-}x^2 - 3x \phantom{\Big|}} \\
- 2x + 8
\end{array}
$$
← Bring down the 8.

Notice that to subtract one polynomial from another, we add its opposite. That is why we change the signs on $x^2 - 3x$ and add what we get to $x^2 - 5x$. (To subtract the second polynomial, simply change the signs and add.)

We perform the same four steps again:

$$
\begin{array}{r}
x - 2 \\
x - 3 \overline{\smash{)}\, x^2 - 5x + 8} \\
\underline{\phantom{-}x^2 - 3x \phantom{\Big|}} \\
- 2x + 8 \\
\underline{\phantom{-}2x - 6} \\
2
\end{array}
$$
← $-2$ is the estimate $(-2x \div x = -2)$.

← Multiply $-2(x - 3) = -2x + 6$.
← Subtract $(-2x + 8) - (-2x + 6) = 2$.

Since there is nothing left to bring down, we have our answer:

$$
\frac{x^2 - 5x + 8}{x - 3} = x - 2 + \frac{2}{x - 3}
$$

To check our answer, we multiply $(x - 3)(x - 2)$ to get $x^2 - 5x + 6$. Then, adding on the remainder, 2, we have $x^2 - 5x + 8$.

---

**EXAMPLE  3**    Divide $\dfrac{6x^2 - 11x - 14}{2x - 5}$.

**Solution**

$$
\begin{array}{r}
3x + 2 \\
2x - 5 \overline{\smash{)}\, 6x^2 - 11x - 14} \\
\underline{\phantom{-}6x^2 - 15x \phantom{\Big|}} \\
4x - 14 \\
\underline{\phantom{-}4x - 10} \\
- 4
\end{array}
$$

$$\frac{6x^2 - 11x - 14}{2x - 5} = 3x + 2 + \frac{-4}{2x - 5}$$

One last step is sometimes necessary. The two polynomials in a division problem must both be in descending powers of the variable and cannot skip any powers from the highest power down to the constant term.

**E X A M P L E   4**   Divide $\dfrac{2x^3 - 3x + 2}{x - 5}$.

**Solution**   The problem will be much less confusing if we write $2x^3 - 3x + 2$ as $2x^3 + 0x^2 - 3x + 2$. Adding $0x^2$ does not change our original problem.

$$
\begin{array}{r}
2x^2 \quad\quad\quad \leftarrow \text{Estimate } 2x^3 \div x = 2x^2. \\
x - 5 \overline{)\ 2x^3 + \ 0x^2 - 3x + 2} \\
\end{array}
$$

← Multiply $2x^2(x - 5) = 2x^3 - 10x^2$.
← Subtract:
$\quad\quad (2x^3 + 0x^2) - (2x^3 - 10x^2) = 10x^2$
— Bring down the next term.

Adding the term $0x^2$ gives us a column in which to write $10x^2$. (Remember, you can add and subtract only similar terms.)

Here is the completed problem:

$$
\begin{array}{r}
2x^2 + 10x + \ 47 \\
x - 5 \overline{)\ 2x^3 + \ 0x^2 - \ 3x + \ \ 2} \\
\underline{2x^3 - 10x^2} \\
+ 10x^2 - \ 3x \\
\underline{10x^2 - 50x} \\
+ 47x + \ \ 2 \\
\underline{47x - 235} \\
237
\end{array}
$$

Our answer is $\dfrac{2x^3 - 3x + 2}{x - 5} = 2x^2 + 10x + 47 + \dfrac{237}{x - 5}$.

As you can see, long division with polynomials is a mechanical process. Once you have done it correctly a couple of times, it becomes very easy to produce the correct answer.

# Getting Ready for Class

*After reading through the preceding section, respond in your own words and in complete sentences.*

**A.** What are the four steps used in long division with whole numbers?

**B.** How is division of two polynomials similar to long division with whole numbers?

**C.** What are the four steps used in long division with polynomials?

**D.** How do we use 0 when dividing the polynomial $2x^3 - 3x + 2$ by $x - 5$?

## PROBLEM SET 5.8

Divide.

**1.** $\dfrac{x^2 - 5x + 6}{x - 3}$

**2.** $\dfrac{x^2 - 5x + 6}{x - 2}$

**3.** $\dfrac{a^2 + 9a + 20}{a + 5}$

**4.** $\dfrac{a^2 + 9a + 20}{a + 4}$

**5.** $\dfrac{x^2 - 6x + 9}{x - 3}$

**6.** $\dfrac{x^2 + 10x + 25}{x + 5}$

**7.** $\dfrac{2x^2 + 5x - 3}{2x - 1}$

**8.** $\dfrac{4x^2 + 4x - 3}{2x - 1}$

**9.** $\dfrac{2a^2 - 9a - 5}{2a + 1}$

**10.** $\dfrac{4a^2 - 8a - 5}{2a + 1}$

**11.** $\dfrac{x^2 + 5x + 8}{x + 3}$

**12.** $\dfrac{x^2 + 5x + 4}{x + 3}$

**13.** $\dfrac{a^2 + 3a + 2}{a + 5}$

**14.** $\dfrac{a^2 + 4a + 3}{a + 5}$

**15.** $\dfrac{x^2 + 2x + 1}{x - 2}$

**16.** $\dfrac{x^2 + 6x + 9}{x - 3}$

**17.** $\dfrac{x^2 + 5x - 6}{x + 1}$

**18.** $\dfrac{x^2 - x - 6}{x + 1}$

**19.** $\dfrac{a^2 + 3a + 1}{a + 2}$

**20.** $\dfrac{a^2 - a + 3}{a + 1}$

**21.** $\dfrac{2x^2 - 2x + 5}{2x + 4}$

**22.** $\dfrac{15x^2 + 19x - 4}{3x + 8}$

**23.** $\dfrac{6a^2 + 5a + 1}{2a + 3}$

**24.** $\dfrac{4a^2 + 4a + 3}{2a + 1}$

**25.** $\dfrac{6a^3 - 13a^2 - 4a + 15}{3a - 5}$

**26.** $\dfrac{2a^3 - a^2 + 3a + 2}{2a + 1}$

Fill in the missing terms in the numerator, and then use long division to find the quotients (see Example 4).

**27.** $\dfrac{x^3 + 4x + 5}{x + 1}$

**28.** $\dfrac{x^3 + 4x^2 - 8}{x + 2}$

**29.** $\dfrac{x^3 - 1}{x - 1}$

**30.** $\dfrac{x^3 + 1}{x + 1}$

**31.** $\dfrac{x^3 - 8}{x - 2}$

**32.** $\dfrac{x^3 + 27}{x + 3}$

## Review Problems

The problems that follow review material we covered in Section 4.4.

Use systems of equations to solve the following word problems.

**33. Number Problem** The sum of two numbers is 25. One of the numbers is 4 times the other. Find the numbers.

**34. Number Problem** The sum of two numbers is 24. One of the numbers is 3 more than twice the other. Find the numbers.

**35. Investing** Suppose you have a total of $1,200 invested in two accounts. One of the accounts pays 8% annual interest, and the other pays 9% annual interest. If your total interest for the year is $100, how much money did you invest in each of the accounts?

36. **Investing**   If you invest twice as much money in an account that pays 12% annual interest as you do in an account that pays 11% annual interest, how much do you have in each account if your total interest for a year is $210?

37. **Money Problem**   If you have a total of $160 in $5 bills and $10 bills, how many of each type of bill do you have if you have 4 more $10 bills than $5 bills?

38. **Coin Problem**   Suppose you have 20 coins worth a total of $2.80. If the coins are all nickels and quarters, how many of each type do you have?

39. **Mixture Problem**   How many gallons of 20% antifreeze solution and 60% antifreeze solution must be mixed to get 16 gallons of 35% antifreeze solution?

40. **Mixture Problem**   A chemist wants to obtain 80 liters of a solution that is 12% hydrochloric acid. How many liters of 10% hydrochloric acid solution and 20% hydrochloric acid solution should he mix to do so?

# CHAPTER 5 SUMMARY

**Examples**

**1.** (a) $2^3 = 2 \cdot 2 \cdot 2 = 8$

(b) $x^5 \cdot x^3 = x^{5+3} = x^8$

(c) $\dfrac{x^5}{x^3} = x^{5-3} = x^2$

(d) $(3x)^2 = 3^2 \cdot x^2 = 9x^2$

(e) $\left(\dfrac{2}{3}\right)^3 = \dfrac{2^3}{3^3} = \dfrac{8}{27}$

(f) $(x^5)^3 = x^{5 \cdot 3} = x^{15}$

(g) $3^{-2} = \dfrac{1}{3^2} = \dfrac{1}{9}$

## Exponents: Definition and Properties [5.1, 5.2]

Integer exponents indicate repeated multiplications.

$a^r \cdot a^s = a^{r+s}$      To multiply with the same base, you add exponents.

$\dfrac{a^r}{a^s} = a^{r-s}$      To divide with the same base, you subtract exponents.

$(ab)^r = a^r \cdot b^r$      Exponents distribute over multiplication.

$\left(\dfrac{a}{b}\right)^r = \dfrac{a^r}{b^r}$      Exponents distribute over division.

$(a^r)^s = a^{r \cdot s}$      A power of a power is the base raised to the product of the powers.

$a^{-r} = \dfrac{1}{a^r}$      Negative exponents imply reciprocals.

**2.** $(5x^2)(3x^4) = 15x^6$

## Multiplication of Monomials [5.3]

To multiply two monomials, multiply coefficients and add exponents.

**3.** $\dfrac{12x^9}{4x^5} = 3x^4$

## Division of Monomials [5.3]

To divide two monomials, divide coefficients and subtract exponents.

**4.** $768{,}000 = 7.68 \times 10^5$

$0.00039 = 3.9 \times 10^{-4}$

## Scientific Notation [5.1, 5.2, 5.3]

A number is in scientific notation when it is written as the product of a number between 1 and 10 and an integer power of 10.

**5.** $(3x^2 - 2x + 1) + (2x^2 + 7x - 3)$
$= 5x^2 + 5x - 2$

## Addition of Polynomials [5.4]

To add two polynomials, add coefficients of similar terms.

**6.** $(3x + 5) - (4x - 3)$
$= 3x + 5 - 4x + 3$
$= -x + 8$

## Subtraction of Polynomials [5.4]

To subtract one polynomial from another, add the opposite of the second to the first.

**7.** (a) $2a^2(5a^2 + 3a - 2)$
$= 10a^4 + 6a^3 - 4a^2$

(b) $(x + 2)(3x - 1)$
$= 3x^2 - x + 6x - 2$
$= 3x^2 + 5x - 2$

(c) $\begin{array}{r} 2x^2 - 3x + 4 \\ 3x - 2 \\ \hline 6x^3 - 9x^2 + 12x \\ -4x^2 + 6x - 8 \\ \hline 6x^3 - 13x^2 + 18x - 8 \end{array}$

## Multiplication of Polynomials [5.5]

To multiply a polynomial by a monomial, we apply the distributive property. To multiply two binomials we use the FOIL method. In other situations we use the COLUMN method. Each method achieves the same result: To multiply any two polynomials, we multiply each term in the first polynomial by each term in the second polynomial.

**8.**
$$(x + 3)^2 = x^2 + 6x + 9$$
$$(x - 3)^2 = x^2 - 6x + 9$$
$$(x + 3)(x - 3) = x^2 - 9$$

**Special Products [5.6]**

$$\left.\begin{array}{l}(a + b)^2 = a^2 + 2ab + b^2 \\ (a - b)^2 = a^2 - 2ab + b^2\end{array}\right\} \quad \text{Binomial squares}$$

$$(a + b)(a - b) = a^2 - b^2 \qquad \text{Difference of two squares}$$

**9.** $\dfrac{12x^3 - 18x^2}{6x} = 2x^2 - 3x$

**Dividing a Polynomial by a Monomial [5.7]**
To divide a polynomial by a monomial, divide each term in the polynomial by the monomial.

**10.**
$$\begin{array}{r} x - 2 \\ x - 3 \overline{\smash{)}\ x^2 - 5x + 8} \\ \underline{\phantom{-}-\ \ +\phantom{xxx}} \\ \cancel{-}\ x^2\ \cancel{-}\ 3x \phantom{xx} \\ \underline{\phantom{xxxx}- 2x + 8} \\ +\phantom{xxx}-\phantom{xx} \\ \underline{\cancel{-}\ 2x\ \cancel{-}\ 6} \\ 2 \end{array}$$

**Long Division With Polynomials [5.8]**
Division with polynomials is similar to long division with whole numbers. The steps in the process are estimate, multiply, subtract, and bring down the next term. The divisors in all the long-division problems in this chapter were binomials.

## COMMON MISTAKES

**1.** If a term contains a variable that is raised to a power, then the exponent on the variable is associated only with that variable, unless there are parentheses. That is, the expression $3x^2$ means $3 \cdot x \cdot x$, not $3x \cdot 3x$. It is a mistake to write $3x^2$ as $9x^2$. The only way to end up with $9x^2$ is to start with $(3x)^2$.

**2.** It is a mistake to add nonsimilar terms. For example, $2x$ and $3x^2$ are nonsimilar terms and therefore cannot be combined. That is, $2x + 3x^2 \neq 5x^3$. If you were to substitute 10 for $x$ in the preceding expression, you would see that the two sides are not equal.

**3.** It is a mistake to distribute exponents over sums and differences. That is, $(a + b)^2 \neq a^2 + b^2$. Convince yourself of this by letting $a = 2$ and $b = 3$ and then simplifying both sides.

**4.** Another common mistake can occur when dividing a polynomial by a monomial. Here is an example:

$$\frac{x + 2}{2} = x + 1 \qquad \text{Mistake}$$

The mistake here is in not dividing both terms in the numerator by 2. The correct way to divide $x + 2$ by 2 looks like this:

$$\frac{x + 2}{2} = \frac{x}{2} + \frac{2}{2} \qquad \text{Correct}$$

$$= \frac{x}{2} + 1$$

# CHAPTER 5  REVIEW

The numbers in brackets refer to the sections of the text in which similar problems can be found.

Simplify. [5.1]

**1.** $(-1)^3$

**2.** $-8^2$

**3.** $\left(\dfrac{3}{7}\right)^2$

**4.** $y^3 \cdot y^9$

**5.** $x^{15} \cdot x^7 \cdot x^5 \cdot x^3$

**6.** $(x^7)^5$

**7.** $(2^6)^4$

**8.** $(3y)^3$

**9.** $(-2xyz)^3$

Simplify each expression. Any answers that contain exponents should contain positive exponents only. [5.2]

**10.** $7^{-2}$

**11.** $4x^{-5}$

**12.** $(3y)^{-3}$

**13.** $\dfrac{a^9}{a^3}$

**14.** $\left(\dfrac{x^3}{x^5}\right)^2$

**15.** $\dfrac{x^9}{x^{-6}}$

**16.** $\dfrac{x^{-7}}{x^{-2}}$

**17.** $(-3xy)^0$

**18.** $3^0 - 5^1 + 5^0$

Simplify. Any answers that contain exponents should contain positive exponents only. [5.1, 5.2]

**19.** $(3x^3y^2)^2$

**20.** $(2a^3b^2)^4(2a^5b^6)^2$

**21.** $(-3xy^2)^{-3}$

**22.** $\dfrac{(b^3)^4(b^2)^5}{(b^7)^3}$

**23.** $\dfrac{(x^{-3})^3(x^6)^{-1}}{(x^{-5})^{-4}}$

Simplify. Write all answers with positive exponents only. [5.3]

**24.** $\dfrac{(2x^4)(15x^9)}{6x^6}$

**25.** $\dfrac{(10x^3y^5)(21x^2y^6)}{(7xy^3)(5x^9y)}$

**26.** $\dfrac{21a^{10}}{3a^4} - \dfrac{18a^{17}}{6a^{11}}$

**27.** $\dfrac{8x^8y^3}{2x^3y} - \dfrac{10x^6y^9}{5xy^7}$

Simplify, and write all answers in scientific notation. [5.3]

**28.** $(3.2 \times 10^3)(2 \times 10^4)$

**29.** $\dfrac{4.6 \times 10^5}{2 \times 10^{-3}}$

**30.** $\dfrac{(4 \times 10^6)(6 \times 10^5)}{3 \times 10^8}$

Perform the following additions and subtractions. [5.4]

**31.** $(3a^2 - 5a + 5) + (5a^2 - 7a - 8)$

**32.** $(-7x^2 + 3x - 6) - (8x^2 - 4x + 7) + (3x^2 - 2x - 1)$

**33.** Subtract $8x^2 + 3x - 2$ from $4x^2 - 3x - 2$.

**34.** Find the value of $2x^2 - 3x + 5$ when $x = 3$.

Multiply. [5.5]

**35.** $3x(4x - 7)$

**36.** $8x^3y(3x^2y - 5xy^2 + 4y^3)$

**37.** $(a + 1)(a^2 + 5a - 4)$

**38.** $(x + 5)(x^2 - 5x + 25)$

**39.** $(3x - 7)(2x - 5)$

**40.** $\left(5y + \dfrac{1}{5}\right)\left(5y - \dfrac{1}{5}\right)$

**41.** $(a^2 - 3)(a^2 + 3)$

Perform the indicated operations. [5.6]

**42.** $(a - 5)^2$

**43.** $(3x + 4)^2$

**44.** $(y^2 + 3)^2$

**45.** Divide $10ab + 20a^2$ by $-5a$. [5.7]

**46.** Divide $40x^5y^4 - 32x^3y^3 - 16x^2y$ by $-8xy$.

Divide using long division. [5.8]

**47.** $\dfrac{x^2 + 15x + 54}{x + 6}$

**48.** $\dfrac{6x^2 + 13x - 5}{3x - 1}$

**49.** $\dfrac{x^3 + 64}{x + 4}$

**50.** $\dfrac{3x^2 - 7x + 10}{3x + 2}$

**51.** $\dfrac{2x^3 - 7x^2 + 6x + 10}{2x + 1}$

**Volume**   A box is in the shape of a rectangular solid and is 3 times as long as it is wide. The height and the width are equal. [5.1, 5.2, 5.3]

**52.** Write an expression for the volume of this box.

**53.** If the width of this box is 2 feet, will it hold as much food as your refrigerator?

**54. Area and Volume of a Sphere**  A discotheque wants to install a new disco ball in the middle of the ceiling over the dance floor. The owners have determined that the radius $r$ of the new ball must be 2 feet. [5.1]

(a) If the surface area of a sphere is $A = 4\pi r^2$, what is the surface area of the new disco ball?

(b) If the volume of a sphere is $V = \dfrac{4}{3}\pi r^3$, what is the volume of the new disco ball?

**55. Area of a Circle**  The Jacobs family is getting a round trampoline. Their square backyard has an area of 225 square feet. The smallest trampoline has a radius of 4 feet. The next largest has a radius of 6 feet, then 8 feet, and the largest has a radius of 10 feet. What is the largest trampoline that will fit in their backyard? [5.1]

# C H A P T E R   5   **TEST**

Simplify each of the following expressions. [5.1]

**1.** $(-3)^4$

**2.** $\left(\dfrac{3}{4}\right)^2$

**3.** $(3x^3)^2(2x^4)^3$

Simplify each expression. Write all answers with positive exponents only. [5.2]

**4.** $3^{-2}$

**5.** $(3a^4b^2)^0$

**6.** $\dfrac{a^{-3}}{a^{-5}}$

**7.** $\dfrac{(x^{-2})^3(x^{-3})^{-5}}{(x^{-4})^{-2}}$

**8.** Write 0.0278 in scientific notation. [5.2]

**9.** Write $2.43 \times 10^5$ in expanded form. [5.2]

Simplify. Write all answers with positive exponents only. [5.3]

**10.** $\dfrac{35x^2y^4z}{70x^6y^2z}$

**11.** $\dfrac{(6a^2b)(9a^3b^2)}{18a^4b^3}$

**12.** $\dfrac{24x^7}{3x^2} + \dfrac{14x^9}{7x^4}$

**13.** $\dfrac{(2.4 \times 10^5)(4.5 \times 10^{-2})}{1.2 \times 10^{-6}}$

Add and subtract as indicated. [5.4]

**14.** $8x^2 - 4x + 6x + 2$

**15.** $(5x^2 - 3x + 4) - (2x^2 - 7x - 2)$

**16.** Subtract $3x - 4$ from $6x - 8$. [5.4]

**17.** Find the value of $2y^2 - 3y - 4$ when $y$ is $-2$. [5.4]

Multiply. [5.5]

**18.** $2a^2(3a^2 - 5a + 4)$

**19.** $\left(x + \dfrac{1}{2}\right)\left(x + \dfrac{1}{3}\right)$

**20.** $(4x - 5)(2x + 3)$

**21.** $(x - 3)(x^2 + 3x + 9)$

Multiply. [5.6]

**22.** $(x + 5)^2$

**23.** $(3a - 2b)^2$

**24.** $(3x - 4y)(3x + 4y)$     **25.** $(a^2 - 3)(a^2 + 3)$

**26.** Divide $10x^3 + 15x^2 - 5x$ by $5x$. [5.7]

Divide. [5.8]

**27.** $\dfrac{8x^2 - 6x - 5}{2x - 3}$     **28.** $\dfrac{3x^3 - 2x + 1}{x - 3}$

**29. Volume**   Find the volume of a cube if the length of a side is 2.5 centimeters. [5.1]

**30. Volume**   Find the volume of a rectangular solid if the length is 5 times the width and the height is $\frac{1}{5}$ the width. [5.3]

# CHAPTERS 1–5 CUMULATIVE REVIEW

Simplify.

**1.** $-\left(-\dfrac{3}{4}\right)$

**2.** $2 \cdot 7 + 10$

**3.** $6 \cdot 7 + 7 \cdot 9$

**4.** $\dfrac{12 - 3}{8 - 8}$

**5.** $6(4a + 2) - 3(5a - 1)$

**6.** $-7\left(-\dfrac{1}{7}\right)$

**7.** $-15 - (-3)$

**8.** $(-9)(-7)$

**9.** $(-9)(-5)$

**10.** $y^{10} \cdot y^6$

**11.** $(2y)^4$

**12.** $(3a^2b^5)^3(2a^6b^7)^2$

**13.** $\dfrac{(12xy^5)^3(16x^2y^2)}{(8x^3y^3)(3x^5y)}$

**14.** $\dfrac{3.5 \times 10^{-7}}{7 \times 10^{-3}}$

**15.** $(5x - 1)^2$

**16.** $(-4x^2 - 5x + 2) + (3x^2 - 6x + 1)$
$\qquad\qquad\qquad - (-x^2 + 2x - 7)$

**17.** $(x - 1)(x^2 + x + 1)$

Solve each equation.

**18.** $8 - 2y + 3y = 12$

**19.** $6a - 5 = 4a$

**20.** $18 = 3(2x - 2)$

**21.** $2(3x + 5) + 8 = 2x + 10$

Solve the inequality, and graph the solution set.

**22.** $-\dfrac{a}{6} > 4$

**23.** $3(2t - 5) - 7 \le 5(3t + 1) + 5$

**24.** $-3 < 2x + 3 < 5$

**25.** $-5 \le 4x + 3 \le 11$

Divide.

**26.** $\dfrac{4x^2 + 8x - 10}{2x - 3}$

**27.** $\dfrac{15x^5 - 10x^2 + 20x}{5x^5}$

Graph on a rectangular coordinate system.

**28.** $x + y = 3$

**29.** $x = 2$

**30.** $3x + 2y = 6$

Solve each system by graphing.

**31.** $x + y = 1$
$\quad\;\; x + 2y = 2$

**32.** $x - y = 3$
$\quad\;\; 2x - 2y = 2$

**33.** $x + 2y = 4$
$\quad\;\; 3x + 6y = 12$

Solve each system.

**34.** $x + 2y = 5$
$\quad\;\; x - y = 2$

**35.** $x + 2y = 5$
$\quad\;\; 3x + 6y = 14$

**36.** $2x + y = -3$
$\quad\;\; x - 3y = -5$

**37.** $\dfrac{1}{6}x + \dfrac{1}{4}y = 1$
$\quad \dfrac{6}{5}x - y = \dfrac{8}{5}$

**38.** $x + y = 9$
$\quad\;\; y = x + 1$

**39.** $4x + 5y = 25$
$\quad\;\;\; 2y = x - 3$

**40.** Subtract 8 from $-9$.

**41.** Find the value of $8x - 3$ when $x = 3$.

**42.** Solve $A = \dfrac{1}{2}bh$ for $h$.

**43.** For the set $\{-3, -\sqrt{2}, 0, \frac{3}{4}, 1.5, \pi\}$ list all the irrational numbers.

**44.** Find $x$ when $y$ is 8 in the equation $y = 3x - 1$.

**45.** Which ordered pairs $(0, 3)$, $(4, 0)$, and $(\frac{16}{3}, 1)$ are solutions to the equation $3x - 4y = 12$?

**46.** Find the $x$- and $y$-intercepts for the equation $y = 2x + 4$.

**47.** Find the $x$- and $y$-intercepts for the line that passes through the points $(-2, 3)$ and $(6, -1)$.

**48.** Find the slope of the line $2x - 5y = 7$.

**49.** Find the equation for the line that passes through $(-2, 4)$ and $(1, 8)$.

**50. Mixture** How many liters of a 40% solution and a 16% solution must be mixed to obtain 20 liters of a 22% solution?

# Factoring

## INTRODUCTION

If you watch professional football on television, you will hear the announcers refer to "hang time" when the punter punts the ball. Hang time is the amount of time the ball is in the air, and it depends on only one thing, the initial vertical velocity imparted to the ball by the kicker's foot. We can find the hang time of a football by solving equations. Table 1 shows the equations to solve for hang time, given various initial vertical velocities. Figure 1 is a visual representation of some equations associated with the ones in Table 1. In Figure 1, you can find hang time on the horizontal axis.

| TABLE 1 Hang Time for a Football | | |
|---|---|---|
| Initial Vertical Velocity (feet/second) | Equation in Factored Form | Hang Time (seconds) |
| 16 | $16t(1 - t) = 0$ | 1 |
| 32 | $16t(2 - t) = 0$ | 2 |
| 48 | $16t(3 - t) = 0$ | 3 |
| 64 | $16t(4 - t) = 0$ | 4 |
| 80 | $16t(5 - t) = 0$ | 5 |

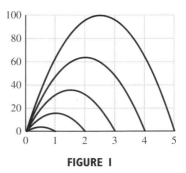

**FIGURE 1**

The equations in the second column of the table are in what is called "factored form." Once the equation is in factored form, hang time can be read from the second factor. In this chapter we develop techniques that allow us to factor a variety of polynomials. Factoring is the key to solving equations like the ones in Table 1.

## STUDY SKILLS

This is the last chapter in which we will mention study skills. You know by now what works best for you and what you have to do to achieve your goals for this course. From now on it is simply a matter of sticking with the things that work for you and avoiding the things that do not work. It seems simple, but as with anything that takes effort, it is up to you to see that you maintain the skills that get you where you want to be in the course.

If you intend to take more classes in mathematics and you want to ensure your success in those classes, then you can work toward this goal: *Become a student who can learn mathematics on your own.* Most people who have degrees in mathematics were students who could learn mathematics on their own. This doesn't mean that you have to learn it all on your own; it simply means that if you have to, you can learn it on your own. Attaining this goal gives you independence and puts you in control of your success in any math class you take.

In Chapter 1 we used the following diagram to illustrate the relationship between multiplication and factoring.

Multiplication

$$\text{Factors} \longrightarrow 3 \cdot 5 = 15 \longleftarrow \text{Product}$$

Factoring

A similar relationship holds for multiplication of polynomials. Reading the following diagram from left to right, we say the product of the binomials $x + 2$ and $x + 3$ is the trinomial $x^2 + 5x + 6$. On the other hand, if we read in the other direction, we can say that $x^2 + 5x + 6$ factors into the product of $x + 2$ and $x + 3$.

Multiplication

$$\text{Factors} \longrightarrow (x + 2)(x + 3) = x^2 + 5x + 6 \longleftarrow \text{Product}$$

Factoring

In this chapter we develop a systematic method of factoring polynomials.

In this section we will apply the distributive property to polynomials to factor from them what is called the greatest common factor.

> **DEFINITION**  The **greatest common factor** for a polynomial is the largest monomial that divides (is a factor of) each term of the polynomial.

We use the term *largest monomial* to mean the monomial with the greatest coefficient and highest power of the variable.

**EXAMPLE 1**  Find the greatest common factor for the polynomial:

$$3x^5 + 12x^2$$

**Solution**  The terms of the polynomial are $3x^5$ and $12x^2$. The largest number that divides the coefficients is 3, and the highest power of $x$ that is a factor of $x^5$ and $x^2$ is $x^2$. Therefore, the greatest common factor for $3x^5 + 12x^2$ is $3x^2$. That is, $3x^2$ is the largest monomial that divides each term of $3x^5 + 12x^2$.

**EXAMPLE 2**  Find the greatest common factor for:

$$8a^3b^2 + 16a^2b^3 + 20a^3b^3$$

**Solution**    The largest number that divides each of the coefficients is 4. The highest power of the variable that is a factor of $a^3b^2$, $a^2b^3$, and $a^3b^3$ is $a^2b^2$. The greatest common factor for $8a^3b^2 + 16a^2b^3 + 20a^3b^3$ is $4a^2b^2$. It is the largest monomial that is a factor of each term.

---

Once we have recognized the greatest common factor of a polynomial, we can apply the distributive property and factor it out of each term. We rewrite the polynomial as the product of its greatest common factor with the polynomial that remains after the greatest common factor has been factored from each term in the original polynomial.

**EXAMPLE  3**    Factor the greatest common factor from $3x - 15$.

**Solution**    The greatest common factor for the terms $3x$ and $15$ is $3$. We can rewrite both $3x$ and $15$ so that the greatest common factor $3$ is showing in each term. It is important to realize that $3x$ means $3 \cdot x$. The 3 and the $x$ are not "stuck" together:

$$3x - 15 = 3 \cdot x - 3 \cdot 5$$

Now, applying the distributive property, we have:

$$3 \cdot x - 3 \cdot 5 = 3(x - 5)$$

To check a factoring problem like this, we can multiply 3 and $x - 5$ to get $3x - 15$, which is what we started with. Factoring is simply a procedure by which we change sums and differences into products. In this case we changed the difference $3x - 15$ into the product $3(x - 5)$. Note, however, that we have not changed the meaning or value of the expression. The expression we end up with is equal to the expression we started with.

**EXAMPLE  4**    Factor the greatest common factor from:

$$5x^3 - 15x^2$$

**Solution**    The greatest common factor is $5x^2$. We rewrite the polynomial as:

$$5x^3 - 15x^2 = 5x^2 \cdot x - 5x^2 \cdot 3$$

Then we apply the distributive property to get:

$$5x^2 \cdot x - 5x^2 \cdot 3 = 5x^2(x - 3)$$

To check our work, we simply multiply $5x^2$ and $(x - 3)$ to get $5x^3 - 15x^2$, which is our original polynomial.

**EXAMPLE  5**    Factor the greatest common factor from:

$$16x^5 - 20x^4 + 8x^3$$

**Solution**   The greatest common factor is $4x^3$. We rewrite the polynomial so that we can see the greatest common factor $4x^3$ in each term; then we apply the distributive property to factor it out.

$$16x^5 - 20x^4 + 8x^3 = 4x^3 \cdot 4x^2 - 4x^3 \cdot 5x + 4x^3 \cdot 2$$
$$= 4x^3(4x^2 - 5x + 2)$$

**EXAMPLE   6**   Factor the greatest common factor from:

$$6x^3y - 18x^2y^2 + 12xy^3$$

**Solution**   The greatest common factor is $6xy$. We rewrite the polynomial in terms of $6xy$ and then apply the distributive property as follows:

$$6x^3y - 18x^2y^2 + 12xy^3 = 6xy \cdot x^2 - 6xy \cdot 3xy + 6xy \cdot 2y^2$$
$$= 6xy(x^2 - 3xy + 2y^2)$$

**EXAMPLE   7**   Factor the greatest common factor from:

$$3a^2b - 6a^3b^2 + 9a^3b^3$$

**Solution**   The greatest common factor is $3a^2b$:

$$3a^2b - 6a^3b^2 + 9a^3b^3 = 3a^2b(1) - 3a^2b(2ab) + 3a^2b(3ab^2)$$
$$= 3a^2b(1 - 2ab + 3ab^2)$$

## FACTORING BY GROUPING

To develop our next method of factoring, called *factoring by grouping,* we start by examining the polynomial $xc + yc$. The greatest common factor for the two terms is $c$. Factoring $c$ from each term we have:

$$xc + yc = c(x + y)$$

But suppose that $c$ itself was a more complicated expression, such as $a + b$, so that the expression we were trying to factor was $x(a + b) + y(a + b)$, instead of $xc + yc$. The greatest common factor for $x(a + b) + y(a + b)$ is $(a + b)$. Factoring this common factor from each term looks like this:

$$x(a + b) + y(a + b) = (a + b)(x + y)$$

To see how all of this applies to factoring polynomials, consider the polynomial

$$xy + 3x + 2y + 6$$

There is no greatest common factor other than the number 1. However, if we group the terms together two at a time, we can factor an $x$ from the first two terms and a 2 from the last two terms:

$$xy + 3x + 2y + 6 = x(y + 3) + 2(y + 3)$$

The expression on the right can be thought of as having two terms: $x(y + 3)$ and $2(y + 3)$. Each of these expressions contains the common factor $y + 3$, which can be factored out using the distributive property:

$$x(y + 3) + 2(y + 3) = (y + 3)(x + 2)$$

This last expression is in factored form. The process we used to obtain it is called factoring by grouping. Here are some additional examples.

**EXAMPLE  8**   Factor $ax + bx + ay + by$.

**Solution**   We begin by factoring $x$ from the first two terms and $y$ from the last two terms:

$$ax + bx + ay + by = x(a + b) + y(a + b)$$
$$= (a + b)(x + y)$$

To convince yourself that this is factored correctly, multiply the two factors $(a + b)$ and $(x + y)$.

**EXAMPLE  9**   Factor by grouping: $3ax - 2a + 15x - 10$.

**Solution**   First, we factor $a$ from the first two terms and 5 from the last two terms. Then, we factor $3x - 2$ from the remaining two expressions:

$$3ax - 2a + 15x - 10 = a(3x - 2) + 5(3x - 2)$$
$$= (3x - 2)(a + 5)$$

Again, multiplying $(3x - 2)$ and $(a + 5)$ will convince you that these are the correct factors.

**EXAMPLE  10**   Factor $2x^2 + 5ax - 2xy - 5ay$.

**Solution**   From the first two terms we factor $x$. From the second two terms we must factor $-y$ so that the binomial that remains after we do so matches the binomial produced by the first two terms:

$$2x^2 + 5ax - 2xy - 5ay = x(2x + 5a) - y(2x + 5a)$$
$$= (2x + 5a)(x - y)$$

Another way to accomplish the same result is to use the commutative property to interchange the middle two terms and then factor by grouping:

$$2x^2 + 5ax - 2xy - 5ay = 2x^2 - 2xy + 5ax - 5ay \qquad \text{Commutative property}$$
$$= 2x(x - y) + 5a(x - y)$$
$$= (x - y)(2x + 5a)$$

This is the same as the result we obtained previously.

**EXAMPLE 11**    Factor $6x^2 - 3x - 4x + 2$ by grouping.

**Solution**    The first two terms have $3x$ in common, and the last two terms have either a 2 or a $-2$ in common. Suppose we factor $3x$ from the first two terms and 2 from the last two terms. We get:

$$6x^2 - 3x - 4x + 2 = 3x(2x - 1) + 2(-2x + 1)$$

We can't go any further because there is no common factor that will allow us to factor further. However, if we factor $-2$, instead of 2, from the last two terms, our problem is solved:

$$6x^2 - 3x - 4x + 2 = 3x(2x - 1) - 2(2x - 1)$$
$$= (2x - 1)(3x - 2)$$

In this case, factoring $-2$ from the last two terms gives us an expression that can be factored further.

# Getting Ready for Class

*After reading through the preceding section, respond in your own words and in complete sentences.*

**A.**  What is the greatest common factor for a polynomial?

**B.**  After factoring a polynomial, how can you check your result?

**C.**  When would you try to factor by grouping?

**D.**  What is the relationship between multiplication and factoring?

# PROBLEM SET 6.1

Factor the following by taking out the greatest common factor.

**1.** $15x + 25$

**2.** $14x + 21$

**3.** $6a + 9$

**4.** $8a + 10$

**5.** $4x - 8y$

**6.** $9x - 12y$

**7.** $3x^2 - 6x - 9$

**8.** $2x^2 + 6x + 4$

**9.** $3a^2 - 3a - 60$

**10.** $2a^2 - 18a + 28$

**11.** $24y^2 - 52y + 24$

**12.** $18y^2 + 48y + 32$

**13.** $9x^2 - 8x^3$

**14.** $7x^3 - 4x^2$

**15.** $13a^2 - 26a^3$

**16.** $5a^2 - 10a^3$

**17.** $21x^2y - 28xy^2$

**18.** $30xy^2 - 25x^2y$

**19.** $22a^2b^2 - 11ab^2$

**20.** $15x^3 - 25x^2 + 30x$

**21.** $7x^3 + 21x^2 - 28x$

**22.** $16x^4 - 20x^2 - 16x$

**23.** $121y^4 - 11x^4$

**24.** $25a^4 - 5b^4$

Factor the following by taking out the greatest common factor.

**25.** $100x^4 - 50x^3 + 25x^2$

**26.** $36x^5 + 72x^3 - 81x^2$

**27.** $8a^2 + 16b^2 + 32c^2$

**28.** $9a^2 - 18b^2 - 27c^2$

**29.** $4a^2b - 16ab^2 + 32a^2b^2$

**30.** $5ab^2 + 10a^2b^2 + 15a^2b$

**31.** $121a^3b^2 - 22a^2b^3 + 33a^3b^3$

**32.** $20a^4b^3 - 18a^3b^4 + 22a^4b^4$

**33.** $12x^2y^3 - 72x^5y^3 - 36x^4y^4$

**34.** $49xy - 21x^2y^2 + 35x^3y^3$

Factor by grouping.

**35.** $xy + 5x + 3y + 15$    **36.** $xy + 2x + 4y + 8$

**37.** $xy + 6x + 2y + 12$    **38.** $xy + 2y + 6x + 12$

**39.** $ab + 7a - 3b - 21$    **40.** $ab + 3b - 7a - 21$

**41.** $ax - bx + ay - by$    **42.** $ax - ay + bx - by$

**43.** $2ax + 6x - 5a - 15$   **44.** $3ax + 21x - a - 7$

**45.** $3xb - 4b - 6x + 8$

**46.** $3xb - 4b - 15x + 20$   **47.** $x^2 + ax + 2x + 2a$

**48.** $x^2 + ax + 3x + 3a$    **49.** $x^2 - ax - bx + ab$

**50.** $x^2 + ax - bx - ab$

Factor by grouping. You can group the terms together two at a time or three at a time. Either way will produce the same result.

**51.** $ax + ay + bx + by + cx + cy$

**52.** $ax + bx + cx + ay + by + cy$

Factor the following polynomials by grouping the terms together two at a time.

**53.** $6x^2 + 9x + 4x + 6$    **54.** $6x^2 - 9x - 4x + 6$

**55.** $20x^2 - 2x + 50x - 5$

**56.** $20x^2 + 25x + 4x + 5$

**57.** $20x^2 + 4x + 25x + 5$

**58.** $20x^2 + 4x - 25x - 5$

**59.** $x^3 + 2x^2 + 3x + 6$

**60.** $x^3 - 5x^2 - 4x + 20$

**61.** $6x^3 - 4x^2 + 15x - 10$

**62.** $8x^3 - 12x^2 + 14x - 21$

**63.** The greatest common factor of the binomial $3x + 6$ is 3. The greatest common factor of the binomial $2x + 4$ is 2. What is the greatest common factor of their product $(3x + 6)(2x + 4)$, when it has been multiplied out?

**64.** The greatest common factors of the binomials $4x + 2$ and $5x + 10$ are 2 and 5, respectively. What is the greatest common factor of their product $(4x + 2)(5x + 10)$, when it has been multiplied out?

**65.** The following factorization is incorrect. Find the mistake, and correct the right-hand side:

$$12x^2 + 6x + 3 = 3(4x^2 + 2x)$$

**66.** Find the mistake in the following factorization, and then rewrite the right-hand side correctly:

$$10x^2 + 2x + 6 = 2(5x^2 + 3)$$

## Applying the Concepts

**67. Investing** If you invest \$1,000 in an account with an annual interest rate of $r$ compounded annually, the amount of money you have in the account after one year is:

$$A = 1,000 + 1,000r$$

Write this formula again with the right side in factored form. Then, find the amount of money in this account at the end of one year if the interest rate is 12%.

**68. Investing** If you invest $P$ dollars in an account with an annual interest rate of 8% compounded annually, then the amount of money in that account after one year is given by the formula:

$$A = P + 0.08P$$

Rewrite this formula with the right side in factored form, and then find the amount of money in the account at the end of one year if \$500 was the initial investment.

**69. Biological Growth** If 1,000,000 bacteria are placed in a Petri dish and the bacteria have a growth rate of $r$ (a percent expressed as a decimal) per hour, then one hour later the amount of bacteria will be $A = 1,000,000 + 1,000,000r$ bacteria.
(a) Factor the right side of the equation.
(b) If $r = 30\%$, find the number of bacteria present after one hour.

**70. Biological Growth** If there are $B$ bacteria present initially in a Petri dish and their growth rate is $r$ (a percent expressed as a decimal) per hour, then after one hour there will be $A = B + Br$ bacteria present.
(a) Factor the right side of this equation.
(b) The following bar graph shows the number of bacteria present initially and the number of bacteria present one hour later. Use the bar chart to find $B$ and $A$ in the preceding equation.

**Bacterial growth**

*Time (hr)*

*Number of bacteria*

$C$ is the original amount and $A$ is the amount after 1,000 years, then $A = C - Cr$, and $r$ is called the decay rate (per thousand years).
(a) Factor and simplify the right side.
(b) From the following chart, find $C$, $A$, and $r$:

|                    | 1,000 Years Ago | Now      |
|--------------------|-----------------|----------|
| Amount of $^{14}C$ | 5.00 kg         | 4.45 kg  |

**71. Negative Growth Rate (Decline)** Old family names in English disappear at a steady rate (and are replaced by new names that come into existence). If there were 7,000 English family names in the year 1500, and if they disappear at the rate of $r$ (a percent expressed as a decimal) every 500 years, then $A = 7,000 - 7,000r$ of them will still be in use in the year 2000.
(a) Factor the right side of the equation.
(b) If $r = 23\%$, how many will still be in use in the year 2000?

**72. Radioactive Decay** Radioactive atoms such as $^{14}C$ (Carbon-14) decay — that is, after a certain amount of time, a certain percentage of the element will be lost by changing into a different element. If

## Review Problems

The following problems review material we covered in Section 5.5. Reviewing these problems will help you with the next section.

Multiply using the FOIL method.

**73.** $(x - 7)(x + 2)$    **74.** $(x - 7)(x - 2)$

**75.** $(x - 3)(x + 2)$    **76.** $(x + 3)(x - 2)$

**77.** $(x + 3)(x^2 - 3x + 9)$

**78.** $(x - 2)(x^2 + 2x + 4)$

**79.** $(2x + 1)(x^2 + 4x - 3)$

**80.** $(3x + 2)(x^2 - 2x - 4)$

---

**6.2    Factoring Trinomials**

In this section we will factor trinomials in which the coefficient of the squared term is 1. The more familiar we are with multiplication of binomials, the easier factoring trinomials will be.

Recall multiplication of binomials from Chapter 5:

$$(x + 3)(x + 4) = x^2 + 7x + 12$$
$$(x - 5)(x + 2) = x^2 - 3x - 10$$

The first term in the answer is the product of the first terms in each binomial. The last term in the answer is the product of the last terms in each binomial. The middle term in the answer comes from adding the product of the outside terms to the product of the inside terms.

Let's have $a$ and $b$ represent real numbers and look at the product of $(x + a)$ and $(x + b)$:

$$(x + a)(x + b) = x^2 + ax + bx + ab$$
$$= x^2 + (a + b)x + ab$$

The coefficient of the middle term is the sum of $a$ and $b$. The last term is the product of $a$ and $b$. Writing this as a factoring problem, we have:

$$x^2 + \underset{\text{Sum}}{(a + b)}x + \underset{\text{Product}}{ab} = (x + a)(x + b)$$

To factor a trinomial in which the coefficient of $x^2$ is 1, we need only find the numbers $a$ and $b$ whose sum is the coefficient of the middle term and whose product is the constant term (last term).

**E X A M P L E   I**    Factor $x^2 + 8x + 12$.

**Solution**    The coefficient of $x^2$ is 1. We need two numbers whose sum is 8 and whose product is 12. The numbers are 6 and 2:

$$x^2 + 8x + 12 = (x + 6)(x + 2).$$

We can easily check our work by multiplying $(x + 6)$ and $(x + 2)$.

$$\text{Check:}\quad (x + 6)(x + 2) = x^2 + 6x + 2x + 12$$
$$= x^2 + 8x + 12$$

**E X A M P L E   2**    Factor $x^2 - 2x - 15$.

**Solution**    The coefficient of $x^2$ is again 1. We need to find a pair of numbers whose sum is $-2$ and whose product is $-15$. Here are all the possibilities for products that are $-15$.

| Products | Sums |
|---|---|
| $-1(15) = -15$ | $-1 + 15 = 14$ |
| $1(-15) = -15$ | $1 + (-15) = -14$ |
| $-5(3) = -15$ | $-5 + 3 = -2$ |
| $5(-3) = -15$ | $5 + (-3) = 2$ |

The third line gives us what we want. The factors of $x^2 - 2x - 15$ are $(x - 5)$ and $(x + 3)$:

$$x^2 - 2x - 15 = (x - 5)(x + 3)$$

**E X A M P L E   3**    Factor $2x^2 + 10x - 28$.

**Solution**    The coefficient of $x^2$ is 2. We begin by factoring out the greatest common factor, which is 2:

$$2x^2 + 10x - 28 = 2(x^2 + 5x - 14)$$

Now, we factor the remaining trinomial by finding a pair of numbers whose sum is 5 and whose product is $-14$. Here are the possibilities:

| Products | Sums |
|---|---|
| $-1(14) = -14$ | $-1 + 14 = 13$ |
| $1(-14) = -14$ | $1 + (-14) = -13$ |
| $-7(2) = -14$ | $-7 + 2 = -5$ |
| $7(-2) = -14$ | $7 + (-2) = 5$ |

From the last line we see that the factors of $x^2 + 5x - 14$ are $(x + 7)$ and $(x - 2)$. Here is the complete problem:

$$2x^2 + 10x - 28 = 2(x^2 + 5x - 14)$$
$$= 2(x + 7)(x - 2)$$

**Note:**  In Example 3 we began by factoring out the greatest common factor. The first step in factoring any trinomial is to look for the greatest common factor. If the trinomial in question has a greatest common factor other than 1, we factor it out first and then try to factor the trinomial that remains.

**E X A M P L E   4**   Factor $3x^3 - 3x^2 - 18x$.

**Solution**   We begin by factoring out the greatest common factor, which is $3x$. Then we factor the remaining trinomial. Without showing the table of products and sums as we did in Examples 2 and 3, here is the complete problem:

$$3x^3 - 3x^2 - 18x = 3x(x^2 - x - 6)$$
$$= 3x(x - 3)(x + 2)$$

**E X A M P L E   5**   Factor $x^2 + 8xy + 12y^2$.

**Solution**   This time we need two expressions whose product is $12y^2$ and whose sum is $8y$. The two expressions are $6y$ and $2y$ (see Example 1 in this section):

$$x^2 + 8xy + 12y^2 = (x + 6y)(x + 2y)$$

You should convince yourself that these factors are correct by finding their product.

## Getting Ready for Class

*After reading through the preceding section, respond in your own words and in complete sentences.*

**A.** When the leading coefficient of a trinomial is 1, what is the relationship between the other two coefficients and the factors of the trinomial?

**B.** When factoring polynomials, what should you look for first?

**C.** How can you check to see that you have factored a trinomial correctly?

**D.** Describe how you would find the factors of $x^2 + 8x + 12$.

# PROBLEM SET 6.2

Factor the following trinomials.

**1.** $x^2 + 7x + 12$

**2.** $x^2 + 7x + 10$

**3.** $x^2 + 3x + 2$

**4.** $x^2 + 7x + 6$

**5.** $a^2 + 10a + 21$

**6.** $a^2 - 7a + 12$

**7.** $x^2 - 7x + 10$

**8.** $x^2 - 3x + 2$

**9.** $y^2 - 10y + 21$

**10.** $y^2 - 7y + 6$

**11.** $x^2 - x - 12$

**12.** $x^2 - 4x - 5$

**13.** $y^2 + y - 12$

**14.** $y^2 + 3y - 18$

**15.** $x^2 + 5x - 14$

**16.** $x^2 - 5x - 24$

**17.** $r^2 - 8r - 9$

**18.** $r^2 - r - 2$

**19.** $x^2 - x - 30$

**20.** $x^2 + 8x + 12$

**21.** $a^2 + 15a + 56$

**22.** $a^2 - 9a + 20$

**23.** $y^2 - y - 42$

**24.** $y^2 + y - 42$

**25.** $x^2 + 13x + 42$

**26.** $x^2 - 13x + 42$

Factor the following problems completely. First, factor out the greatest common factor, and then factor the remaining trinomial.

**27.** $2x^2 + 6x + 4$

**28.** $3x^2 - 6x - 9$

**29.** $3a^2 - 3a - 60$

**30.** $2a^2 - 18a + 28$

**31.** $100x^2 - 500x + 600$

**32.** $100x^2 - 900x + 2,000$

**33.** $100p^2 - 1,300p + 4,000$

**34.** $100p^2 - 1,200p + 3,200$

**35.** $x^4 - x^3 - 12x^2$

**36.** $x^4 - 11x^3 + 24x^2$

**37.** $2r^3 + 4r^2 - 30r$

**38.** $5r^3 + 45r^2 + 100r$

**39.** $2y^4 - 6y^3 - 8y^2$

**40.** $3r^3 - 3r^2 - 6r$

**41.** $x^5 + 4x^4 + 4x^3$

**42.** $x^5 + 13x^4 + 42x^3$

**43.** $3y^4 - 12y^3 - 15y^2$

**44.** $5y^4 - 10y^3 + 5y^2$

**45.** $4x^4 - 52x^3 + 144x^2$

**46.** $3x^3 - 3x^2 - 18x$

Factor the following trinomials.

**47.** $x^2 + 5xy + 6y^2$

**48.** $x^2 - 5xy + 6y^2$

**49.** $x^2 - 9xy + 20y^2$

**50.** $x^2 + 9xy + 20y^2$

**51.** $a^2 + 2ab - 8b^2$

**52.** $a^2 - 2ab - 8b^2$

**53.** $a^2 - 10ab + 25b^2$

**54.** $a^2 + 6ab + 9b^2$

**55.** $a^2 + 10ab + 25b^2$

**56.** $a^2 - 6ab + 9b^2$

**57.** $x^2 + 2xa - 48a^2$

**58.** $x^2 - 3xa - 10a^2$

**59.** $x^2 - 5xb - 36b^2$

**60.** $x^2 - 13xb + 36b^2$

Factor completely.

**61.** $x^4 - 5x^2 + 6$

**62.** $x^6 - 2x^3 - 15$

**63.** $x^2 - 80x - 2,000$

**64.** $x^2 - 190x - 2,000$

**65.** $x^2 - x + \dfrac{1}{4}$

**66.** $x^2 - \dfrac{2}{3}x + \dfrac{1}{9}$

**67.** $x^2 + 0.6x + 0.08$

**68.** $x^2 + 0.8x + 0.15$

**69.** If one of the factors of $x^2 + 24x + 128$ is $x + 8$, what is the other factor?

**70.** If one factor of $x^2 + 260x + 2,500$ is $x + 10$, what is the other factor?

**71.** What polynomial, when factored, gives $(4x + 3)(x - 1)$?

**72.** What polynomial factors to $(4x - 3)(x + 1)$?

## Review Problems

The problems that follow review material we covered in Sections 5.4 and 5.5. Reviewing the problems from Section 5.5 will help you with the next section.

Multiply using the FOIL method. [5.5]

**73.** $(6a + 1)(a + 2)$

**74.** $(6a - 1)(a - 2)$

**75.** $(3a + 2)(2a + 1)$

**76.** $(3a - 2)(2a - 1)$

**77.** $(6a + 2)(a + 1)$

**78.** $(3a + 1)(2a + 2)$

Subtract. [5.4]

**79.** $(5x^2 + 5x - 4) - (3x^2 - 2x + 7)$

**80.** $(7x^4 - 4x^2 - 5) - (2x^4 - 4x^2 + 5)$

**81.** Subtract $4x - 5$ from $7x + 3$.

**82.** Subtract $3x + 2$ from $-6x + 1$.

**83.** Subtract $2x^2 - 4x$ from $5x^2 - 5$.

**84.** Subtract $6x^2 + 3$ from $2x^2 - 4x$.

We will now consider trinomials whose greatest common factor is 1 and whose leading coefficient (the coefficient of the squared term) is a number other than 1. We present two methods for factoring trinomials of this type. The first method involves listing possible factors until the correct pair of factors is found. This requires a certain amount of trial and error. The second method is based on the factoring by grouping process, which we covered in Section 6.1. Either method can be used to factor trinomials whose leading coefficient is a number other than 1.

## METHOD 1: FACTORING $ax^2 + bx + c$ BY TRIAL AND ERROR

Suppose we want to factor the trinomial $2x^2 - 5x - 3$. We know the factors (if they exist) will be a pair of binomials. The product of their first terms is $2x^2$, and the product of the last terms is $-3$. Let us list all the possible factors along with the trinomial that would result if we were to multiply them together. Remember, the middle term comes from the product of the inside terms plus the product of the outside terms.

| Binomial Factors | First Term | Middle Term | Last Term |
|---|---|---|---|
| $(2x - 3)(x + 1)$ | $2x^2$ | $-x$ | $-3$ |
| $(2x + 3)(x - 1)$ | $2x^2$ | $+x$ | $-3$ |
| $(2x - 1)(x + 3)$ | $2x^2$ | $+5x$ | $-3$ |
| $(2x + 1)(x - 3)$ | $2x^2$ | $-5x$ | $-3$ |

We can see from the last line that the factors of $2x^2 - 5x - 3$ are $(2x + 1)$ and $(x - 3)$. We call this the trial and error method of factoring trinomials. We look for possible factors that, when multiplied, will give the correct first and last terms, and then we see if we can adjust them to give the correct middle term.

**EXAMPLE 1**  Factor $6a^2 + 7a + 2$.

**Solution**  We list all the possible pairs of factors that, when multiplied together, give a trinomial whose first term is $6a^2$ and whose last term is $+2$.

| Binomial Factors | First Term | Middle Term | Last Term |
|---|---|---|---|
| $(6a + 1)(a + 2)$ | $6a^2$ | $+13a$ | $+2$ |
| $(6a - 1)(a - 2)$ | $6a^2$ | $-13a$ | $+2$ |
| $(3a + 2)(2a + 1)$ | $6a^2$ | $+7a$ | $+2$ |
| $(3a - 2)(2a - 1)$ | $6a^2$ | $-7a$ | $+2$ |

The factors of $6a^2 + 7a + 2$ are $(3a + 2)$ and $(2a + 1)$.

Check:      $(3a + 2)(2a + 1) = 6a^2 + 7a + 2$

---

Notice that in the preceding list we did not include the factors $(6a + 2)$ and $(a + 1)$. We do not need to try these, because the first factor has a 2 common to each term and so could be factored again, giving $2(3a + 1)(a + 1)$. Since our original trinomial, $6a^2 + 7a + 2$, did *not* have a greatest common factor of 2, neither of its factors will.

**EXAMPLE 2**   Factor $4x^2 - x - 3$.

**Solution**   We list all the possible factors that, when multiplied, give a trinomial whose first term is $4x^2$ and whose last term is $-3$.

| Binomial Factors | First Term | Middle Term | Last Term |
|---|---|---|---|
| $(4x + 1)(x - 3)$ | $4x^2$ | $-11x$ | $-3$ |
| $(4x - 1)(x + 3)$ | $4x^2$ | $+11x$ | $-3$ |
| $(4x + 3)(x - 1)$ | $4x^2$ | $-x$ | $-3$ |
| $(4x - 3)(x + 1)$ | $4x^2$ | $+x$ | $-3$ |
| $(2x + 1)(2x - 3)$ | $4x^2$ | $-4x$ | $-3$ |
| $(2x - 1)(2x + 3)$ | $4x^2$ | $+4x$ | $-3$ |

The third line shows that the factors are $(4x + 3)$ and $(x - 1)$.

Check:      $(4x + 3)(x - 1) = 4x^2 - x - 3$

---

You will find that the more practice you have at factoring this type of trinomial, the faster you will get the correct factors. You will pick up some shortcuts along the way or may come across a system of eliminating some factors as possibilities. Whatever works best for you is the method you should use. Factoring is a very important tool, and you must be good at it.

**EXAMPLE 3**   Factor $12y^3 + 10y^2 - 12y$.

**Solution**   We begin by factoring out the greatest common factor, $2y$:

$$12y^3 + 10y^2 - 12y = 2y(6y^2 + 5y - 6)$$

We now list all possible factors of a trinomial with the first term $6y^2$ and last term $-6$, along with the associated middle terms.

| Possible Factors | Middle Term When Multiplied |
|---|---|
| $(3y + 2)(2y - 3)$ | $-5y$ |
| $(3y - 2)(2y + 3)$ | $+5y$ |
| $(6y + 1)(y - 6)$ | $-35y$ |
| $(6y - 1)(y + 6)$ | $+35y$ |

The second line gives the correct factors. The complete problem is:

$$12y^3 + 10y^2 - 12y = 2y(6y^2 + 5y - 6)$$
$$= 2y(3y - 2)(2y + 3)$$

**EXAMPLE   4**   Factor $30x^2y - 5xy^2 - 10y^3$.

**Solution**   The greatest common factor is $5y$:

$$30x^2y - 5xy^2 - 10y^3 = 5y(6x^2 - xy - 2y^2)$$
$$= 5y(2x + y)(3x - 2y)$$

## METHOD 2: FACTORING $ax^2 + bx + c$ BY GROUPING

Recall from Section 6.1 that we can use factoring by grouping to factor the polynomial $6x^2 - 3x - 4x + 2$. We begin by factoring $3x$ from the first two terms and $-2$ from the last two terms. For review, here is the complete problem:

$$6x^2 - 3x - 4x + 2 = 3x(2x - 1) - 2(2x - 1)$$
$$= (2x - 1)(3x - 2)$$

Now, let's back up a little and notice that our original polynomial $6x^2 - 3x - 4x + 2$ can be simplified to $6x^2 - 7x + 2$ by adding $-3x$ and $-4x$. This means that $6x^2 - 7x + 2$ can be factored to $(2x - 1)(3x - 2)$ by the grouping method shown in Section 6.1. The key to using this process is to rewrite the middle term $-7x$ as $-3x - 4x$.

To generalize this discussion, here are the steps we use to factor trinomials by grouping.

### STRATEGY FOR FACTORING $ax^2 + bx + c$ BY GROUPING

*Step 1:*  Form the product $ac$.

*Step 2:*  Find a pair of numbers whose product is $ac$ and whose sum is $b$.

*Step 3:*  Rewrite the polynomial to be factored so that the middle term $bx$ is written as the sum of two terms whose coefficients are the two numbers found in step 2.

*Step 4:*  Factor by grouping.

**EXAMPLE  5**   Factor $3x^2 - 10x - 8$ using these steps.

**Solution**   The trinomial $3x^2 - 10x - 8$ has the form $ax^2 + bx + c$, where $a = 3$, $b = -10$, and $c = -8$.

Step 1: The product $ac$ is $3(-8) = -24$.

Step 2: We need to find two numbers whose product is $-24$ and whose sum is $-10$. Let's systematically begin to list all the pairs of numbers whose product is $-24$ to find the pair whose sum is $-10$.

| Product | Sum |
|---------|-----|
| $-24(1) = -24$ | $-24 + 1 = -23$ |
| $-12(2) = -24$ | $-12 + 2 = -10$ |

We stop here because we have found the pair of numbers whose product is $-24$ and whose sum is $-10$. The numbers are $-12$ and 2.

Step 3: We now rewrite our original trinomial so that the middle term, $-10x$, is written as the sum of $-12x$ and $2x$:

$$3x^2 - 10x - 8 = 3x^2 - 12x + 2x - 8$$

Step 4: Factoring by grouping, we have:

$$3x^2 - 12x + 2x - 8 = 3x(x - 4) + 2(x - 4)$$
$$= (x - 4)(3x + 2)$$

We can check our work by multiplying $x - 4$ and $3x + 2$ to get $3x^2 - 10x - 8$.

**EXAMPLE  6**   Factor $4x^2 - x - 3$.

**Solution**   In this case, $a = 4$, $b = -1$, and $c = -3$. The product $ac$ is $4(-3) = -12$. We need a pair of numbers whose product is $-12$ and whose sum is $-1$. We begin by listing pairs of numbers whose product is $-12$ in order to find the pair whose sum is $-1$.

| Product | Sum |
|---------|-----|
| $-12(1) = -12$ | $-12 + 1 = -11$ |
| $-6(2) = -12$ | $-6 + 2 = -4$ |
| $-4(3) = -12$ | $-4 + 3 = -1$ |

We stop here because we have found the pair of numbers for which we are looking. They are $-4$ and 3. Next, we rewrite the middle term $-x$ as the sum $-4x + 3x$ and proceed to factor by grouping.

$$4x^2 - x - 3 = 4x^2 - 4x + 3x - 3$$
$$= 4x(x - 1) + 3(x - 1)$$
$$= (x - 1)(4x + 3)$$

Compare this procedure and the result with those shown in Example 2 of this section.

**E X A M P L E   7**    Factor $8x^2 - 2x - 15$.

**Solution**    The product $ac$ is $8(-15) = -120$. There are many pairs of numbers whose product is $-120$. We are looking for the pair whose sum is also $-2$. The numbers are $-12$ and $10$. Writing $-2x$ as $-12x + 10x$ and then factoring by grouping, we have:

$$8x^2 - 2x - 15 = 8x^2 - 12x + 10x - 15$$
$$= 4x(2x - 3) + 5(2x - 3)$$
$$= (2x - 3)(4x + 5)$$

**E X A M P L E   8**    A ball is tossed into the air with an upward velocity of 16 feet per second from the top of a building 32 feet high. The equation that gives the height of the ball above the ground at any time $t$ is

$$h = 32 + 16t - 16t^2$$

Factor the right side of this equation and then find $h$ when $t$ is 2.

**Solution**    We begin by factoring out the greatest common factor, 16. Then, we factor the trinomial that remains:

$$h = 32 + 16t - 16t^2$$
$$h = 16(2 + t - t^2)$$
$$h = 16(2 - t)(1 + t)$$

Letting $t = 2$ in the equation, we have

$$h = 16(2 - 2)(1 + 2)$$
$$= 16(0)(3)$$
$$= 0$$

When $t$ is 2, $h$ is 0.

**Getting Ready for Class**

*After reading through the preceding section, respond in your own words and in complete sentences.*

**A.** What is the first step in factoring a trinomial?

**B.** Describe the criteria you would use to set up a table of possible factors of a trinomial.

**C.** What does it mean if you factor a trinomial and one of your factors has a greatest common factor of 3?

**D.** Describe how you would look for possible factors of $6a^2 + 7a + 2$.

# PROBLEM SET 6.3

Factor the following trinomials.

**1.** $2x^2 + 7x + 3$

**2.** $2x^2 + 5x + 3$

**3.** $2a^2 - a - 3$

**4.** $2a^2 + a - 3$

**5.** $3x^2 + 2x - 5$

**6.** $3x^2 - 2x - 5$

**7.** $3y^2 - 14y - 5$

**8.** $3y^2 + 14y - 5$

**9.** $6x^2 + 13x + 6$

**10.** $6x^2 - 13x + 6$

**11.** $4x^2 - 12xy + 9y^2$

**12.** $4x^2 + 12xy + 9y^2$

**13.** $4y^2 - 11y - 3$

**14.** $4y^2 + y - 3$

**15.** $20x^2 - 41x + 20$

**16.** $20x^2 + 9x - 20$

**17.** $20a^2 + 48ab - 5b^2$

**18.** $20a^2 + 29ab + 5b^2$

**19.** $20x^2 - 21x - 5$

**20.** $20x^2 - 48x - 5$

**21.** $12m^2 + 16m - 3$

**22.** $12m^2 + 20m + 3$

**23.** $20x^2 + 37x + 15$

**24.** $20x^2 + 13x - 15$

**25.** $12a^2 - 25ab + 12b^2$

**26.** $12a^2 + 7ab - 12b^2$

**27.** $3x^2 - xy - 14y^2$

**28.** $3x^2 + 19xy - 14y^2$

**29.** $14x^2 + 29x - 15$

**30.** $14x^2 + 11x - 15$

**31.** $6x^2 - 43x + 55$

**32.** $6x^2 - 7x - 55$

**33.** $15t^2 - 67t + 38$

**34.** $15t^2 - 79t - 34$

Factor each of the following completely. Look first for the greatest common factor.

**35.** $4x^2 + 2x - 6$

**36.** $6x^2 - 51x + 63$

**37.** $24a^2 - 50a + 24$

**38.** $18a^2 + 48a + 32$

**39.** $10x^3 - 23x^2 + 12x$

**40.** $10x^4 + 7x^3 - 12x^2$

**41.** $6x^4 - 11x^3 - 10x^2$

**42.** $6x^3 + 19x^2 + 10x$

**43.** $10a^3 - 6a^2 - 4a$

**44.** $6a^3 + 15a^2 + 9a$

**45.** $15x^3 - 102x^2 - 21x$

**46.** $2x^4 - 24x^3 + 64x^2$

**47.** $35y^3 - 60y^2 - 20y$

**48.** $14y^4 - 32y^3 + 8y^2$

**49.** $15a^4 - 2a^3 - a^2$

**50.** $10a^5 - 17a^4 + 3a^3$

**51.** $24x^2y - 6xy - 45y$

**52.** $8x^2y^2 + 26xy^2 + 15y^2$

**53.** $12x^2y - 34xy^2 + 14y^3$

**54.** $12x^2y - 46xy^2 + 14y^3$

**55.** Evaluate the expression $2x^2 + 7x + 3$ and the expression $(2x + 1)(x + 3)$ for $x = 2$.

**56.** Evaluate the expression $2a^2 - a - 3$ and the expression $(2a - 3)(a + 1)$ for $a = 5$.

**57.** What polynomial factors to $(2x + 3)(2x - 3)$?

**58.** What polynomial factors to $(5x + 4)(5x - 4)$?

**59.** What polynomial factors to $(x + 3)(x - 3)(x^2 + 9)$?

**60.** What polynomial factors to $(x + 2)(x - 2)(x^2 + 4)$?

## Applying the Concepts

**61. Archery** Margaret shoots an arrow into the air. The equation for the height (in feet) of the tip of the arrow is:

$$h = 8 + 62t - 16t^2$$

Factor the right side of this equation. Then fill in

the table for various heights of the arrow, using the factored form of the equation.

| Time $t$ (seconds) | Height $h$ (feet) |
|---|---|
| 0 | |
| 1 | |
| 2 | |
| 3 | |
| 4 | |

**62. Coin Toss** At the beginning of every football game, the referee flips a coin to see who will kick off. The equation that gives the height (in feet) of

the coin tossed in the air is:

$$h = 6 + 29t - 16t^2$$

(a) Factor this equation.

(b) Use the factored form of the equation to find the height of the quarter after 0 second, 1 second, and 2 seconds.

**63. Constructing a Box** Yesterday I was experimenting with how to cut and fold a certain piece of cardboard to make a box with different volumes. Unfortunately, today I have lost both the cardboard and most of my notes. I remember that I made the box by cutting equal squares from the corners, then folding up the side flaps:

I don't remember how big the cardboard was, and I can only find the last page of notes, which says that, if $x$ is the length of a side of a small square (in inches), then the volume is $V = 99x - 40x^2 + 4x^3$.

(a) Factor the right side of this expression completely.

(b) What were the dimensions of the original piece of cardboard?

**64. Constructing a Box** Repeat Problem 63 if the remaining formula is $V = 15x - 16x^2 + 4x^3$.

## Review Problems

The following problems review material we covered in Section 5.6. Reviewing these problems will help you understand the next section.

Multiply.

**65.** $(x + 3)(x - 3)$      **66.** $(x + 5)(x - 5)$

**67.** $(6a + 1)(6a - 1)$      **68.** $(4a + 5)(4a - 5)$

**69.** $(x + 4)^2$      **70.** $(x - 5)^2$

**71.** $(2x + 3)^2$      **72.** $(2x - 3)^2$

In Chapter 5 we listed the following three special products:

$$(a + b)^2 = (a + b)(a + b) = a^2 + 2ab + b^2$$
$$(a - b)^2 = (a - b)(a - b) = a^2 - 2ab + b^2$$
$$(a + b)(a - b) = a^2 - b^2$$

Since factoring is the reverse of multiplication, we can also consider the three special products as three special factorings:

$$a^2 + 2ab + b^2 = (a + b)^2$$
$$a^2 - 2ab + b^2 = (a - b)^2$$
$$a^2 - b^2 = (a + b)(a - b)$$

Any trinomial of the form $a^2 + 2ab + b^2$ or $a^2 - 2ab + b^2$ can be factored by the methods of Section 6.3. The last line is the factorization for the difference of two squares. The difference of two squares always factors in this way. Again, these are patterns you must be able to recognize on sight.

**E X A M P L E  1**  Factor $16x^2 - 25$.

**Solution**  We can see that the first term is a perfect square, and the last term is also. This fact becomes even more obvious if we rewrite the problem as:

$$16x^2 - 25 = (4x)^2 - (5)^2$$

The first term is the square of the quantity $4x$, and the last term is the square of 5. The completed problem looks like this:

$$16x^2 - 25 = (4x)^2 - (5)^2$$
$$= (4x + 5)(4x - 5)$$

To check our results, we multiply:

$$(4x + 5)(4x - 5) = 16x^2 + 20x - 20x - 25$$
$$= 16x^2 - 25$$

**E X A M P L E  2**  Factor $36a^2 - 1$.

**Solution**  We rewrite the two terms to show they are perfect squares and then factor. Remember, 1 is its own square, $1^2 = 1$.

$$36a^2 - 1 = (6a)^2 - (1)^2$$
$$= (6a + 1)(6a - 1)$$

**335**

To check our results, we multiply:

$$(6a + 1)(6a - 1) = 36a^2 + 6a - 6a - 1$$
$$= 36a^2 - 1$$

**E X A M P L E   3**   Factor $x^4 - y^4$.

**Solution**   $x^4$ is the perfect square $(x^2)^2$, and $y^4$ is $(y^2)^2$:

$$x^4 - y^4 = (x^2)^2 - (y^2)^2$$
$$= (x^2 - y^2)(x^2 + y^2)$$

The factor $(x^2 - y^2)$ is itself the difference of two squares and therefore can be factored again. The factor $(x^2 + y^2)$ is the *sum* of two squares and cannot be factored again. The complete problem is this:

$$x^4 - y^4 = (x^2)^2 - (y^2)^2$$
$$= (x^2 - y^2)(x^2 + y^2)$$
$$= (x + y)(x - y)(x^2 + y^2)$$

**Note:**   If you think the sum of two squares $x^2 + y^2$ factors, you should try it. Write down the factors you think it has, and then multiply them using the FOIL method. You won't get $x^2 + y^2$.

**E X A M P L E   4**   Factor $25x^2 - 60x + 36$.

**Solution**   Although this trinomial can be factored by the method we used in Section 6.3, we notice that the first and last terms are the perfect squares $(5x)^2$ and $(6)^2$. Before going through the method for factoring trinomials by listing all possible factors, we can check to see if $25x^2 - 60x + 36$ factors to $(5x - 6)^2$. We need only multiply to check:

$$(5x - 6)^2 = (5x - 6)(5x - 6)$$
$$= 25x^2 - 30x - 30x + 36$$
$$= 25x^2 - 60x + 36$$

The trinomial $25x^2 - 60x + 36$ factors to $(5x - 6)(5x - 6) = (5x - 6)^2$.

**E X A M P L E   5**   Factor $m^2 + 14m + 49$.

**Solution**   Since the first and last terms are perfect squares, we can try the factors $(m + 7)(m + 7)$:

$$(m + 7)^2 = (m + 7)(m + 7)$$
$$= m^2 + 7m + 7m + 49$$
$$= m^2 + 14m + 49$$

The factors of $m^2 + 14m + 49$ are $(m + 7)(m + 7) = (m + 7)^2$.

**Note:**  As we have indicated before, perfect square trinomials like the ones in Examples 4 and 5 can be factored by the methods developed in previous sections. Recognizing that they factor to binomial squares simply saves time in factoring.

**EXAMPLE 6**  Factor $5x^2 + 30x + 45$.

**Solution**  We begin by factoring out the greatest common factor, which is 5. Then we notice that the trinomial that remains is a perfect square trinomial:

$$5x^2 + 30x + 45 = 5(x^2 + 6x + 9)$$
$$= 5(x + 3)^2$$

**EXAMPLE 7**  Factor $(x - 3)^2 - 25$.

**Solution**  This example has the form $a^2 - b^2$, where $a$ is $x - 3$ and $b$ is 5. We factor it according to the formula for the difference of two squares:

$$(x - 3)^2 - 25 = (x - 3)^2 - 5^2 \qquad \text{Write 25 as } 5^2.$$
$$= [(x - 3) - 5][(x - 3) + 5] \qquad \text{Factor.}$$
$$= (x - 8)(x + 2) \qquad \text{Simplify.}$$

Notice in this example we could have expanded $(x - 3)^2$, subtracted 25, and then factored to obtain the same result:

$$(x - 3)^2 - 25 = x^2 - 6x + 9 - 25 \qquad \text{Expand } (x - 3)^2.$$
$$= x^2 - 6x - 16 \qquad \text{Simplify.}$$
$$= (x - 8)(x + 2) \qquad \text{Factor.}$$

# Getting Ready for Class

*After reading through the preceding section, respond in your own words and in complete sentences.*

**A.** Describe how you factor the difference of two squares.

**B.** What is a perfect square trinomial?

**C.** How do you know when you've factored completely?

**D.** Describe how you would factor $25x^2 - 60x + 36$.

# PROBLEM SET 6.4

Factor the following.

**1.** $x^2 - 9$

**2.** $x^2 - 25$

**3.** $a^2 - 36$

**4.** $a^2 - 64$

**5.** $x^2 - 49$

**6.** $x^2 - 121$

**7.** $4a^2 - 16$

**8.** $4a^2 + 16$

**9.** $9x^2 + 25$

**10.** $16x^2 - 36$

**11.** $25x^2 - 169$

**12.** $x^2 - y^2$

**13.** $9a^2 - 16b^2$

**14.** $49a^2 - 25b^2$

**15.** $9 - m^2$

**16.** $16 - m^2$

**17.** $25 - 4x^2$

**18.** $36 - 49y^2$

**19.** $2x^2 - 18$

**20.** $3x^2 - 27$

**21.** $32a^2 - 128$

**22.** $3a^3 - 48a$

**23.** $8x^2y - 18y$

**24.** $50a^2b - 72b$

**25.** $a^4 - b^4$

**26.** $a^4 - 16$

**27.** $16m^4 - 81$

**28.** $81 - m^4$

**29.** $3x^3y - 75xy^3$

**30.** $2xy^3 - 8x^3y$

Factor the following.

**31.** $x^2 - 2x + 1$

**32.** $x^2 - 6x + 9$

**33.** $x^2 + 2x + 1$

**34.** $x^2 + 6x + 9$

**35.** $a^2 - 10a + 25$

**36.** $a^2 + 10a + 25$

**37.** $y^2 + 4y + 4$

**38.** $y^2 - 8y + 16$

**39.** $x^2 - 4x + 4$

**40.** $x^2 + 8x + 16$

**41.** $m^2 - 12m + 36$

**42.** $m^2 + 12m + 36$

**43.** $4a^2 + 12a + 9$

**44.** $9a^2 - 12a + 4$

**45.** $49x^2 - 14x + 1$

**46.** $64x^2 - 16x + 1$

**47.** $9y^2 - 30y + 25$

**48.** $25y^2 + 30y + 9$

**49.** $x^2 + 10xy + 25y^2$

**50.** $25x^2 + 10xy + y^2$

**51.** $9a^2 + 6ab + b^2$

**52.** $9a^2 - 6ab + b^2$

Factor the following by first factoring out the greatest common factor.

**53.** $3a^2 + 18a + 27$

**54.** $4a^2 - 16a + 16$

**55.** $2x^2 + 20xy + 50y^2$

**56.** $3x^2 + 30xy + 75y^2$

**57.** $5x^3 + 30x^2y + 45xy^2$

**58.** $12x^2y - 36xy^2 + 27y^3$

Factor by grouping the first three terms together.

**59.** $x^2 + 6x + 9 - y^2$

**60.** $x^2 + 10x + 25 - y^2$   **61.** $x^2 + 2xy + y^2 - 9$

**62.** $a^2 + 2ab + b^2 - 25$

**63.** Find a value for $b$ so that the polynomial $x^2 + bx + 49$ factors to $(x + 7)^2$.

**64.** Find a value of $b$ so that the polynomial $x^2 + bx + 81$ factors to $(x + 9)^2$.

**65.** Find the value of $c$ for which the polynomial $x^2 + 10x + c$ factors to $(x + 5)^2$.

**66.** Find the value of $a$ for which the polynomial $ax^2 + 12x + 9$ factors to $(2x + 3)^2$.

## Applying the Concepts

**67. Area**

(a) What is the area of the following figure?

(b) Factor the answer from part (a).

(c) Find a way to cut the figure into two pieces and put them back together to show that the factorization in part (b) is correct.

**68. Area**

(a) What is the area of the following figure?

(b) Factor the expression from part (a).

(c) Cut and rearrange the figure to show that the factorization is correct.

Find the area for the shaded regions; then write your result in factored form.

**69.**

**70.**

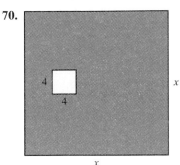

## Review Problems

The following problems review material we covered in Section 5.8.

Use long division to divide.

**71.** $\dfrac{x^2 - 5x + 8}{x - 3}$

**72.** $\dfrac{x^2 + 7x + 12}{x + 4}$

**73.** $\dfrac{6x^2 + 5x + 3}{2x + 3}$

**74.** $\dfrac{x^3 + 27}{x + 3}$

---

**6.5**   **Factoring: A General Review**

In this section we will review the different methods of factoring that we presented in the previous sections of the chapter. This section is important because it will give you an opportunity to factor a variety of polynomials. Prior to this section, the polynomials you worked with were grouped together according to the method used to factor them. That is, in Section 6.4 all the polynomials you factored were either the difference of two squares or perfect square trinomials. What usually happens in a situation like this is that you become proficient at factoring the kind of polynomial you are working with at the time, but have trouble when given a variety of polynomials to factor.

We begin this section with a checklist that can be used in factoring polynomials of any type. When you have finished this section and the problem set that follows, you want to be proficient enough at factoring so that the checklist is second nature to you.

### STRATEGY FOR FACTORING A POLYNOMIAL

*Step 1:* If the polynomial has a greatest common factor other than 1, then factor out the greatest common factor.

*Step 2:* If the polynomial has two terms (it is a binomial), then see if it is the difference of two squares. Remember, if it is the sum of two squares, it will not factor.

*Step 3:* If the polynomial has three terms (a trinomial), then either it is a perfect square trinomial, which will factor into the square of a binomial, or it is not a perfect square trinomial, in which case you use one of the methods developed in Section 6.3.

*Step 4:* If the polynomial has more than three terms, try to factor it by grouping.

*Step 5:* As a final check, see if any of the factors you have written can be factored further. If you have overlooked a common factor, you can catch it here.

Here are some examples illustrating how we use the checklist.

**EXAMPLE 1**　Factor $2x^5 - 8x^3$.

**Solution**　First, we check to see if the greatest common factor is other than 1. Since the greatest common factor is $2x^3$, we begin by factoring it out. Once we have done so, we notice that the binomial that remains is the difference of two squares:

$$2x^5 - 8x^3 = 2x^3(x^2 - 4) \qquad \text{Factor out the greatest} \\ \text{common factor, } 2x^3.$$

$$= 2x^3(x + 2)(x - 2) \qquad \text{Factor the difference} \\ \text{of two squares.}$$

Note that the greatest common factor $2x^3$ that we factored from each term in the first step of Example 1 remains as part of the answer to the problem. That is because it is one of the factors of the original binomial. Remember, the expression we end up with when factoring must be equal to the expression we start with. We can't just drop a factor and expect the resulting expression to equal the original expression.

**EXAMPLE 2**　Factor $3x^4 - 18x^3 + 27x^2$.

**Solution**　Step 1 is to factor out the greatest common factor, $3x^2$. After we have done so, we notice that the trinomial that remains is a perfect square trinomial, which will factor as the square of a binomial:

$$3x^4 - 18x^3 + 27x^2 = 3x^2(x^2 - 6x + 9) \qquad \text{Factor out } 3x^2.$$

$$= 3x^2(x - 3)^2 \qquad x^2 - 6x + 9 \text{ is the} \\ \text{square of } x - 3.$$

**EXAMPLE 3**　Factor $y^3 + 25y$.

**Solution**　We begin by factoring out the $y$ that is common to both terms. The binomial that remains after we have done so is the sum of two squares, which does not factor, so after the first step we are finished:

$$y^3 + 25y = y(y^2 + 25) \qquad \text{Factor out the greatest common factor, } y; \\ \text{then notice that } y^2 + 25 \text{ cannot be} \\ \text{factored further.}$$

**EXAMPLE 4**　Factor $6a^2 - 11a + 4$.

**Solution**　Here we have a trinomial that does not have a greatest common factor other than 1. Since it is not a perfect square trinomial, we factor it by trial and error. That is, we look for binomial factors, the product of whose first terms is $6a^2$ and the product of whose last terms is 4. Then we look for the combination of these types of binomials whose product gives us a middle term of $-11a$. Without showing all the different possibilities, here is the answer:

$$6a^2 - 11a + 4 = (3a - 4)(2a - 1)$$

**EXAMPLE  5**  Factor $6x^3 - 12x^2 - 48x$.

**Solution**  This trinomial has a greatest common factor of $6x$. The trinomial that remains after the $6x$ has been factored from each term may be factored by trial and error:

$$6x^3 - 12x^2 - 48x = 6x(x^2 - 2x - 8)$$
$$= 6x(x - 4)(x + 2)$$

**EXAMPLE  6**  Factor $2ab^5 + 8ab^4 + 2ab^3$.

**Solution**  The greatest common factor is $2ab^3$. We begin by factoring it from each term. After that we find that the trinomial that remains cannot be factored further:

$$2ab^5 + 8ab^4 + 2ab^3 = 2ab^3(b^2 + 4b + 1)$$

**EXAMPLE  7**  Factor $xy + 8x + 3y + 24$.

**Solution**  Since our polynomial has four terms, we try factoring by grouping:

$$xy + 8x + 3y + 24 = x(y + 8) + 3(y + 8)$$
$$= (y + 8)(x + 3)$$

 Getting Ready for Class

*After reading through the preceding section, respond in your own words and in complete sentences.*

**A.** What is the first step in factoring any polynomial?

**B.** If a polynomial has four terms, what method of factoring should you try?

**C.** If a polynomial has two terms, what method of factoring should you try?

**D.** What is the last step in factoring any polynomial?

## PROBLEM SET 6.5

Factor each of the following polynomials completely. That is, once you are finished factoring, none of the factors you obtain should be factorable. Also, note that the even-numbered problems are not necessarily similar to the odd-numbered problems that precede them in this problem set.

**1.** $x^2 - 81$

**2.** $x^2 - 18x + 81$

**3.** $x^2 + 2x - 15$

**4.** $15x^2 + 11x - 6$

**5.** $x^2 + 6x + 9$

**6.** $12x^2 - 11x + 2$

**7.** $y^2 - 10y + 25$

**8.** $21y^2 - 25y - 4$

**9.** $2a^3b + 6a^2b + 2ab$

**10.** $6a^2 - ab - 15b^2$

**11.** $x^2 + x + 1$

**12.** $2x^2 - 4x + 2$

**13.** $12a^2 - 75$

**14.** $18a^2 - 50$

**15.** $9x^2 - 12xy + 4y^2$

**16.** $x^3 - x^2$

**17.** $4x^3 + 16xy^2$

**18.** $16x^2 + 49y^2$

**19.** $2y^3 + 20y^2 + 50y$

**20.** $3y^2 - 9y - 30$

**21.** $a^6 + 4a^4b^2$

**22.** $5a^2 - 45b^2$

**23.** $xy + 3x + 4y + 12$

**24.** $xy + 7x + 6y + 42$

**25.** $x^4 - 16$

**26.** $x^4 - 81$

**27.** $xy - 5x + 2y - 10$

**28.** $xy - 7x + 3y - 21$

**29.** $5a^2 + 10ab + 5b^2$

**30.** $3a^3b^2 + 15a^2b^2 + 3ab^2$

**31.** $x^2 + 49$

**32.** $16 - x^4$

**33.** $3x^2 + 15xy + 18y^2$

**34.** $3x^2 + 27xy + 54y^2$

**35.** $2x^2 + 15x - 38$

**36.** $2x^2 + 7x - 85$

**37.** $100x^2 - 300x + 200$

**38.** $100x^2 - 400x + 300$

**39.** $x^2 - 64$

**40.** $9x^2 - 4$

**41.** $x^2 + 3x + ax + 3a$

**42.** $x^2 + 4x + bx + 4b$

**43.** $49a^7 - 9a^5$

**44.** $a^4 - 1$

**45.** $49x^2 + 9y^2$

**46.** $12x^4 - 62x^3 + 70x^2$

**47.** $25a^3 + 20a^2 + 3a$

**48.** $36a^4 - 100a^2$

**49.** $xa - xb + ay - by$

**50.** $xy - bx + ay - ab$

**51.** $48a^4b - 3a^2b$

**52.** $18a^4b^2 - 12a^3b^3 + 8a^2b^4$

**53.** $20x^4 - 45x^2$

**54.** $16x^3 + 16x^2 + 3x$

**55.** $3x^2 + 35xy - 82y^2$

**56.** $3x^2 + 37xy - 86y^2$

**57.** $16x^5 - 44x^4 + 30x^3$

**58.** $16x^2 + 16x - 1$

**59.** $2x^2 + 2ax + 3x + 3a$

**60.** $2x^2 + 2ax + 5x + 5a$

**61.** $y^4 - 1$

**62.** $25y^7 - 16y^5$

**63.** $12x^4y^2 + 36x^3y^3 + 27x^2y^4$

**64.** $16x^3y^2 - 4xy^2$

## Review Problems

The problems that follow review material we covered in Sections 2.3 and 5.1. Reviewing the problems from Section 2.3 will help you understand the next section.

Solve each equation. [2.3]

**65.** $3x - 6 = 9$

**66.** $5x - 1 = 14$

**67.** $2x + 3 = 0$

**68.** $4x - 5 = 0$

**69.** $4x + 3 = 0$

**70.** $3x - 1 = 0$

Simplify, using the properties of exponents. [5.1]

**71.** $x^8 \cdot x^7$

**72.** $(x^5)^2$

**73.** $(3x^3)^2(2x^4)^3$

**74.** $(5x^2y)^2(4xy^3)^2$

**75.** Write the number 57,600 in scientific notation.

**76.** Write the number $4.3 \times 10^5$ in expanded form.

---

**6.6** **Solving Equations by Factoring**

In this section we will use the methods of factoring developed in previous sections, along with a special property of 0, to solve quadratic equations.

> **DEFINITION**   Any equation that can be put in the form $ax^2 + bx + c = 0$, where $a, b$, and $c$ are real numbers $(a \neq 0)$, is called a **quadratic equation**. The equation $ax^2 + bx + c = 0$ is called **standard form** for a quadratic equation:
>
> an $x^2$ term          an $x$ term          and a constant term
>
> $$a(\text{variable})^2 + b(\text{variable}) + (\text{absence of the variable}) = 0$$

The number 0 has a special property. If we multiply two numbers and the product is 0, then one or both of the original two numbers must be 0. In symbols, this property looks like this.

## ZERO-FACTOR PROPERTY

Let $a$ and $b$ represent real numbers. If $a \cdot b = 0$, then $a = 0$ or $b = 0$.

Suppose we want to solve the quadratic equation $x^2 + 5x + 6 = 0$. We can factor the left side into $(x + 2)(x + 3)$. Then we have:

$$x^2 + 5x + 6 = 0$$

$$(x + 2)(x + 3) = 0$$

Now, $(x + 2)$ and $(x + 3)$ both represent real numbers. Their product is 0; therefore, either $(x + 3)$ is 0 or $(x + 2)$ is 0. Either way we have a solution to our equation. We use the property of 0 stated to finish the problem:

$$x^2 + 5x + 6 = 0$$

$$(x + 2)(x + 3) = 0$$

$$x + 2 = 0 \qquad \text{or} \qquad x + 3 = 0$$

$$x = -2 \qquad \text{or} \qquad x = -3$$

Our solution set is $\{-2, -3\}$. Our equation has two solutions. To check our solutions we have to check each one separately to see that they both produce a true statement when used in place of the variable:

When                                             $x = -3$

the equation                          $x^2 + 5x + 6 = 0$

becomes                           $(-3)^2 + 5(-3) + 6 \stackrel{?}{=} 0$

$$9 + (-15) + 6 = 0$$

$$0 = 0$$

When                                             $x = -2$

the equation                          $x^2 + 5x + 6 = 0$

becomes
$$(-2)^2 + 5(-2) + 6 \stackrel{?}{=} 0$$
$$4 + (-10) + 6 = 0$$
$$0 = 0$$

We have solved a quadratic equation by replacing it with two linear equations in one variable.

### STRATEGY FOR SOLVING A QUADRATIC EQUATION BY FACTORING

*Step 1:* Put the equation in standard form, that is, 0 on one side and decreasing powers of the variable on the other.

*Step 2:* Factor completely.

*Step 3:* Use the zero-factor property to set each variable factor from step 2 to 0.

*Step 4:* Solve each equation produced in step 3.

*Step 5:* Check each solution, if necessary.

**EXAMPLE 1** Solve the equation $2x^2 - 5x = 12$.

**Solution**

*Step 1:* We begin by adding $-12$ to both sides, so the equation is in standard form:
$$2x^2 - 5x = 12$$
$$2x^2 - 5x - 12 = 0$$

*Step 2:* We factor the left side completely:
$$(2x + 3)(x - 4) = 0$$

*Step 3:* We set each factor to 0:
$$2x + 3 = 0 \qquad \text{or} \qquad x - 4 = 0$$

*Step 4:* Solve each of the equations from step 3:
$$2x + 3 = 0 \qquad x - 4 = 0$$
$$2x = -3 \qquad x = 4$$
$$x = -\frac{3}{2}$$

*Step 5:* Substitute each solution into $2x^2 - 5x = 12$ to check:

Check: $-\dfrac{3}{2}$ 　　　　　　Check: $4$

$$2\left(-\frac{3}{2}\right)^2 - 5\left(-\frac{3}{2}\right) \stackrel{?}{=} 12 \qquad 2(4)^2 - 5(4) \stackrel{?}{=} 12$$

$$2\left(\frac{9}{4}\right) + 5\left(\frac{3}{2}\right) = 12 \qquad 2(16) - 20 = 12$$

$$\frac{9}{2} + \frac{15}{2} = 12 \qquad 32 - 20 = 12$$

$$\frac{24}{2} = 12 \qquad 12 = 12$$

$$12 = 12$$

**E X A M P L E   2**   Solve for $a$: $16a^2 - 25 = 0$

**Solution**   The equation is already in standard form:

$$16a^2 - 25 = 0$$

$$(4a - 5)(4a + 5) = 0 \qquad \text{Factor left side.}$$

$$4a - 5 = 0 \quad \text{or} \quad 4a + 5 = 0 \qquad \text{Set each factor to 0.}$$

$$4a = 5 \qquad\qquad 4a = -5 \qquad \text{Solve the resulting equations.}$$

$$a = \frac{5}{4} \qquad\qquad a = -\frac{5}{4}$$

**E X A M P L E   3**   Solve $4x^2 = 8x$.

**Solution**   We begin by adding $-8x$ to each side of the equation to put it in standard form. Then we factor the left side of the equation by factoring out the greatest common factor.

$$4x^2 = 8x$$

$$4x^2 - 8x = 0 \qquad \text{Add } -8x \text{ to each side.}$$

$$4x(x - 2) = 0 \qquad \text{Factor the left side.}$$

$$4x = 0 \quad \text{or} \quad x - 2 = 0 \qquad \text{Set each factor to 0.}$$

$$x = 0 \quad \text{or} \quad x = 2 \qquad \text{Solve the resulting equations.}$$

The solutions are 0 and 2.

**E X A M P L E   4**   Solve $x(2x + 3) = 44$.

**Solution**   We must multiply out the left side first and then put the equation in standard form:

$$x(2x + 3) = 44$$

$$2x^2 + 3x = 44 \qquad \text{Multiply out the left side.}$$

$$2x^2 + 3x - 44 = 0 \qquad \text{Add } -44 \text{ to each side.}$$

$$(2x + 11)(x - 4) = 0 \qquad \text{Factor the left side.}$$

$$2x + 11 = 0 \qquad \text{or} \qquad x - 4 = 0 \qquad \text{Set each factor to 0.}$$

$$2x = -11 \qquad \text{or} \qquad x = 4 \qquad \text{Solve the resulting equations.}$$

$$x = -\frac{11}{2}$$

The two solutions are $-\frac{11}{2}$ and 4.

**E X A M P L E   5**   Solve for $x$: $5^2 = x^2 + (x + 1)^2$

**Solution**   Before we can put this equation in standard form we must square the binomial. Remember, to square a binomial, we use the formula $(a + b)^2 = a^2 + 2ab + b^2$:

$$5^2 = x^2 + (x + 1)^2$$

$$25 = x^2 + x^2 + 2x + 1 \qquad \text{Expand } 5^2 \text{ and } (x + 1)^2.$$

$$25 = 2x^2 + 2x + 1 \qquad \text{Simplify the right side.}$$

$$0 = 2x^2 + 2x - 24 \qquad \text{Add } -25 \text{ to each side.}$$

$$0 = 2(x^2 + x - 12) \qquad \text{Begin factoring.}$$

$$0 = 2(x + 4)(x - 3) \qquad \text{Factor completely.}$$

$$x + 4 = 0 \qquad \text{or} \qquad x - 3 = 0 \qquad \text{Set each variable factor to 0.}$$

$$x = -4 \qquad \text{or} \qquad x = 3$$

Note, in the second to the last line, that we do not set 2 equal to 0. That is because 2 can never be 0. It is always 2. We only use the zero-factor property to set variable factors to 0 because they are the only factors that can possibly be 0.

Also notice that it makes no difference which side of the equation is 0 when we write the equation in standard form.

Although the equation in the next example is not a quadratic equation, it can be solved by the method shown in the first five examples.

**E X A M P L E   6**   Solve $24x^3 = -10x^2 + 6x$ for $x$.

**Solution**   First, we write the equation in standard form:

$$24x^3 + 10x^2 - 6x = 0 \qquad \text{Standard form}$$

$$2x(12x^2 + 5x - 3) = 0 \qquad \text{Factor out } 2x.$$

$$2x(3x - 1)(4x + 3) = 0 \qquad \text{Factor remaining trinomial.}$$

$$2x = 0 \quad \text{or} \quad 3x - 1 = 0 \quad \text{or} \quad 4x + 3 = 0 \qquad \text{Set factors to 0.}$$

$$x = 0 \quad \text{or} \quad x = \frac{1}{3} \quad \text{or} \quad x = -\frac{3}{4} \qquad \text{Solutions}$$

 ## Getting Ready for Class

*After reading through the preceding section, respond in your own words and in complete sentences.*

**A.** When is an equation in standard form?

**B.** What is the first step in solving an equation by factoring?

**C.** Describe the zero-factor property in your own words.

**D.** Describe how you would solve the equation $2x^2 - 5x = 12$.

## PROBLEM SET 6.6

The following equations are already in factored form. Use the special zero-factor property to set the factors to 0 and solve.

**1.** $(x + 2)(x - 1) = 0$     **2.** $(x + 3)(x + 2) = 0$

**3.** $(a - 4)(a - 5) = 0$     **4.** $(a + 6)(a - 1) = 0$

**5.** $x(x + 1)(x - 3) = 0$

**6.** $x(2x + 1)(x - 5) = 0$

**7.** $(3x + 2)(2x + 3) = 0$

**8.** $(4x - 5)(x - 6) = 0$

**9.** $m(3m + 4)(3m - 4) = 0$

**10.** $m(2m - 5)(3m - 1) = 0$

**11.** $2y(3y + 1)(5y + 3) = 0$

**12.** $3y(2y - 3)(3y - 4) = 0$

Solve the following equations.

**13.** $x^2 + 3x + 2 = 0$     **14.** $x^2 - x - 6 = 0$

**15.** $x^2 - 9x + 20 = 0$     **16.** $x^2 + 2x - 3 = 0$

**17.** $a^2 - 2a - 24 = 0$     **18.** $a^2 - 11a + 30 = 0$

**19.** $100x^2 - 500x + 600 = 0$

**20.** $100x^2 - 300x + 200 = 0$

**21.** $x^2 = -6x - 9$     **22.** $x^2 = 10x - 25$

**23.** $a^2 - 16 = 0$     **24.** $a^2 - 36 = 0$

**25.** $2x^2 + 5x - 12 = 0$     **26.** $3x^2 + 14x - 5 = 0$

**27.** $9x^2 + 12x + 4 = 0$

**28.** $12x^2 - 24x + 9 = 0$

**29.** $a^2 + 25 = 10a$     **30.** $a^2 + 16 = 8a$

**31.** $2x^2 = 3x + 20$     **32.** $6x^2 = x + 2$

**33.** $3m^2 = 20 - 7m$     **34.** $2m^2 = -18 + 15m$

**35.** $4x^2 - 49 = 0$     **36.** $16x^2 - 25 = 0$

**37.** $x^2 + 6x = 0$     **38.** $x^2 - 8x = 0$

**39.** $x^2 - 3x = 0$     **40.** $x^2 + 5x = 0$

**41.** $2x^2 = 8x$     **42.** $2x^2 = 10x$

**43.** $3x^2 = 15x$     **44.** $5x^2 = 15x$

**45.** $1{,}400 = 400 + 700x - 100x^2$

**46.** $2{,}700 = 700 + 900x - 100x^2$

**47.** $6x^2 = -5x + 4$     **48.** $9x^2 = 12x - 4$

**49.** $x(2x - 3) = 20$     **50.** $x(3x - 5) = 12$

**51.** $t(t + 2) = 80$     **52.** $t(t + 2) = 99$

**53.** $4{,}000 = (1{,}300 - 100p)p$

**54.** $3{,}200 = (1{,}200 - 100p)p$

**55.** $x(14 - x) = 48$     **56.** $x(12 - x) = 32$

**57.** $(x + 5)^2 = 2x + 9$     **58.** $(x + 7)^2 = 2x + 13$

**59.** $(y - 6)^2 = y - 4$     **60.** $(y + 4)^2 = y + 6$

**61.** $10^2 = (x + 2)^2 + x^2$

**62.** $15^2 = (x + 3)^2 + x^2$

**63.** $2x^3 + 11x^2 + 12x = 0$

**64.** $3x^3 + 17x^2 + 10x = 0$

**65.** $4y^3 - 2y^2 - 30y = 0$

**66.** $9y^3 + 6y^2 - 24y = 0$

**67.** $8x^3 + 16x^2 = 10x$

**68.** $24x^3 - 22x^2 = -4x$

**69.** $20a^3 = -18a^2 + 18a$

**70.** $12a^3 = -2a^2 + 10a$

Use factoring by grouping to solve the following equations.

**71.** $x^3 + 3x^2 - 4x - 12 = 0$

**72.** $x^3 + 5x^2 - 9x - 45 = 0$

**73.** $x^3 + x^2 - 16x - 16 = 0$

**74.** $4x^3 + 12x^2 - 9x - 27 = 0$

## Review Problems

The following problems review material we covered in Sections 4.4 and 5.2.

The following word problems are taken from the book *Academic Algebra*, written by William J. Milne and published by the American Book Company in 1901.

Solve each problem. [4.4]

**75. Cost of a Bicycle and a Suit**  A bicycle and a suit cost $90. How much did each cost, if the bicycle cost 5 times as much as the suit?

**76. Cost of a Cow and a Calf**  A man bought a cow and a calf for $36, paying 8 times as much for the cow as for the calf. What was the cost of each?

**77. Cost of a House and a Lot**  A house and a lot cost $3,000. If the house cost 4 times as much as the lot, what was the cost of each?

**78. Daily Wages**  A plumber and two helpers together earned $7.50 per day. How much did each earn per day, if the plumber earned 4 times as much as each helper?

Use the properties of exponents to simplify each expression. [5.2]

**79.** $2^{-3}$

**80.** $5^{-2}$

**81.** $\dfrac{x^5}{x^{-3}}$

**82.** $\dfrac{x^{-2}}{x^{-5}}$

**83.** $\dfrac{(x^2)^3}{(x^{-3})^4}$

**84.** $\dfrac{(x^2)^{-4}(x^{-2})^3}{(x^{-3})^{-5}}$

**85.** Write the number 0.0056 in scientific notation.

**86.** Write the number $2.34 \times 10^{-4}$ in expanded form.

 **6.7**  **Applications**

In this section we will look at some application problems, the solutions to which require solving a quadratic equation. We will also introduce the Pythagorean theorem, one of the oldest theorems in the history of mathematics. The person whose name we associate with the theorem, Pythagoras (of Samos), was a Greek philosopher and mathematician who lived from about 560 B.C. to 480 B.C. According to the British philosopher Bertrand Russell, Pythagoras was "intellectually one of the most important men that ever lived."

Also in this section, the solutions to the examples show only the essential steps from our Blueprint for Problem Solving. Recall that step 1 is done mentally. We read the problem and mentally list the items that are known and the items that are unknown. This is an essential part of problem solving. However, now that you have had experience with application problems, you are doing step 1 automatically.

## NUMBER PROBLEMS

**EXAMPLE  1**   The product of two consecutive odd integers is 63. Find the integers.

**Solution**   Let $x$ = the first odd integer; then $x + 2$ = the second odd integer. An equation that describes the situation is:

$$x(x + 2) = 63 \qquad \text{(Their product is 63.)}$$

We solve the equation:

$$x(x + 2) = 63$$
$$x^2 + 2x = 63$$
$$x^2 + 2x - 63 = 0$$
$$(x - 7)(x + 9) = 0$$
$$x - 7 = 0 \qquad \text{or} \qquad x + 9 = 0$$
$$x = 7 \qquad \text{or} \qquad x = -9$$

If the first odd integer is 7, the next odd integer is $7 + 2 = 9$. If the first odd integer is $-9$, the next consecutive odd integer is $-9 + 2 = -7$. We have two pairs of consecutive odd integers that are solutions. They are 7, 9 and $-9$, $-7$.
   We check to see that their products are 63:

$$7(9) = 63$$
$$-7(-9) = 63$$

   Suppose we know that the sum of two numbers is 50. We want to find a way to represent each number using only one variable. If we let $x$ represent one of the two numbers, how can we represent the other? Let's suppose for a moment that $x$ turns out to be 30. Then the other number will be 20, because their sum is 50. That is, if two numbers add up to 50 and one of them is 30, then the other must be $50 - 30 = 20$. Generalizing this to any number $x$, we see that if two numbers have a sum of 50 and one of the numbers is $x$, then the other must be $50 - x$. The table that follows shows some additional examples.

| If Two Numbers Have a Sum of | And One of Them Is | Then the Other Must Be |
|:---:|:---:|:---:|
| 50 | $x$ | $50 - x$ |
| 100 | $x$ | $100 - x$ |
| 10 | $y$ | $10 - y$ |
| 12 | $n$ | $12 - n$ |

Now, let's look at an example that uses this idea.

**EXAMPLE 2**   The sum of two numbers is 13. Their product is 40. Find the numbers.

**Solution**   If we let $x$ represent one of the numbers, then $13 - x$ must be the other number, because their sum is 13. Since their product is 40, we can write:

$$x(13 - x) = 40 \qquad \text{The product of the two numbers is } 40.$$

$$13x - x^2 = 40 \qquad \text{Multiply left side.}$$

$$x^2 - 13x = -40 \qquad \begin{array}{l}\text{Multiply both sides by } -1 \text{ and} \\ \text{reverse the order of the terms of} \\ \text{the left side.}\end{array}$$

$$x^2 - 13x + 40 = 0 \qquad \text{Add 40 to each side.}$$

$$(x - 8)(x - 5) = 0 \qquad \text{Factor the left side.}$$

$$x - 8 = 0 \quad \text{or} \quad x - 5 = 0$$

$$x = 8 \quad \text{or} \quad x = 5$$

The two solutions are 8 and 5. If $x$ is 8, then the other number is $13 - x = 13 - 8 = 5$. Likewise, if $x$ is 5, the other number is $13 - x = 13 - 5 = 8$. Therefore, the two numbers we are looking for are 8 and 5. Their sum is 13 and their product is 40.

## GEOMETRY PROBLEMS

Many word problems dealing with area can best be described algebraically by quadratic equations.

**EXAMPLE 3**   The length of a rectangle is 3 more than twice the width. The area is 44 square inches. Find the dimensions (find the length and width).

**Solution**   As shown in Figure 1, let $x =$ the width of the rectangle. Then $2x + 3 =$ the length of the rectangle, because the length is three more than twice the width.

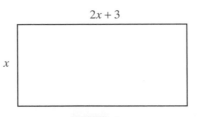

**FIGURE 1**

Since the area is 44 square inches, an equation that describes the situation is

$$x(2x + 3) = 44 \qquad (\text{Length} \cdot \text{width} = \text{area})$$

We now solve the equation:

$$x(2x + 3) = 44$$
$$2x^2 + 3x = 44$$
$$2x^2 + 3x - 44 = 0$$
$$(2x + 11)(x - 4) = 0$$

$$2x + 11 = 0 \qquad \text{or} \qquad x - 4 = 0$$

$$x = -\frac{11}{2} \qquad \text{or} \qquad x = 4$$

The solution $x = -\frac{11}{2}$ cannot be used, since length and width are always given in positive units. The width is 4. The length is 3 more than twice the width or $2(4) + 3 = 11$.

$$\text{Width} = 4 \text{ inches}$$

$$\text{Length} = 11 \text{ inches}$$

The solutions check in the original problem, since $4(11) = 44$.

**EXAMPLE  4**   The numerical value of the area of a square is twice its perimeter. What is the length of its side?

**Solution**   As shown in Figure 2, let $x =$ the length of its side. Then $x^2 =$ the area of the square and $4x =$ the perimeter of the square:

**FIGURE  2**

An equation that describes the situation is

$$x^2 = 2(4x) \qquad \text{The area is 2 times the perimeter.}$$
$$x^2 = 8x$$
$$x^2 - 8x = 0$$
$$x(x - 8) = 0$$

$$x = 0 \qquad \text{or} \qquad x = 8$$

Since $x = 0$ does not make sense in our original problem, we use $x = 8$. If the side has length 8, then the perimeter is $4(8) = 32$, and the area is $8^2 = 64$. Since 64 is twice 32, our solution is correct.

FACTS FROM
## Geometry

### The Pythagorean Theorem

Next, we will work some problems involving the Pythagorean theorem, which we mentioned in the introduction to this section. It may interest you to know that Pythagoras formed a secret society around the year 540 B.C. Known as the Pythagoreans, members kept no written record of their work; everything was handed down by spoken word. They influenced not only mathematics, but religion, science, medicine, and music as well. Among other things, they discovered the correlation between musical notes and the reciprocals of counting numbers, $\frac{1}{2}, \frac{1}{3}, \frac{1}{4}$, and so on. In their daily lives, they followed strict dietary and moral rules to achieve a higher rank in future lives.

### PYTHAGOREAN THEOREM

In any right triangle (Figure 3), the square of the longest side (called the hypotenuse) is equal to the sum of the squares of the other two sides (called legs).

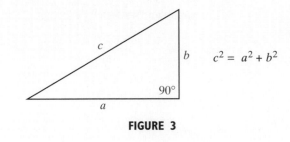

**FIGURE 3**

**EXAMPLE 5**   The three sides of a right triangle are three consecutive integers. Find the lengths of the three sides.

**Solution**   Let $x =$ the first integer (shortest side)

then $x + 1 =$ the next consecutive integer

and $x + 2 =$ the last consecutive integer (longest side)

A diagram of the triangle is shown in Figure 4.

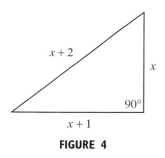

**FIGURE 4**

The Pythagorean theorem tells us that the square of the longest side $(x + 2)^2$ is equal to the sum of the squares of the two shorter sides, $(x + 1)^2 + x^2$. Here is the equation:

$$(x + 2)^2 = (x + 1)^2 + x^2$$

$$x^2 + 4x + 4 = x^2 + 2x + 1 + x^2 \qquad \text{Expand squares.}$$

$$x^2 - 2x - 3 = 0 \qquad \text{Standard form}$$

$$(x - 3)(x + 1) = 0 \qquad \text{Factor.}$$

$$x - 3 = 0 \quad \text{or} \quad x + 1 = 0 \qquad \text{Set factors to 0.}$$

$$x = 3 \quad \text{or} \quad x = -1$$

Since a triangle cannot have a side with a negative number for its length, we must not use $-1$ for a solution to our original problem; therefore, the shortest side is 3. The other two sides are the next two consecutive integers, 4 and 5.

---

**E X A M P L E   6** The hypotenuse of a right triangle is 5 inches, and the lengths of the two legs (the other two sides) are given by two consecutive integers. Find the lengths of the two legs.

**Solution** If we let $x =$ the length of the shortest side, then the other side must be $x + 1$. A diagram of the triangle is shown in Figure 5.

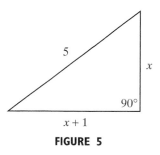

**FIGURE 5**

The Pythagorean theorem tells us that the square of the longest side, $5^2$, is equal to the sum of the squares of the two shorter sides, $x^2 + (x + 1)^2$. Here is the equation:

$$5^2 = x^2 + (x + 1)^2 \qquad \text{Pythagorean theorem}$$
$$25 = x^2 + x^2 + 2x + 1 \qquad \text{Expand } 5^2 \text{ and } (x+1)^2.$$
$$25 = 2x^2 + 2x + 1 \qquad \text{Simplify the right side.}$$
$$0 = 2x^2 + 2x - 24 \qquad \text{Add } -25 \text{ to each side.}$$
$$0 = 2(x^2 + x - 12) \qquad \text{Begin factoring.}$$
$$0 = 2(x + 4)(x - 3) \qquad \text{Factor completely.}$$
$$x + 4 = 0 \quad \text{or} \quad x - 3 = 0 \qquad \text{Set variable factors to 0.}$$
$$x = -4 \quad \text{or} \quad x = 3$$

Since a triangle cannot have a side with a negative number for its length, we cannot use $-4$; therefore, the shortest side must be 3 inches. The next side is $x + 1 = 3 + 1 = 4$ inches. Since the hypotenuse is 5, we can check our solutions with the Pythagorean theorem as shown in Figure 6.

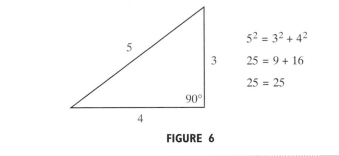

**FIGURE 6**

## BUSINESS PROBLEMS

Our next two examples are from the world of business. If you are planning on taking finite mathematics, statistics, or business calculus in the future, these examples will give you a head start on some of the problems you will see in those classes.

**E X A M P L E 7** A company can manufacture $x$ hundred items for a total cost of $C = 300 + 500x - 100x^2$. How many items were manufactured if the total cost is \$900?

**Solution** We are looking for $x$ when $C$ is 900. We begin by substituting 900 for $C$ in the cost equation. Then we solve for $x$:

When $\qquad\qquad C = 900$

the equation $\qquad C = 300 + 500x - 100x^2$

becomes $\qquad 900 = 300 + 500x - 100x^2$

We can write this equation in standard form by adding $-300$, $-500x$, and $100x^2$ to each side. The result looks like this:

$$100x^2 - 500x + 600 = 0$$

$$100(x^2 - 5x + 6) = 0 \qquad \text{Begin factoring.}$$

$$100(x - 2)(x - 3) = 0 \qquad \text{Factor completely.}$$

$$x - 2 = 0 \quad \text{or} \quad x - 3 = 0 \qquad \text{Set variable factors to 0.}$$

$$x = 2 \quad \text{or} \quad x = 3$$

Our solutions are 2 and 3, which means that the company can manufacture 2 hundred items or 3 hundred items for a total cost of $900.

**EXAMPLE 8**    A manufacturer of small portable radios knows that the number of radios she can sell each week is related to the price of the radios by the equation $x = 1,300 - 100p$ ($x$ is the number of radios, and $p$ is the price per radio). What price should she charge for the radios in order to have a weekly revenue of $4,000?

**Solution**    First, we must find the revenue equation. The equation for total revenue is $R = xp$, where $x$ is the number of units sold and $p$ is the price per unit. Since we want $R$ in terms of $p$, we substitute $1,300 - 100p$ for $x$ in the equation $R = xp$:

$$\text{If} \qquad R = xp$$

$$\text{and} \qquad x = 1,300 - 100p$$

$$\text{then} \qquad R = (1,300 - 100p)p$$

We want to find $p$ when $R$ is 4,000. Substituting 4,000 for $R$ in the equation gives us:

$$4,000 = (1,300 - 100p)p$$

If we multiply out the right side, we have:

$$4,000 = 1,300p - 100p^2$$

To write this equation in standard form, we add $100p^2$ and $-1,300p$ to each side:

$$100p^2 - 1,300p + 4,000 = 0 \qquad \text{Add } 100p^2 \text{ and } -1,300p \text{ to each side.}$$

$$100(p^2 - 13p + 40) = 0 \qquad \text{Begin factoring.}$$

$$100(p - 5)(p - 8) = 0 \qquad \text{Factor completely.}$$

$$p - 5 = 0 \quad \text{or} \quad p - 8 = 0 \qquad \text{Set variable factors to 0.}$$

$$p = 5 \quad \text{or} \quad p = 8$$

If she sells the radios for $5 each or for $8 each, she will have a weekly revenue of $4,000.

# Getting Ready for Class

*After reading through the preceding section, respond in your own words and in complete sentences.*

**A.** What are consecutive integers?

**B.** Explain the Pythagorean theorem in words.

**C.** Write an application problem for which the solution depends on solving the equation $x(x + 1) = 12$.

**D.** Write an application problem for which the solution depends on solving the equation $x(2x - 3) = 44$.

## PROBLEM SET 6.7

Solve the following word problems. Be sure to show the equation used.

### Number Problems

**1.** The product of two consecutive even integers is 80. Find the two integers.

**2.** The product of two consecutive integers is 72. Find the two integers.

**3.** The product of two consecutive odd integers is 99. Find the two integers.

**4.** The product of two consecutive integers is 132. Find the two integers.

**5.** The product of two consecutive even integers is 10 less than 5 times their sum. Find the two integers.

**6.** The product of two consecutive odd integers is 1 less than 4 times their sum. Find the two integers.

**7.** The sum of two numbers is 14. Their product is 48. Find the numbers.

**8.** The sum of two numbers is 12. Their product is 32. Find the numbers.

**9.** One number is 2 more than 5 times another. Their product is 24. Find the numbers.

**10.** One number is 1 more than twice another. Their product is 55. Find the numbers.

**11.** One number is 4 times another. Their product is 4 times their sum. Find the numbers.

**12.** One number is 2 more than twice another. Their product is 2 more than twice their sum. Find the numbers.

### Geometry Problems

**13.** The length of a rectangle is 1 more than the width. The area is 12 square inches. Find the dimensions.

**14.** The length of a rectangle is 3 more than twice the width. The area is 44 square inches. Find the dimensions.

**15.** The height of a triangle is twice the base. The area is 9 square inches. Find the base.

**16.** The height of a triangle is 2 more than twice the base. The area is 20 square feet. Find the base.

**17.** The hypotenuse of a right triangle is 10 inches. The lengths of the two legs are given by two consecutive even integers. Find the lengths of the two legs.

**18.** The hypotenuse of a right triangle is 15 inches. One of the legs is 3 inches longer than the other. Find the lengths of the two legs.

**19.** The shorter leg of a right triangle is 5 meters. The hypotenuse is 1 meter longer than the longer leg. Find the length of the longer leg.

**20.** The shorter leg of a right triangle is 12 yards. If the hypotenuse is 20 yards, how long is the other leg?

### Business Problems

**21.** A company can manufacture $x$ hundred items for a total cost of $C = 400 + 700x - 100x^2$. Find $x$ if the total cost is $1,400.

**22.** If the total cost $C$ of manufacturing $x$ hundred items is given by the equation $C = 700 + 900x - 100x^2$, find $x$ when $C$ is $2,700.

**23.** The total cost $C$ of manufacturing $x$ hundred video-tapes is given by the equation

$$C = 600 + 1,000x - 100x^2$$

Find $x$ if the total cost is $2,200.

**24.** The total cost $C$ of manufacturing $x$ hundred pen and pencil sets is given by the equation $C = 500 + 800x - 100x^2$. Find $x$ when $C$ is $1,700.

**25.** A company that manufactures typewriter ribbons knows that the number of ribbons it can sell each week, $x$, is related to the price $p$ per ribbon by the equation $x = 1,200 - 100p$. At what price should the company sell the ribbons if it wants the weekly revenue to be $3,200? (*Remember:* The equation for revenue is $R = xp$.)

**26.** A company manufactures diskettes for home computers. It knows from past experience that the number of diskettes it can sell each day, $x$, is related to the price $p$ per diskette by the equation $x = 800 - 100p$. At what price should the company sell the diskettes if it wants the daily revenue to be $1,200?

**27.** The relationship between the number of calculators a company sells per week, $x$, and the price $p$ of each calculator is given by the equation $x = 1,700 - 100p$. At what price should the calculators be sold if the weekly revenue is to be $7,000?

**28.** The relationship between the number of pencil sharpeners a company can sell each week, $x$, and the price $p$ of each sharpener is given by the equation $x = 1,800 - 100p$. At what price should the sharpeners be sold if the weekly revenue is to be $7,200?

**29. Pythagorean Theorem** A 13-foot ladder is placed so that it reaches to a point on the wall that is 2 feet higher than twice the distance from the base of the wall to the base of the ladder.
(a) How far from the wall is the base of the ladder?
(b) How high does the ladder reach?

**Wall**     **13 ft**

**Ground**

**30. Height of a Projectile** If a rocket is fired vertically into the air with a speed of 240 feet per second, its height at time $t$ seconds is given by $h(t) = -16t^2 + 240t$. Here is a graph of its height at various times, with the details left out:

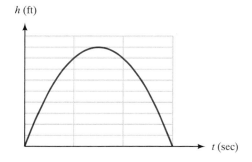

$h$ (ft)

$t$ (sec)

At what time(s) will the rocket be the following number of feet above the ground?
(a) 704 feet
(b) 896 feet
(c) Why do parts (a) and (b) each have two answers?
(d) How long will the rocket be in the air? (*Hint:* How high is it when it hits the ground?)
(e) When the equation for part (d) is solved, one of the answers is $t = 0$ second. What does this represent?

**31. Projectile Motion** A gun fires a bullet almost straight up from the edge of a 100-foot cliff. If the bullet leaves the gun with a speed of 396 feet per second, its height at time $t$ is given by $h(t) =$

$-16t^2 + 396t + 100$, measured from the ground below the cliff.

(a) When will the bullet land on the ground below the cliff? (*Hint:* What is its height when it lands? Remember that we are measuring from the ground below, not from the cliff.)

(b) Make a table showing the bullet's height every five seconds, from the time it is fired ($t = 0$) to the time it lands. (*Note:* It is faster to substitute into the factored form.)

**32. Constructing a Box** I have a piece of cardboard that is twice as long as it is wide. If I cut a 2-inch by 2-inch square from each corner and fold up the resulting flaps, I get a box with a volume of 32 cubic inches. What are the dimensions of the cardboard?

## Review Problems

The problems that follow review material we covered in Chapter 5.

Simplify each expression. (Write all answers with positive exponents only.)

**33.** $(5x^3)^2(2x^6)^3$

**34.** $2^{-3}$

**35.** $\dfrac{x^4}{x^{-3}}$

**36.** $\dfrac{(20x^2y^3)(5x^4y)}{(2xy^5)(10x^2y^3)}$

**37.** $(2 \times 10^{-4})(4 \times 10^5)$

**38.** $\dfrac{9 \times 10^{-3}}{3 \times 10^{-2}}$

**39.** $20ab^2 - 16ab^2 + 6ab^2$

**40.** Subtract $6x^2 - 5x - 7$ from $9x^2 + 3x - 2$.

Multiply.

**41.** $2x^2(3x^2 + 3x - 1)$

**42.** $(2x + 3)(5x - 2)$

**43.** $(3y - 5)^2$

**44.** $(a - 4)(a^2 + 4a + 16)$

**45.** $(2a^2 + 7)(2a^2 - 7)$

**46.** Divide $15x^{10} - 10x^8 + 25x^6$ by $5x^6$.

# CHAPTER 6    SUMMARY

**Examples**

**1.** $8x^4 - 10x^3 + 6x^2$

$= 2x^2 \cdot 4x^2 - 2x^2 \cdot 5x + 2x^2 \cdot 3$

$= 2x^2(4x^2 - 5x + 3)$

## Greatest Common Factor [6.1]

The largest monomial that divides each term of a polynomial is called the greatest common factor for that polynomial. We begin all factoring by factoring out the greatest common factor.

**2.** $x^2 + 5x + 6 = (x + 2)(x + 3)$

$x^2 - 5x + 6 = (x - 2)(x - 3)$

$6x^2 - x - 2 = (2x + 1)(3x - 2)$

$6x^2 + 7x + 2 = (2x + 1)(3x + 2)$

## Factoring Trinomials [6.2, 6.3]

One method of factoring a trinomial is to list all pairs of binomials, the product of whose first terms gives the first term of the trinomial and the product of whose last terms gives the last term of the trinomial. We then choose the pair that gives the correct middle term for the original trinomial.

**3.** $x^2 + 10x + 25 = (x + 5)^2$

$x^2 - 10x + 25 = (x - 5)^2$

$x^2 - 25 = (x + 5)(x - 5)$

## Special Factorings [6.4]

$$a^2 + 2ab + b^2 = (a + b)^2$$

$$a^2 - 2ab + b^2 = (a - b)^2$$

$$a^2 - b^2 = (a + b)(a - b)$$

**4.** (a) $2x^5 - 8x^3 = 2x^3(x^2 - 4)$

$= 2x^3(x + 2)(x - 2)$

(b) $3x^4 - 18x^3 + 27x^2$

$= 3x^2(x^2 - 6x + 9)$

$= 3x^2(x - 3)^2$

(c) $6x^3 - 12x^2 - 48x$

$= 6x(x^2 - 2x - 8)$

$= 6x(x - 4)(x + 2)$

(d) $x^2 + ax + bx + ab$

$= x(x + a) + b(x + a)$

$= (x + a)(x + b)$

## Strategy for Factoring a Polynomial [6.5]

**Step 1:** If the polynomial has a greatest common factor other than 1, then factor out the greatest common factor.

**Step 2:** If the polynomial has two terms (it is a binomial), then see if it is the difference of two squares. Remember, if it is the sum of two squares, it will not factor.

**Step 3:** If the polynomial has three terms (a trinomial), then it is either a perfect square trinomial that will factor into the square of a binomial, or it is not a perfect square trinomial, in which case you use one of the methods developed in Section 6.3.

**Step 4:** If the polynomial has more than three terms, then try to factor it by grouping.

**Step 5:** As a final check, see if any of the factors you have written can be factored further. If you have overlooked a common factor, you can catch it here.

**5.** Solve $x^2 - 6x = -8$.

$$x^2 - 6x + 8 = 0$$
$$(x - 4)(x - 2) = 0$$
$$x - 4 = 0 \quad \text{or} \quad x - 2 = 0$$
$$x = 4 \quad \text{or} \quad x = 2$$

Both solutions check.

**6.** The hypotenuse of a right triangle is 5 inches, and the lengths of the two legs (the other two sides) are given by two consecutive integers. Find the lengths of the two legs.

If we let $x =$ the length of the shortest side, then the other side must be $x + 1$. The Pythagorean theorem tells us that

$$5^2 = x^2 + (x + 1)^2$$

$$25 = x^2 + x^2 + 2x + 1$$

$$25 = 2x^2 + 2x + 1$$

$$0 = 2x^2 + 2x - 24$$

$$0 = 2(x^2 + x - 12)$$

$$0 = 2(x + 4)(x - 3)$$

$$x + 4 = 0 \quad \text{or} \quad x - 3 = 0$$

$$x = -4 \quad \text{or} \quad x = 3$$

Since a triangle cannot have a side with a negative number for its length, we cannot use $-4$. The two sides are $x = 3$ and $x + 1 = 3 + 1 = 4$.

**Strategy for Solving a Quadratic Equation [6.6]**

> ***Step 1:*** Write the equation in standard form:
>
> $$ax^2 + bx + c = 0$$
>
> ***Step 2:*** Factor completely.
> ***Step 3:*** Set each variable factor equal to 0.
> ***Step 4:*** Solve the equations found in step 3.
> ***Step 5:*** Check solutions, if necessary.

**The Pythagorean Theorem [6.7]**
In any right triangle, the square of the longest side (called the hypotenuse) is equal to the sum of the squares of the other two sides (called legs).

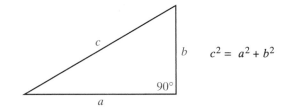

$$c^2 = a^2 + b^2$$

**COMMON MISTAKE**

It is a mistake to apply the zero-factor property to numbers other than zero. For example, consider the equation $(x - 3)(x + 4) = 18$. A fairly common mistake is to attempt to solve it with the following steps:

$$(x - 3)(x + 4) = 18$$

$$x - 3 = 18 \quad \text{or} \quad x + 4 = 18 \leftarrow \text{Mistake}$$

$$x = 21 \quad \text{or} \quad x = 14$$

These are obviously not solutions, as a quick check will verify:

Check: $x = 21$

$$(21 - 3)(21 + 4) \overset{?}{=} 18$$
$$18 \cdot 25 = 18$$
$$450 = 18 \quad \xleftarrow{\text{False statements}}$$

Check: $x = 14$

$$(14 - 3)(14 + 4) \overset{?}{=} 18$$
$$11 \cdot 18 = 18$$
$$198 = 18$$

The mistake is in setting each factor equal to 18. It is not necessarily true, when the product of two numbers is 18, that either one of them is itself 18. The correct solution looks like this:

$$(x - 3)(x + 4) = 18$$
$$x^2 + x - 12 = 18$$
$$x^2 + x - 30 = 0$$
$$(x + 6)(x - 5) = 0$$

$$x + 6 = 0 \quad\quad \text{or} \quad\quad x - 5 = 0$$
$$x = -6 \quad\quad \text{or} \quad\quad x = 5$$

To avoid this mistake, remember that before you factor a quadratic equation, you must write it in standard form. It is in standard form only when 0 is on one side and decreasing powers of the variable are on the other.

# CHAPTER 6   REVIEW

The numbers in brackets refer to the sections of the text in which similar problems can be found.

Factor the following by factoring out the greatest common factor. [6.1]

**1.** $10x - 20$

**2.** $4x^3 - 9x^2$

**3.** $5x - 5y$

**4.** $7x^3 + 2x$

**5.** $8x + 4$

**6.** $2x^2 + 14x + 6$

**7.** $24y^2 - 40y + 48$

**8.** $30xy^3 - 45x^3y^2$

**9.** $49a^3 - 14b^3$

**10.** $6ab^2 + 18a^3b^3 - 24a^2b$

Factor by grouping. [6.1]

**11.** $xy + bx + ay + ab$

**12.** $xy + 4x - 5y - 20$

**13.** $2xy + 10x - 3y - 15$

**14.** $5x^2 - 4ax - 10bx + 8ab$

Factor the following trinomials. [6.2]

**15.** $y^2 + 9y + 14$

**16.** $w^2 + 15w + 50$

**17.** $a^2 - 14a + 48$

**18.** $r^2 - 18r + 72$

**19.** $y^2 + 20y + 99$

**20.** $y^2 + 8y + 12$

Factor the following trinomials. [6.3]

**21.** $2x^2 + 13x + 15$

**22.** $4y^2 - 12y + 5$

**23.** $5y^2 + 11y + 6$

**24.** $20a^2 - 27a + 9$

**25.** $6r^2 + 5rt - 6t^2$

**26.** $10x^2 - 29x - 21$

Factor the following if possible. [6.4]

**27.** $n^2 - 81$

**28.** $4y^2 - 9$

**29.** $x^2 + 49$

**30.** $36y^2 - 121x^2$

**31.** $64a^2 - 121b^2$

**32.** $64 - 9m^2$

Factor the following. [6.4]

**33.** $y^2 + 20y + 100$

**34.** $m^2 - 16m + 64$

**35.** $64t^2 + 16t + 1$

**36.** $16n^2 - 24n + 9$

**37.** $4r^2 - 12rt + 9t^2$

**38.** $9m^2 + 30mn + 25n^2$

Factor the following. [6.2]

**39.** $2x^2 + 20x + 48$

**40.** $a^3 - 10a^2 + 21a$

**41.** $3m^3 - 18m^2 - 21m$

**42.** $5y^4 + 10y^3 - 40y^2$

Factor the following trinomials. [6.3]

**43.** $8x^2 + 16x + 6$

**44.** $3a^3 - 14a^2 - 5a$

**45.** $20m^3 - 34m^2 + 6m$

**46.** $30x^2y - 55xy^2 + 15y^3$

Factor the following. [6.4]

**47.** $4x^2 + 40x + 100$

**48.** $4x^3 + 12x^2 + 9x$

**49.** $5x^2 - 45$

**50.** $12x^3 - 27xy^2$

Factor the following polynomials completely. [6.5]

**51.** $6a^3b + 33a^2b^2 + 15ab^3$

**52.** $x^5 - x^3$    **53.** $4y^6 + 9y^4$

**54.** $12x^5 + 20x^4y - 8x^3y^2$

**55.** $30a^4b + 35a^3b^2 - 15a^2b^3$

**56.** $18a^3b^2 + 3a^2b^3 - 6ab^4$

Solve. [6.6]

**57.** $(x - 5)(x + 2) = 0$

**58.** $3(2y + 5)(2y - 5) = 0$

**59.** $m^2 + 3m = 10$

**60.** $a^2 - 49 = 0$    **61.** $m^2 - 9m = 0$

**62.** $6y^2 = -13y - 6$    **63.** $9x^4 + 9x^3 = 10x^2$

Solve the following word problems. [6.7]

**64. Number Problem**  The product of two consecutive even integers is 120. Find the two integers.

**65. Number Problem**  The product of two consecutive integers is 110. Find the two integers.

**66. Number Problem**  The product of two consecutive odd integers is 1 less than 3 times their sum. Find the integers.

**67. Number Problem**  The sum of two numbers is 20. Their product is 75. Find the numbers.

**68. Number Problem**  One number is 1 less than twice another. Their product is 66. Find the numbers.

**69. Geometry**  The height of a triangle is 8 times the base. The area is 16 square inches. Find the base.

# CHAPTER 6    TEST

Factor out the greatest common factor. [6.1]

**1.** $5x - 10$

**2.** $18x^2y - 9xy - 36xy^2$

Factor by grouping. [6.1]

**3.** $x^2 + 2ax - 3bx - 6ab$

**4.** $xy + 4x - 7y - 28$

Factor the following completely. [6.2–6.6]

**5.** $x^2 - 5x + 6$    **6.** $x^2 - x - 6$

**7.** $a^2 - 16$    **8.** $x^2 + 25$

**9.** $x^4 - 81$    **10.** $27x^2 - 75y^2$

**11.** $x^3 + 5x^2 - 9x - 45$    **12.** $x^2 - bx + 5x - 5b$

**13.** $4a^2 + 22a + 10$    **14.** $3m^2 - 3m - 18$

**15.** $6y^2 + 7y - 5$    **16.** $12x^3 - 14x^2 - 10x$

Solve the following quadratic equations. [6.6]

**17.** $x^2 + 7x + 12 = 0$    **18.** $x^2 - 4x + 4 = 0$

**19.** $x^2 - 36 = 0$    **20.** $x^2 = x + 20$

**21.** $x^2 - 11x = -30$    **22.** $y^3 = 16y$

**23.** $2a^2 = a + 15$    **24.** $30x^3 - 20x^2 = 10x$

Solve the following word problems. Be sure to show the equation used. [6.7]

**25. Number Problem**  Two numbers have a sum of 20. Their product is 64. Find the numbers.

**26. Consecutive Integers**  The product of two consecutive odd integers is 7 more than their sum. Find the integers.

**27. Geometry**  The length of a rectangle is 5 more than 3 times the width. The area is 42 square feet. Find the dimensions.

**28. Geometry**  One leg of a right triangle is 2 more than twice the other. The hypotenuse is 13 meters. Find the lengths of the two legs.

**29. Production Cost**  A company can manufacture $x$ hundred items for a total cost $C$ if $C = 200 + 500x - 100x^2$. How many items can be manufactured if the total cost is to be $800?

**30. Price and Revenue**  A manufacturer knows that the number of items he can sell each week, $x$, is related to the price $p$ of each item by the equation $x = 900 - 100p$. What price should he charge for each item in order to have a weekly revenue of $1,800? (*Remember: R = xp.*)

# CHAPTERS 1 – 6   CUMULATIVE REVIEW

Simplify.

**1.** $-|-9|$

**2.** $9 + (-7) + (-8)$

**3.** $20 - (-9)$

**4.** $\dfrac{-63}{-7}$

**5.** $\dfrac{9(-2)}{-2}$

**6.** $(-4)^3$

**7.** $\dfrac{-3(4-7) - 5(7-2)}{-5 - 2 - 1}$

**8.** $-a + 3 + 6a - 8$

**9.** $6 - 2(4a + 2) - 5$

**10.** $(x^4)^{10}$

**11.** $(9xy)^0$

**12.** $\dfrac{x^{-9}}{x^{-13}}$

**13.** $\dfrac{50x^8y^8}{25x^4y^2} + \dfrac{28x^7y^7}{14x^3y}$

**14.** $(5x - 2)(3x + 4)$

Solve each equation.

**15.** $3x = -18$

**16.** $\dfrac{x}{2} = 5$

**17.** $-\dfrac{x}{3} = 7$

**18.** $\dfrac{1}{2}(4t - 1) + \dfrac{1}{3} = -\dfrac{25}{6}$

**19.** $4m(m - 7)(2m - 7) = 0$

**20.** $16x^2 - 81 = 0$

Solve each inequality.

**21.** $-2x > -8$

**22.** $-3x \ge -6$ or $2x - 7 \ge 7$

Graph on a rectangular coordinate system.

**23.** $y = -3x$

**24.** $y = 2$

**25.** $2x + 3y \ge 6$

**26.** Find the $x$- and $y$-intercepts for the equation $10x - 3y = 30$.

**27.** Find the slope of the line that passes through the points $(7, 3)$ and $(-2, -4)$.

**28.** Find the equation for the line with slope $\frac{2}{3}$ that passes through $(-6, 4)$.

**29.** Find the equation for the line with slope $-\frac{2}{5}$ and $y$-intercept $-\frac{2}{3}$.

**30.** Find the slope and $y$-intercept for the equation $3x - 4y = -16$.

**31.** Solve the following system by graphing:

$$y = x + 3$$
$$x + y = -1$$

Solve each system.

**32.** $-x + y = 3$
$\phantom{-}x + y = 7$

**33.** $5x + 7y = -18$
$8x + 3y = 4$

**34.** $2x + y = 4$
$\phantom{2x +} x = y - 1$

**35.** $\phantom{0.0}x + y = 5{,}000$
$0.04x + 0.06y = 270$

**36.** $4x + 7y = -1$
$3x - 2y = -8$

Factor completely.

**37.** $n^2 - 5n - 36$

**38.** $14x^2 + 31xy - 10y^2$

**39.** $16 - a^2$

**40.** $49x^2 - 14x + 1$

**41.** $45x^2y - 30xy^2 + 5y^3$

**42.** $18x^3 - 3x^2y - 3xy^2$

**43.** $3xy + 15x - 2y - 10$

**44.** Give the opposite and reciprocal for $-2$.

**45.** What property justifies the statement $(2 + x) + 3 = (x + 2) + 3$?

**46.** Multiply and write your answer in scientific notation: $(5 \times 10^5)(2.1 \times 10^3)$.

**47.** Divide $28x^4y^4 - 14x^2y^3 + 21xy^2$ by $-7xy^2$.

**48.** Divide using long division $\dfrac{x^3 - 27}{x - 3}$.

**49. Cutting a Board Into Two Pieces**   A 72-inch board is to be cut into two pieces. One piece is to be 4 inches longer than the other. How long is each piece?

**50. Hamburgers and Fries**   Sheila bought burgers and fries for her children and some friends. The burgers cost $2.05 each, and the fries are $0.85 each. She bought a total of 14 items, for a total cost of $19.10. How many of each did she buy?

# Rational Expressions

(Jon Riley/© Tony Stone Images)

## INTRODUCTION

First Bank of San Luis Obispo charges $2.00 per month and $0.15 per check for a regular checking account. If we write $x$ checks in one month, the total monthly cost of the checking account will be $C = 2.00 + 0.15x$. From this formula we see that the more checks we write in a month, the more we pay for the account. But, it is also true that the more checks we write in a month, the lower the cost per check. To find the average cost per check, we divide the total cost by the number of checks written:

$$\text{Average cost} = A = \frac{C}{x} = \frac{2.00 + 0.15x}{x}$$

We can use this formula to create Table 1 and Figure 1, giving us a visual interpretation of the relationship between the number of checks written and the average cost per check.

| TABLE I  Average Cost | |
|---|---|
| Number of Checks | Average Cost Per Check |
| 1 | 2.15 |
| 2 | 1.15 |
| 5 | 0.55 |
| 10 | 0.35 |
| 15 | 0.28 |
| 20 | 0.25 |

**FIGURE I**

As you can see, if we write one check per month, the cost per check is relatively high, $2.15. On the other hand, if we write 20 checks per month, each check costs us only $0.25. Using average cost per check is a good way to compare different checking accounts. The expression $\dfrac{2.00 + 0.15x}{x}$ in the average cost formula is a rational expression. When you have finished this chapter you will have a good working knowledge of rational expressions.

In Chapter 1 we defined the set of rational numbers to be the set of all numbers that could be put in the form $\frac{a}{b}$, where $a$ and $b$ are integers ($b \neq 0$):

$$\text{Rational numbers} = \left\{ \frac{a}{b} \,\middle|\, a \text{ and } b \text{ are integers, } b \neq 0 \right\}$$

A *rational expression* is any expression that can be put in the form $\frac{P}{Q}$, where $P$ and $Q$ are polynomials and $Q \neq 0$:

$$\text{Rational expressions} = \left\{ \frac{P}{Q} \,\middle|\, P \text{ and } Q \text{ are polynomials, } Q \neq 0 \right\}$$

Each of the following is an example of a rational expression:

$$\frac{2x + 3}{x} \qquad \frac{x^2 - 6x + 9}{x^2 - 4} \qquad \frac{5}{x^2 + 6} \qquad \frac{2x^2 + 3x + 4}{2}$$

For the rational expression

$$\frac{x^2 - 6x + 9}{x^2 - 4}$$

the polynomial on top, $x^2 - 6x + 9$, is called the numerator, and the polynomial on the bottom, $x^2 - 4$, is called the denominator. The same is true of the other rational expressions.

We must be careful that we do not use a value of the variable that will give us a denominator of zero. Remember, division by zero is not defined.

**EXAMPLES** State the restrictions on the variable in the following rational expressions:

**1.** $\dfrac{x + 2}{x - 3}$

**Solution** The variable $x$ can be any real number except $x = 3$, since, when $x = 3$, the denominator is $3 - 3 = 0$. We state this restriction by writing $x \neq 3$.

**2.** $\dfrac{5}{x^2 - x - 6}$

**Solution** If we factor the denominator, we have $x^2 - x - 6 = (x - 3)(x + 2)$. If either of the factors is zero, the whole denominator is zero. Our restrictions are $x \neq 3$ and $x \neq -2$, since either one makes $x^2 - x - 6 = 0$.

We will not always list each restriction on a rational expression, but we should be aware of them and keep in mind that no rational expression can have a denominator of zero.

The two fundamental properties of rational expressions are listed next. We will use these two properties many times in this chapter.

## PROPERTIES OF RATIONAL EXPRESSIONS

### PROPERTY I
*Multiplying* the numerator and denominator of a rational expression by the same nonzero quantity will not change the value of the rational expression.

### PROPERTY 2
*Dividing* the numerator and denominator of a rational expression by the same nonzero quantity will not change the value of the rational expression.

We can use Property 2 to reduce rational expressions to lowest terms. Since this process is almost identical to the process of reducing fractions to lowest terms, let's recall how the fraction $\frac{6}{15}$ is reduced to lowest terms:

$$\frac{6}{15} = \frac{2 \cdot 3}{5 \cdot 3} \qquad \text{Factor numerator and denominator.}$$

$$= \frac{2 \cdot \cancel{3}}{5 \cdot \cancel{3}} \qquad \text{Divide out the common factor, 3.}$$

$$= \frac{2}{5} \qquad \text{Reduce to lowest terms.}$$

The same procedure applies to reducing rational expressions to lowest terms. The process is summarized in the following rule.

## RULE

To reduce a rational expression to lowest terms, first factor the numerator and denominator completely and then divide both the numerator and denominator by any factors they have in common.

**EXAMPLE 3**    Reduce $\dfrac{x^2 - 9}{x^2 + 5x + 6}$ to lowest terms.

**Solution**    We begin by factoring:

$$\frac{x^2 - 9}{x^2 + 5x + 6} = \frac{(x - 3)(x + 3)}{(x + 2)(x + 3)}$$

Notice that both polynomials contain the factor $(x + 3)$. If we divide the numerator by $(x + 3)$, we are left with $(x - 3)$. If we divide the denominator by $(x + 3)$, we are left with $(x + 2)$. The complete problem looks like this:

$$\frac{x^2 - 9}{x^2 + 5x + 6} = \frac{(x - 3)\cancel{(x + 3)}}{(x + 2)\cancel{(x + 3)}}$$    Factor the numerator and
    denominator completely.

$$= \frac{x - 3}{x + 2}$$    Divide out the common
    factor, $x + 3$.

It is convenient to draw a line through the factors as we divide them out. It is especially helpful when the problems become longer.

**EXAMPLE 4**    Reduce to lowest terms $\dfrac{10a + 20}{5a^2 - 20}$.

**Solution**    We begin by factoring out the greatest common factor from the numerator and denominator:

$$\frac{10a + 20}{5a^2 - 20} = \frac{10(a + 2)}{5(a^2 - 4)}$$    Factor the greatest common factor
    from the numerator and denominator.

$$= \frac{10\cancel{(a + 2)}}{5\cancel{(a + 2)}(a - 2)}$$    Factor the denominator as the
    difference of two squares.

$$= \frac{2}{a - 2}$$    Divide out the common
    factors 5 and $a + 2$.

**EXAMPLE 5**    Reduce $\dfrac{2x^3 + 2x^2 - 24x}{x^3 + 2x^2 - 8x}$ to lowest terms.

**Solution**    We begin by factoring the numerator and denominator completely. Then we divide out all factors common to the numerator and denominator. Here is what it looks like:

$$\frac{2x^3 + 2x^2 - 24x}{x^3 + 2x^2 - 8x} = \frac{2x(x^2 + x - 12)}{x(x^2 + 2x - 8)}$$    Factor out the greatest
    common factor first.

$$= \frac{2\cancel{x}(x - 3)\cancel{(x + 4)}}{\cancel{x}(x - 2)\cancel{(x + 4)}}$$    Factor the remaining trinomials.

$$= \frac{2(x - 3)}{x - 2}$$    Divide out the factors common to
    the numerator and denominator.

**EXAMPLE 6**    Reduce $\dfrac{x - 5}{x^2 - 25}$ to lowest terms.

**Solution**

$$\frac{x-5}{x^2-25} = \frac{\cancel{x-5}}{(\cancel{x-5})(x+5)}$$     Factor numerator and denominator completely.

$$= \frac{1}{x+5}$$     Divide out the common factor, $x-5$.

---

## RATIOS

For the rest of this section we will concern ourselves with ratios, a topic closely related to reducing fractions and rational expressions to lowest terms. Let's start with a definition.

> **DEFINITION**   If $a$ and $b$ are any two numbers, $b \neq 0$, then the ratio of $a$ and $b$ is
>
> $$\frac{a}{b}$$

As you can see, ratios are another name for fractions or rational numbers. They are a way of comparing quantities. Since we can also think of $a/b$ as the quotient of $a$ and $b$, ratios are also quotients. The following table gives some ratios in words and as fractions.

| Ratio | As a Fraction | In Lowest Terms | |
|-------|---------------|-----------------|--|
| 25 to 75 | $\frac{25}{75}$ | $\frac{1}{3}$ | |
| 8 to 2 | $\frac{8}{2}$ | $\frac{4}{1}$ | With ratios it is common to leave the 1 in the denominator. |
| 20 to 16 | $\frac{20}{16}$ | $\frac{5}{4}$ | |

**EXAMPLE 7**   A solution of hydrochloric acid (HCl) and water contains 49 milliliters of water and 21 milliliters of HCl. Find the ratio of HCl to water and of HCl to the total volume of the solution.

**Solution**   The ratio of HCl to water is 21 to 49, or

$$\frac{21}{49} = \frac{3}{7}$$

The amount of total solution volume is $49 + 21 = 70$ milliliters. Therefore, the ratio of HCl to total solution is 21 to 70, or

$$\frac{21}{70} = \frac{3}{10}$$

## RATE EQUATION

Many of the problems in this chapter will use what is called the *rate equation*. You use this equation on an intuitive level when you are estimating how long it will take you to drive long distances. For example, if you drive at 50 miles per hour for 2 hours, you will travel 100 miles. Here is the rate equation:

$$\text{Distance} = \text{rate} \cdot \text{time}$$

$$d = r \cdot t$$

The rate equation has two equivalent forms, the most common of which is obtained by solving for *r*. Here it is:

$$r = \frac{d}{t}$$

The rate *r* in the rate equation is the ratio of distance to time and is also referred to as *average speed*. The units for rate are miles per hour, feet per second, kilometers per hour, and so on.

**E X A M P L E  8**   The Forest chair lift at the Northstar ski resort in Lake Tahoe is 5,603 feet long. If a ride on this chair lift takes 11 minutes, what is the average speed of the lift in feet per minute?

**Solution**   To find the speed of the lift, we find the ratio of distance covered to time. (Our answer is rounded to the nearest whole number.)

$$\text{Rate} = \frac{\text{distance}}{\text{time}} = \frac{5{,}603 \text{ feet}}{11 \text{ minutes}} = \frac{5{,}603}{11} \text{ feet/minute} = 509 \text{ feet/minute}$$

Note how we separate the numerical part of the problem from the units. In the next section, we will convert this rate to miles per hour.

L = 5,603 ft

Getting Ready for Class

*After reading through the preceding section, respond in your own words and in complete sentences.*

**A.** How do you reduce a rational expression to lowest terms?

**B.** What are the properties we use to manipulate rational expressions?

**C.** For what values of the variable is a rational expression undefined?

**D.** What is a ratio?

## PROBLEM SET 7.1

Reduce the following rational expressions to lowest terms, if possible. Also, specify any restrictions on the variable in Problems 1 through 10.

**1.** $\dfrac{5}{5x - 10}$

**2.** $\dfrac{-4}{2x - 8}$

**3.** $\dfrac{a - 3}{a^2 - 9}$

**4.** $\dfrac{a + 4}{a^2 - 16}$

**5.** $\dfrac{x + 5}{x^2 - 25}$

**6.** $\dfrac{x - 2}{x^2 - 4}$

**7.** $\dfrac{2x^2 - 8}{4}$

**8.** $\dfrac{5x - 10}{x - 2}$

**9.** $\dfrac{2x - 10}{3x - 6}$

**10.** $\dfrac{4x - 8}{x - 2}$

**11.** $\dfrac{10a + 20}{5a + 10}$

**12.** $\dfrac{11a + 33}{6a + 18}$

**13.** $\dfrac{5x^2 - 5}{4x + 4}$

**14.** $\dfrac{7x^2 - 28}{2x + 4}$

**15.** $\dfrac{x - 3}{x^2 - 6x + 9}$

**16.** $\dfrac{x^2 - 10x + 25}{x - 5}$

**17.** $\dfrac{3x + 15}{3x^2 + 24x + 45}$

**18.** $\dfrac{5x + 15}{5x^2 + 40x + 75}$

**19.** $\dfrac{a^2 - 3a}{a^3 - 8a^2 + 15a}$

**20.** $\dfrac{a^2 + 3a}{a^3 - 2a^2 - 15a}$

**21.** $\dfrac{3x - 2}{9x^2 - 4}$

**22.** $\dfrac{2x - 3}{4x^2 - 9}$

**23.** $\dfrac{x^2 + 8x + 15}{x^2 + 5x + 6}$

**24.** $\dfrac{x^2 - 8x + 15}{x^2 - x - 6}$

**25.** $\dfrac{2m^3 - 2m^2 - 12m}{m^2 - 5m + 6}$

**26.** $\dfrac{2m^3 + 4m^2 - 6m}{m^2 - m - 12}$

**27.** $\dfrac{x^3 + 3x^2 - 4x}{x^3 - 16x}$

**28.** $\dfrac{3a^2 - 8a + 4}{9a^3 - 4a}$

**29.** $\dfrac{4x^3 - 10x^2 + 6x}{2x^3 + x^2 - 3x}$

**30.** $\dfrac{3a^3 - 8a^2 + 5a}{4a^3 - 5a^2 + 1a}$

**31.** $\dfrac{4x^2 - 12x + 9}{4x^2 - 9}$

**32.** $\dfrac{5x^2 + 18x - 8}{5x^2 + 13x - 6}$

**33.** $\dfrac{x + 3}{x^4 - 81}$

**34.** $\dfrac{x^2 + 9}{x^4 - 81}$

**35.** $\dfrac{3x^2 + x - 10}{x^4 - 16}$

**36.** $\dfrac{5x^2 - 26x + 24}{x^4 - 64}$

**37.** $\dfrac{42x^3 - 20x^2 - 48x}{6x^2 - 5x - 4}$

**38.** $\dfrac{36x^3 + 132x^2 - 135x}{6x^2 + 25x - 9}$

To reduce each of the following rational expressions to lowest terms, you will have to use factoring by grouping. Be sure to factor each numerator and denominator completely before dividing out any common factors. (Remember, factoring by grouping takes two steps.)

**39.** $\dfrac{xy + 3x + 2y + 6}{xy + 3x + 5y + 15}$

**40.** $\dfrac{xy + 7x + 4y + 28}{xy + 3x + 4y + 12}$

**41.** $\dfrac{x^2 - 3x + ax - 3a}{x^2 - 3x + bx - 3b}$

**42.** $\dfrac{x^2 - 6x + ax - 6a}{x^2 - 7x + ax - 7a}$

**43.** $\dfrac{xy + bx + ay + ab}{xy + bx + 3y + 3b}$

**44.** $\dfrac{x^2 + 5x + ax + 5a}{x^2 + 5x + bx + 5b}$

Write each ratio as a fraction in lowest terms.

**45.** 8 to 6

**46.** 6 to 8

**47.** 200 to 250

**48.** 250 to 200

**49.** 32 to 4

**50.** 4 to 32

## Applying the Concepts

**51. Cost and Average Cost** As we mentioned in the introduction to this chapter, if a bank charges

$2.00 per month and $0.15 per check for one of its checking accounts, then the total monthly cost to write $x$ checks is $C = 2.00 + 0.15x$, and the average cost of each of the $x$ checks written is $A = \dfrac{2.00 + 0.15x}{x}$. Compare these two formulas by filling in the following table. Round to the nearest cent.

| Checks Written $x$ | Total Cost $2.00 + 0.15x$ | Cost per Check $\dfrac{2.00 + 0.15x}{x}$ |
|---|---|---|
| 0 | | |
| 5 | | |
| 10 | | |
| 15 | | |
| 20 | | |

**52. Cost and Average Cost**   A rewritable CD-ROM drive for a computer costs $250. An individual CD for the drive costs $5.00 and can store 640 megabytes of information. The total cost of filling $x$ CDs with information is $C = 250 + 5x$ dollars. The average cost per megabyte of information is given by $A = \dfrac{5x + 250}{640x}$. Compare the total cost and average cost per megabyte of storage by completing the following table. Round all answers to the nearest cent.

| CDs Purchased $x$ | Total Cost $250 + 5x$ | Cost per Megabyte $\dfrac{5x + 250}{640x}$ |
|---|---|---|
| 0 | | |
| 5 | | |
| 10 | | |
| 15 | | |
| 20 | | |

**53. Speed of a Car**   A car travels 122 miles in 3 hours. Find the average speed of the car in miles per hour. Round to the nearest tenth.

**54. Speed of a Bullet**   A bullet fired from a gun trav-
els 4,500 feet in 3 seconds. Find the average speed of the bullet in feet per second.

**55. Ferris Wheel**   The first Ferris wheel was designed and built by George Ferris in 1893. It was a large wheel with a circumference of 785 feet. If one trip around the circumference of the wheel took 20 minutes, find the average speed of a rider in feet per minute.

**56. Ferris Wheel**   In 1897 a large Ferris wheel was built in Vienna; it is still in operation today. Known as *The Great Wheel*, it has a circumference of 618 feet. If one trip around the wheel takes 15 minutes, find the average speed of a rider on this wheel in feet per minute.

**57. Ferris Wheel**   A Ferris wheel called *Colossus* was built in St. Louis in 1986. It has a circumference of 518 feet. If a trip around the circumference of *Colossus* takes 40 seconds, find the average speed of a rider in feet per second.

**58. Ferris Wheel**   A person riding a Ferris wheel travels once around the wheel, a distance of 188 feet, in 30 seconds. What is the average speed of the rider in feet per second? Round to the nearest tenth.

**59. Average Speed**   Tina is training for a biathlon. As part of her training, she runs an 8-mile course, 2 miles of which is on level ground and 6 miles of which is downhill. It takes her 20 minutes to run the level part of the course and 40 minutes to run the downhill part of the course. Find her average speed in minutes per mile and in miles per minute for each part of the course. Round to the nearest hundredth, if rounding is necessary.

**60. Jogging**   A jogger covers a distance of 3 miles in 24 minutes. Find the average speed of the jogger in miles per minute.

**61. Fuel Consumption** An economy car travels 168 miles on 3.5 gallons of gas. Give the average fuel consumption of the car in miles per gallon.

**62. Fuel Consumption** A luxury car travels 100 miles on 8 gallons of gas. Give the average fuel consumption of the car in miles per gallon.

**63. Comparing Expressions** Replace $x$ with 5 and $y$ with 4 in the expression

$$\frac{x^2 - y^2}{x - y}$$

and simplify the result. Is the result equal to $5 - 4$ or $5 + 4$?

**64. Comparing Expressions** Replace $x$ with 2 in the expression

$$\frac{x^3 - 1}{x - 1}$$

and simplify the result. Your answer should be equal to what you would get if you replaced $x$ with 2 in $x^2 + x + 1$.

**65. Comparing Expressions** Complete the following table; then show why the table turns out as it does.

| $x$ | $\dfrac{x - 3}{3 - x}$ |
|-----|-----|
| $-2$ | |
| $-1$ | |
| $0$ | |
| $1$ | |
| $2$ | |

**66. Comparing Expressions** Complete the following table; then show why the table turns out as it does.

| $x$ | $\dfrac{25 - x^2}{x^2 - 25}$ |
|-----|-----|
| $-4$ | |
| $-2$ | |
| $0$ | |
| $2$ | |
| $4$ | |

**67. Comparing Expressions** You know from reading through Example 6 in this section that $\dfrac{x - 5}{x^2 - 25} = \dfrac{1}{x + 5}$. Compare these expressions by completing the following table. (Be careful, not all the rows have equal entries.)

| $x$ | $\dfrac{x - 5}{x^2 - 25}$ | $\dfrac{1}{x + 5}$ |
|-----|-----|-----|
| $0$ | | |
| $-2$ | | |
| $2$ | | |
| $-5$ | | |
| $5$ | | |

**68. Comparing Expressions** You know from your work in this section that $\dfrac{x^2 - 6x + 9}{x^2 - 9} = \dfrac{x - 3}{x + 3}$. Compare these expressions by completing the following table. (Be careful, not all the rows have equal entries.)

| $x$ | $\dfrac{x^2 - 6x + 9}{x^2 - 9}$ | $\dfrac{x - 3}{x + 3}$ |
|-----|-----|-----|
| $-3$ | | |
| $-2$ | | |
| $-1$ | | |
| $0$ | | |
| $1$ | | |
| $2$ | | |
| $3$ | | |

## Review Problems

The following problems review material we covered in Sections 5.3 and 5.7.

Simplify. [5.3]

**69.** $\dfrac{27x^5}{9x^2} - \dfrac{45x^8}{15x^5}$

**70.** $\dfrac{36x^9}{4x} - \dfrac{45x^3}{5x^{-5}}$

**71.** $\dfrac{72a^3b^7}{9ab^5} + \dfrac{64a^5b^3}{8a^3b}$

**72.** $\dfrac{80a^5b^{11}}{10a^2b} + \dfrac{33a^6b^{12}}{11a^3b^2}$

Divide. [5.7]

**73.** $\dfrac{38x^7 + 42x^5 - 84x^3}{2x^3}$

**74.** $\dfrac{49x^6 - 63x^4 - 35x^2}{7x^2}$

**75.** $\dfrac{28a^5b^5 + 36ab^4 - 44a^4b}{4ab}$

**76.** $\dfrac{30a^3b - 12a^2b^2 + 6ab^3}{6ab}$

---

**7.2   Multiplication and Division of Rational Expressions**

Recall that to multiply two fractions we simply multiply numerators and multiply denominators and then reduce to lowest terms, if possible:

$$\frac{3}{4} \cdot \frac{10}{21} = \frac{30}{84} \quad \leftarrow \text{Multiply numerators.}$$
$$\qquad\qquad\qquad \leftarrow \text{Multiply denominators.}$$

$$= \frac{5}{14} \quad \leftarrow \text{Reduce to lowest terms.}$$

Recall also that the same result can be achieved by factoring numerators and denominators first and then dividing out the factors they have in common:

$$\frac{3}{4} \cdot \frac{10}{21} = \frac{3}{2 \cdot 2} \cdot \frac{2 \cdot 5}{3 \cdot 7} \qquad \text{Factor.}$$

$$= \frac{3 \cdot 2 \cdot 5}{2 \cdot 2 \cdot 3 \cdot 7} \qquad \begin{array}{l}\text{Multiply numerators.}\\ \text{Multiply denominators.}\end{array}$$

$$= \frac{5}{14} \qquad\qquad \text{Divide out common factors.}$$

We can apply the second process to the product of two rational expressions, as the following example illustrates.

---

**EXAMPLE 1**   Multiply $\dfrac{x - 2}{x + 3} \cdot \dfrac{x^2 - 9}{2x - 4}$.

**Solution**   We begin by factoring numerators and denominators as much as possible. Then we multiply the numerators and denominators. The last step consists of dividing out all factors common to the numerator and denominator:

$$\frac{x-2}{x+3} \cdot \frac{x^2-9}{2x-4} = \frac{x-2}{x+3} \cdot \frac{(x-3)(x+3)}{2(x-2)}$$    Factor completely.

$$= \frac{(x-2)(x-3)(x+3)}{(x+3)(2)(x-2)}$$    Multiply numerators and denominators.

$$= \frac{x-3}{2}$$    Divide out common factors.

In Chapter 1 we defined division as the equivalent of multiplication by the reciprocal. This is how it looks with fractions:

$$\frac{4}{5} \div \frac{8}{9} = \frac{4}{5} \cdot \frac{9}{8}$$    Division as multiplication by the reciprocal

$$= \frac{2 \cdot 2 \cdot 3 \cdot 3}{5 \cdot 2 \cdot 2 \cdot 2}$$

$$= \frac{9}{10}$$    Factor and divide out common factors.

The same idea holds for division with rational expressions. The rational expression that follows the division symbol is called the *divisor;* to divide, we multiply by the reciprocal of the divisor.

**EXAMPLE 2**    Divide $\dfrac{3x-9}{x^2-x-20} \div \dfrac{x^2+2x-15}{x^2-25}$.

**Solution**    We begin by taking the reciprocal of the divisor and writing the problem again in terms of multiplication. We then factor, multiply, and, finally, divide out all factors common to the numerator and denominator of the resulting expression. The complete problem looks like this:

$$\frac{3x-9}{x^2-x-20} \div \frac{x^2+2x-15}{x^2-25}$$

$$= \frac{3x-9}{x^2-x-20} \cdot \frac{x^2-25}{x^2+2x-15}$$    Multiply by the reciprocal of the divisor.

$$= \frac{3(x-3)}{(x+4)(x-5)} \cdot \frac{(x-5)(x+5)}{(x+5)(x-3)}$$    Factor.

$$= \frac{3(x-3)(x-5)(x+5)}{(x+4)(x-5)(x+5)(x-3)}$$    Divide out common factors.

$$= \frac{3}{x+4}$$

As you can see, factoring is the single most important tool we use in working with rational expressions. Most of the work we have done or will do with rational expressions is most easily accomplished if the rational expressions are in factored form. Here are some more examples of multiplication and division with rational expressions.

## EXAMPLES

**3.** Multiply $\dfrac{3a + 6}{a^2} \cdot \dfrac{a}{2a + 4}$.

**Solution**

$$\dfrac{3a + 6}{a^2} \cdot \dfrac{a}{2a + 4}$$

$$= \dfrac{3(a + 2)}{a^2} \cdot \dfrac{a}{2(a + 2)} \qquad \text{Factor completely.}$$

$$= \dfrac{3(a + 2)a}{a^2(2)(a + 2)} \qquad \text{Multiply.}$$

$$= \dfrac{3}{2a} \qquad\qquad \begin{array}{l}\text{Divide numerator and denominator}\\ \text{by common factors } a(a + 2).\end{array}$$

**4.** Divide $\dfrac{x^2 + 7x + 12}{x^2 - 16} \div \dfrac{x^2 + 6x + 9}{2x - 8}$.

**Solution**

$$\dfrac{x^2 + 7x + 12}{x^2 - 16} \div \dfrac{x^2 + 6x + 9}{2x - 8}$$

$$= \dfrac{x^2 + 7x + 12}{x^2 - 16} \cdot \dfrac{2x - 8}{x^2 + 6x + 9} \qquad \begin{array}{l}\text{Division is multiplication}\\ \text{by the reciprocal.}\end{array}$$

$$= \dfrac{(x + 3)(x + 4)(2)(x - 4)}{(x - 4)(x + 4)(x + 3)(x + 3)} \qquad \text{Factor and multiply.}$$

$$= \dfrac{2}{x + 3} \qquad\qquad \text{Divide out common factors.}$$

In Example 4 we factored and multiplied the two expressions in a single step. This saves writing the problem one extra time.

**EXAMPLE 5**    Multiply $(x^2 - 49)\left(\dfrac{x + 4}{x + 7}\right)$.

**Solution**    We can think of the polynomial $x^2 - 49$ as having a denominator of 1. Thinking of $x^2 - 49$ in this way allows us to proceed as we did in previous examples:

$$(x^2 - 49)\left(\frac{x + 4}{x + 7}\right) = \frac{x^2 - 49}{1} \cdot \frac{x + 4}{x + 7}$$    Write $x^2 - 49$ with denominator 1.

$$= \frac{\cancel{(x + 7)}(x - 7)(x + 4)}{\cancel{x + 7}}$$    Factor and multiply.

$$= (x - 7)(x + 4)$$    Divide out common factors.

We can leave the answer in this form or multiply to get $x^2 - 3x - 28$. In this section let's agree to leave our answers in factored form.

**EXAMPLE 6**    Multiply $a(a + 5)(a - 5)\left(\dfrac{a + 4}{a^2 + 5a}\right)$.

**Solution**    We can think of the expression $a(a + 5)(a - 5)$ as having a denominator of 1:

$$a(a + 5)(a - 5)\left(\frac{a + 4}{a^2 + 5a}\right)$$

$$= \frac{a(a + 5)(a - 5)}{1} \cdot \frac{a + 4}{a^2 + 5a}$$

$$= \frac{\cancel{a(a + 5)}(a - 5)(a + 4)}{\cancel{a(a + 5)}}$$    Factor and multiply.

$$= (a - 5)(a + 4)$$    Divide out common factors.

## UNIT ANALYSIS

*Unit analysis* is a method of converting between units of measure by multiplying by the number 1. Here is our first illustration: Suppose you are flying in a commercial airliner and the pilot tells you the plane has reached its cruising altitude of 35,000 feet. How many miles is the plane above the ground?

If you know that 1 mile is 5,280 feet, then it is simply a matter of deciding what to do with the two numbers, 5,280 and 35,000. By using unit analysis, this decision is unnecessary:

$$35{,}000 \text{ feet} = \frac{35{,}000 \text{ feet}}{1} \cdot \frac{1 \text{ mile}}{5{,}280 \text{ feet}}$$

We treat the units common to the numerator and denominator in the same way we treat factors common to the numerator and denominator; common units can be divided out, just as common factors are. In the previous expression, we have feet common to the numerator and denominator. Dividing them out leaves us with miles only. Here is the complete problem:

$$35{,}000 \text{ feet} = \frac{35{,}000 \text{ feet}}{1} \cdot \frac{1 \text{ mile}}{5{,}280 \text{ feet}}$$

$$= \frac{35{,}000}{5{,}280} \text{ miles}$$

$$= 6.6 \text{ miles to the nearest tenth of a mile}$$

The expression $\dfrac{1 \text{ mile}}{5{,}280 \text{ feet}}$ is called a *conversion factor*. It is simply the number 1 written in a convenient form. Because it is the number 1, we can multiply any other number by it and always be sure we have not changed that number. The key to unit analysis is choosing the right conversion factors.

---

**E X A M P L E   7**    The Mall of America in the Twin Cities covers 78 acres of land. If 1 square mile = 640 acres, how many square miles does the Mall of America cover? Round your answer to the nearest hundredth of a square mile.

**Solution**    We are starting with acres and want to end up with square miles. We need to multiply by a conversion factor that will allow acres to divide out and leave us with square miles:

$$78 \text{ acres} = \frac{78 \text{ acres}}{1} \cdot \frac{1 \text{ square mile}}{640 \text{ acres}}$$

$$= \frac{78}{640} \text{ square mile}$$

$$= 0.12 \text{ square mile to the nearest hundredth}$$

---

The next example is a continuation of Example 8 from Section 7.1.

---

**E X A M P L E   8**    The Forest chair lift at the Northstar ski resort in Lake Tahoe is 5,603 feet long. If a ride on this chair lift takes 11 minutes, what is the average speed of the lift in miles per hour?

**Solution**    First, we find the speed of the lift in feet per second, as we did in Example 8 of Section 7.1, by taking the ratio of distance to time.

$$\text{Rate} = \frac{\text{distance}}{\text{time}} = \frac{5{,}603 \text{ feet}}{11 \text{ minutes}} = \frac{5{,}603}{11} \text{ feet per minute}$$

$$= 509 \text{ feet per minute}$$

L = 5,603 ft

Next, we convert feet per minute to miles per hour. To do this, we need to know that

1 mile = 5,280 feet
1 hour = 60 minutes

$$\text{Speed} = 509 \text{ feet per minute} = \frac{509 \text{ feet}}{1 \text{ minute}} \cdot \frac{1 \text{ mile}}{5,280 \text{ feet}} \cdot \frac{60 \text{ minutes}}{1 \text{ hour}}$$

$$= \frac{509 \cdot 60}{5,280} \text{ miles per hour}$$

$$= 5.8 \text{ miles per hour to the nearest tenth}$$

 ## Getting Ready for Class

*After reading through the preceding section, respond in your own words and in complete sentences.*

**A.** How do we multiply rational expressions?

**B.** Explain the steps used to divide rational expressions.

**C.** What part does factoring play in multiplying and dividing rational expressions?

**D.** Why are all conversion factors the same as the number 1?

# PROBLEM SET 7.2

Multiply or divide as indicated. Be sure to reduce all answers to lowest terms. (That is, the numerator and denominator of the answer should not have any factors in common.)

**1.** $\dfrac{x+y}{3} \cdot \dfrac{6}{x+y}$

**2.** $\dfrac{x-1}{x+1} \cdot \dfrac{5}{x-1}$

**3.** $\dfrac{2x+10}{x^2} \cdot \dfrac{x^3}{4x+20}$

**4.** $\dfrac{3x^4}{3x-6} \cdot \dfrac{x-2}{x^2}$

**5.** $\dfrac{9}{2a-8} \div \dfrac{3}{a-4}$

**6.** $\dfrac{8}{a^2-25} \div \dfrac{16}{a+5}$

**7.** $\dfrac{x+1}{x^2-9} \div \dfrac{2x+2}{x+3}$

**8.** $\dfrac{11}{x-2} \div \dfrac{22}{2x^2-8}$

**9.** $\dfrac{a^2+5a}{7a} \cdot \dfrac{4a^2}{a^2+4a}$

**10.** $\dfrac{4a^2+4a}{a^2-25} \cdot \dfrac{a^2-5a}{8a}$

**11.** $\dfrac{y^2-5y+6}{2y+4} \div \dfrac{2y-6}{y+2}$

**12.** $\dfrac{y^2-7y}{3y^2-48} \div \dfrac{y^2-9}{y^2-7y+12}$

**13.** $\dfrac{2x-8}{x^2-4} \cdot \dfrac{x^2+6x+8}{x-4}$

**14.** $\dfrac{x^2+5x+1}{7x-7} \cdot \dfrac{x-1}{x^2+5x+1}$

**15.** $\dfrac{x-1}{x^2-x-6} \cdot \dfrac{x^2+5x+6}{x^2-1}$

**16.** $\dfrac{x^2-3x-10}{x^2-4x+3} \cdot \dfrac{x^2-5x+6}{x^2-3x-10}$

**17.** $\dfrac{a^2+10a+25}{a+5} \div \dfrac{a^2-25}{a-5}$

**18.** $\dfrac{a^2+a-2}{a^2+5a+6} \div \dfrac{a-1}{a}$

**19.** $\dfrac{y^3-5y^2}{y^4+3y^3+2y^2} \div \dfrac{y^2-5y+6}{y^2-2y-3}$

**20.** $\dfrac{y^2-5y}{y^2+7y+12} \div \dfrac{y^3-7y^2+10y}{y^2+9y+18}$

**21.** $\dfrac{2x^2+17x+21}{x^2+2x-35} \cdot \dfrac{x^2-25}{2x^2-7x-15}$

**22.** $\dfrac{x^2-13x+42}{4x^2+31x+21} \cdot \dfrac{4x^2-5x-6}{x^2-4}$

**23.** $\dfrac{2x^2+10x+12}{4x^2+24x+32} \cdot \dfrac{2x^2+18x+40}{x^2+8x+15}$

**24.** $\dfrac{3x^2-3}{6x^2+18x+12} \cdot \dfrac{2x^2-8}{x^2-3x+2}$

**25.** $\dfrac{2a^2+7a+3}{a^2-16} \div \dfrac{4a^2+8a+3}{2a^2-5a-12}$

**26.** $\dfrac{3a^2+7a-20}{a^2+3a-4} \div \dfrac{3a^2-2a-5}{a^2-2a+1}$

**27.** $\dfrac{4y^2-12y+9}{y^2-36} \div \dfrac{2y^2-5y+3}{y^2+5y-6}$

**28.** $\dfrac{5y^2-6y+1}{y^2-1} \div \dfrac{16y^2-9}{4y^2+7y+3}$

**29.**
$$\dfrac{x^2-1}{6x^2+42x+60} \cdot \dfrac{7x^2+17x+6}{x+1} \cdot \dfrac{6x+30}{7x^2-11x-6}$$

**30.** $\dfrac{4x^2-1}{3x-15} \cdot \dfrac{4x^2-17x-15}{4x^2-9x-9} \cdot \dfrac{3x-3}{x^2-9}$

**31.** $\dfrac{18x^3+21x^2-60x}{21x^2-25x-4} \cdot \dfrac{28x^2-17x-3}{16x^3+28x^2-30x}$

**32.** $\dfrac{56x^3+54x^2-20x}{8x^2-2x-15} \cdot \dfrac{6x^2+5x-21}{63x^3+129x^2-42x}$

Multiply the following expressions using the method shown in Examples 5 and 6 in this section.

**33.** $(x^2-9)\left(\dfrac{2}{x+3}\right)$

**34.** $(x^2-9)\left(\dfrac{-3}{x-3}\right)$

**35.** $a(a+5)(a-5)\left(\dfrac{2}{a^2-25}\right)$

**36.** $a(a^2-4)\left(\dfrac{a}{a+2}\right)$

**37.** $(x^2-x-6)\left(\dfrac{x+1}{x-3}\right)$

**38.** $(x^2-2x-8)\left(\dfrac{x+3}{x-4}\right)$

**39.** $(x^2-4x-5)\left(\dfrac{-2x}{x+1}\right)$

**40.** $(x^2-6x+8)\left(\dfrac{4x}{x-2}\right)$

Each of the following problems involves some factoring by grouping. Remember, before you can divide out factors common to the numerators and denominators of a product, you must factor completely.

**41.** $\dfrac{x^2 - 9}{x^2 - 3x} \cdot \dfrac{2x + 10}{xy + 5x + 3y + 15}$

**42.** $\dfrac{x^2 - 16}{x^2 - 4x} \cdot \dfrac{3x + 18}{xy + 6x + 4y + 24}$

**43.** $\dfrac{2x^2 + 4x}{x^2 - y^2} \cdot \dfrac{x^2 + 3x + xy + 3y}{x^2 + 5x + 6}$

**44.** $\dfrac{x^2 - 25}{3x^2 + 3xy} \cdot \dfrac{x^2 + 4x + xy + 4y}{x^2 + 9x + 20}$

**45.** $\dfrac{x^3 - 3x^2 + 4x - 12}{x^4 - 16} \cdot \dfrac{3x^2 + 5x - 2}{3x^2 - 10x + 3}$

**46.** $\dfrac{x^3 - 5x^2 + 9x - 45}{x^4 - 81} \cdot \dfrac{5x^2 + 18x + 9}{5x^2 - 22x - 15}$

Simplify each expression. Work inside parentheses first, and then divide out common factors.

**47.** $\left(1 - \dfrac{1}{2}\right)\left(1 - \dfrac{1}{3}\right)\left(1 - \dfrac{1}{4}\right)\left(1 - \dfrac{1}{5}\right)$

**48.** $\left(1 + \dfrac{1}{2}\right)\left(1 + \dfrac{1}{3}\right)\left(1 + \dfrac{1}{4}\right)\left(1 + \dfrac{1}{5}\right)$

The dots in the following problems represent factors not written that are in the same pattern as the surrounding factors. Simplify.

**49.**

$\left(1 - \dfrac{1}{2}\right)\left(1 - \dfrac{1}{3}\right)\left(1 - \dfrac{1}{4}\right) \cdots \left(1 - \dfrac{1}{99}\right)\left(1 - \dfrac{1}{100}\right)$

**50.**

$\left(1 - \dfrac{1}{3}\right)\left(1 - \dfrac{1}{4}\right)\left(1 - \dfrac{1}{5}\right) \cdots \left(1 - \dfrac{1}{98}\right)\left(1 - \dfrac{1}{99}\right)$

## Applying the Concepts

**51. Mount Whitney** The top of Mount Whitney, the highest point in California, is 14,494 feet above sea level. Give this height in miles to the nearest tenth of a mile.

**52. Motor Displacement** The relationship between liters and cubic inches, both of which are measures of volume, is 0.0164 liter = 1 cubic inch. If a Ford Mustang has a motor with a displacement of 4.9 liters, what is the displacement in cubic inches? Round your answer to the nearest cubic inch.

**53. Speed of Sound** The speed of sound is 1,088 feet per second. Convert the speed of sound to miles per hour. Round your answer to the nearest whole number.

**54. Average Speed** A car travels 122 miles in 3 hours. Find the average speed of the car in feet per second. Round to the nearest whole number.

**55. Ferris Wheel** As we mentioned in Problem Set 7.1, the first Ferris wheel was built in 1893. It was a large wheel with a circumference of 785 feet. If one trip around the circumference of the wheel took 20 minutes, find the average speed of a rider in miles per hour. Round to the nearest hundredth.

**56. Ferris Wheel** In 1897 a large Ferris wheel was built in Vienna. Known as *The Great Wheel*, it has a circumference of 618 feet. If one trip around the wheel takes 15 minutes, find the average speed of a rider on this wheel in miles per hour. Round to the nearest hundredth.

**57. Ferris Wheel** A Ferris wheel called *Colossus* has a circumference of 518 feet. If a trip around the circumference of *Colossus* takes 40 seconds, find the average speed of a rider in miles per hour. Round to the nearest tenth.

**58. Ferris Wheel** A person riding a Ferris wheel travels once around the wheel, a distance of 188 feet, in 30 seconds. What is the average speed of the rider in miles per hour?

**59. Average Speed** Tina is training for a biathlon. As part of her training, she runs an 8-mile course, 2 miles of which is on level ground and 6 miles of which is downhill. It takes her 20 minutes to run the level part of the course and 40 minutes to run the downhill part of the course. Find her average speed in miles per hour for each part of the course.

**60. Jogging**   A jogger covers a distance of 3 miles in 24 minutes. Find the average speed of the jogger in miles per hour.

## Review Problems

The following problems review material we covered in Sections 1.9 and 5.3. Reviewing these problems will help you in the next section.

Add the following fractions. [1.9]

**61.** $\dfrac{1}{2} + \dfrac{5}{2}$  **62.** $\dfrac{2}{3} + \dfrac{8}{3}$  **63.** $2 + \dfrac{3}{4}$

**64.** $1 + \dfrac{4}{7}$  **65.** $\dfrac{1}{10} + \dfrac{3}{14}$  **66.** $\dfrac{1}{12} + \dfrac{11}{30}$

Simplify each term, then add. [5.3]

**67.** $\dfrac{10x^4}{2x^2} + \dfrac{12x^6}{3x^4}$  **68.** $\dfrac{32x^8}{8x^3} + \dfrac{27x^7}{3x^2}$

**69.** $\dfrac{12a^2b^5}{3ab^3} + \dfrac{14a^4b^7}{7a^3b^5}$  **70.** $\dfrac{16a^3b^2}{4ab} + \dfrac{25a^6b^5}{5a^4b^4}$

---

**7.3**   **Addition and Subtraction of Rational Expressions**

In Chapter 1 we combined fractions having the same denominator by combining their numerators and putting the result over the common denominator. We use the same process to add two rational expressions with the same denominator.

**EXAMPLES**

**1.**  Add $\dfrac{5}{x} + \dfrac{3}{x}$.

**Solution**   Adding numerators, we have:

$$\frac{5}{x} + \frac{3}{x} = \frac{8}{x}$$

**2.**  Add $\dfrac{x}{x^2 - 9} + \dfrac{3}{x^2 - 9}$.

**Solution**   Since both expressions have the same denominator, we add numerators and reduce to lowest terms:

$$\frac{x}{x^2 - 9} + \frac{3}{x^2 - 9} = \frac{x + 3}{x^2 - 9}$$

$$\left.\begin{array}{l} = \dfrac{\cancel{x + 3}}{\cancel{(x + 3)}(x - 3)} \\[2ex] = \dfrac{1}{x - 3} \end{array}\right\} \quad \begin{array}{l}\text{Reduce to lowest terms by} \\ \text{factoring the denominator} \\ \text{and then dividing out com-} \\ \text{mon factor } x + 3.\end{array}$$

Remember, it is the distributive property that allows us to add rational expressions by simply adding numerators. Because of this, we must begin all addition problems involving rational expressions by first making sure all the expressions have the same denominator.

---

**DEFINITION**   The **least common denominator** (LCD) for a set of denominators is the simplest quantity that is exactly divisible by all the denominators.

---

**EXAMPLE 3**   Add $\frac{1}{10} + \frac{3}{14}$.

**Solution**

*Step 1:* Find the LCD for 10 and 14. To do so, we factor each denominator and build the LCD from the factors:

$$\left.\begin{array}{l} 10 = 2 \cdot 5 \\ 14 = 2 \cdot 7 \end{array}\right\} \quad \text{LCD} = 2 \cdot 5 \cdot 7 = 70$$

We know the LCD is divisible by 10 because it contains the factors 2 and 5. It is also divisible by 14 because it contains the factors 2 and 7.

*Step 2:* Change to equivalent fractions that each have denominator 70. To accomplish this task, we multiply the numerator and denominator of each fraction by the factor of the LCD that is not also a factor of its denominator:

| Original Fractions | | Denominators in Factored Form | | Multiply by Factor Needed to Obtain LCD | | These Have the Same Value as the Original Fractions |
|---|---|---|---|---|---|---|
| $\dfrac{1}{10}$ | $=$ | $\dfrac{1}{2 \cdot 5}$ | $=$ | $\dfrac{1}{2 \cdot 5} \cdot \dfrac{7}{7}$ | $=$ | $\dfrac{7}{70}$ |
| $\dfrac{3}{14}$ | $=$ | $\dfrac{3}{2 \cdot 7}$ | $=$ | $\dfrac{3}{2 \cdot 7} \cdot \dfrac{5}{5}$ | $=$ | $\dfrac{15}{70}$ |

The fraction $\frac{7}{70}$ has the same value as the fraction $\frac{1}{10}$. Likewise, the fractions $\frac{15}{70}$ and $\frac{3}{14}$ are equivalent; they have the same value.

*Step 3:* Add numerators and put the result over the LCD:

$$\frac{7}{70} + \frac{15}{70} = \frac{7 + 15}{70} = \frac{22}{70}$$

*Step 4:* Reduce to lowest terms:

$$\frac{22}{70} = \frac{11}{35} \qquad \text{Divide numerator and denominator by 2.}$$

The main idea in adding fractions is to write each fraction again with the LCD for a denominator. Once we have done that, we simply add numerators. The same process can be used to combine rational expressions, as the next example illustrates.

**EXAMPLE 4**    Subtract $\dfrac{3}{x} - \dfrac{1}{2}$.

**Solution**

*Step 1:* The LCD for $x$ and 2 is $2x$. It is the smallest expression divisible by $x$ and by 2.

*Step 2:* To change to equivalent expressions with the denominator $2x$, we multiply the first fraction by $\frac{2}{2}$ and the second by $\dfrac{x}{x}$:

$$\frac{3}{x} \cdot \frac{2}{2} = \frac{6}{2x}$$

$$\frac{1}{2} \cdot \frac{x}{x} = \frac{x}{2x}$$

*Step 3:* Subtracting numerators of the rational expressions in step 2, we have:

$$\frac{6}{2x} - \frac{x}{2x} = \frac{6-x}{2x}$$

*Step 4:* Since $6 - x$ and $2x$ do not have any factors in common, we cannot reduce any further. Here is the complete problem:

$$\frac{3}{x} - \frac{1}{2} = \frac{3}{x} \cdot \frac{2}{2} - \frac{1}{2} \cdot \frac{x}{x}$$

$$= \frac{6}{2x} - \frac{x}{2x}$$

$$= \frac{6-x}{2x}$$

**EXAMPLE 5**    Add $\dfrac{5}{2x-6} + \dfrac{x}{x-3}$.

**Solution**    If we factor $2x - 6$, we have $2x - 6 = 2(x - 3)$. We need only multiply the second rational expression in our problem by $\frac{2}{2}$ to have two expressions with the same denominator:

$$\frac{5}{2x-6} + \frac{x}{x-3} = \frac{5}{2(x-3)} + \frac{x}{x-3}$$

$$= \frac{5}{2(x-3)} + \frac{2}{2}\left(\frac{x}{x-3}\right)$$

$$= \frac{5}{2(x-3)} + \frac{2x}{2(x-3)}$$

$$= \frac{2x+5}{2(x-3)}$$

---

**EXAMPLE 6**    Add $\dfrac{1}{x+4} + \dfrac{8}{x^2-16}$.

**Solution**    After writing each denominator in factored form, we find that the least common denominator is $(x+4)(x-4)$. To change the first rational expression to an equivalent rational expression with the common denominator, we multiply its numerator and denominator by $x-4$:

$$\frac{1}{x+4} + \frac{8}{x^2-16}$$

$$= \frac{1}{x+4} + \frac{8}{(x+4)(x-4)} \qquad \text{Factor each denominator.}$$

$$= \frac{1}{x+4} \cdot \frac{x-4}{x-4} + \frac{8}{(x+4)(x-4)} \qquad \begin{array}{l}\text{Change to equivalent} \\ \quad \text{rational expressions.}\end{array}$$

$$= \frac{x-4}{(x+4)(x-4)} + \frac{8}{(x+4)(x-4)} \qquad \text{Simplify.}$$

$$= \frac{x+4}{(x+4)(x-4)} \qquad \text{Add numerators.}$$

$$= \frac{1}{x-4} \qquad \begin{array}{l}\text{Divide out common} \\ \quad \text{factor } x+4.\end{array}$$

**Note:**    In the last step we reduced the rational expression to lowest terms by dividing out the common factor of $x+4$.

---

**EXAMPLE 7**    Add $\dfrac{2}{x^2+5x+6} + \dfrac{x}{x^2-9}$.

**Solution**

**Step 1:** We factor each denominator and build the LCD from the factors:

$$\left.\begin{array}{l} x^2+5x+6 = (x+2)(x+3) \\ x^2-9 = (x+3)(x-3) \end{array}\right\} \qquad \text{LCD} = (x+2)(x+3)(x-3)$$

**Step 2:** Change to equivalent rational expressions:

$$\frac{2}{x^2+5x+6} = \frac{2}{(x+2)(x+3)} \cdot \frac{(x-3)}{(x-3)} = \frac{2x-6}{(x+2)(x+3)(x-3)}$$

$$\frac{x}{x^2-9} = \frac{x}{(x+3)(x-3)} \cdot \frac{(x+2)}{(x+2)} = \frac{x^2+2x}{(x+2)(x+3)(x-3)}$$

**Step 3:** Add numerators of the rational expressions produced in step 2:

$$\frac{2x - 6}{(x + 2)(x + 3)(x - 3)} + \frac{x^2 + 2x}{(x + 2)(x + 3)(x - 3)}$$

$$= \frac{x^2 + 4x - 6}{(x + 2)(x + 3)(x - 3)}$$

The numerator and denominator do not have any factors in common.

**E X A M P L E   8**   Subtract $\dfrac{x + 4}{2x + 10} - \dfrac{5}{x^2 - 25}$.

**Solution**   We begin by factoring each denominator:

$$\frac{x + 4}{2x + 10} - \frac{5}{x^2 - 25} = \frac{x + 4}{2(x + 5)} - \frac{5}{(x + 5)(x - 5)}$$

The LCD is $2(x + 5)(x - 5)$. Completing the problem, we have:

$$= \frac{x + 4}{2(x + 5)} \cdot \frac{(x - 5)}{(x - 5)} + \frac{-5}{(x + 5)(x - 5)} \cdot \frac{2}{2}$$

$$= \frac{x^2 - x - 20}{2(x + 5)(x - 5)} + \frac{-10}{2(x + 5)(x - 5)}$$

$$= \frac{x^2 - x - 30}{2(x + 5)(x - 5)}$$

To see if this expression will reduce, we factor the numerator into $(x - 6)(x + 5)$:

$$= \frac{(x - 6)\cancel{(x + 5)}}{2\cancel{(x + 5)}(x - 5)}$$

$$= \frac{x - 6}{2(x - 5)}$$

**Note:**   In the first step we replaced subtraction by addition of the opposite. There seems to be less chance for error when this is done on longer problems.

**E X A M P L E   9**   Write an expression for the sum of a number and its reciprocal, and then simplify that expression.

**Solution**   If we let $x$ = the number, then its reciprocal is $\dfrac{1}{x}$. To find the sum of the number and its reciprocal, we add them:

$$x + \frac{1}{x}$$

The first term $x$ can be thought of as having a denominator of 1. Since the denominators are 1 and $x$, the least common denominator is $x$.

$$x + \frac{1}{x} = \frac{x}{1} + \frac{1}{x} \qquad \text{Write } x \text{ as } \frac{x}{1}.$$

$$= \frac{x}{1} \cdot \frac{x}{x} + \frac{1}{x} \qquad \text{The LCD is } x.$$

$$= \frac{x^2}{x} + \frac{1}{x}$$

$$= \frac{x^2 + 1}{x} \qquad \text{Add numerators.}$$

 ## Getting Ready for Class

*After reading through the preceding section, respond in your own words and in complete sentences.*

**A.** How do we add two rational expressions that have the same denominator?

**B.** What is the least common denominator for two fractions?

**C.** What role does factoring play in finding a least common denominator?

**D.** Explain how to find a common denominator for two rational expressions.

## PROBLEM SET 7.3

Find the following sums and differences.

**1.** $\dfrac{3}{x} + \dfrac{4}{x}$

**2.** $\dfrac{5}{x} + \dfrac{3}{x}$

**3.** $\dfrac{9}{a} - \dfrac{5}{a}$

**4.** $\dfrac{8}{a} - \dfrac{7}{a}$

**5.** $\dfrac{1}{x+1} + \dfrac{x}{x+1}$

**6.** $\dfrac{x}{x-3} - \dfrac{3}{x-3}$

**7.** $\dfrac{y^2}{y-1} - \dfrac{1}{y-1}$

**8.** $\dfrac{y^2}{y+3} - \dfrac{9}{y+3}$

**9.** $\dfrac{x^2}{x+2} + \dfrac{4x+4}{x+2}$

**10.** $\dfrac{x^2-6x}{x-3} + \dfrac{9}{x-3}$

**11.** $\dfrac{x^2}{x-2} - \dfrac{4x-4}{x-2}$

**12.** $\dfrac{x^2}{x-5} - \dfrac{10x-25}{x-5}$

**13.** $\dfrac{x+2}{x+6} - \dfrac{x-4}{x+6}$

**14.** $\dfrac{x+5}{x+2} - \dfrac{x+3}{x+2}$

**15.** $\dfrac{y}{2} - \dfrac{2}{y}$

**16.** $\dfrac{3}{y} + \dfrac{y}{3}$

**17.** $\dfrac{1}{2} + \dfrac{a}{3}$

**18.** $\dfrac{2}{3} + \dfrac{2a}{5}$

**19.** $\dfrac{x}{x+1} + \dfrac{3}{4}$

**20.** $\dfrac{x}{x-3} + \dfrac{1}{3}$

**21.** $\dfrac{x+1}{x-2} - \dfrac{4x+7}{5x-10}$

**22.** $\dfrac{3x+1}{2x-6} - \dfrac{x+2}{x-3}$

**23.** $\dfrac{4x-2}{3x+12} - \dfrac{x-2}{x+4}$

**24.** $\dfrac{6x+5}{5x-25} - \dfrac{x+2}{x-5}$

**25.** $\dfrac{6}{x(x-2)} + \dfrac{3}{x}$

**26.** $\dfrac{10}{x(x+5)} - \dfrac{2}{x}$

**27.** $\dfrac{4}{a} - \dfrac{12}{a^2+3a}$

**28.** $\dfrac{5}{a} + \dfrac{20}{a^2-4a}$

**29.** $\dfrac{2}{x+5} - \dfrac{10}{x^2-25}$

**30.** $\dfrac{6}{x^2-1} + \dfrac{3}{x+1}$

**31.** $\dfrac{x-4}{x-3} + \dfrac{6}{x^2-9}$       **32.** $\dfrac{x+1}{x-1} - \dfrac{4}{x^2-1}$

**33.** $\dfrac{a-4}{a-3} + \dfrac{5}{a^2-a-6}$

**34.** $\dfrac{a+2}{a+1} + \dfrac{7}{a^2-5a-6}$

**35.** $\dfrac{8}{x^2-16} - \dfrac{7}{x^2-x-12}$

**36.** $\dfrac{6}{x^2-9} - \dfrac{5}{x^2-x-6}$

**37.** $\dfrac{4y}{y^2+6y+5} - \dfrac{3y}{y^2+5y+4}$

**38.** $\dfrac{3y}{y^2+7y+10} - \dfrac{2y}{y^2+6y+8}$

**39.** $\dfrac{4x+1}{x^2+5x+4} - \dfrac{x+3}{x^2+4x+3}$

**40.** $\dfrac{2x-1}{x^2+x-6} - \dfrac{x+2}{x^2+5x+6}$

**41.** $\dfrac{1}{x} + \dfrac{x}{3x+9} - \dfrac{3}{x^2+3x}$

**42.** $\dfrac{1}{x} + \dfrac{x}{2x+4} - \dfrac{2}{x^2+2x}$

Complete the following tables.

**43.**

| Number $x$ | Reciprocal $\dfrac{1}{x}$ | Sum $1+\dfrac{1}{x}$ | Sum $\dfrac{x+1}{x}$ |
|---|---|---|---|
| 1 | | | |
| 2 | | | |
| 3 | | | |
| 4 | | | |

**44.**

| Number $x$ | Reciprocal $\dfrac{1}{x}$ | Difference $1-\dfrac{1}{x}$ | Difference $\dfrac{x-1}{x}$ |
|---|---|---|---|
| 1 | | | |
| 2 | | | |
| 3 | | | |
| 4 | | | |

**45.**

| $x$ | $x+\dfrac{4}{x}$ | $\dfrac{x^2+4}{x}$ | $x+4$ |
|---|---|---|---|
| 1 | | | |
| 2 | | | |
| 3 | | | |
| 4 | | | |

**46.**

| $x$ | $2x+\dfrac{6}{x}$ | $\dfrac{2x^2+6}{x}$ | $2x+6$ |
|---|---|---|---|
| 1 | | | |
| 2 | | | |
| 3 | | | |
| 4 | | | |

Add and subtract as indicated.

**47.** $1 + \dfrac{1}{x+2}$       **48.** $1 - \dfrac{1}{x+2}$

**49.** $1 - \dfrac{1}{x+3}$       **50.** $1 + \dfrac{1}{x+3}$

## Applying the Concepts

**51. Number Problem**   Write an expression for the sum of a number and twice its reciprocal. Then, simplify that expression. (If the reciprocal of a number is $\dfrac{1}{x}$, then twice that is $\dfrac{2}{x}$, not $\dfrac{1}{2x}$.)

**52. Number Problem**   Write an expression for the sum of a number and 3 times its reciprocal. Then, simplify that expression.

**53. Number Problem**   One number is twice another. Write an expression for the sum of their reciprocals. Then, simplify that expression. (*Hint:* The numbers are $x$ and $2x$. Their reciprocals are, respectively, $\dfrac{1}{x}$ and $\dfrac{1}{2x}$.)

**54. Number Problem**   One number is three times another. Write an expression for the sum of their reciprocals. Then, simplify that expression.

## Review Problems

The following problems review material we covered in Sections 2.4 and 6.6. Reviewing these problems will help you understand the next section.

Solve each equation. [2.4]

**55.** $2x + 3(x - 3) = 6$     **56.** $4x - 2(x - 5) = 6$

**57.** $x - 3(x + 3) = x - 3$

**58.** $x - 4(x + 4) = x - 4$

**59.** $7 - 2(3x + 1) = 4x + 3$

**60.** $8 - 5(2x - 1) = 2x + 4$

Solve each quadratic equation. [6.6]

**61.** $x^2 + 5x + 6 = 0$     **62.** $x^2 - 5x + 6 = 0$

**63.** $x^2 - x = 6$     **64.** $x^2 + x = 6$

**65.** $x^2 - 5x = 0$     **66.** $x^2 - 6x = 0$

---

### 7.4  Equations Involving Rational Expressions

The first step in solving an equation that contains one or more rational expressions is to find the LCD for all denominators in the equation. Once the LCD has been found, we multiply both sides of the equation by it. The resulting equation should be equivalent to the original one (unless we inadvertently multiplied by zero) and free from any denominators except the number 1.

**EXAMPLE  I**   Solve $\dfrac{x}{3} + \dfrac{5}{2} = \dfrac{1}{2}$ for $x$.

**Solution**   The LCD for 3 and 2 is 6. If we multiply both sides by 6, we have:

$$6\left(\frac{x}{3} + \frac{5}{2}\right) = 6\left(\frac{1}{2}\right) \qquad \text{Multiply both sides by 6.}$$

$$6\left(\frac{x}{3}\right) + 6\left(\frac{5}{2}\right) = 6\left(\frac{1}{2}\right) \qquad \text{Distributive property}$$

$$2x + 15 = 3$$

$$2x = -12$$

$$x = -6$$

We can check our solution by replacing $x$ with $-6$ in the original equation:

$$-\frac{6}{3} + \frac{5}{2} \overset{?}{=} \frac{1}{2}$$

$$\frac{1}{2} = \frac{1}{2}$$

---

Multiplying both sides of an equation containing fractions by the LCD clears the equation of all denominators, because the LCD has the property that all denominators will divide it evenly.

**E X A M P L E   2**    Solve $\dfrac{3}{x-1} = \dfrac{3}{5}$ for $x$.

**Solution**    The LCD for $(x - 1)$ and 5 is $5(x - 1)$. Multiplying both sides by $5(x - 1)$, we have:

$$5(x-1) \cdot \frac{3}{x-1} = 5(x-1) \cdot \frac{3}{5}$$

$$5 \cdot 3 = (x-1) \cdot 3$$

$$15 = 3x - 3$$

$$18 = 3x$$

$$6 = x$$

If we substitute $x = 6$ into the original equation, we have:

$$\frac{3}{6-1} \overset{?}{=} \frac{3}{5}$$

$$\frac{3}{5} = \frac{3}{5}$$

The solution set is $\{6\}$.

**E X A M P L E   3**    Solve $1 - \dfrac{5}{x} = \dfrac{-6}{x^2}$.

**Solution**    The LCD is $x^2$. Multiplying both sides by $x^2$, we have:

$$x^2\left(1 - \frac{5}{x}\right) = x^2\left(\frac{-6}{x^2}\right) \qquad \text{Multiply both sides by } x^2.$$

$$x^2(1) - x^2\left(\frac{5}{x}\right) = x^2\left(\frac{-6}{x^2}\right) \qquad \begin{array}{l}\text{Apply distributive property}\\ \text{to the left side.}\end{array}$$

$$x^2 - 5x = -6 \qquad \text{Simplify each side.}$$

We have a quadratic equation, which we write in standard form, factor, and solve as we did in Section 6.6:

$$x^2 - 5x + 6 = 0 \qquad \text{Standard form}$$

$$(x - 2)(x - 3) = 0 \qquad \text{Factor.}$$

$$x - 2 = 0 \quad \text{or} \quad x - 3 = 0 \qquad \text{Set factors equal to 0.}$$

$$x = 2 \quad \text{or} \quad x = 3$$

The two possible solutions are 2 and 3. Checking each in the original equation, we find they both give true statements. They are both solutions to the original equation:

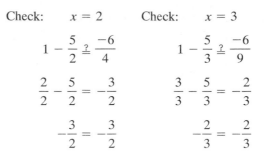

$$\text{EXAMPLE } 4 \quad \text{Solve } \frac{x}{x^2 - 9} - \frac{3}{x - 3} = \frac{1}{x + 3}.$$

**Solution**   The factors of $x^2 - 9$ are $(x + 3)(x - 3)$. The LCD, then, is $(x + 3)(x - 3)$:

$$(x + 3)(x - 3) \cdot \frac{x}{(x + 3)(x - 3)} + (x + 3)(x - 3) \cdot \frac{-3}{x - 3}$$

$$= (x + 3)(x - 3) \cdot \frac{1}{x + 3}$$

$$x + (x + 3)(-3) = (x - 3)1$$

$$x + (-3x) + (-9) = x - 3$$

$$-2x - 9 = x - 3$$

$$-3x = 6$$

$$x = -2$$

The solution is $x = -2$. It checks when replaced for $x$ in the original equation.

$$\text{EXAMPLE } 5 \quad \text{Solve } \frac{x}{x - 3} + \frac{3}{2} = \frac{3}{x - 3}.$$

**Solution**   We begin by multiplying each term on both sides of the equation by $2(x - 3)$:

$$2(x - 3) \cdot \frac{x}{x - 3} + 2(x - 3) \cdot \frac{3}{2} = 2(x - 3) \cdot \frac{3}{x - 3}$$

$$2x + (x - 3) \cdot 3 = 2 \cdot 3$$

$$2x + 3x - 9 = 6$$

$$5x - 9 = 6$$

$$5x = 15$$

$$x = 3$$

Our only possible solution is $x = 3$. If we substitute $x = 3$ into our original equation, we get:

$$\frac{3}{3-3} + \frac{3}{2} \overset{?}{=} \frac{3}{3-3}$$

$$\frac{3}{0} + \frac{3}{2} = \frac{3}{0}$$

Two of the terms are undefined, so the equation is meaningless. What has happened is that we have multiplied both sides of the original equation by zero. The equation produced by doing this is not equivalent to our original equation. We must always check our solution when we multiply both sides of an equation by an expression containing the variable in order to make sure we have not multiplied both sides by zero.

Our original equation has no solutions. That is, there is no real number $x$ such that:

$$\frac{x}{x-3} + \frac{3}{2} = \frac{3}{x-3}$$

The solution set is $\varnothing$.

**EXAMPLE 6**    Solve $\dfrac{a+4}{a^2+5a} = \dfrac{-2}{a^2-25}$ for $a$.

**Solution**    Factoring each denominator, we have:

$$a^2 + 5a = a(a + 5)$$

$$a^2 - 25 = (a + 5)(a - 5)$$

The LCD is $a(a + 5)(a - 5)$. Multiplying both sides of the equation by the LCD gives us:

$$a(a+5)(a-5) \cdot \frac{a+4}{a(a+5)} = \frac{-2}{(a+5)(a-5)} \cdot a(a+5)(a-5)$$

$$(a - 5)(a + 4) = -2a$$

$$a^2 - a - 20 = -2a$$

The result is a quadratic equation, which we write in standard form, factor, and solve:

$$a^2 + a - 20 = 0 \qquad \text{Add } 2a \text{ to both sides.}$$

$$(a + 5)(a - 4) = 0 \qquad \text{Factor.}$$

$$a + 5 = 0 \qquad \text{or} \qquad a - 4 = 0 \qquad \text{Set each factor to 0.}$$

$$a = -5 \qquad \text{or} \qquad a = 4$$

The two possible solutions are $-5$ and 4. There is no problem with the 4. It checks when substituted for $a$ in the original equation. However, $-5$ is not a solution. Substituting $-5$ into the original equation gives:

$$\frac{-5 + 4}{(-5)^2 + 5(-5)} \overset{?}{=} \frac{-2}{(-5)^2 - 25}$$

$$\frac{-1}{0} = \frac{-2}{0}$$

This indicates $-5$ is not a solution. The solution is 4.

## Getting Ready for Class

*After reading through the preceding section, respond in your own words and in complete sentences.*

**A.** What is the first step in solving an equation that contains rational expressions?

**B.** Explain how to find the LCD used to clear an equation of fractions.

**C.** When will a possible solution to an equation containing rational expressions not be an actual solution?

**D.** When do we check for extraneous solutions to an equation containing rational expressions?

## PROBLEM SET 7.4

Solve the following equations. Be sure to check each answer in the original equation if you multiply both sides by an expression that contains the variable.

**1.** $\dfrac{x}{3} + \dfrac{1}{2} = -\dfrac{1}{2}$

**2.** $\dfrac{x}{2} + \dfrac{4}{3} = -\dfrac{2}{3}$

**3.** $\dfrac{4}{a} = \dfrac{1}{5}$

**4.** $\dfrac{2}{3} = \dfrac{6}{a}$

**5.** $\dfrac{3}{x} + 1 = \dfrac{2}{x}$

**6.** $\dfrac{4}{x} + 3 = \dfrac{1}{x}$

**7.** $\dfrac{3}{a} - \dfrac{2}{a} = \dfrac{1}{5}$

**8.** $\dfrac{7}{a} + \dfrac{1}{a} = 2$

**9.** $\dfrac{3}{x} + 2 = \dfrac{1}{2}$

**10.** $\dfrac{5}{x} + 3 = \dfrac{4}{3}$

**11.** $\dfrac{1}{y} - \dfrac{1}{2} = -\dfrac{1}{4}$

**12.** $\dfrac{3}{y} - \dfrac{4}{5} = -\dfrac{1}{5}$

**13.** $1 - \dfrac{8}{x} = \dfrac{-15}{x^2}$

**14.** $1 - \dfrac{3}{x} = \dfrac{-2}{x^2}$

**15.** $\dfrac{x}{2} - \dfrac{4}{x} = -\dfrac{7}{2}$

**16.** $\dfrac{x}{2} - \dfrac{5}{x} = -\dfrac{3}{2}$

**17.** $\dfrac{x - 3}{2} + \dfrac{2x}{3} = \dfrac{5}{6}$

**18.** $\dfrac{x - 2}{3} + \dfrac{5x}{2} = 5$

**19.** $\dfrac{x + 1}{3} + \dfrac{x - 3}{4} = \dfrac{1}{6}$

**20.** $\dfrac{x + 2}{3} + \dfrac{x - 1}{5} = -\dfrac{3}{5}$

**21.** $\dfrac{6}{x + 2} = \dfrac{3}{5}$

**22.** $\dfrac{4}{x + 3} = \dfrac{1}{2}$

**23.** $\dfrac{3}{y - 2} = \dfrac{2}{y - 3}$

**24.** $\dfrac{5}{y + 1} = \dfrac{4}{y + 2}$

**25.** $\dfrac{x}{x-2} + \dfrac{2}{3} = \dfrac{2}{x-2}$

**26.** $\dfrac{x}{x-5} + \dfrac{1}{5} = \dfrac{5}{x-5}$

**27.** $\dfrac{x}{x-2} + \dfrac{3}{2} = \dfrac{9}{2(x-2)}$

**28.** $\dfrac{x}{x+1} + \dfrac{4}{5} = \dfrac{-14}{5(x+1)}$

**29.** $\dfrac{5}{x+2} + \dfrac{1}{x+3} = \dfrac{-1}{x^2+5x+6}$

**30.** $\dfrac{3}{x-1} + \dfrac{2}{x+3} = \dfrac{-3}{x^2+2x-3}$

**31.** $\dfrac{8}{x^2-4} + \dfrac{3}{x+2} = \dfrac{1}{x-2}$

**32.** $\dfrac{10}{x^2-25} - \dfrac{1}{x-5} = \dfrac{3}{x+5}$

**33.** $\dfrac{a}{2} + \dfrac{3}{a-3} = \dfrac{a}{a-3}$

**34.** $\dfrac{a}{2} + \dfrac{4}{a-4} = \dfrac{a}{a-4}$

**35.** $\dfrac{6}{y^2-4} = \dfrac{4}{y^2+2y}$    **36.** $\dfrac{2}{y^2-9} = \dfrac{5}{y^2-3y}$

**37.** $\dfrac{2}{a^2-9} = \dfrac{3}{a^2+a-12}$

**38.** $\dfrac{2}{a^2-1} = \dfrac{6}{a^2-2a-3}$

**39.** $\dfrac{3x}{x-5} - \dfrac{2x}{x+1} = \dfrac{-42}{x^2-4x-5}$

**40.** $\dfrac{4x}{x-4} - \dfrac{3x}{x-2} = \dfrac{-3}{x^2-6x+8}$

**41.** $\dfrac{2x}{x+2} = \dfrac{x}{x+3} - \dfrac{3}{x^2+5x+6}$

**42.** $\dfrac{3x}{x-4} = \dfrac{2x}{x-3} + \dfrac{6}{x^2-7x+12}$

**43.** The following table is the answer to Problem 45 in Problem Set 7.3. Solve the equation

$$x + \dfrac{4}{x} = 5$$

and compare your solutions with the entries in the table.

| $x$ | $x + \dfrac{4}{x}$ | $\dfrac{x^2+4}{x}$ | $x + 4$ |
|-----|--------------------|--------------------|---------|
| 1   | 5                  | 5                  | 5       |
| 2   | 4                  | 4                  | 6       |
| 3   | $\dfrac{13}{3}$    | $\dfrac{13}{3}$    | 7       |
| 4   | 5                  | 5                  | 8       |

**44.** The following table is the answer to Problem 46 in Problem Set 7.3. Solve the equation

$$2x + \dfrac{6}{x} = 8$$

and compare your solutions with the entries in the table.

| $x$ | $2x + \dfrac{6}{x}$ | $\dfrac{2x^2+6}{x}$ | $2x + 6$ |
|-----|---------------------|---------------------|----------|
| 1   | 8                   | 8                   | 8        |
| 2   | 7                   | 7                   | 10       |
| 3   | 8                   | 8                   | 12       |
| 4   | $\dfrac{19}{2}$     | $\dfrac{19}{2}$     | 14       |

## Review Problems

The problems that follow review material we covered in Sections 2.6 and 6.7. Reviewing these problems will help you with the next section.

Solve each word problem. [2.6]

**45. Number Problem**    If twice the difference of a number and 3 were decreased by 5, the result would be 3. Find the number.

**46. Number Problem**    If 3 times the sum of a number and 2 were increased by 6, the result would be 27. Find the number.

**47. Geometry**    The length of a rectangle is 5 more than twice the width. The perimeter is 34 inches. Find the length and width.

**48. Geometry**    The length of a rectangle is 2 more than 3 times the width. The perimeter is 44 feet. Find the length and width.

Solve each problem. Be sure to show the equation that describes the situation. [6.7]

**49. Number Problem** The product of two consecutive even integers is 48. Find the two integers.

**50. Number Problem** The product of two consecutive odd integers is 35. Find the two integers.

**51. Geometry** The hypotenuse (the longest side) of a

right triangle is 10 inches, and the lengths of the two legs (the other two sides) are given by two consecutive even integers. Find the lengths of the two legs.

**52. Geometry** One leg of a right triangle is 2 more than twice the other. If the hypotenuse is 13 feet, find the lengths of the two legs.

---

## 7.5  Applications

In this section we will solve some word problems whose equations involve rational expressions. Like the other word problems we have encountered, the more you work with them, the easier they become.

**EXAMPLE 1** One number is twice another. The sum of their reciprocals is $\frac{9}{2}$. Find the two numbers.

**Solution** Let $x =$ the smaller number. The larger then must be $2x$. Their reciprocals are $\dfrac{1}{x}$ and $\dfrac{1}{2x}$, respectively. An equation that describes the situation is:

$$\frac{1}{x} + \frac{1}{2x} = \frac{9}{2}$$

We can multiply both sides by the LCD $2x$ and then solve the resulting equation:

$$2x\left(\frac{1}{x}\right) + 2x\left(\frac{1}{2x}\right) = 2x\left(\frac{9}{2}\right)$$

$$2 + 1 = 9x$$

$$3 = 9x$$

$$x = \frac{3}{9} = \frac{1}{3}$$

The smaller number is $\frac{1}{3}$. The other number is twice as large, or $\frac{2}{3}$. If we add their reciprocals, we have:

$$\frac{3}{1} + \frac{3}{2} = \frac{6}{2} + \frac{3}{2} = \frac{9}{2}$$

The solutions check with the original problem.

**EXAMPLE 2** A boat travels 30 miles up a river in the same amount of time it takes to travel 50 miles down the same river. If the current is 5 miles per hour, what is the speed of the boat in still water?

**Solution**    The easiest way to work a problem like this is with a table. The top row of the table is labeled with $d$ for distance, $r$ for rate, and $t$ for time. The left column of the table is labeled with the two trips: upstream and downstream. Here is what the table looks like:

|            | $d$ | $r$ | $t$ |
|------------|-----|-----|-----|
| Upstream   |     |     |     |
| Downstream |     |     |     |

The next step is to read the problem over again and fill in as much of the table as we can with the information in the problem. The distance the boat travels upstream is 30 and the distance downstream is 50. Since we are asked for the speed of the boat in still water, we will let that be $x$. If the speed of the boat in still water is $x$, then its speed upstream (against the current) must be $x - 5$, and its speed downstream (with the current) must be $x + 5$. Putting these four quantities into the appropriate positions in the table, we have

|            | $d$ | $r$     | $t$ |
|------------|-----|---------|-----|
| Upstream   | 30  | $x - 5$ |     |
| Downstream | 50  | $x + 5$ |     |

The last positions in the table are filled in by using the equation $t = \dfrac{d}{r}$.

|            | $d$ | $r$     | $t$               |
|------------|-----|---------|-------------------|
| Upstream   | 30  | $x - 5$ | $\dfrac{30}{x - 5}$ |
| Downstream | 50  | $x + 5$ | $\dfrac{50}{x + 5}$ |

Reading the problem again, we find that the time for the trip upstream is equal to the time for the trip downstream. Setting these two quantities equal to each other, we have our equation:

$$\text{Time (downstream)} = \text{time (upstream)}$$

$$\frac{50}{x + 5} = \frac{30}{x - 5}$$

The LCD is $(x + 5)(x - 5)$. We multiply both sides of the equation by the LCD to clear it of all denominators. Here is the solution:

$$(x + 5)(x - 5) \cdot \frac{50}{x + 5} = (x + 5)(x - 5) \cdot \frac{30}{x - 5}$$

$$50x - 250 = 30x + 150$$

$$20x = 400$$

$$x = 20$$

The speed of the boat in still water is 20 miles per hour.

**EXAMPLE 3**    Tina is training for a biathlon. To train for the bicycle portion, she rides her bike 15 miles up a hill and then 15 miles back down the same hill. The complete trip takes her 2 hours. If her downhill speed is 20 miles per hour faster than her uphill speed, how fast does she ride uphill?

**Up and back**
**total time = 2 hours**

**Solution**    Again, we make a table. As in the previous example, we label the top row with distance, rate, and time. We label the left column with the two trips, uphill and downhill.

|          | *d* | *r* | *t* |
|----------|-----|-----|-----|
| Uphill   |     |     |     |
| Downhill |     |     |     |

Next, we fill in the table with as much information as we can from the problem. We know the distance traveled is 15 miles uphill and 15 miles downhill, which allows us to fill in the distance column. To fill in the rate column, we first note that she rides 20 miles per hour faster downhill than uphill. Therefore, if we let $x$ equal her rate uphill, then her rate downhill is $x + 20$. Filling in the table with this information gives us

|          | d  | r      | t |
|----------|----|--------|---|
| Uphill   | 15 | $x$    |   |
| Downhill | 15 | $x + 20$ |   |

Since time is distance divided by rate, $t = \dfrac{d}{r}$, we can fill in the last column in the table.

|          | d  | r        | t                    |
|----------|----|----------|----------------------|
| Uphill   | 15 | $x$      | $\dfrac{15}{x}$      |
| Downhill | 15 | $x + 20$ | $\dfrac{15}{x + 20}$ |

Rereading the problem, we find that the total time (the time riding uphill plus the time riding downhill) is 2 hours. We write our equation as follows:

$$\text{Time (uphill)} + \text{time (downhill)} = 2$$

$$\frac{15}{x} + \frac{15}{x + 20} = 2$$

We solve this equation for $x$ by first finding the LCD and then multiplying each term in the equation by it in order to clear the equation of all denominators. Our LCD is $x(x + 20)$. Here is our solution:

$$x(x + 20)\frac{15}{x} + x(x + 20)\frac{15}{x + 20} = 2 \cdot [x(x + 20)]$$

$$15(x + 20) + 15x = 2x(x + 20)$$

$$15x + 300 + 15x = 2x^2 + 40x$$

$$0 = 2x^2 + 10x - 300$$

$$0 = x^2 + 5x - 150 \qquad \text{Divide both}$$

$$0 = (x + 15)(x - 10) \qquad \text{sides by 2.}$$

$$x = -15 \qquad \text{or} \qquad x = 10$$

Since we cannot have a negative speed, our only solution is $x = 10$. Tina rides her bike at a rate of 10 miles per hour when going uphill. (Her downhill speed is $x + 20 = 30$ miles per hour.)

**EXAMPLE 4**    An inlet pipe can fill a water tank in 10 hours, and an outlet pipe can empty the same tank in 15 hours. By mistake, both pipes are left open. How long will it take to fill the water tank with both pipes open?

**Solution**    Let $x$ = amount of time to fill the tank with both pipes open.

One method of solving this type of problem is to think in terms of how much of the job is done by a pipe in 1 hour.

**1.** If the inlet pipe fills the tank in 10 hours, then in 1 hour the inlet pipe fills $\frac{1}{10}$ of the tank.

**2.** If the outlet pipe empties the tank in 15 hours, then in 1 hour the outlet pipe empties $\frac{1}{15}$ of the tank.

**3.** If it takes $x$ hours to fill the tank with both pipes open, then in 1 hour the tank is $\frac{1}{x}$ full.

Here is how we set up the equation. In one hour,

$$\frac{1}{10} \quad - \quad \frac{1}{15} \quad = \quad \frac{1}{x}$$

Amount of water let      Amount of water let      Total amount of
in by inlet pipe          out by outlet pipe       water in tank

The LCD for our equation is $30x$. We multiply both sides by the LCD and solve:

$$30x\left(\frac{1}{10}\right) - 30x\left(\frac{1}{15}\right) = 30x\left(\frac{1}{x}\right)$$

$$3x - 2x = 30$$

$$x = 30$$

It takes 30 hours with both pipes open to fill the tank.

**Note:**    In solving a problem of this type, we have to assume that the thing doing the work (whether it is a pipe, a person, or a machine) is working at a constant rate. That is, as much work gets done in the first hour as is done in the last hour and any other hour in between.

**EXAMPLE 5**    Graph the equation $y = \dfrac{1}{x}$.

**Solution**    Since this is the first time we have graphed an equation of this form, we will make a table of values for $x$ and $y$ that satisfy the equation. Before we do, let's make some generalizations about the graph (Figure 1).

First, notice that, since $y$ is equal to 1 divided by $x$, $y$ will be positive when $x$ is positive. (The quotient of two positive numbers is a positive number.) Likewise, when $x$ is negative, $y$ will be negative. In other words, $x$ and $y$ will always have the same sign. Thus, our graph will appear in quadrants 1 and 3 only, because in those quadrants $x$ and $y$ have the same sign.

Next, notice that the expression $\dfrac{1}{x}$ will be undefined when $x$ is 0, meaning that there is no value of $y$ corresponding to $x = 0$. Because of this, the graph will not cross the $y$-axis. Further, the graph will not cross the $x$-axis either. If we try to find the $x$-intercept by letting $y = 0$, we have

$$0 = \frac{1}{x}$$

But there is no value of $x$ to divide into 1 to obtain 0. Therefore, since there is no solution to this equation, our graph will not cross the $x$-axis.

To summarize, we can expect to find the graph in quadrants I and III only, and the graph will cross neither axis.

| $x$ | $y$ |
|-----|-----|
| $-3$ | $-\frac{1}{3}$ |
| $-2$ | $-\frac{1}{2}$ |
| $-1$ | $-1$ |
| $-\frac{1}{2}$ | $-2$ |
| $-\frac{1}{3}$ | $-3$ |
| $0$ | Undefined |
| $\frac{1}{3}$ | $3$ |
| $\frac{1}{2}$ | $2$ |
| $1$ | $1$ |
| $2$ | $\frac{1}{2}$ |
| $3$ | $\frac{1}{3}$ |

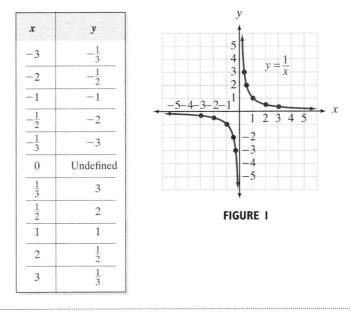

**FIGURE 1**

**EXAMPLE 6**    Graph the equation $y = \dfrac{-6}{x}$.

**Solution**    Since $y$ is $-6$ divided by $x$, when $x$ is positive, $y$ will be negative (a negative divided by a positive is negative), and when $x$ is negative, $y$ will be positive (a negative divided by a negative). Thus, the graph (Figure 2) will appear in quadrants II and IV only. As was the case in Example 5, the graph will not cross either axis.

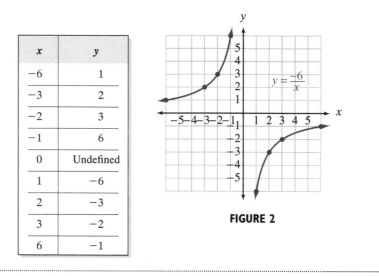

| $x$ | $y$ |
|-----|-----|
| $-6$ | 1 |
| $-3$ | 2 |
| $-2$ | 3 |
| $-1$ | 6 |
| 0 | Undefined |
| 1 | $-6$ |
| 2 | $-3$ |
| 3 | $-2$ |
| 6 | $-1$ |

**FIGURE 2**

## Getting Ready for Class

*After reading through the preceding section, respond in your own words and in complete sentences.*

**A.** Write an application problem for which the solution depends on solving the equation $\dfrac{1}{2} + \dfrac{1}{3} = \dfrac{1}{x}$.

**B.** How does the current of a river affect the speed of a motor boat traveling against the current?

**C.** How does the current of a river affect the speed of a motor boat traveling in the same direction as the current?

**D.** What is the relationship between the total number of minutes it takes for a drain to empty a sink and the amount of water that drains out of the sink in 1 minute?

# PROBLEM SET 7.5

## Number Problems

**1.** One number is 3 times as large as another. The sum of their reciprocals is $\dfrac{16}{3}$. Find the two numbers.

**2.** If $\dfrac{3}{5}$ is added to twice the reciprocal of a number, the result is 1. Find the number.

**3.** The sum of a number and its reciprocal is $\dfrac{13}{6}$. Find the number.

**4.** The sum of a number and 10 times its reciprocal is 7. Find the number.

**5.** If a certain number is added to both the numerator and denominator of the fraction $\dfrac{7}{9}$, the result is $\dfrac{5}{7}$. Find the number.

**6.** The numerator of a certain fraction is 2 more than the denominator. If $\dfrac{1}{3}$ is added to the fraction, the result is 2. Find the fraction.

**7.** The sum of the reciprocals of two consecutive even integers is $\dfrac{5}{12}$. Find the integers.

**8.** The sum of the reciprocals of two consecutive integers is $\dfrac{7}{12}$. Find the two integers.

## Motion Problems

9. A boat travels 26 miles up a river in the same amount of time it takes to travel 38 miles down the same river. If the current is 3 miles per hour, what is the speed of the boat in still water?

|  | *d* | *r* | *t* |
|---|---|---|---|
| Upstream |  |  |  |
| Downstream |  |  |  |

10. A boat can travel 9 miles up a river in the same amount of time it takes to travel 11 miles down the same river. If the current is 2 miles per hour, what is the speed of the boat in still water?

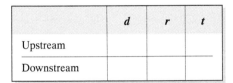

|  | *d* | *r* | *t* |
|---|---|---|---|
| Upstream |  |  |  |
| Downstream |  |  |  |

11. An airplane flying against the wind travels 140 miles in the amount of time it would take the same plane to travel 160 miles with the wind. If the wind speed is a constant 20 miles per hour, how fast would the plane travel in still air?

12. An airplane flying against the wind travels 500 miles in the amount of time that it would take to travel 600 miles with the wind. If the speed of the wind is 50 miles per hour, what is the speed of the plane in still air?

13. One plane can travel 20 miles per hour faster than another. One of them goes 285 miles in the time it takes the other to go 255 miles. What are their speeds?

14. One car travels 300 miles in the amount of time it takes a second car traveling 5 miles per hour slower than the first to go 275 miles. What are the speeds of the cars?

15. Tina, whom we mentioned in Example 3 of this section, is training for a biathlon. To train for the running portion of the race, she runs 8 miles each day over the same course. The first 2 miles of the course are on level ground; the last 6 miles are downhill. She runs 3 miles per hour slower on level ground than she runs downhill. If the complete

course takes 1 hour, how fast does she run on the downhill part of the course?

16. Jerri is training for the same biathlon as Tina (Example 3 and Problem 15). To train for the bicycle portion of the race, she rides 24 miles out a straight road, and then turns around and rides 24 miles back. The trip out is against the wind; the trip back is with the wind. If she rides 10 miles per hour faster with the wind than she does against the wind and the complete trip out and back takes 2 hours, how fast does she ride when she rides against the wind?

17. To train for the running of a triathlon, Jerri jogs 1 hour each day over the same 9-mile course. Five miles of the course are downhill, while the other 4 miles are on level ground. Jerri figures that she runs 2 miles per hour faster downhill than she runs on level ground. Find the rate at which Jerri runs on level ground.

18. Travis paddles his kayak in the harbor at Morro Bay, California, where the incoming tide has caused a current in the water. From the point where he enters the water, he paddles 1 mile against the current, and then turns around and paddles 1 mile back to where he started. His average speed when paddling with the current is 4 miles per hour faster than his speed against the current. If the complete trip (out and back) takes him 1.2 hours, find his average speed when he paddles against the current.

## Work Problems

19. An inlet pipe can fill a pool in 12 hours, and an outlet pipe can empty it in 15 hours. If both pipes are left open, how long will it take to fill the pool?

20. A water tank can be filled in 20 hours by an inlet pipe and emptied in 25 hours by an outlet pipe. How long will it take to fill the tank if both pipes are left open?

21. A bathtub can be filled by the cold-water faucet in 10 minutes and by the hot-water faucet in 12 minutes. How long does it take to fill the tub if both faucets are open?

22. A water faucet can fill a sink in 6 minutes, and the drain can empty it in 4 minutes. If the sink is full, how long will it take to empty if both the faucet and the drain are open?

23. A sink can be filled by the cold-water faucet in 3 minutes. The drain can empty a full sink in 4 min-

utes. If the sink is empty and both the cold-water faucet and the drain are open, how long will it take the sink to overflow?

**24.** A bathtub can be filled by the cold-water faucet in 9 minutes and by the hot-water faucet in 10 minutes. The drain can empty the tub in 5 minutes. Can the tub be filled if both faucets and the drain are open?

**25. Average Cost** As we mentioned in the introduction to this chapter, if a bank charges $2.00 per month and $0.15 per check for one of its checking accounts, then the average cost of each of the $x$ checks written is $A = \dfrac{2.00 + 0.15x}{x}$. The following table was constructed using this formula. Use the template to construct a line graph using the information in the table.

| Checks Written $x$ | Average Cost per Check $A = \dfrac{2.00 + 0.15x}{x}$ |
|---|---|
| 1 | 2.15 |
| 5 | 0.55 |
| 10 | 0.35 |
| 15 | 0.28 |
| 20 | 0.25 |
| 25 | 0.23 |

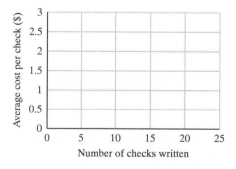

**26. Average Cost** A rewritable CD-ROM drive for a computer costs $250. An individual CD for the drive costs $5.00 and can store 640 megabytes of information. The average cost per megabyte to fill $x$ CDs with information is $A = \dfrac{5x + 250}{640x}$. The following table was constructed using this formula. Use the template to construct a line graph using the information in the table.

| CDs Purchased $x$ | Cost per Megabyte $\dfrac{5x + 250}{640x}$ |
|---|---|
| 1 | 0.40 |
| 5 | 0.09 |
| 10 | 0.05 |
| 15 | 0.03 |
| 20 | 0.03 |
| 25 | 0.02 |

Graph each of the following equations.

**27.** $y = \dfrac{-4}{x}$          **28.** $y = \dfrac{4}{x}$

**29.** $y = \dfrac{8}{x}$          **30.** $y = \dfrac{-8}{x}$

**31.** Graph $y = \dfrac{3}{x}$ and $x + y = 4$ on the same coordinate system. At what points do the two graphs intersect?

**32.** Graph $y = \dfrac{4}{x}$ and $x - y = 3$ on the same coordinate system. At what points do the two graphs intersect?

## Review Problems

The following problems review material we covered in Chapter 6.

**33.** Factor out the greatest common factor for
$$15a^3b^3 - 20a^2b - 35ab^2$$

**34.** Factor by grouping: $3ax - 2a + 15x - 10$

Factor completely.

**35.** $x^2 - 4x - 12$

**36.** $4x^2 - 20xy + 25y^2$

**37.** $x^4 - 16$

**38.** $2x^2 + xy - 21y^2$

**39.** $5x^3 - 25x^2 - 30x$

Solve each equation.

**40.** $x^2 - 9x + 18 = 0$

**41.** $x^2 - 6x = 0$

**42.** $8x^2 = -2x + 15$

**43.** $x(x + 2) = 80$

**44. Number Problem**   The product of two consecutive even integers is 4 more than twice their sum. Find the two integers.

**45. Geometry**   The hypotenuse of a right triangle is 15 inches. One of the legs is 3 inches more than the other. Find the lengths of the two legs.

## 7.6    Complex Fractions

A complex fraction is a fraction or rational expression that contains other fractions in its numerator or denominator. Each of the following is a complex fraction:

$$\frac{\dfrac{1}{2}}{\dfrac{2}{3}} \qquad \frac{x + \dfrac{1}{y}}{y + \dfrac{1}{x}} \qquad \frac{\dfrac{a + 1}{a^2 - 9}}{\dfrac{2}{a + 3}}$$

We will begin this section by simplifying the first of these complex fractions. Before we do, though, let's agree on some vocabulary. So that we won't have to use phrases such as the numerator of the denominator, let's call the numerator of a complex fraction the *top* and the denominator of a complex fraction the *bottom*.

**EXAMPLE 1**   Simplify $\dfrac{\dfrac{1}{2}}{\dfrac{2}{3}}$.

**Solution**   There are two methods we can use to solve this problem.

*Method 1*   We can multiply the top and bottom of this complex fraction by the LCD for both fractions. In this case the LCD is 6:

$$\frac{\dfrac{1}{2}}{\dfrac{2}{3}} = \frac{6 \cdot \dfrac{1}{2}}{6 \cdot \dfrac{2}{3}} = \frac{3}{4}$$

***Method 2***    We can treat this as a division problem. Instead of dividing by $\frac{2}{3}$, we can multiply by its reciprocal $\frac{3}{2}$:

$$\frac{\dfrac{1}{2}}{\dfrac{2}{3}} = \frac{1}{2} \cdot \frac{3}{2} = \frac{3}{4}$$

Using either method, we obtain the same result.

**E X A M P L E   2**    Simplify:

$$\frac{\dfrac{2x^3}{y^2}}{\dfrac{4x}{y^5}}$$

**Solution**

***Method 1***    The LCD for each rational expression is $y^5$. Multiplying the top and bottom of the complex fraction by $y^5$, we have:

$$\frac{\dfrac{2x^3}{y^2}}{\dfrac{4x}{y^5}} = \frac{y^5 \cdot \dfrac{2x^3}{y^2}}{y^5 \cdot \dfrac{4x}{y^5}} = \frac{2x^3 y^3}{4x} = \frac{x^2 y^3}{2}$$

***Method 2***    Instead of dividing by $\dfrac{4x}{y^5}$, we can multiply by its reciprocal, $\dfrac{y^5}{4x}$:

$$\frac{\dfrac{2x^3}{y^2}}{\dfrac{4x}{y^5}} = \frac{2x^3}{y^2} \cdot \frac{y^5}{4x} = \frac{x^2 y^3}{2}$$

Again the result is the same, whether we use Method 1 or Method 2.

**E X A M P L E   3**    Simplify:

$$\frac{x + \dfrac{1}{y}}{y + \dfrac{1}{x}}$$

**Solution**    To apply Method 2 as we did in the first two examples, we would have to simplify the top and bottom separately to obtain a single rational expression for both before we could multiply by the reciprocal. It is much easier, in this case, to multiply the top and bottom by the LCD $xy$:

$$\frac{x + \dfrac{1}{y}}{y + \dfrac{1}{x}} = \frac{xy\left(x + \dfrac{1}{y}\right)}{xy\left(y + \dfrac{1}{x}\right)} \qquad \text{Multiply top and bottom by } \mathbf{xy.}$$

$$= \frac{xy \cdot x + xy \cdot \dfrac{1}{y}}{xy \cdot y + xy \cdot \dfrac{1}{x}} \qquad \text{Distributive property}$$

$$= \frac{x^2 y + x}{xy^2 + y} \qquad \text{Simplify.}$$

We can factor an $x$ from $x^2 y + x$ and a $y$ from $xy^2 + y$ and then reduce to lowest terms:

$$= \frac{x(\cancel{xy + 1})}{y(\cancel{xy + 1})}$$

$$= \frac{x}{y}$$

**EXAMPLE 4**   Simplify:

$$\frac{1 - \dfrac{4}{x^2}}{1 - \dfrac{1}{x} - \dfrac{6}{x^2}}$$

**Solution**   The simplest way to simplify this complex fraction is to multiply the top and bottom by the LCD, $x^2$:

$$\frac{1 - \dfrac{4}{x^2}}{1 - \dfrac{1}{x} - \dfrac{6}{x^2}} = \frac{x^2\left(1 - \dfrac{4}{x^2}\right)}{x^2\left(1 - \dfrac{1}{x} - \dfrac{6}{x^2}\right)} \qquad \begin{array}{l}\text{Multiply top and} \\ \text{bottom by } x^2.\end{array}$$

$$= \frac{x^2 \cdot 1 - x^2 \cdot \dfrac{4}{x^2}}{x^2 \cdot 1 - x^2 \cdot \dfrac{1}{x} - x^2 \cdot \dfrac{6}{x^2}} \qquad \text{Distributive property}$$

$$= \frac{x^2 - 4}{x^2 - x - 6} \qquad \text{Simplify.}$$

$$= \frac{(x-2)(x+2)}{(x-3)(x+2)} \qquad \text{Factor.}$$

$$= \frac{x-2}{x-3} \qquad \text{Reduce.}$$

In our next example, we examine the relationship between the numbers in the Fibonacci sequence and a sequence involving complex frations.

**E X A M P L E    5**    Simplify each term in the following sequence, and then explain how this sequence is related to the Fibonacci sequence:

$$1 + \frac{1}{1+1}, 1 + \frac{1}{1 + \dfrac{1}{1+1}}, 1 + \frac{1}{1 + \dfrac{1}{1 + \dfrac{1}{1+1}}}, \ldots$$

**Solution**    We can simplify our work somewhat if we notice that the first term $1 + \dfrac{1}{1+1}$ is the larger denominator in the second term, and that the second term is the largest denominator in the third term:

First term:    $1 + \dfrac{1}{1+1} = 1 + \dfrac{1}{2} = \dfrac{2}{2} + \dfrac{1}{2} = \dfrac{3}{2}$

Second term:    $1 + \dfrac{1}{1 + \dfrac{1}{1+1}} = 1 + \dfrac{1}{\dfrac{3}{2}} = 1 + \dfrac{2}{3} = \dfrac{3}{3} + \dfrac{2}{3} = \dfrac{5}{3}$

Third term:    $1 + \dfrac{1}{1 + \dfrac{1}{1 + \dfrac{1}{1+1}}} = 1 + \dfrac{1}{\dfrac{5}{3}} = 1 + \dfrac{3}{5} = \dfrac{5}{5} + \dfrac{3}{5} = \dfrac{8}{5}$

Here are the simplified numbers for the first three terms in our sequence:

$$\frac{3}{2}, \frac{5}{3}, \frac{8}{5}, \ldots$$

Recall the Fibonacci sequence:

$$1, 1, 2, 3, 5, 8, 13, 21, \ldots$$

As you can see, each term in the sequence we have simplified is the ratio of two consecutive numbers in the Fibonacci sequence. If the pattern continues in this manner, the next number in our sequence will be $\dfrac{13}{8}$.

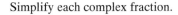 **Getting Ready for Class**

*After reading through the preceding section, respond in your own words and in complete sentences.*

**A.** What is a complex fraction?

**B.** Explain one method of simplifying complex fractions.

**C.** How is a least common denominator used to simplify a complex fraction?

**D.** What types of complex fractions can be rewritten as division problems?

## PROBLEM SET 7.6

Simplify each complex fraction.

**1.** $\dfrac{\frac{3}{4}}{\frac{1}{8}}$

**2.** $\dfrac{\frac{1}{3}}{\frac{5}{6}}$

**3.** $\dfrac{\frac{2}{3}}{\frac{3}{4}}$

**4.** $\dfrac{\frac{5}{1}}{\frac{1}{2}}$

**5.** $\dfrac{\frac{x^2}{y}}{\frac{x}{y^3}}$

**6.** $\dfrac{\frac{x^5}{y^3}}{\frac{x^2}{y^8}}$

**7.** $\dfrac{\frac{4x^3}{y^6}}{\frac{8x^2}{y^7}}$

**8.** $\dfrac{\frac{6x^4}{y}}{\frac{2x}{y^5}}$

**9.** $\dfrac{y + \frac{1}{x}}{x + \frac{1}{y}}$

**10.** $\dfrac{y - \frac{1}{x}}{x - \frac{1}{y}}$

**11.** $\dfrac{1 + \frac{1}{a}}{1 - \frac{1}{a}}$

**12.** $\dfrac{\frac{1}{a} - 1}{\frac{1}{a} + 1}$

**13.** $\dfrac{\frac{x + 1}{x^2 - 9}}{\frac{2}{x + 3}}$

**14.** $\dfrac{\frac{3}{x - 5}}{\frac{x + 1}{x^2 - 25}}$

**15.** $\dfrac{\frac{1}{a + 2}}{\frac{1}{a^2 - a - 6}}$

**16.** $\dfrac{\frac{1}{a^2 + 5a + 6}}{\frac{1}{a + 3}}$

**17.** $\dfrac{1 - \frac{9}{y^2}}{1 - \frac{1}{y} - \frac{6}{y^2}}$

**18.** $\dfrac{1 - \frac{4}{y^2}}{1 - \frac{2}{y} - \frac{8}{y^2}}$

**19.** $\dfrac{\frac{1}{y} + \frac{1}{x}}{\frac{1}{xy}}$

**20.** $\dfrac{\frac{1}{xy}}{\frac{1}{y} - \frac{1}{x}}$

**21.** $\dfrac{1 - \frac{1}{a^2}}{1 - \frac{1}{a}}$

**22.** $\dfrac{1 + \frac{1}{a}}{1 - \frac{1}{a^2}}$

**23.** $\dfrac{\frac{1}{10x} - \frac{y}{10x^2}}{\frac{1}{10} - \frac{y}{10x}}$

**24.** $\dfrac{\frac{1}{2x} + \frac{y}{2x^2}}{\frac{1}{4} + \frac{y}{4x}}$

**25.** $\dfrac{\frac{1}{a + 1} + 2}{\frac{1}{a + 1} + 3}$

**26.** $\dfrac{\frac{2}{a + 1} + 3}{\frac{3}{a + 1} + 4}$

Although the following problems do not contain complex fractions, they do involve more than one operation. Simplify inside the parentheses first, and then multiply.

**27.** $\left(1 - \dfrac{1}{x}\right)\left(1 - \dfrac{1}{x + 1}\right)\left(1 - \dfrac{1}{x + 2}\right)$

**28.** $\left(1 + \dfrac{1}{x}\right)\left(1 + \dfrac{1}{x + 1}\right)\left(1 + \dfrac{1}{x + 2}\right)$

**29.** $\left(1 + \dfrac{1}{x + 3}\right)\left(1 + \dfrac{1}{x + 2}\right)\left(1 + \dfrac{1}{x + 1}\right)$

**30.** $\left(1 - \dfrac{1}{x + 3}\right)\left(1 - \dfrac{1}{x + 2}\right)\left(1 - \dfrac{1}{x + 1}\right)$

**31.** Simplify each term in the following sequence.

$$2 + \frac{1}{2+1}, \; 2 + \frac{1}{2+\dfrac{1}{2+1}}, \; 2 + \frac{1}{2+\dfrac{1}{2+\dfrac{1}{2+1}}}, \; \ldots$$

**32.** Simplify each term in the following sequence.

$$2 + \frac{3}{2+3}, \; 2 + \frac{3}{2+\dfrac{3}{2+3}}, \; 2 + \frac{3}{2+\dfrac{3}{2+\dfrac{3}{2+3}}}, \; \ldots$$

Complete the following tables.

**33.**

| Number $x$ | Reciprocal $\dfrac{1}{x}$ | Quotient $\dfrac{x}{\dfrac{1}{x}}$ | Square $x^2$ |
|---|---|---|---|
| 1 | | | |
| 2 | | | |
| 3 | | | |
| 4 | | | |

**34.**

| Number $x$ | Reciprocal $\dfrac{1}{x}$ | Quotient $\dfrac{\dfrac{1}{x}}{x}$ | Square $x^2$ |
|---|---|---|---|
| 1 | | | |
| 2 | | | |
| 3 | | | |
| 4 | | | |

**35.**

| Number $x$ | Reciprocal $\dfrac{1}{x}$ | Sum $1+\dfrac{1}{x}$ | Quotient $\dfrac{1+\dfrac{1}{x}}{\dfrac{1}{x}}$ |
|---|---|---|---|
| 1 | | | |
| 2 | | | |
| 3 | | | |
| 4 | | | |

**36.**

| Number $x$ | Reciprocal $\dfrac{1}{x}$ | Difference $1-\dfrac{1}{x}$ | Quotient $\dfrac{1-\dfrac{1}{x}}{\dfrac{1}{x}}$ |
|---|---|---|---|
| 1 | | | |
| 2 | | | |
| 3 | | | |
| 4 | | | |

## Review Problems

The following problems review material we covered in Section 2.8.

Solve each inequality.

**37.** $2x + 3 < 5$      **38.** $3x - 2 > 7$

**39.** $-3x \leq 21$      **40.** $-5x \geq -10$

**41.** $-2x + 8 > -4$      **42.** $-4x - 1 < 11$

**43.** $4 - 2(x + 1) \geq -2$      **44.** $6 - 2(x + 3) \leq -8$

A proportion is two equal ratios. That is, if $\dfrac{a}{b}$ and $\dfrac{c}{d}$ are ratios, then:

$$\frac{a}{b} = \frac{c}{d}$$

is a proportion.

    Each of the four numbers in a proportion is called a *term* of the proportion. We number the terms as follows:

$$\text{First term} \;\longrightarrow\; \frac{a}{b} = \frac{c}{d} \;\longleftarrow\; \text{Third term}$$
$$\text{Second term} \;\longrightarrow\; \phantom{\frac{a}{b}} \phantom{=} \phantom{\frac{c}{d}} \;\longleftarrow\; \text{Fourth term}$$

The first and fourth terms are called the *extremes,* and the second and third terms are called the *means:*

$$\text{Means} \quad \frac{a}{b} = \frac{c}{d} \quad \text{Extremes}$$

For example, in the proportion:

$$\frac{3}{8} = \frac{12}{32}$$

the extremes are 3 and 32, and the means are 8 and 12.

## MEANS-EXTREMES PROPERTY

If $a$, $b$, $c$, and $d$ are real numbers with $b \neq 0$ and $d \neq 0$, then

$$\text{if} \qquad \frac{a}{b} = \frac{c}{d}$$
$$\text{then} \qquad ad = bc$$

*In words:*    In any proportion, the product of the extremes is equal to the product of the means.

    This property of proportions comes from the multiplication property of equality. We can use it to solve for a missing term in a proportion.

**E X A M P L E   I**    Solve the proportion $\dfrac{3}{x} = \dfrac{6}{7}$ for $x$.

**Solution**    We could solve for $x$ by using the method developed in Section 7.4, that is, multiplying both sides by the LCD $7x$. Instead, let's use our new means-extremes property:

$$\frac{3}{x} = \frac{6}{7} \qquad \text{Extremes are 3 and 7;}$$
$$\text{means are } x \text{ and 6.}$$

$$21 = 6x \qquad \text{Product of extremes} =$$
$$\text{product of means}$$

$$\frac{21}{6} = x \qquad \text{Divide both sides by 6.}$$

$$x = \frac{7}{2} \qquad \text{Reduce to lowest terms.}$$

**E X A M P L E   2**   Solve for $x$: $\dfrac{x+1}{2} = \dfrac{3}{x}$

**Solution**   Again, we want to point out that we could solve for $x$ by using the method we used in Section 7.5. Using the means-extremes property is simply an alternative to the method developed in Section 7.5:

$$\frac{x+1}{2} = \frac{3}{x} \qquad \text{Extremes are } x + 1 \text{ and } x;$$
$$\text{means are 2 and 3.}$$

$$x^2 + x = 6 \qquad \text{Product of extremes} =$$
$$\text{product of means}$$

$$x^2 + x - 6 = 0 \qquad \text{Standard form for a}$$
$$\text{quadratic equation}$$

$$(x+3)(x-2) = 0 \qquad \text{Factor.}$$

$$x + 3 = 0 \qquad \text{or} \qquad x - 2 = 0 \qquad \text{Set factors equal to 0.}$$

$$x = -3 \qquad \text{or} \qquad x = 2$$

This time we have two solutions: $-3$ and 2.

**E X A M P L E   3**   A manufacturer knows that during a production run, 8 out of every 100 parts produced by a certain machine will be defective. If the machine produces 1,450 parts, how many can be expected to be defective?

**Solution**   The ratio of defective parts to total parts produced is $\frac{8}{100}$. If we let $x$ represent the number of defective parts out of the total of 1,450 parts, then we can write this ratio again as $\dfrac{x}{1,450}$. This gives us a proportion to solve:

Defective parts
in numerator——

Total parts
in denominator——

$$\frac{x}{1,450} = \frac{8}{100} \qquad \text{Extremes are } x \text{ and 100;}$$
$$\text{means are 1,450 and 8.}$$

$$100x = 11,600 \qquad \text{Product of extremes} =$$
$$\text{product of means}$$

$$x = 116$$

The manufacturer can expect 116 defective parts out of the total of 1,450 parts if the machine usually produces 8 defective parts for every 100 parts it produces.

## Getting Ready for Class

*After reading through the preceding section, respond in your own words and in complete sentences.*

**A.** What is a proportion?

**B.** What are the means and extremes of a proportion?

**C.** What is the relationship between the means and the extremes in a proportion? (It is called the means-extremes property of proportions.)

**D.** How are ratios and proportions related?

## PROBLEM SET 7.7

Solve each of the following proportions.

**1.** $\dfrac{x}{2} = \dfrac{6}{12}$

**2.** $\dfrac{x}{4} = \dfrac{6}{8}$

**3.** $\dfrac{2}{5} = \dfrac{4}{x}$

**4.** $\dfrac{3}{8} = \dfrac{9}{x}$

**5.** $\dfrac{10}{20} = \dfrac{20}{x}$

**6.** $\dfrac{15}{60} = \dfrac{60}{x}$

**7.** $\dfrac{a}{3} = \dfrac{5}{12}$

**8.** $\dfrac{a}{2} = \dfrac{7}{20}$

**9.** $\dfrac{2}{x} = \dfrac{6}{7}$

**10.** $\dfrac{4}{x} = \dfrac{6}{7}$

**11.** $\dfrac{x+1}{3} = \dfrac{4}{x}$

**12.** $\dfrac{x+1}{6} = \dfrac{7}{x}$

**13.** $\dfrac{x}{2} = \dfrac{8}{x}$

**14.** $\dfrac{x}{9} = \dfrac{4}{x}$

**15.** $\dfrac{4}{a+2} = \dfrac{a}{2}$

**16.** $\dfrac{3}{a+2} = \dfrac{a}{5}$

**17.** $\dfrac{1}{x} = \dfrac{x-5}{6}$

**18.** $\dfrac{1}{x} = \dfrac{x-6}{7}$

## Applying the Concepts

**19. Baseball** A baseball player gets 6 hits in the first 18 games of the season. If he continues hitting at the same rate, how many hits will he get in the first 45 games?

**20. Basketball** A basketball player makes 8 of 12 free throws in the first game of the season. If she shoots with the same accuracy in the second game, how many of the 15 free throws she attempts will she make?

**21. Mixture Problem** A solution contains 12 milliliters of alcohol and 16 milliliters of water. If another solution is to have the same concentration of alcohol in water but is to contain 28 milliliters of water, how much alcohol must it contain?

**22. Mixture Problem** A solution contains 15 milliliters of HCl and 42 milliliters of water. If another solution is to have the same concentration of HCl in water but is to contain 140 milliliters of water, how much HCl must it contain?

**23. Nutrition** If 100 grams of ice cream contain 13 grams of fat, how much fat is in 350 grams of ice cream?

**24. Nutrition** A 6-ounce serving of grapefruit juice contains 159 grams of water. How many grams of water are in 20 ounces of grapefruit juice?

**25. Map Reading** A map is drawn so that every 3.5 inches on the map corresponds to an actual distance of 100 miles. If the actual distance between two cities is 420 miles, how far apart are they on the map?

26. **Map Reading**  The scale on a map indicates that 1 inch on the map corresponds to an actual distance of 105 miles. Two cities are 4.5 inches apart on the map. What is the actual distance between the two cities?

27. **Distance**  A man drives his car 245 miles in 5 hours. At this rate, how far will he travel in 7 hours?

28. **Distance**  An airplane flies 1,380 miles in 3 hours. How far will it fly in 5 hours?

## Review Problems

The following problems review material we covered in Sections 7.1, 7.2, and 7.3.

Reduce to lowest terms. [7.1]

**29.** $\dfrac{x^2 - x - 6}{x^2 - 9}$    **30.** $\dfrac{xy + 5x + 3y + 15}{x^2 + ax + 3x + 3a}$

Multiply and divide. [7.2]

**31.** $\dfrac{x^2 - 25}{x + 4} \cdot \dfrac{2x + 8}{x^2 - 9x + 20}$

**32.** $\dfrac{3x + 6}{x^2 + 4x + 3} \div \dfrac{x^2 + x - 2}{x^2 + 2x - 3}$

Add and subtract. [7.3]

**33.** $\dfrac{x}{x^2 - 16} + \dfrac{4}{x^2 - 16}$

**34.** $\dfrac{2}{x^2 - 1} - \dfrac{5}{x^2 + 3x - 4}$

---

### 7.8    Variation

Two variables are said to *vary directly* if one is a constant multiple of the other. For instance, $y$ varies directly as $x$ if $y = Kx$, where $K$ is a constant. The constant $K$ is called the constant of variation. The following table gives the relation between direct variation statements and their equivalent algebraic equations.

| Statement | Equation ($K$ = constant of variation) |
|---|---|
| $y$ varies directly as $x$. | $y = Kx$ |
| $y$ varies directly as the square of $x$. | $y = Kx^2$ |
| $s$ varies directly as the square root of $t$. | $s = K\sqrt{t}$ |
| $r$ varies directly as the cube of $s$. | $r = Ks^3$ |

Any time we run across a statement similar to those in the table, we can immediately write an equivalent expression involving variables and a constant of variation $K$.

**EXAMPLE 1**    Suppose $y$ varies directly as $x$. When $y$ is 15, $x$ is 3. Find $y$ when $x$ is 4.

**Solution**    From the first sentence we can write the relationship between $x$ and $y$ as:

$$y = Kx$$

We now use the second sentence to find the value of $K$. Since $y$ is 15 when $x$ is 3, we have:

$$15 = K(3) \quad \text{or} \quad K = 5$$

Now we can rewrite the relationship between $x$ and $y$ more specifically as:

$$y = 5x$$

To find the value of $y$ when $x$ is 4 we simply substitute $x = 4$ into our last equation. Substituting:

$$x = 4$$

into:

$$y = 5x$$

we have:

$$y = 5(4)$$

$$y = 20$$

**E X A M P L E   2**    Suppose $y$ varies directly as the square of $x$. When $x$ is 4, $y$ is 32. Find $x$ when $y$ is 50.

**Solution**    The first sentence gives us:

$$y = Kx^2$$

Since $y$ is 32 when $x$ is 4, we have:

$$32 = K(4)^2$$

$$32 = 16K$$

$$K = 2$$

The equation now becomes:

$$y = 2x^2$$

When $y$ is 50, we have:

$$50 = 2x^2$$

$$25 = x^2$$

$$x = \pm 5$$

There are two possible solutions, $x = 5$ or $x = -5$.

**E X A M P L E   3**    The cost of a certain kind of candy varies directly with the weight of the candy. If 12 ounces of the candy cost \$1.68, how much will 16 ounces cost?

**Solution**    Let $x =$ the number of ounces of candy and $y =$ the cost of the candy. Then $y = Kx$. Since $y$ is 1.68 when $x$ is 12, we have:

$$1.68 = K \cdot 12$$

$$K = \frac{1.68}{12}$$

$$= 0.14$$

The equation must be:

$$y = 0.14x$$

When $x$ is 16, we have:

$$y = 0.14(16)$$

$$= 2.24$$

The cost of 16 ounces of candy is \$2.24.

## INVERSE VARIATION

Two variables are said to *vary inversely* if one is a constant multiple of the reciprocal of the other. For example, $y$ varies inversely as $x$ if $y = \dfrac{K}{x}$, where $K$ is a real number constant. Again, $K$ is called the constant of variation. The table that follows gives some examples of inverse variation statements and their associated algebraic equations.

| Statement | Equation<br>($K$ = constant of variation) |
|---|---|
| $y$ varies inversely as $x$. | $y = \dfrac{K}{x}$ |
| $y$ varies inversely as the square of $x$. | $y = \dfrac{K}{x^2}$ |
| $F$ varies inversely as the square root of $t$. | $F = \dfrac{K}{\sqrt{t}}$ |
| $r$ varies inversely as the cube of $s$. | $r = \dfrac{K}{s^3}$ |

Every inverse variation statement has an associated inverse variation equation.

**EXAMPLE 4**    Suppose $y$ varies inversely as $x$. When $y$ is 4, $x$ is 5. Find $y$ when $x$ is 10.

**Solution**    The first sentence gives us the relationship between $x$ and $y$:

$$y = \frac{K}{x}$$

We use the second sentence to find the value of the constant $K$:

$$4 = \frac{K}{5}$$

or:

$$K = 20$$

We can now write the relationship between $x$ and $y$ more specifically as:

$$y = \frac{20}{x}$$

We use this equation to find the value of $y$ when $x$ is 10. Substituting:

$$x = 10$$

into:

$$y = \frac{20}{x}$$

we have:

$$y = \frac{20}{10}$$

$$y = 2$$

**EXAMPLE  5**    The intensity ($I$) of light from a source varies inversely as the square of the distance ($d$) from the source. Ten feet away from the source the intensity is 200 candlepower. What is the intensity 5 feet from the source?

**Solution**

$$I = \frac{K}{d^2}$$

Since $I = 200$ when $d = 10$, we have:

$$200 = \frac{K}{10^2}$$

$$200 = \frac{K}{100}$$

$$K = 20,000$$

The equation becomes:

$$I = \frac{20,000}{d^2}$$

When $d = 5$, we have:

$$I = \frac{20,000}{5^2}$$

$$= \frac{20,000}{25}$$

$$= 800 \text{ candlepower}$$

**Getting Ready for Class**

*After reading through the preceding section, respond in your own words and in complete sentences.*

**A.** What does it mean when we say "$y$ varies directly with $x$"?

**B.** Give an example of a sentence that is a direct variation statement.

**C.** Translate the equation $y = \dfrac{k}{x}$ into words.

**D.** Give an example of an everyday situation where one quantity varies inversely with another.

## PROBLEM SET 7.8

For each of the following problems, $y$ varies directly as $x$.

**1.** If $y = 10$ when $x = 5$, find $y$ when $x$ is 4.

**2.** If $y = 20$ when $x = 4$, find $y$ when $x$ is 11.

**3.** If $y = 39$ when $x = 3$, find $y$ when $x$ is 10.

**4.** If $y = -18$ when $x = 6$, find $y$ when $x$ is 3.

**5.** If $y = -24$ when $x = 4$, find $x$ when $y$ is $-30$.

**6.** If $y = 30$ when $x = -15$, find $x$ when $y$ is 8.

**7.** If $y = -7$ when $x = -1$, find $x$ when $y$ is $-21$.

**8.** If $y = 30$ when $x = 4$, find $y$ when $x$ is 7.

For each of the following problems, $y$ varies directly as the square of $x$.

**9.** If $y = 75$ when $x = 5$, find $y$ when $x$ is 1.

**10.** If $y = -72$ when $x = 6$, find $y$ when $x$ is 3.

**11.** If $y = 48$ when $x = 4$, find $y$ when $x$ is 9.

**12.** If $y = 27$ when $x = 3$, find $x$ when $y$ is 75.

For each of the following problems, $y$ varies inversely with $x$.

**13.** If $y = 5$ when $x = 2$, find $y$ when $x$ is 5.

**14.** If $y = 2$ when $x = 10$, find $y$ when $x$ is 4.

**15.** If $y = 2$ when $x = 1$, find $y$ when $x$ is 4.

**16.** If $y = 4$ when $x = 3$, find $y$ when $x$ is 6.

**17.** If $y = 5$ when $x = 3$, find $x$ when $y$ is 15.

**18.** If $y = 12$ when $x = 10$, find $x$ when $y$ is 60.

**19.** If $y = 10$ when $x = 10$, find $x$ when $y$ is 20.

**20.** If $y = 15$ when $x = 2$, find $x$ when $y$ is 6.

For each of the following problems, $y$ varies inversely as the square of $x$.

**21.** If $y = 4$ when $x = 5$, find $y$ when $x$ is 2.

**22.** If $y = 5$ when $x = 2$, find $y$ when $x$ is 6.

**23.** If $y = 4$ when $x = 3$, find $y$ when $x$ is 2.

**24.** If $y = 9$ when $x = 4$, find $y$ when $x$ is 3.

## Applying the Concepts

Solve the following problems.

**25. Tension in a Spring**  The tension $t$ in a spring varies directly with the distance $d$ the spring is stretched. If the tension is 42 pounds when the spring is stretched 2 inches, find the tension when the spring is stretched twice as far.

**26. Fill Time**  The time $t$ it takes to fill a bucket varies directly with the volume $g$ of the bucket. If it takes 1 minute to fill a 4-gallon bucket, how long will it take to fill a 6-gallon bucket?

**27. Electricity**  The power $P$ in an electric circuit varies directly with the square of the current $I$. If $P = 30$ when $I = 2$, find $P$ when $I = 7$.

**28. Electricity**  The resistance $R$ in an electric circuit varies directly with the voltage $V$. If $R = 20$ when $V = 120$, find $R$ when $V = 240$.

**29. Wages**  The amount of money $M$ a woman makes per week varies directly with the number of hours $h$ she works per week. If she works 20 hours and earns $157, how much does she make if she works 30 hours?

**30. Volume**  The volume $V$ of a gas varies directly as the temperature $T$. If $V = 3$ when $T = 150$, find $V$ when $T$ is 200.

**31. Weight**  The weight $F$ of a body varies inversely with the square of the distance $d$ between the body and the center of the Earth. If a man weighs 150 pounds 4,000 miles from the center of the Earth, how much will he weigh at a distance of 5,000 miles from the center of the Earth?

**32. Light Intensity**  The intensity $I$ of a light source varies inversely with the square of the distance $d$ from the source. Four feet from the source, the intensity is 9 foot-candles. What is the intensity 3 feet from the source?

**33. Electricity**  The current $I$ in an electric circuit varies inversely with the resistance $R$. If a current of 30 amperes is produced by a resistance of 2 ohms, what current will be produced by a resistance of 5 ohms?

**34. Pressure**  The pressure exerted by a gas on the container in which it is held varies inversely with the volume of the container. A pressure of 40 pounds per square inch is exerted on a container of volume 2 cubic feet. What is the pressure on a container whose volume is 8 cubic feet?

## Review Problems

The following problems review material we covered in Sections 4.2 and 4.3.

Solve each system of equations by the elimination method. [4.2]

**35.** $2x + y = 3$
    $3x - y = 7$

**36.** $3x - y = -6$
    $4x + y = -8$

**37.** $4x - 5y = 1$
    $x - 2y = -2$

**38.** $6x - 4y = 2$
    $2x + y = 10$

Solve by the substitution method. [4.3]

**39.** $5x + 2y = 7$
    $y = 3x - 2$

**40.** $-7x - 5y = -1$
    $y = x + 5$

**41.** $2x - 3y = 4$
    $x = 2y + 1$

**42.** $4x - 5y = 2$
    $x = 2y - 1$

## C H A P T E R   7   SUMMARY

**Examples**

**1.** We can reduce $\frac{6}{8}$ to lowest terms by dividing the numerator and denominator by their greatest common factor 2:

$$\frac{6}{8} = \frac{2 \cdot 3}{2 \cdot 4} = \frac{3}{4}$$

### Rational Numbers [7.1]

Any number that can be put in the form $\frac{a}{b}$, where $a$ and $b$ are integers ($b \neq 0$), is called a rational number.

Multiplying or dividing the numerator and denominator of a rational number by the same nonzero number never changes the value of the rational number.

**2.** We reduce rational expressions to lowest terms by factoring the numerator and denominator and then dividing out any factors they have in common:

$$\frac{x - 3}{x^2 - 9} = \frac{\cancel{x - 3}}{(\cancel{x - 3})(x + 3)} = \frac{1}{x + 3}$$

### Rational Expressions [7.1]

Any expression of the form $\frac{P}{Q}$, where $P$ and $Q$ are polynomials ($Q \neq 0$), is a rational expression.

Multiplying or dividing the numerator and denominator of a rational expression by the same nonzero quantity always produces a rational expression equivalent to the original one.

**3.** $\dfrac{x - 1}{x^2 + 2x - 3} \cdot \dfrac{x^2 - 9}{x - 2}$

$$= \frac{\cancel{x - 1}}{(x + 3)(\cancel{x - 1})} \cdot \frac{(x - 3)(\cancel{x + 3})}{x - 2}$$

$$= \frac{x - 3}{x - 2}$$

### Multiplication [7.2]

To multiply two rational numbers or two rational expressions, multiply numerators, multiply denominators, and divide out any factors common to the numerator and denominator:

For rational numbers $\dfrac{a}{b}$ and $\dfrac{c}{d}$,  $\dfrac{a}{b} \cdot \dfrac{c}{d} = \dfrac{ac}{bd}$

For rational expressions $\dfrac{P}{Q}$ and $\dfrac{R}{S}$,  $\dfrac{P}{Q} \cdot \dfrac{R}{S} = \dfrac{PR}{QS}$

**4.** $\dfrac{2x}{x^2 - 25} \div \dfrac{4}{x - 5}$

$$= \frac{2x}{(\cancel{x - 5})(x + 5)} \cdot \frac{(\cancel{x - 5})}{4}$$

$$= \frac{x}{2(x + 5)}$$

### Division [7.2]

To divide by a rational number or rational expression, simply multiply by its reciprocal:

For rational numbers $\dfrac{a}{b}$ and $\dfrac{c}{d}$,  $\dfrac{a}{b} \div \dfrac{c}{d} = \dfrac{a}{b} \cdot \dfrac{d}{c}$

For rational expressions $\dfrac{P}{Q}$ and $\dfrac{R}{S}$,  $\dfrac{P}{Q} \div \dfrac{R}{S} = \dfrac{P}{Q} \cdot \dfrac{S}{R}$

**5.** $\dfrac{3}{x-1} + \dfrac{x}{2}$

$= \dfrac{3}{x-1} \cdot \dfrac{2}{2} + \dfrac{x}{2} \cdot \dfrac{x-1}{x-1}$

$= \dfrac{6}{2(x-1)} + \dfrac{x^2 - x}{2(x-1)}$

$= \dfrac{x^2 - x + 6}{2(x-1)}$

**Addition [7.3]**

To add two rational numbers or rational expressions, find a common denominator, change each expression to an equivalent expression having the common denominator, then add numerators, and reduce if possible:

For rational numbers $\dfrac{a}{c}$ and $\dfrac{b}{c}$,   $\dfrac{a}{c} + \dfrac{b}{c} = \dfrac{a+b}{c}$

For rational expressions $\dfrac{P}{S}$ and $\dfrac{Q}{S}$,   $\dfrac{P}{S} + \dfrac{Q}{S} = \dfrac{P+Q}{S}$

**6.** $\dfrac{x}{x^2 - 4} - \dfrac{2}{x^2 - 4}$

$= \dfrac{x-2}{x^2 - 4}$

$= \dfrac{\cancel{x-2}}{(\cancel{x-2})(x+2)}$

$= \dfrac{1}{x+2}$

**Subtraction [7.3]**

To subtract a rational number or rational expression, simply add its opposite:

For rational numbers $\dfrac{a}{c}$ and $\dfrac{b}{c}$,   $\dfrac{a}{c} - \dfrac{b}{c} = \dfrac{a}{c} + \left(\dfrac{-b}{c}\right)$

For rational expressions $\dfrac{P}{S}$ and $\dfrac{Q}{S}$,   $\dfrac{P}{S} - \dfrac{Q}{S} = \dfrac{P}{S} + \left(\dfrac{-Q}{S}\right)$

**7.** Solve $\dfrac{1}{2} + \dfrac{3}{x} = 5$.

$2x\left(\dfrac{1}{2}\right) + 2x\left(\dfrac{3}{x}\right) = 2x(5)$

$x + 6 = 10x$

$6 = 9x$

$x = \dfrac{2}{3}$

**Equations [7.4]**

To solve equations involving rational expressions, first find the least common denominator (LCD) for all denominators. Then multiply both sides by the LCD and solve as usual. Check all solutions in the original equation to be sure there are no undefined terms.

**8.** $\dfrac{1 - \dfrac{4}{x}}{x - \dfrac{16}{x}} = \dfrac{x\left(1 - \dfrac{4}{x}\right)}{x\left(x - \dfrac{16}{x}\right)}$

$= \dfrac{x - 4}{x^2 - 16}$

$= \dfrac{\cancel{x-4}}{(\cancel{x-4})(x+4)}$

$= \dfrac{1}{x+4}$

**Complex Fractions [7.6]**

A rational expression that contains a fraction in its numerator or denominator is called a complex fraction. The most common method of simplifying a complex fraction is to multiply the top and bottom by the LCD for all denominators.

**9.** Solve for $x$: $\dfrac{3}{x} = \dfrac{5}{20}$

$$3 \cdot 20 = 5 \cdot x$$
$$60 = 5x$$
$$x = 12$$

**Ratio and Proportion** [7.1, 7.7]

The ratio of $a$ to $b$ is:

$$\frac{a}{b}$$

Two equal ratios form a proportion. In the proportion

$$\frac{a}{b} = \frac{c}{d}$$

$a$ and $d$ are the *extremes,* and $b$ and $c$ are the *means.* In any proportion the product of the extremes is equal to the product of the means.

**10.** If $y$ varies directly with the square of $x$, then:

$$y = Kx^2$$

**Direct Variation** [7.8]

The variable $y$ is said to vary directly with the variable $x$ if $y = Kx$, where $K$ is a real number.

**11.** If $y$ varies inversely with the cube of $x$, then:

$$y = \frac{K}{x^3}$$

**Inverse Variation** [7.8]

The variable $y$ is said to vary inversely with the variable $x$ if $y = \dfrac{K}{x}$, where $K$ is a real number.

# CHAPTER 7   REVIEW

The numbers in brackets refer to the sections of the text in which similar problems can be found.

Reduce to lowest terms. Also specify any restriction on the variable. [7.1]

**1.** $\dfrac{7}{14x - 28}$

**2.** $\dfrac{a + 6}{a^2 - 36}$

**3.** $\dfrac{8x - 4}{4x + 12}$

**4.** $\dfrac{x + 4}{x^2 + 8x + 16}$

**5.** $\dfrac{3x^3 + 16x^2 - 12x}{2x^3 + 9x^2 - 18x}$

**6.** $\dfrac{x + 2}{x^4 - 16}$

**7.** $\dfrac{x^2 + 5x - 14}{x + 7}$

**8.** $\dfrac{a^2 + 16a + 64}{a + 8}$

**9.** $\dfrac{xy + bx + ay + ab}{xy + 5x + ay + 5a}$

Multiply or divide as indicated. [7.2]

**10.** $\dfrac{3x + 9}{x^2} \cdot \dfrac{x^3}{6x + 18}$

**11.** $\dfrac{x^2 + 8x + 16}{x^2 + x - 12} \div \dfrac{x^2 - 16}{x^2 - x - 6}$

**12.** $(a^2 - 4a - 12)\left(\dfrac{a - 6}{a + 2}\right)$

**13.** $\dfrac{3x^2 - 2x - 1}{x^2 + 6x + 8} \div \dfrac{3x^2 + 13x + 4}{x^2 + 8x + 16}$

Find the following sums and differences. [7.3]

**14.** $\dfrac{2x}{2x + 3} + \dfrac{3}{2x + 3}$

**15.** $\dfrac{x^2}{x - 9} - \dfrac{18x - 81}{x - 9}$

**16.** $\dfrac{a + 4}{a + 8} - \dfrac{a - 9}{a + 8}$

**17.** $\dfrac{x}{x + 9} + \dfrac{5}{x}$

**18.** $\dfrac{5}{4x + 20} + \dfrac{x}{x + 5}$

**19.** $\dfrac{3}{x^2 - 36} - \dfrac{2}{x^2 - 4x - 12}$

**20.** $\dfrac{3a}{a^2 + 8a + 15} - \dfrac{2}{a + 5}$

Solve each equation. [7.4]

**21.** $\dfrac{3}{x} + \dfrac{1}{2} = \dfrac{5}{x}$

**22.** $\dfrac{a}{a - 3} = \dfrac{3}{2}$

**23.** $1 - \dfrac{7}{x} = \dfrac{-6}{x^2}$

**24.** $\dfrac{3}{x + 6} - \dfrac{1}{x - 2} = \dfrac{-8}{x^2 + 4x - 12}$

**25.** $\dfrac{2}{y^2 - 16} = \dfrac{10}{y^2 + 4y}$

**26. Number Problem**   The sum of a number and 7 times its reciprocal is $\frac{16}{3}$. Find the number. [7.5]

**27. Distance, Rate, and Time**   A boat travels 48 miles up a river in the same amount of time it takes to travel 72 miles down the same river. If the current is 3 miles per hour, what is the speed of the boat in still water? [7.5]

**28. Filling a Pool**   An inlet pipe can fill a pool in 21 hours, while an outlet pipe can empty it in 28

hours. If both pipes are left open, how long will it take to fill the pool? [7.5]

Simplify each complex fraction. [7.6]

**29.** $\dfrac{\dfrac{x + 4}{x^2 - 16}}{\dfrac{2}{x - 4}}$

**30.** $\dfrac{1 - \dfrac{9}{y^2}}{1 + \dfrac{4}{y} - \dfrac{21}{y^2}}$

**31.** $\dfrac{\dfrac{1}{a - 2} + 4}{\dfrac{1}{a - 2} + 1}$

**32.** Write the ratio of 40 to 100 as a fraction in lowest terms. [7.7]

**33.** If there are 60 seconds in 1 minute, what is the ratio of 40 seconds to 3 minutes? [7.7]

Solve each proportion. [7.7]

**34.** $\dfrac{x}{9} = \dfrac{4}{3}$

**35.** $\dfrac{a}{3} = \dfrac{12}{a}$

**36.** $\dfrac{8}{x - 2} = \dfrac{x}{6}$

Work the following problems involving variation. [7.8]

**37.** $y$ varies directly as $x$. If $y = -20$ when $x = 4$, find $y$ when $x = 7$.

**38.** $y$ varies inversely with $x$. If $y = 3$ when $x = 2$, find $y$ when $x = 12$.

# C H A P T E R   7   **TEST**

Reduce to lowest terms. [7.1]

**1.** $\dfrac{x^2 - 16}{x^2 - 8x + 16}$

**2.** $\dfrac{10a + 20}{5a^2 + 20a + 20}$

**3.** $\dfrac{xy + 7x + 5y + 35}{x^2 + ax + 5x + 5a}$

Multiply or divide as indicated. [7.2]

**4.** $\dfrac{3x - 12}{4} \cdot \dfrac{8}{2x - 8}$

**5.** $\dfrac{x^2 - 49}{x + 1} \div \dfrac{x + 7}{x^2 - 1}$

**6.** $\dfrac{x^2 - 3x - 10}{x^2 - 8x + 15} \div \dfrac{3x^2 + 2x - 8}{x^2 + x - 12}$

**7.** $(x^2 - 9)\left(\dfrac{x + 2}{x + 3}\right)$

Add or subtract as indicated. [7.3]

**8.** $\dfrac{3}{x - 2} - \dfrac{6}{x - 2}$

**9.** $\dfrac{x}{x^2 - 9} + \dfrac{4}{4x - 12}$

10. $\dfrac{2x}{x^2 - 1} + \dfrac{x}{x^2 - 3x + 2}$

Solve the following equations. [7.4]

11. $\dfrac{7}{5} = \dfrac{x + 2}{3}$

12. $\dfrac{10}{x + 4} = \dfrac{6}{x} - \dfrac{4}{x}$

13. $\dfrac{3}{x - 2} - \dfrac{4}{x + 1} = \dfrac{5}{x^2 - x - 2}$

Solve the following problems. [7.5]

14. **Speed of a Boat** A boat travels 26 miles up a river in the same amount of time it takes to travel 34 miles down the same river. If the current is 2 miles per hour, what is the speed of the boat in still water?

15. **Emptying a Pool** An inlet pipe can fill a pool in 15 hours, and an outlet pipe can empty it in 12 hours. If the pool is full and both pipes are open, how long will it take to empty?

Solve the following problems involving ratio and proportion. [7.7]

16. **Ratio** A solution of alcohol and water contains 27 milliliters of alcohol and 54 milliliters of water. What is the ratio of alcohol to water and the ratio of alcohol to total volume?

17. **Ratio** A manufacturer knows that during a production run 8 out of every 100 parts produced by a certain machine will be defective. If the machine produces 1,650 parts, how many can be expected to be defective?

Simplify each complex fraction. [7.6]

18. $\dfrac{1 + \dfrac{1}{x}}{1 - \dfrac{1}{x}}$

19. $\dfrac{1 - \dfrac{16}{x^2}}{1 - \dfrac{2}{x} - \dfrac{8}{x^2}}$

20. **Direct Variation** Suppose $y$ varies directly with the square of $x$. If $x$ is 3 when $y$ is 36, find $y$ when $x$ is 5. [7.8]

21. **Inverse Variation** If $y$ varies inversely with $x$, and $x$ is 3 when $y$ is 6, find $y$ when $x$ is 9. [7.8]

# CHAPTERS 1-7 CUMULATIVE REVIEW

Simplify.

**1.** $8 - 11$

**2.** $-20 + 14$

**3.** $\dfrac{-48}{12}$

**4.** $\dfrac{1}{6}(-18)$

**5.** $5x - 4 - 9x$

**6.** $8 - x - 4$

**7.** $9^{-2}$

**8.** $\left(\dfrac{x^5}{x^3}\right)^{-2}$

**9.** $4^1 + 9^0 + (-7)^0$

**10.** $\dfrac{(x^{-4})^{-3}(x^{-3})^4}{x^0}$

**11.** $(4a^3 - 10a^2 + 6) - (6a^3 + 5a - 7)$

**12.** $(a^2 + 7)(a^2 - 7)$

**13.** $\dfrac{x^2}{x - 7} - \dfrac{14x - 49}{x - 7}$

**14.** $\dfrac{6}{10x + 30} + \dfrac{x}{x + 3}$

**15.** $\dfrac{\dfrac{x - 2}{x^2 + 6x + 8}}{\dfrac{4}{x + 4}}$

Solve each equation.

**16.** $\dfrac{3}{4}(8x - 12) = \dfrac{1}{2}(4x + 4)$

**17.** $x - \dfrac{3}{4} = \dfrac{5}{6}$

**18.** $4x - 3 = 8x + 5$

**19.** $98r^2 - 18 = 0$

**20.** $6x^4 = 33x^3 - 42x^2$

**21.** $\dfrac{5}{x} - \dfrac{1}{3} = \dfrac{3}{x}$

**22.** $\dfrac{4}{x - 5} - \dfrac{3}{x + 2} = \dfrac{28}{x^2 - 3x - 10}$

**23.** $\dfrac{x}{3} = \dfrac{6}{x - 3}$

Solve each system.

**24.** $x + 2y = 1$
$\quad\ x - y = 4$

**25.** $\ \ 9x + 14y = -4$
$\quad 15x - 8y = 9$

**26.** $\quad x = 4y - 3$
$\quad 2x - 5y = 9$

**27.** $\dfrac{1}{2}x + \dfrac{1}{3}y = -1$
$\quad \dfrac{1}{3}x = \dfrac{1}{4}y + 5$

Graph each equation on a rectangular coordinate system.

**28.** $y = -3x + 2$

**29.** $y < 2$

**30.** Solve the inequality $-\dfrac{a}{3} \le -2$.

Factor completely.

**31.** $xy + 5x + ay + 5a$

**32.** $a^2 + 2a - 35$

**33.** $20y^2 - 27y + 9$

**34.** $4r^2 - 9t^2$

**35.** $16x^2 + 72xy + 81y^2$

**36.** Solve the following system by graphing.
$2x + \ y = 4$
$3x - 2y = 6$

**37.** Which of the ordered pairs $(3, -1)$, $(1, -3)$, $(-2, 9)$ are solutions to the equation $y = 2x - 5$?

**38.** For the equation $2x + 5y = 10$, find the $x$- and $y$-intercepts.

**39.** For the equation $3x - y = 4$, find the slope and $y$-intercept.

**40.** Find the equation for the line with slope $-5$ and $y$-intercept $-1$.

**41.** Find the equation for the line with slope $-\dfrac{2}{5}$ that passes through $(-2, -1)$.

**42.** Add $-2$ to the product of $-3$ and 4.

**43.** What property justifies the statement
$(2 + x) + 3 = 2 + (x + 3)$?

**44.** For the set $\{-3, -\sqrt{2}, 0, \frac{3}{4}, 1.5, \pi\}$ list all the rational numbers.

**45.** Multiply $\dfrac{6x - 12}{6x + 12} \cdot \dfrac{3x + 3}{12x - 24}$.

**46.** Divide $\dfrac{x^2 - 3x - 28}{x + 4}$.

**47.** Reduce to lowest terms $\dfrac{2xy + 10x + 3y + 15}{3xy + 15x + 2y + 10}$.

**48.** Solve $P = 2l + 2w$ for $l$.

**49. Number Problem** The sum of two numbers is 40. The difference of the two numbers is 18. Find the numbers.

**50. Investing** Ms. Jones invested $18,000 in two accounts. One account pays 6% simple interest, and the other pays 8%. Her total interest for the year was $1,290. How much did she have in each account?

# Roots and Radicals

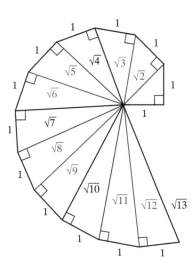

## INTRODUCTION

The diagram shown here is called the *spiral of roots*. It is constructed using the Pythagorean theorem, which we introduced in Chapter 6. The spiral of roots gives us a way to visualize positive square roots, one of the topics we will cover in this chapter. If we take each of the diagonals in the spiral of roots and place it above the corresponding whole numbers on the *x*-axis, and then connect the tops of all these segments with a smooth curve, we have the curve shown in Figure 1. Table 1 gives the lengths of the diagonals in the spiral of roots, accurate to the nearest hundredth. That curve in Figure 1 is also the graph of the equation $y = \sqrt{x}$.

| TABLE I | Approximate Length of Diagonals | |
|---|---|
| **Number** | **Positive Square Root** |
| 1 | 1 |
| 2 | 1.41 |
| 3 | 1.73 |
| 4 | 2 |
| 5 | 2.24 |
| 6 | 2.45 |
| 7 | 2.65 |
| 8 | 2.83 |
| 9 | 3 |
| 10 | 3.16 |

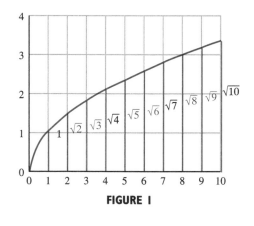

**FIGURE I**

In Chapter 5 we developed notation (exponents) that would take us from a number to its square. If we wanted the square of 5, we wrote $5^2 = 25$. In this section we will use another type of notation that will take us in the reverse direction — from the square of a number back to the number itself.

In general we are interested in going from a number — say, 49 — back to the number we squared to get 49. Since the square of 7 is 49, we say 7 is a square root of 49. The notation we use looks like this:

$$\sqrt{49}$$

**Notation**   In the expression $\sqrt{49}$, 49 is called the *radicand,* $\sqrt{\phantom{x}}$ is the *radical sign,* and the complete expression, $\sqrt{49}$, is called the *radical.*

> **DEFINITION**   If $x$ represents any positive real number, then the expression $\sqrt{x}$ is the **positive square root** of $x$. It is the *positive* number we square to get $x$.
> The expression $-\sqrt{x}$ is the **negative square root** of $x$. It is the *negative* number we square to get $x$.

## SQUARE ROOTS OF POSITIVE NUMBERS

Every positive number has two square roots, one positive and the other negative. Some books refer to the positive square root of a number as the principal root.

**EXAMPLE   I**   The positive square root of 25 is 5 and can be written $\sqrt{25} = 5$. The negative square root of 25 is $-5$ and can be written $-\sqrt{25} = -5$.

If we want to consider the negative square root of a number, we must put a negative sign in front of the radical. It is a common mistake to think of $\sqrt{25}$ as meaning either 5 or $-5$. The expression $\sqrt{25}$ means the *positive* square root of 25, which is 5. If we want the negative square root, we write $-\sqrt{25}$ to begin with.

**EXAMPLES**   Find the following roots.

**2.** $\sqrt{49} = 7$          7 is the positive number we square to get 49.
**3.** $-\sqrt{49} = -7$       $-7$ is the negative number we square to get 49.
**4.** $\sqrt{121} = 11$       11 is the positive number we square to get 121.

**EXAMPLE   5**   The positive square root of 17 is written $\sqrt{17}$. The negative square root of 17 is written $-\sqrt{17}$.

We have no other exact representation for the two roots in Example 5. Since 17 itself is not a perfect square (the square of an integer), its two square roots, $\sqrt{17}$ and $-\sqrt{17}$, are irrational numbers. They have a place on the real number line but cannot be written as the ratio of two integers. The square roots of any number that is not itself a perfect square are irrational numbers.

### EXAMPLE 6

| Number | Positive Square Root | Negative Square Root | Roots Are |
|---|---|---|---|
| 9 | 3 | $-3$ | Rational numbers |
| 36 | 6 | $-6$ | Rational numbers |
| 7 | $\sqrt{7}$ | $-\sqrt{7}$ | Irrational numbers |
| 22 | $\sqrt{22}$ | $-\sqrt{22}$ | Irrational numbers |
| 100 | 10 | $-10$ | Rational numbers |

## SQUARE ROOT OF ZERO

The number 0 is the only real number with one square root. It is also its own square root:

$$\sqrt{0} = 0$$

## SQUARE ROOTS OF NEGATIVE NUMBERS

Negative numbers have square roots, but their square roots are not real numbers. They do not have a place on the real number line. We will consider square roots of negative numbers later in the book.

### EXAMPLE 7    The expression $\sqrt{-4}$ does not represent a real number, since there is no real number we can square and end up with $-4$. The same is true of square roots of any negative number.

## OTHER ROOTS

There are many other roots of numbers besides square roots, although square roots seem to be the most commonly used. The cube root of a number is the number we cube (raise to the third power) to get the original number. The cube root of 8 is 2, since $2^3 = 8$. The cube root of 27 is 3, since $3^3 = 27$. The notation for cube roots looks like this:

$$\text{The 3 is called the index} \rightarrow \sqrt[3]{8} = 2$$
$$\rightarrow \sqrt[3]{27} = 3$$

We can go as high as we want with roots. The fourth root of 16 is 2, because $2^4 = 16$. We can write this in symbols as $\sqrt[4]{16} = 2$.

Here is a list of the most common roots. They are the roots that will come up most often in the remainder of the book and should be memorized.

| Square Roots | | Cube Roots | Fourth Roots |
|---|---|---|---|
| $\sqrt{1} = 1$ | $\sqrt{49} = 7$ | $\sqrt[3]{1} = 1$ | $\sqrt[4]{1} = 1$ |
| $\sqrt{4} = 2$ | $\sqrt{64} = 8$ | $\sqrt[3]{8} = 2$ | $\sqrt[4]{16} = 2$ |
| $\sqrt{9} = 3$ | $\sqrt{81} = 9$ | $\sqrt[3]{27} = 3$ | $\sqrt[4]{81} = 3$ |
| $\sqrt{16} = 4$ | $\sqrt{100} = 10$ | $\sqrt[3]{64} = 4$ | $\sqrt[4]{256} = 4$ |
| $\sqrt{25} = 5$ | $\sqrt{121} = 11$ | $\sqrt[3]{125} = 5$ | $\sqrt[4]{625} = 5$ |
| $\sqrt{36} = 6$ | $\sqrt{144} = 12$ | | |

With even roots — square roots, fourth roots, sixth roots, and so on — we cannot have negative numbers *under* the radical sign. With odd roots, negative numbers under the radical sign do not cause problems.

**EXAMPLES**   Find the following roots, if possible.

**8.** $\sqrt[3]{-8} = -2$        Because $(-2)^3 = -8$

**9.** $\sqrt[3]{-27} = -3$       Because $(-3)^3 = -27$

**10.** $\sqrt{-4}$              Not a real number, since there is no real number whose square is $-4$

**11.** $-\sqrt{4} = -2$         Because $-2$ is the negative number we square to get 4

**12.** $\sqrt[4]{-16}$          Not a real number, since there is no real number that can be raised to the fourth power to obtain $-16$

**13.** $-\sqrt[4]{16} = -2$      Because $-2$ is the negative number we raise to the fourth power to get 16

**Note:** At first it may be difficult to see the difference in some of these examples. Generally, we have to be careful with even roots of negative numbers. That is, if the index on the radical is an even number, then we cannot have a negative number under the radical sign. That is why $\sqrt{-4}$ and $\sqrt[4]{-16}$ are not real numbers.

## VARIABLES UNDER THE RADICAL SIGN

In this chapter, unless we say otherwise, we will assume that all variables that appear under a radical sign represent positive numbers. That way we can simplify expressions involving radicals that contain variables. Here are some examples.

**EXAMPLE 14**   Simplify $\sqrt{49x^2}$.

**Solution**   We are looking for the expression we square to get $49x^2$. Since the square of 7 is 49 and the square of $x$ is $x^2$, we can square $7x$ and get $49x^2$:

$$\sqrt{49x^2} = 7x \qquad \text{Because } (7x)^2 = 49x^2$$

**EXAMPLE 15**   Simplify $\sqrt{16a^2b^2}$.

**Solution**   We want an expression whose square is $16a^2b^2$. That expression is $4ab$:

$$\sqrt{16a^2b^2} = 4ab \qquad \text{Because } (4ab)^2 = 16a^2b^2$$

**EXAMPLE 16**   Simplify $\sqrt[3]{125a^3}$.

**Solution**   We are looking for the expression we cube to get $125a^3$. That expression is $5a$:

$$\sqrt[3]{125a^3} = 5a \qquad \text{Because } (5a)^3 = 125a^3$$

**EXAMPLE 17**   Simplify $\sqrt{x^6}$.

**Solution**   The number we square to obtain $x^6$ is $x^3$.

$$\sqrt{x^6} = x^3 \qquad \text{Because } (x^3)^2 = x^6$$

**EXAMPLE 18**   Simplify $\sqrt[4]{16a^8b^4}$.

**Solution**   The number we raise to the fourth power to obtain $16a^8b^4$ is $2a^2b$.

$$\sqrt[4]{16a^8b^4} = 2a^2b \qquad \text{Because } (2a^2b)^4 = 16a^8b^4$$

Many application problems involving radicals require decimal approximations. When we need a decimal approximation to a square root, we can use a calculator.

**EXAMPLE 19**   If you invest $P$ dollars in an account and after 2 years the account has $A$ dollars in it, then the annual rate of return $r$ on the money you originally invested is given by the formula

$$r = \frac{\sqrt{A} - \sqrt{P}}{\sqrt{P}}$$

Suppose you pay $65 for a coin collection and find that the same coins sell for $84 two years later. Find the annual rate of return on your investment.

**Solution**   Substituting $A = 84$ and $P = 65$ in the formula, we have

$$r = \frac{\sqrt{84} - \sqrt{65}}{\sqrt{65}}$$

From either the table of square roots or a calculator, we find that $\sqrt{84} = 9.165$ and $\sqrt{65} = 8.062$. Using these numbers in our formula gives us

$$r = \frac{9.165 - 8.062}{8.062}$$

$$= \frac{1.103}{8.062}$$

$$= 0.137, \text{ or } 13.7\%$$

To earn as much as this in a savings account that compounds interest once a year, you have to find an account that pays 13.7% in annual interest.

## THE PYTHAGOREAN THEOREM AGAIN

Now that we have had some experience working with square roots, we can rewrite the Pythagorean theorem using a square root. In Figure 1, if triangle $ABC$ is a right triangle with $C = 90°$, then the length of the longest side is the *square root* of the sum of the squares of the other two sides.

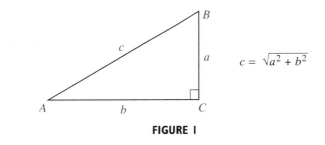

$$c = \sqrt{a^2 + b^2}$$

**FIGURE 1**

**EXAMPLE 20**    A tent pole is 8 feet in length and makes an angle of 90° with the ground. One end of a rope is attached to the top of the pole, and the other end of the rope is anchored to the ground 6 feet from the bottom of the pole. Find the length of the rope.

**Solution**    The diagram in Figure 2 is a visual representation of the situation. To find the length of the pole, we apply the Pythagorean theorem.

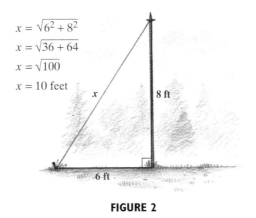

$$x = \sqrt{6^2 + 8^2}$$
$$x = \sqrt{36 + 64}$$
$$x = \sqrt{100}$$
$$x = 10 \text{ feet}$$

**FIGURE 2**

**Getting Ready for Class**

*After reading through the preceding section, respond in your own words and in complete sentences.*

**A.** Every real number has two square roots. Explain the notation we use to tell them apart. Use the square roots of 3 for examples.

**B.** Explain why a square root of $-4$ is not a real number.

**C.** We use the notation $\sqrt{17}$ to represent the positive square root of 17. Explain why there isn't a simpler way to express the positive square root of 17.

**D.** Write a list of the notation and terms associated with radicals.

## PROBLEM SET 8.1

Find the following roots. If the root does not exist as a real number, write "not a real number."

**1.** $\sqrt{9}$    **2.** $\sqrt{16}$    **3.** $-\sqrt{9}$

**4.** $-\sqrt{16}$    **5.** $\sqrt{-25}$    **6.** $\sqrt{-36}$

**7.** $-\sqrt{144}$    **8.** $\sqrt{256}$    **9.** $\sqrt{625}$

**10.** $-\sqrt{625}$    **11.** $\sqrt{-49}$    **12.** $\sqrt{-169}$

**13.** $-\sqrt{64}$    **14.** $-\sqrt{25}$    **15.** $-\sqrt{100}$

**16.** $\sqrt{121}$    **17.** $\sqrt{1,225}$    **18.** $-\sqrt{1,681}$

**19.** $\sqrt[4]{1}$    **20.** $-\sqrt[4]{81}$    **21.** $\sqrt[3]{-8}$

**22.** $\sqrt[3]{125}$    **23.** $-\sqrt[3]{125}$    **24.** $-\sqrt[3]{-8}$

**25.** $\sqrt[3]{-1}$    **26.** $-\sqrt[3]{-1}$    **27.** $\sqrt[3]{-27}$

**28.** $-\sqrt[3]{27}$    **29.** $-\sqrt[4]{16}$    **30.** $\sqrt[4]{-16}$

Assume all variables are positive, and find the following roots.

**31.** $\sqrt{x^2}$    **32.** $\sqrt{a^2}$    **33.** $\sqrt{9x^2}$

**34.** $\sqrt{25x^2}$    **35.** $\sqrt{x^2y^2}$    **36.** $\sqrt{a^2b^2}$

**37.** $\sqrt{(a+b)^2}$    **38.** $\sqrt{(x+y)^2}$    **39.** $\sqrt{49x^2y^2}$

**40.** $\sqrt{81x^2y^2}$    **41.** $\sqrt[3]{x^3}$    **42.** $\sqrt[3]{a^3}$

**43.** $\sqrt[3]{8x^3}$    **44.** $\sqrt[3]{27x^3}$    **45.** $\sqrt{x^4}$

**46.** $\sqrt{x^6}$    **47.** $\sqrt{36a^6}$    **48.** $\sqrt{64a^4}$

**49.** $\sqrt{25a^8b^4}$    **50.** $\sqrt{16a^4b^8}$    **51.** $\sqrt[3]{x^6}$

**52.** $\sqrt[3]{x^9}$    **53.** $\sqrt[3]{27a^{12}}$    **54.** $\sqrt[3]{8a^6}$

**55.** $\sqrt[4]{x^8}$    **56.** $\sqrt[4]{x^{12}}$

Simplify each expression.

**57.** $\sqrt{9} + \sqrt{16}$    **58.** $\sqrt{64} + \sqrt{36}$

**59.** $\sqrt{9+16}$    **60.** $\sqrt{64+36}$

**61.** $\sqrt{144} + \sqrt{25}$    **62.** $\sqrt{25} - \sqrt{16}$

**63.** $\sqrt{144+25}$    **64.** $\sqrt{25-16}$

Simplify each of the following pairs of expressions.

**65.** $\dfrac{5+\sqrt{49}}{2}$ and $\dfrac{5-\sqrt{49}}{2}$

**66.** $\dfrac{3+\sqrt{25}}{4}$ and $\dfrac{3-\sqrt{25}}{4}$

**67.** $\dfrac{2+\sqrt{16}}{2}$ and $\dfrac{2-\sqrt{16}}{2}$

**68.** $\dfrac{3+\sqrt{9}}{2}$ and $\dfrac{3-\sqrt{9}}{2}$

**69.** We know that the trinomial $x^2 + 6x + 9$ is the square of the binomial $x + 3$. That is,

$$x^2 + 6x + 9 = (x + 3)^2$$

Use this fact to find $\sqrt{x^2 + 6x + 9}$.

**70.** Use the fact that $x^2 + 10x + 25 = (x + 5)^2$ to find $\sqrt{x^2 + 10x + 25}$.

## Applying the Concepts

**71. Pendulum Problem**   The time (in seconds) it takes for the pendulum on a clock to swing through one complete cycle is given by the formula

$$T = \frac{11}{7}\sqrt{\frac{L}{2}}$$

where $L$ is the length (in feet) of the pendulum. Use this formula and a calculator to complete the following table. Round your answers to the nearest hundredth.

| Length $L$ (feet) | Time $T$ (seconds) |
|---|---|
| 1 | |
| 2 | |
| 3 | |
| 4 | |
| 5 | |
| 6 | |

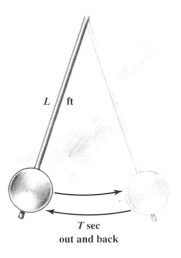

$L$ ft

$T$ sec
out and back

**72. Lighthouse Problem**   The higher you are above the ground, the farther you can see. If your view is unobstructed, then the distance in miles that you can see from $h$ feet above the ground is given by the formula

$$d = \sqrt{\frac{3h}{2}}$$

$h$   $d$

The following figure shows a lighthouse with a door and windows at various heights. The preced-

ing formula can be used to find the distance to the ocean horizon from these heights. Use the formula and a calculator to complete the following table. Round your answers to the nearest whole number.

| Height $h$ (feet) | Distance $d$ (miles) |
|---|---|
| 10 | |
| 50 | |
| 90 | |
| 130 | |
| 170 | |
| 190 | |

190 ft

170 ft

130 ft

90 ft

50 ft

10 ft

**Return on an Investment**   In the next four problems, you will need the following formula. The formula gives the annual rate of return $r$ on an investment of $P$ dollars that grows to $A$ dollars over a 2-year period of time:

$$r = \frac{\sqrt{A} - \sqrt{P}}{\sqrt{P}}$$

**73.** If you invest \$50 in silver coins and 2 years later you find the coins are worth \$65, what is the annual rate of return on your investment?

**74.** If you pay \$35 for some rare stamps and 2 years later find that the same stamps cost \$45, what is the annual rate of return on your investment?

**75.** Suppose you invest \$500 in the stock market and 2 years later your stocks are worth \$600. Find the annual rate of return on your stocks.

**76.** Suppose you invest $700 in the stock market and 2 years later your stocks are worth $800. Find the annual rate of return on your stocks.

Find $x$ in each of the following right triangles.

**77.**

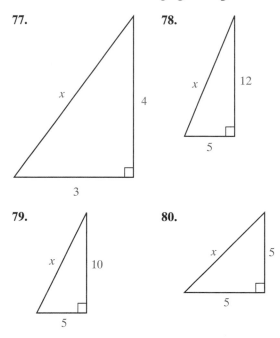

**78.**

12

$x$

5

3

**79.**

$x$

10

5

**80.**

$x$

5

5

**81. Geometry** One end of a wire is attached to the top of a 24-foot pole; the other end of the wire is anchored to the ground 18 feet from the bottom of the pole. If the pole makes an angle of 90° with the ground, find the length of the wire.

24 ft

90°

18 ft

**82. Geometry** The screen on a television set is in the shape of a rectangle. If the length is 20 inches and the width is 12 inches, how many inches is it from one corner of the screen to the opposite corner?

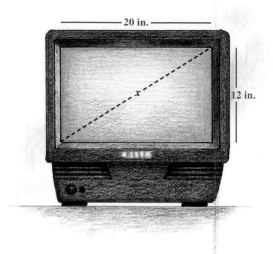

— 20 in. —

$x$

12 in.

**83. Geometry** Two children are trying to cross a stream. They want to use a log that goes from one bank to the other. If the left bank is 5 feet higher than the right bank and the stream is 12 feet wide, how long must a log be to just barely reach?

left bank

5 ft

right bank

12 ft

**84. Geometry, Surveying** A surveying team wants to calculate how long a straight tunnel through a mountain will be. They find a place where the lines

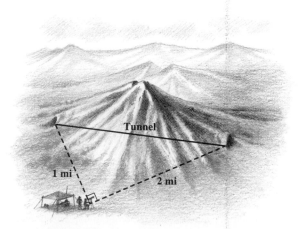

Tunnel

1 mi

2 mi

to the two ends of the proposed tunnel form a right angle. The distances are shown. How long will the tunnel be? Give the exact answer, and, if you have a calculator, an approximate answer to two decimal places.

The following diagram is a rectangular solid with length $l$, width $w$, and height $h$. Diagonal $c$ is in the bottom of the solid; diagonal $d$ runs through the center of the solid.

**85. Geometry**   If the length of the solid is 3 centimeters, the width is 4 centimeters, and the height is 12 centimeters, find the length of both diagonals.

**86. Geometry**   Find a formula for diagonal $d$ that involves only $l$, $w$, and $h$.

## Review Problems

The following problems review material we covered in Section 7.1.

Reduce each expression to lowest terms.

**87.** $\dfrac{x^2 - 16}{x + 4}$

**88.** $\dfrac{x - 5}{x^2 - 25}$

**89.** $\dfrac{10a + 20}{5a^2 - 20}$

**90.** $\dfrac{8a - 16}{4a^2 - 16}$

**91.** $\dfrac{2x^2 - 5x - 3}{x^2 - 3x}$

**92.** $\dfrac{x^2 - 5x}{3x^2 - 13x - 10}$

**93.** $\dfrac{xy + 3x + 2y + 6}{xy + 3x + ay + 3a}$

**94.** $\dfrac{xy + 5x + 4y + 20}{x^2 + bx + 4x + 4b}$

---

## 8.2    Properties of Radicals

In this section we will consider the first part of what is called *simplified form* for radical expressions. A radical expression is any expression containing a radical, whether it is a square root, a cube root, or a higher root. Simplified form for a radical expression is the form that is easiest to work with. The first step in putting a radical expression in simplified form is to take as much out from under the radical sign as possible. To do this, we must first develop two properties of radicals in general.

Consider the following two problems:

$$\sqrt{9 \cdot 16} = \sqrt{144} = 12$$

$$\sqrt{9} \cdot \sqrt{16} = 3 \cdot 4 = 12$$

Since the answers to both are equal, the original problems must also be equal. That is, $\sqrt{9 \cdot 16} = \sqrt{9} \cdot \sqrt{16}$. We can generalize this property as follows.

### PROPERTY I FOR RADICALS

If $x$ and $y$ represent nonnegative real numbers, then it is always true that

$$\sqrt{xy} = \sqrt{x}\,\sqrt{y}$$

*In words:*   The square root of a product is the product of the square roots.

We can use this property to simplify radical expressions.

**E X A M P L E   1**   Simplify $\sqrt{20}$.

**Solution**   To simplify $\sqrt{20}$, we want to take as much out from under the radical sign as possible. We begin by looking for the largest perfect square that is a factor of 20. The largest perfect square that divides 20 is 4, so we write 20 as $4 \cdot 5$:

$$\sqrt{20} = \sqrt{4 \cdot 5}$$

Next, we apply the first property of radicals and write

$$\sqrt{4 \cdot 5} = \sqrt{4}\,\sqrt{5}$$

And since $\sqrt{4} = 2$, we have

$$\sqrt{4}\,\sqrt{5} = 2\sqrt{5}$$

The expression $2\sqrt{5}$ is the simplified form of $\sqrt{20}$, since we have taken as much out from under the radical sign as possible.

**Note:**   Working a problem like the one in Example 1 depends on recognizing the largest perfect square that divides (is a factor of) the number under the radical sign. The set of perfect squares is the set

$$\{1, 4, 9, 16, 25, 36, \ldots\}$$

To simplify an expression like $\sqrt{20}$, we must first find the largest number in this set that is a factor of the number under the radical sign.

**E X A M P L E   2**   Simplify $\sqrt{75}$.

**Solution**   Since 25 is the largest perfect square that divides 75, we have

$$\sqrt{75} = \sqrt{25 \cdot 3} \qquad \text{Factor 75 into } 25 \cdot 3.$$
$$= \sqrt{25}\,\sqrt{3} \qquad \text{Property 1 for radicals}$$
$$= 5\sqrt{3} \qquad \sqrt{25} = 5$$

The expression $5\sqrt{3}$ is the simplified form for $\sqrt{75}$, since we have taken as much out from under the radical sign as possible.

The next two examples involve square roots of expressions that contain variables. Remember, we are assuming that all variables that appear under a radical sign represent positive numbers.

**E X A M P L E   3**   Simplify $\sqrt{25x^3}$.

**Solution**   The largest perfect square that is a factor of $25x^3$ is $25x^2$. We write $25x^3$ as $25x^2 \cdot x$ and apply Property 1:

$$\sqrt{25x^3} = \sqrt{25x^2 \cdot x} \qquad \text{Factor } 25x^3 \text{ into } 25x^2 \cdot x.$$
$$= \sqrt{25x^2}\,\sqrt{x} \qquad \text{Property 1 for radicals}$$
$$= 5x\sqrt{x} \qquad \sqrt{25x^2} = 5x$$

**E X A M P L E   4**   Simplify $\sqrt{18y^4}$.

**Solution**   The largest perfect square that is a factor of $18y^4$ is $9y^4$. We write $18y^4$ as $9y^4 \cdot 2$ and apply Property 1:

$$\sqrt{18y^4} = \sqrt{9y^4 \cdot 2} \qquad \text{Factor } 18y^4 \text{ into } 9y^4 \cdot 2.$$
$$= \sqrt{9y^4}\,\sqrt{2} \qquad \text{Property 1 for radicals}$$
$$= 3y^2\sqrt{2} \qquad \sqrt{9y^4} = 3y^2$$

**E X A M P L E   5**   Simplify $3\sqrt{32}$.

**Solution**   We want to get as much out from under $\sqrt{32}$ as possible. Since 16 is the largest perfect square that divides 32, we have:

$$3\sqrt{32} = 3\sqrt{16 \cdot 2} \qquad \text{Factor 32 into } 16 \cdot 2.$$
$$= 3\sqrt{16}\,\sqrt{2} \qquad \text{Property 1 for radicals}$$
$$= 3 \cdot 4\sqrt{2} \qquad \sqrt{16} = 4$$
$$= 12\sqrt{2} \qquad 3 \cdot 4 = 12$$

Although we have stated Property 1 for radicals in terms of square roots only, it holds for higher roots as well. If we were to state Property 1 again for cube roots, it would look like this:

$$\sqrt[3]{xy} = \sqrt[3]{x}\,\sqrt[3]{y}$$

**E X A M P L E   6**   Simplify $\sqrt[3]{24x^3}$.

**Solution**   Since we are simplifying a cube root, we look for the largest perfect cube that is a factor of $24x^3$. Since 8 is a perfect cube, the largest perfect cube that is a factor of $24x^3$ is $8x^3$.

$$\sqrt[3]{24x^3} = \sqrt[3]{8x^3 \cdot 3} \qquad \text{Factor } 24x^3 \text{ into } 8x^3 \cdot 3.$$
$$= \sqrt[3]{8x^3}\,\sqrt[3]{3} \qquad \text{Property 1 for radicals}$$
$$= 2x\sqrt[3]{3} \qquad \sqrt[3]{8x^3} = 2x$$

The second property of radicals has to do with division. The property becomes apparent when we consider the following two problems:

$$\sqrt{\frac{64}{16}} = \sqrt{4} = 2$$

$$\frac{\sqrt{64}}{\sqrt{16}} = \frac{8}{4} = 2$$

Since the answers in each case are equal, the original problem must be also:

$$\sqrt{\frac{64}{16}} = \frac{\sqrt{64}}{\sqrt{16}}$$

Here is the property in general.

## PROPERTY 2 FOR RADICALS

If $x$ and $y$ both represent nonnegative real numbers and $y \neq 0$, then it is always true that

$$\sqrt{\frac{x}{y}} = \frac{\sqrt{x}}{\sqrt{y}}$$

*In words:*   The square root of a quotient is the quotient of the square roots.

Although we have stated Property 2 for square roots only, it holds for higher roots as well.

We can use Property 2 for radicals in much the same way as we used Property 1 to simplify radical expressions.

**EXAMPLE 7**   Simplify $\sqrt{\frac{49}{81}}$.

**Solution**   We begin by applying Property 2 for radicals to separate the fraction into two separate radicals. Then we simplify each radical separately:

$$\sqrt{\frac{49}{81}} = \frac{\sqrt{49}}{\sqrt{81}} \qquad \text{Property 2 for radicals}$$

$$= \frac{7}{9} \qquad \sqrt{49} = 7 \text{ and } \sqrt{81} = 9$$

**EXAMPLE 8**   Simplify $\sqrt[4]{\frac{81}{16}}$.

**Solution**   Remember, although Property 2 has been stated in terms of square roots, it holds for higher roots as well. Proceeding as we did in Example 7, we have

$$\sqrt[4]{\frac{81}{16}} = \frac{\sqrt[4]{81}}{\sqrt[4]{16}} \qquad \text{Property 2 for radicals}$$

$$= \frac{3}{2} \qquad \sqrt[4]{81} = 3 \text{ and } \sqrt[4]{16} = 2$$

**EXAMPLE 9**  Simplify $\sqrt{\dfrac{50}{49}}$.

**Solution**  Applying Property 2 for radicals and then simplifying each resulting radical separately, we have

$$\sqrt{\dfrac{50}{49}} = \dfrac{\sqrt{50}}{\sqrt{49}} \qquad \text{Property 2 for radicals}$$

$$= \dfrac{\sqrt{25 \cdot 2}}{7} \qquad \text{Factor } 50 = 25 \cdot 2 \text{ and } \sqrt{49} = 7.$$

$$= \dfrac{\sqrt{25}\,\sqrt{2}}{7} \qquad \text{Property 1 for radicals}$$

$$= \dfrac{5\sqrt{2}}{7} \qquad \sqrt{25} = 5$$

**EXAMPLE 10**  Simplify $\sqrt{\dfrac{12x^2}{25}}$.

**Solution**  Proceeding as we have in the previous three examples, we use Property 2 for radicals to separate the numerator and denominator into two separate radicals. Then, we simplify each radical separately:

$$\sqrt{\dfrac{12x^2}{25}} = \dfrac{\sqrt{12x^2}}{\sqrt{25}} \qquad \text{Property 2 for radicals}$$

$$= \dfrac{\sqrt{4x^2 \cdot 3}}{5} \qquad \text{Factor } 12x^2 = 4x^2 \cdot 3 \text{ and } \sqrt{25} = 5.$$

$$= \dfrac{\sqrt{4x^2}\,\sqrt{3}}{5} \qquad \text{Property 1 for radicals}$$

$$= \dfrac{2x\sqrt{3}}{5} \qquad \sqrt{4x^2} = 2x$$

**EXAMPLE 11**  Simplify $\sqrt{\dfrac{50x^3y^2}{49}}$.

**Solution**  We begin by taking the square roots of $50x^3y^2$ and 49 separately and then writing $\sqrt{49}$ as 7:

$$\sqrt{\dfrac{50x^3y^2}{49}} = \dfrac{\sqrt{50x^3y^2}}{\sqrt{49}} \qquad \text{Property 2 for radicals}$$

$$= \dfrac{\sqrt{50x^3y^2}}{7} \qquad \sqrt{49} = 7$$

To simplify the numerator of this last expression, we determine that the largest perfect square that is a factor of $50x^3y^2$ is $25x^2y^2$. Continuing, we have

$$= \frac{\sqrt{25x^2y^2 \cdot 2x}}{7} \qquad \text{Factor } 50x^3y^2 \text{ into } 25x^2y^2 \cdot 2x.$$

$$= \frac{\sqrt{25x^2y^2}\,\sqrt{2x}}{7} \qquad \text{Property 1 for radicals}$$

$$= \frac{5xy\sqrt{2x}}{7} \qquad \sqrt{25x^2y^2} = 5xy$$

## Getting Ready for Class

*After reading through the preceding section, respond in your own words and in complete sentences.*

**A.** Describe Property 1 for radicals.

**B.** Describe Property 2 for radicals.

**C.** Explain why this statement is false: "The square root of a sum is the sum of the square roots."

**D.** Describe how you would apply Property 1 for radicals to $\sqrt{20}$.

## PROBLEM SET 8.2

Use Property 1 for radicals to simplify the following radical expressions as much as possible. Assume all variables represent positive numbers.

**1.** $\sqrt{8}$      **2.** $\sqrt{18}$      **3.** $\sqrt{12}$

**4.** $\sqrt{27}$      **5.** $\sqrt[3]{24}$      **6.** $\sqrt[3]{54}$

**7.** $\sqrt{50x^2}$      **8.** $\sqrt{32x^2}$      **9.** $\sqrt{45a^2b^2}$

**10.** $\sqrt{128a^2b^2}$      **11.** $\sqrt[3]{54x^3}$      **12.** $\sqrt[3]{128x^3}$

**13.** $\sqrt{32x^4}$      **14.** $\sqrt{48x^4}$      **15.** $5\sqrt{80}$

**16.** $3\sqrt{125}$      **17.** $\frac{1}{2}\sqrt{28x^3}$      **18.** $\frac{2}{3}\sqrt{54x^3}$

**19.** $x\sqrt[3]{8x^4}$      **20.** $x\sqrt[3]{8x^5}$      **21.** $2a\sqrt[3]{27a^5}$

**22.** $3a\sqrt[3]{27a^4}$      **23.** $\frac{4}{3}\sqrt{45a^3}$      **24.** $\frac{3}{5}\sqrt{300a^3}$

**25.** $3\sqrt{50xy^2}$      **26.** $4\sqrt{18xy^2}$      **27.** $7\sqrt{12x^2y}$

**28.** $6\sqrt{20x^2y}$

Use Property 2 for radicals to simplify each of the following. Assume all variables represent positive numbers.

**29.** $\sqrt{\dfrac{16}{25}}$      **30.** $\sqrt{\dfrac{81}{64}}$      **31.** $\sqrt{\dfrac{4}{9}}$

**32.** $\sqrt{\dfrac{49}{16}}$      **33.** $\sqrt[3]{\dfrac{8}{27}}$      **34.** $\sqrt[3]{\dfrac{64}{27}}$

**35.** $\sqrt[4]{\dfrac{16}{81}}$      **36.** $\sqrt[4]{\dfrac{81}{16}}$      **37.** $\sqrt{\dfrac{100x^2}{25}}$

**38.** $\sqrt{\dfrac{100x^2}{4}}$      **39.** $\sqrt{\dfrac{81a^2b^2}{9}}$      **40.** $\sqrt{\dfrac{64a^2b^2}{16}}$

**41.** $\sqrt[3]{\dfrac{27x^3}{8y^3}}$      **42.** $\sqrt[3]{\dfrac{125x^3}{64y^3}}$

Use combinations of Properties 1 and 2 for radicals to simplify the following problems as much as possible. Assume all variables represent positive numbers.

**43.** $\sqrt{\dfrac{50}{9}}$      **44.** $\sqrt{\dfrac{32}{49}}$      **45.** $\sqrt{\dfrac{75}{25}}$

**46.** $\sqrt{\dfrac{300}{4}}$   **47.** $\sqrt{\dfrac{128}{49}}$   **48.** $\sqrt{\dfrac{32}{64}}$

**49.** $\sqrt{\dfrac{288x}{25}}$   **50.** $\sqrt{\dfrac{28y}{81}}$   **51.** $\sqrt{\dfrac{54a^2}{25}}$

**52.** $\sqrt{\dfrac{243a^2}{49}}$   **53.** $\dfrac{3\sqrt{50}}{2}$   **54.** $\dfrac{5\sqrt{48}}{3}$

**55.** $\dfrac{7\sqrt{28y^2}}{3}$   **56.** $\dfrac{9\sqrt{243x^2}}{2}$   **57.** $\dfrac{5\sqrt{72a^2b^2}}{\sqrt{36}}$

**58.** $\dfrac{2\sqrt{27a^2b^2}}{\sqrt{9}}$   **59.** $\dfrac{6\sqrt{8x^2y}}{\sqrt{4}}$   **60.** $\dfrac{5\sqrt{32xy^2}}{\sqrt{25}}$

Use a calculator to help complete the following tables. If an answer needs rounding, round to the nearest thousandth.

**61.**

| $x$ | $\sqrt{x}$ | $2\sqrt{x}$ | $\sqrt{4x}$ |
|---|---|---|---|
| 1 | | | |
| 2 | | | |
| 3 | | | |
| 4 | | | |

**62.**

| $x$ | $\sqrt{x}$ | $2\sqrt{x}$ | $\sqrt{4x}$ |
|---|---|---|---|
| 1 | | | |
| 4 | | | |
| 9 | | | |
| 16 | | | |

**63.**

| $x$ | $\sqrt{x}$ | $3\sqrt{x}$ | $\sqrt{9x}$ |
|---|---|---|---|
| 1 | | | |
| 2 | | | |
| 3 | | | |
| 4 | | | |

**64.**

| $x$ | $\sqrt{x}$ | $3\sqrt{x}$ | $\sqrt{9x}$ |
|---|---|---|---|
| 1 | | | |
| 4 | | | |
| 9 | | | |
| 16 | | | |

## Applying the Concepts

**65. Falling Time**   The formula that gives the number of seconds $t$ it takes for an object to reach the ground when dropped from the top of a building $h$ feet high is

$$t = \sqrt{\dfrac{h}{16}}$$

If a rock is dropped from the top of a building 25 feet high, how long will it take for the rock to hit the ground?

25 ft

**66. Falling Time**   Using the formula given in Problem 65, how long will it take for an object dropped from a building 100 feet high to reach the ground?

## Review Problems

The following problems review material we covered in Section 7.2.

Multiply or divide as indicated.

**67.** $\dfrac{8x}{x^2 - 5x} \cdot \dfrac{x^2 - 25}{4x^2 + 4x}$   **68.** $\dfrac{x^2 + 4x}{4x^2} \cdot \dfrac{7x}{x^2 + 5x}$

**69.** $\dfrac{x^2 + 3x - 4}{3x^2 + 7x - 20} \div \dfrac{x^2 - 2x + 1}{3x^2 - 2x - 5}$

**70.** $\dfrac{x^2 - 16}{2x^2 + 7x + 3} \div \dfrac{2x^2 - 5x - 12}{4x^2 + 8x + 3}$

**71.** $(x^2 - 36)\left(\dfrac{x + 3}{x - 6}\right)$   **72.** $(x^2 - 49)\left(\dfrac{x + 5}{x + 7}\right)$

A radical expression is in simplified form if it has three special characteristics. Radical expressions that are in simplified form are generally easier to work with.

---

**DEFINITION** A radical expression is in **simplified form** if

1. There are no perfect squares that are factors of the quantity under the square root sign, no perfect cubes that are factors of the quantity under the cube root sign, and so on. We want as little as possible under the radical sign.
2. There are no fractions under the radical sign.
3. There are no radicals in the denominator.

---

A radical expression that has these three characteristics is said to be in simplified form. As we will see, simplified form is not always the least complicated expression. In many cases, the simplified expression looks more complicated than the original expression. The important thing about simplified form for radicals is that simplified expressions are easier to work with.

The tools we will use to put radical expressions into simplified form are the properties of radicals. We list the properties again for clarity.

### PROPERTIES OF RADICALS

If $a$ and $b$ represent any two nonnegative real numbers, then it is always true that

1. $\sqrt{a}\,\sqrt{b} = \sqrt{a \cdot b}$
2. $\dfrac{\sqrt{a}}{\sqrt{b}} = \sqrt{\dfrac{a}{b}}$ $\qquad b \neq 0$
3. $\sqrt{a}\,\sqrt{a} = (\sqrt{a})^2 = a$ $\qquad$ This property comes directly from the definition of radicals.

The following examples illustrate how we put a radical expression into simplified form using the three properties of radicals. Although the properties are stated for square roots only, they hold for all roots. [Property 3 written for cube roots would be $\sqrt[3]{a}\,\sqrt[3]{a}\,\sqrt[3]{a} = (\sqrt[3]{a})^3 = a$.]

---

**EXAMPLE 1** Put $\sqrt{\dfrac{1}{2}}$ into simplified form.

**Solution** The expression $\sqrt{\dfrac{1}{2}}$ is not in simplified form because there is a fraction under the radical sign. We can change this by applying Property 2 for radicals:

$$\sqrt{\frac{1}{2}} = \frac{\sqrt{1}}{\sqrt{2}} \qquad \text{Property 2 for radicals}$$

$$= \frac{1}{\sqrt{2}} \qquad \sqrt{1} = 1$$

The expression $\dfrac{1}{\sqrt{2}}$ is not in simplified form because there is a radical sign in the denominator. If we multiply the numerator and denominator of $\dfrac{1}{\sqrt{2}}$ by $\sqrt{2}$, the denominator becomes $\sqrt{2} \cdot \sqrt{2} = 2$:

$$\frac{1}{\sqrt{2}} = \frac{1}{\sqrt{2}} \cdot \frac{\sqrt{2}}{\sqrt{2}} \qquad \text{Multiply numerator and denominator by } \sqrt{2}.$$

$$= \frac{\sqrt{2}}{2} \qquad \begin{array}{l} 1 \cdot \sqrt{2} = \sqrt{2} \\ \sqrt{2} \cdot \sqrt{2} = \sqrt{4} = 2 \end{array}$$

If we check the expression $\dfrac{\sqrt{2}}{2}$ against our definition of simplified form for radicals, we find that all three rules hold. There are no perfect squares that are factors of 2. There are no fractions under the radical sign. No radicals appear in the denominator. The expression $\dfrac{\sqrt{2}}{2}$, therefore, must be in simplified form.

**EXAMPLE 2** Write $\sqrt{\frac{2}{3}}$ in simplified form.

**Solution** We proceed as we did in Example 1:

$$\sqrt{\frac{2}{3}} = \frac{\sqrt{2}}{\sqrt{3}} \qquad \text{Use Property 2 to separate radicals.}$$

$$= \frac{\sqrt{2}}{\sqrt{3}} \cdot \frac{\sqrt{3}}{\sqrt{3}} \qquad \text{Multiply by } \frac{\sqrt{3}}{\sqrt{3}} \text{ to remove the radical from the denominator.}$$

$$= \frac{\sqrt{6}}{3} \qquad \begin{array}{l} \sqrt{2} \cdot \sqrt{3} = \sqrt{6} \\ \sqrt{3} \cdot \sqrt{3} = \sqrt{9} = 3 \end{array}$$

**EXAMPLE 3** Put the expression $\dfrac{6\sqrt{20}}{2\sqrt{5}}$ into simplified form.

**Solution** Although there are many ways to begin this problem, we notice that 20 is divisible by 5. Using Property 2 for radicals as the first step, we can quickly put the expression into simplified form:

$$\frac{6\sqrt{20}}{2\sqrt{5}} = \frac{6}{2}\sqrt{\frac{20}{5}} \qquad \text{Property 2 for radicals}$$

$$= 3\sqrt{4} \qquad \frac{20}{5} = 4$$

$$= 3 \cdot 2 \qquad \sqrt{4} = 2$$

$$= 6$$

**Note:**  When working with square roots of variable quantities, we will always assume that the variables represent positive numbers. That way we can say that $\sqrt{x^2} = x$.

---

**E X A M P L E  4**   Simplify $\sqrt{\dfrac{4x^3y^2}{3}}$.

**Solution**   We begin by separating the numerator and denominator, and then taking the perfect squares out of the numerator:

$$\sqrt{\frac{4x^3y^2}{3}} = \frac{\sqrt{4x^3y^2}}{\sqrt{3}} \qquad \text{Property 2 for radicals}$$

$$= \frac{\sqrt{4x^2y^2}\,\sqrt{x}}{\sqrt{3}} \qquad \text{Property 1 for radicals}$$

$$= \frac{2xy\,\sqrt{x}}{\sqrt{3}} \qquad \sqrt{4x^2y^2} = 2xy$$

The only thing keeping our expression from being in simplified form is the $\sqrt{3}$ in the denominator. We can take care of this by multiplying the numerator and denominator by $\sqrt{3}$:

$$\frac{2xy\sqrt{x}}{\sqrt{3}} = \frac{2xy\sqrt{x}}{\sqrt{3}} \cdot \frac{\sqrt{3}}{\sqrt{3}} \qquad \text{Multiply numerator and denominator by } \sqrt{\mathbf{3}}.$$

$$= \frac{2xy\sqrt{3x}}{3} \qquad \sqrt{3} \cdot \sqrt{3} = \sqrt{9} = 3$$

---

Although the final expression may look more complicated than the original expression, it is in simplified form. The last step is called *rationalizing the denominator*. We have taken the radical out of the denominator and replaced it with a rational number.

---

**E X A M P L E  5**   Simplify $\sqrt[3]{\dfrac{2}{3}}$.

**Solution**   We can apply Property 2 first to separate the cube roots:

$$\sqrt[3]{\frac{2}{3}} = \frac{\sqrt[3]{2}}{\sqrt[3]{3}}$$

To write this expression in simplified form, we must remove the radical from the denominator. Since the radical is a cube root, we will need to multiply it by an expression that will give us a perfect cube under that cube root. We can accomplish this by multiplying the numerator and denominator by $\sqrt[3]{9}$. Here is what it looks like:

$$\frac{\sqrt[3]{2}}{\sqrt[3]{3}} = \frac{\sqrt[3]{2}}{\sqrt[3]{3}} \cdot \frac{\sqrt[3]{9}}{\sqrt[3]{9}}$$

$$= \frac{\sqrt[3]{18}}{\sqrt[3]{27}} \qquad \sqrt[3]{2}\cdot\sqrt[3]{9} = \sqrt[3]{18}$$
$$\qquad\qquad\qquad \sqrt[3]{3}\cdot\sqrt[3]{9} = \sqrt[3]{27}$$

$$= \frac{\sqrt[3]{18}}{3} \qquad \sqrt[3]{27} = 3$$

To see why multiplying numerator and denominator by $\sqrt[3]{9}$ works in this example, you must first convince yourself that multiplying the numerator and denominator by $\sqrt[3]{3}$ would not have worked.

**EXAMPLE 6**   Simplify $\sqrt[3]{\frac{1}{4}}$.

**Solution**   We begin by separating the numerator and denominator:

$$\sqrt[3]{\frac{1}{4}} = \frac{\sqrt[3]{1}}{\sqrt[3]{4}} \qquad \text{Property 2 for radicals}$$

$$= \frac{1}{\sqrt[3]{4}} \qquad \sqrt[3]{1} = 1$$

To rationalize the denominator, we need to have a perfect cube under the cube root sign. If we multiply the numerator and denominator by $\sqrt[3]{2}$, we will have $\sqrt[3]{4}\cdot\sqrt[3]{2} = \sqrt[3]{8}$ in the denominator:

$$\frac{1}{\sqrt[3]{4}} = \frac{1}{\sqrt[3]{4}} \cdot \frac{\sqrt[3]{2}}{\sqrt[3]{2}} \qquad \text{Multiply numerator and denominator by } \sqrt[3]{2}.$$

$$= \frac{\sqrt[3]{2}}{\sqrt[3]{8}} \qquad \sqrt[3]{4}\cdot\sqrt[3]{2} = \sqrt[3]{8}$$

$$= \frac{\sqrt[3]{2}}{2} \qquad \sqrt[3]{8} = 2$$

The final expression has no radical sign in the denominator and therefore is in simplified form.

## THE SPIRAL OF ROOTS

To visualize the square roots of the positive integers, we can construct the spiral of roots, which we mentioned in the introduction to this chapter. To begin, we draw two line segments, each of length 1, at right angles to each other. Then we use the Pythagorean theorem to find the length of the diagonal. Figure 1 illustrates:

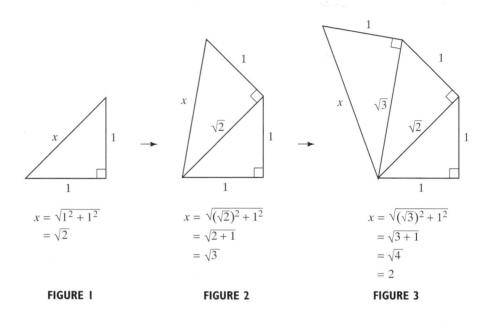

$$x = \sqrt{1^2 + 1^2}$$
$$= \sqrt{2}$$

**FIGURE 1**

$$x = \sqrt{(\sqrt{2})^2 + 1^2}$$
$$= \sqrt{2 + 1}$$
$$= \sqrt{3}$$

**FIGURE 2**

$$x = \sqrt{(\sqrt{3})^2 + 1^2}$$
$$= \sqrt{3 + 1}$$
$$= \sqrt{4}$$
$$= 2$$

**FIGURE 3**

Next, we construct a second triangle by connecting a line segment of length 1 to the end of the first diagonal so that the angle formed is a right angle. We find the length of the second diagonal using the Pythagorean theorem. Figure 2 illustrates this procedure. As we continue to draw new triangles by connecting line segments of length 1 to the end of each new diagonal, so that the angle formed is a right angle, the spiral of roots begins to appear (see Figure 3).

# Getting Ready for Class

*After reading through the preceding section, respond in your own words and in complete sentences.*

**A.** Describe how you would put $\sqrt{\frac{1}{2}}$ in simplified form.

**B.** What is simplified form for an expression that contains a square root?

**C.** What does it mean to rationalize the denominator in an expression?

**D.** What is useful about the spiral of roots?

# PROBLEM SET 8.3

Put each of the following radical expressions into simplified form. Assume all variables represent positive numbers.

**1.** $\sqrt{\dfrac{1}{2}}$  **2.** $\sqrt{\dfrac{1}{5}}$  **3.** $\sqrt{\dfrac{1}{3}}$

**4.** $\sqrt{\dfrac{1}{6}}$  **5.** $\sqrt{\dfrac{2}{5}}$  **6.** $\sqrt{\dfrac{3}{7}}$

**7.** $\sqrt{\dfrac{3}{2}}$  **8.** $\sqrt{\dfrac{5}{3}}$  **9.** $\sqrt{\dfrac{20}{3}}$

**10.** $\sqrt{\dfrac{32}{5}}$  **11.** $\sqrt{\dfrac{45}{6}}$  **12.** $\sqrt{\dfrac{48}{7}}$

**13.** $\sqrt{\dfrac{20}{5}}$  **14.** $\sqrt{\dfrac{12}{3}}$  **15.** $\dfrac{\sqrt{21}}{\sqrt{3}}$

**16.** $\dfrac{\sqrt{21}}{\sqrt{7}}$  **17.** $\dfrac{\sqrt{35}}{\sqrt{7}}$  **18.** $\dfrac{\sqrt{35}}{\sqrt{5}}$

**19.** $\dfrac{10\sqrt{15}}{5\sqrt{3}}$  **20.** $\dfrac{4\sqrt{12}}{8\sqrt{3}}$  **21.** $\dfrac{6\sqrt{21}}{3\sqrt{7}}$

**22.** $\dfrac{8\sqrt{50}}{16\sqrt{2}}$  **23.** $\dfrac{6\sqrt{35}}{12\sqrt{5}}$  **24.** $\dfrac{8\sqrt{35}}{16\sqrt{7}}$

**25.** $\sqrt{\dfrac{4x^2y^2}{2}}$  **26.** $\sqrt{\dfrac{9x^2y^2}{3}}$  **27.** $\sqrt{\dfrac{5x^2y}{3}}$

**28.** $\sqrt{\dfrac{7x^2y}{5}}$  **29.** $\sqrt{\dfrac{16a^4}{5}}$  **30.** $\sqrt{\dfrac{25a^4}{7}}$

**31.** $\sqrt{\dfrac{72a^5}{5}}$  **32.** $\sqrt{\dfrac{12a^5}{5}}$  **33.** $\sqrt{\dfrac{20x^2y^3}{3}}$

**34.** $\sqrt{\dfrac{27x^2y^3}{2}}$  **35.** $\dfrac{2\sqrt{20x^2y^3}}{3}$  **36.** $\dfrac{5\sqrt{27x^3y^2}}{2}$

**37.** $\dfrac{6\sqrt{54a^2b^3}}{5}$  **38.** $\dfrac{7\sqrt{75a^3b^2}}{6}$  **39.** $\dfrac{3\sqrt{72x^4}}{\sqrt{2x}}$

**40.** $\dfrac{2\sqrt{45x^4}}{\sqrt{5x}}$  **41.** $\sqrt[3]{\dfrac{1}{2}}$  **42.** $\sqrt[3]{\dfrac{1}{4}}$

**43.** $\sqrt[3]{\dfrac{1}{9}}$  **44.** $\sqrt[3]{\dfrac{1}{3}}$  **45.** $\sqrt[3]{\dfrac{3}{2}}$

**46.** $\sqrt[3]{\dfrac{7}{9}}$

**47.** Use a calculator to help complete the following table. If an answer needs rounding, round to the nearest thousandth.

| $x$ | $\sqrt{x}$ | $\dfrac{1}{\sqrt{x}}$ | $\dfrac{\sqrt{x}}{x}$ |
|---|---|---|---|
| 1 | | | |
| 2 | | | |
| 3 | | | |
| 4 | | | |
| 5 | | | |
| 6 | | | |

**48.** Complete the following table. Write answers as whole numbers or fractions in lowest terms.

| $x$ | $\sqrt{x}$ | $\dfrac{1}{\sqrt{x}}$ | $\dfrac{\sqrt{x}}{x}$ |
|---|---|---|---|
| 1 | | | |
| 4 | | | |
| 9 | | | |
| 16 | | | |
| 25 | | | |
| 36 | | | |

Use a calculator to help complete the following tables. If an answer needs rounding, round to the nearest thousandth.

**49.**

| $x$ | $\sqrt{x^2}$ | $\sqrt{x^3}$ | $x\sqrt{x}$ |
|---|---|---|---|
| 1 | | | |
| 2 | | | |
| 3 | | | |
| 4 | | | |
| 5 | | | |
| 6 | | | |

**50.**

| $x$ | $\sqrt{x^2}$ | $\sqrt{x^3}$ | $x\sqrt{x}$ |
|-----|------|------|------|
| 1 | | | |
| 4 | | | |
| 9 | | | |
| 16 | | | |
| 25 | | | |
| 36 | | | |

## Applying the Concepts

**51. How Far Can You See?** The higher you are above the surface of the Earth, the farther you can see. The formula that gives the approximate distance in miles, $d$, that you can see from a height of $h$ feet above the surface of the Earth is

$$d = \sqrt{\frac{3h}{2}}$$

How far can you see from a window that is 24 feet above the ground? (This is assuming that your view is unobstructed.)

**52. How Far Can You See?** Using the formula given in Problem 51, how far can you see from the window of an airplane flying at 6,000 feet?

**53. Spiral of Roots** Construct your own spiral of roots by using a ruler. Draw the first triangle by using two 1-inch lines. The first diagonal will have a length of $\sqrt{2}$ inches. Each new triangle will be formed by drawing a 1-inch line segment at the end of the previous diagonal so that the angle formed is 90°.

**54. Spiral of Roots** Construct a spiral of roots by using the line segments of length 2 inches. The length

of the first diagonal will be $2\sqrt{2}$ inches. The length of the second diagonal will be $2\sqrt{3}$ inches.

**55. Number Sequence** Simplify the terms in the following sequence. The result will be a sequence that gives the lengths of the diagonals in the spiral of roots you constructed in Problem 53.

$$\sqrt{1^2 + 1}, \sqrt{(\sqrt{2})^2 + 1}, \sqrt{(\sqrt{3})^2 + 1}, \ldots$$

**56. Number Sequence** Simplify the terms in the following sequence. The result will be a sequence that gives the lengths of the diagonals in the spiral of roots you constructed in Problem 54.

$$\sqrt{2^2 + 4}, \sqrt{(2\sqrt{2})^2 + 4}, \sqrt{(2\sqrt{3})^2 + 4}, \ldots$$

## Review Problems

The following problems review material we covered in Sections 2.1 and 7.3. Reviewing the problems from Section 2.1 will help you understand the next section.

Use the distributive property to combine the following. [2.1]

**57.** $3x + 7x$

**58.** $3x - 7x$

**59.** $15x + 8x$

**60.** $15x - 8x$

**61.** $7a - 3a + 6a$

**62.** $25a + 3a - a$

Add or subtract as indicated. [7.3]

**63.** $\dfrac{x^2}{x + 5} + \dfrac{10x + 25}{x + 5}$

**64.** $\dfrac{x^2}{x - 3} - \dfrac{9}{x - 3}$

**65.** $\dfrac{a}{3} + \dfrac{2}{5}$

**66.** $\dfrac{4}{a} + \dfrac{2}{3}$

**67.** $\dfrac{6}{a^2 - 9} - \dfrac{5}{a^2 - a - 6}$

**68.** $\dfrac{4a}{a^2 + 6a + 5} - \dfrac{3a}{a^2 + 5a + 4}$

**8.4   Addition and Subtraction of Radical Expressions**

To add two or more radical expressions, we apply the distributive property. Adding radical expressions is similar to adding similar terms of polynomials.

**EXAMPLE  1**   Combine terms in the expression $3\sqrt{5} - 7\sqrt{5}$.

**Solution**   The two terms $3\sqrt{5}$ and $7\sqrt{5}$ each have $\sqrt{5}$ in common. Since $3\sqrt{5}$ means 3 times $\sqrt{5}$, or $3 \cdot \sqrt{5}$, we apply the distributive property:

$$3\sqrt{5} - 7\sqrt{5} = (3 - 7)\sqrt{5} \qquad \text{Distributive property}$$
$$= -4\sqrt{5} \qquad\qquad 3 - 7 = -4$$

Since we use the distributive property to add radical expressions, each expression must contain exactly the same radical.

**EXAMPLE  2**   Combine terms in the expression $7\sqrt{2} - 3\sqrt{2} + 6\sqrt{2}$.

**Solution**   $7\sqrt{2} - 3\sqrt{2} + 6\sqrt{2} = (7 - 3 + 6)\sqrt{2} \qquad \text{Distributive property}$
$$= 10\sqrt{2} \qquad\qquad\quad \text{Addition}$$

In Examples 1 and 2, each term was a radical expression in simplified form. If one or more terms are not in simplified form, we must put them into simplified form and then combine terms, if possible.

**RULE**

To combine two or more radical expressions, put each expression in simplified form, and then apply the distributive property, if possible.

**EXAMPLE  3**   Combine terms in the expression $3\sqrt{50} + 2\sqrt{32}$.

**Solution**   We begin by putting each term into simplified form:

$$3\sqrt{50} + 2\sqrt{32} = 3\sqrt{25}\,\sqrt{2} + 2\sqrt{16}\,\sqrt{2} \qquad \text{Property 1 for radicals}$$
$$= 3\cdot 5\sqrt{2} + 2\cdot 4\sqrt{2} \qquad\quad \sqrt{25} = 5 \text{ and } \sqrt{16} = 4$$
$$= 15\sqrt{2} + 8\sqrt{2} \qquad\qquad \text{Multiplication}$$

Applying the distributive property to the last line, we have

$$15\sqrt{2} + 8\sqrt{2} = (15 + 8)\sqrt{2} \qquad \text{Distributive property}$$
$$= 23\sqrt{2} \qquad\qquad 15 + 8 = 23$$

**EXAMPLE  4**   Combine terms in the expression $5\sqrt{75} + \sqrt{27} - \sqrt{3}$.

**Solution**

$$5\sqrt{75} + \sqrt{27} - \sqrt{3} = 5\sqrt{25}\,\sqrt{3} + \sqrt{9}\,\sqrt{3} - \sqrt{3} \qquad \text{Property 1 for radicals}$$
$$= 5 \cdot 5\sqrt{3} + 3\sqrt{3} - \sqrt{3} \qquad \sqrt{25} = 5 \text{ and } \sqrt{9} = 3$$
$$= 25\sqrt{3} + 3\sqrt{3} - \sqrt{3} \qquad 5 \cdot 5 = 25$$
$$= (25 + 3 - 1)\sqrt{3} \qquad \text{Distributive property}$$
$$= 27\sqrt{3} \qquad \text{Addition}$$

The most time-consuming part of combining most radical expressions is simplifying each term in the expression. Once this has been done, applying the distributive property is simple and fast.

**E X A M P L E   5**   Simplify $a\sqrt{12} + 5\sqrt{3a^2}$.

**Solution**   We must assume that $a$ represents a positive number. Then we simplify each term in the expression by putting it in simplified form for radicals:

$$a\sqrt{12} + 5\sqrt{3a^2} = a\sqrt{4}\,\sqrt{3} + 5\sqrt{a^2}\,\sqrt{3} \qquad \text{Property 1 for radicals}$$
$$= a \cdot 2\sqrt{3} + 5 \cdot a\sqrt{3} \qquad \sqrt{4} = 2 \text{ and } \sqrt{a^2} = a$$
$$= 2a\sqrt{3} + 5a\sqrt{3} \qquad \text{Commutative property}$$
$$= (2a + 5a)\sqrt{3} \qquad \text{Distributive property}$$
$$= 7a\sqrt{3} \qquad \text{Addition}$$

**E X A M P L E   6**   Combine terms in the expression:

$$\sqrt{20x^3} - 3x\sqrt{45x} + 10\sqrt{25x^2}$$

(Assume $x$ is a positive real number.)

**Solution**

$$\sqrt{20x^3} - 3x\sqrt{45x} + 10\sqrt{25x^2} = \sqrt{4x^2}\,\sqrt{5x} - 3x\sqrt{9}\,\sqrt{5x} + 10\sqrt{25x^2}$$
$$= 2x\sqrt{5x} - 3x \cdot 3\sqrt{5x} + 10 \cdot 5x$$
$$= 2x\sqrt{5x} - 9x\sqrt{5x} + 50x$$

Each term is now in simplified form. The best we can do next is to combine the first two terms. The last term does not have the common radical $\sqrt{5x}$.

$$2x\sqrt{5x} - 9x\sqrt{5x} + 50x = (2x - 9x)\sqrt{5x} + 50x$$
$$= -7x\sqrt{5x} + 50x$$

We have, in any case, succeeded in reducing the number of terms in our original problem.

Our next example involves an expression that is similar to many of the expressions we will find when we solve quadratic equations in Chapter 9.

---

**EXAMPLE 7** Simplify $\dfrac{6 + \sqrt{12}}{4}$.

**Solution**   We begin by writing $\sqrt{12}$ as $2\sqrt{3}$:

$$\frac{6 + \sqrt{12}}{4} = \frac{6 + 2\sqrt{3}}{4}$$

Factor 2 from the numerator and denominator and then reduce to lowest terms.

$$\frac{6 + 2\sqrt{3}}{4} = \frac{2(3 + \sqrt{3})}{2 \cdot 2}$$

$$= \frac{3 + \sqrt{3}}{2}$$

---

# Getting Ready for Class

*After reading through the preceding section, respond in your own words and in complete sentences.*

**A.** What are similar radicals?

**B.** When can we add two radical expressions?

**C.** What is the first step when adding or subtracting expressions containing radicals?

**D.** It is not possible to add $\sqrt{2}$ and $\sqrt{3}$. Explain why.

---

## PROBLEM SET 8.4

In each of the following problems, simplify each term, if necessary, and then use the distributive property to combine terms, if possible.

**1.** $3\sqrt{2} + 4\sqrt{2}$

**2.** $7\sqrt{3} + 2\sqrt{3}$

**3.** $9\sqrt{5} - 7\sqrt{5}$

**4.** $6\sqrt{7} - 10\sqrt{7}$

**5.** $\sqrt{3} + 6\sqrt{3}$

**6.** $\sqrt{2} + 10\sqrt{2}$

**7.** $\dfrac{5}{8}\sqrt{5} - \dfrac{3}{7}\sqrt{5}$

**8.** $\dfrac{5}{6}\sqrt{11} - \dfrac{7}{9}\sqrt{11}$

**9.** $14\sqrt{13} - \sqrt{13}$

**10.** $-2\sqrt{6} - 9\sqrt{6}$

**11.** $-3\sqrt{10} + 9\sqrt{10}$

**12.** $11\sqrt{11} + \sqrt{11}$

**13.** $5\sqrt{5} + \sqrt{5}$

**14.** $\sqrt{6} - 10\sqrt{6}$

**15.** $\sqrt{8} + 2\sqrt{2}$

**16.** $\sqrt{20} + 3\sqrt{5}$

**17.** $3\sqrt{3} - \sqrt{27}$

**18.** $4\sqrt{5} - \sqrt{80}$

**19.** $5\sqrt{12} - 10\sqrt{48}$

**20.** $3\sqrt{300} - 5\sqrt{27}$

**21.** $-\sqrt{75} - \sqrt{3}$

**22.** $5\sqrt{20} + 8\sqrt{80}$

**23.** $\dfrac{1}{5}\sqrt{75} - \dfrac{1}{2}\sqrt{12}$

**24.** $\dfrac{1}{2}\sqrt{24} + \dfrac{1}{5}\sqrt{150}$

**25.** $\dfrac{3}{4}\sqrt{8} + \dfrac{3}{10}\sqrt{75}$

**26.** $\dfrac{5}{6}\sqrt{54} - \dfrac{3}{4}\sqrt{24}$

**27.** $\sqrt{27} - 2\sqrt{12} + \sqrt{3}$

**28.** $\sqrt{20} + 3\sqrt{45} - \sqrt{5}$

**29.** $\dfrac{5}{6}\sqrt{72} - \dfrac{3}{8}\sqrt{8} + \dfrac{3}{10}\sqrt{50}$

**30.** $\dfrac{3}{4}\sqrt{24} - \dfrac{5}{6}\sqrt{54} - \dfrac{7}{10}\sqrt{150}$

**31.** $5\sqrt{7} + 2\sqrt{28} - 4\sqrt{63}$

**32.** $3\sqrt{3} - 5\sqrt{27} + 8\sqrt{75}$

**33.** $6\sqrt{48} - 2\sqrt{12} + 5\sqrt{27}$

**34.** $5\sqrt{50} + 8\sqrt{12} - \sqrt{32}$

**35.** $6\sqrt{48} - \sqrt{72} - 3\sqrt{300}$

**36.** $7\sqrt{44} - 8\sqrt{99} + \sqrt{176}$

All variables in the following problems represent positive real numbers. Simplify each term, and combine, if possible.

**37.** $\sqrt{x^3} + x\sqrt{x}$     **38.** $2\sqrt{x} - 2\sqrt{4x}$

**39.** $5\sqrt{3a^2} - a\sqrt{3}$     **40.** $6a\sqrt{a} + 7\sqrt{a^3}$

**41.** $5\sqrt{8x^3} + x\sqrt{50x}$     **42.** $2\sqrt{27x^2} - x\sqrt{48}$

**43.** $3\sqrt{75x^3y} - 2x\sqrt{3xy}$

**44.** $9\sqrt{24x^3y^2} - 5x\sqrt{54xy^2}$   **45.** $\sqrt{20ab^2} - b\sqrt{45a}$

**46.** $4\sqrt{a^3b^2} - 5a\sqrt{ab^2}$    **47.** $9\sqrt{18x^3} - 2x\sqrt{48x}$

**48.** $8\sqrt{72x^2} - x\sqrt{8}$

**49.** $7\sqrt{50x^2y} + 8x\sqrt{8y} - 7\sqrt{32x^2y}$

**50.** $6\sqrt{44x^3y^3} - 8x\sqrt{99xy^3} - 6y\sqrt{176x^3y}$

Simplify each expression.

**51.** $\dfrac{8 - \sqrt{24}}{6}$   **52.** $\dfrac{8 + \sqrt{48}}{8}$   **53.** $\dfrac{6 + \sqrt{8}}{2}$

**54.** $\dfrac{4 - \sqrt{12}}{2}$     **55.** $\dfrac{-10 + \sqrt{50}}{10}$

**56.** $\dfrac{-12 + \sqrt{20}}{6}$

Use a calculator to help complete the following tables. If an answer needs rounding, round to the nearest thousandth.

**57.**

| $x$ | $\sqrt{x^2 + 9}$ | $x + 3$ |
|---|---|---|
| 1 | | |
| 2 | | |
| 3 | | |
| 4 | | |
| 5 | | |
| 6 | | |

**58.**

| $x$ | $\sqrt{x^2 + 16}$ | $x + 4$ |
|---|---|---|
| 1 | | |
| 2 | | |
| 3 | | |
| 4 | | |
| 5 | | |
| 6 | | |

**59.**

| $x$ | $\sqrt{x + 3}$ | $\sqrt{x} + \sqrt{3}$ |
|---|---|---|
| 1 | | |
| 2 | | |
| 3 | | |
| 4 | | |
| 5 | | |
| 6 | | |

**60.**

| $x$ | $\sqrt{x + 4}$ | $\sqrt{x} + 2$ |
|---|---|---|
| 1 | | |
| 2 | | |
| 3 | | |
| 4 | | |
| 5 | | |
| 6 | | |

**61. Comparing Expressions** The following statement is false. Correct the right side to make the statement true.

$$4\sqrt{3} + 5\sqrt{3} = 9\sqrt{6}$$

**62. Comparing Expressions** The following statement is false. Correct the right side to make the statement true.

$$7\sqrt{5} - 3\sqrt{5} = 4\sqrt{25}$$

## Review Problems

The following problems review material we covered in Sections 5.6 and 7.4. Reviewing the problems from Section 5.6 will help you in the next section.

Multiply. [5.6]

**63.** $(3x + y)^2$          **64.** $(2x - 3y)^2$

**65.** $(3x - 4y)(3x + 4y)$

**66.** $(7x + 2y)(7x - 2y)$

Solve each equation. [7.4]

**67.** $\dfrac{x}{3} - \dfrac{1}{2} = \dfrac{5}{2}$          **68.** $\dfrac{3}{x} + \dfrac{1}{5} = \dfrac{4}{5}$

**69.** $1 - \dfrac{5}{x} = \dfrac{-6}{x^2}$          **70.** $1 - \dfrac{1}{x} = \dfrac{6}{x^2}$

**71.** $\dfrac{a}{a - 4} - \dfrac{a}{2} = \dfrac{4}{a - 4}$

**72.** $\dfrac{a}{a - 3} - \dfrac{a}{2} = \dfrac{3}{a - 3}$

## 8.5   Multiplication and Division of Radicals

In this section we will look at multiplication and division of expressions that contain radicals. As you will see, multiplication of expressions that contain radicals is very similar to multiplication of polynomials. The division problems in this section are just an extension of the work we did previously when we rationalized denominators.

**E X A M P L E   1**   Multiply $(3\sqrt{5})(2\sqrt{7})$.

**Solution**   We can rearrange the order and grouping of the numbers in this product by applying the commutative and associative properties. Following that, we apply Property 1 for radicals and multiply:

$$(3\sqrt{5})(2\sqrt{7}) = (3 \cdot 2)(\sqrt{5} \ \sqrt{7}) \qquad \text{Commutative and associative properties}$$
$$= (3 \cdot 2)(\sqrt{5 \cdot 7}) \qquad \text{Property 1 for radicals}$$
$$= 6\sqrt{35} \qquad \text{Multiplication}$$

In actual practice, it is not necessary to show either of the first two steps, although you may want to show them on the first few problems you work, just to be sure you understand them.

**E X A M P L E   2**   Multiply $\sqrt{5}(\sqrt{2} + \sqrt{5})$.

**Solution**

$$\sqrt{5}(\sqrt{2} + \sqrt{5}) = \sqrt{5} \cdot \sqrt{2} + \sqrt{5} \cdot \sqrt{5} \qquad \text{Distributive property}$$
$$= \sqrt{10} + 5 \qquad \text{Multiplication}$$

**E X A M P L E   3**   Multiply $3\sqrt{2}(2\sqrt{5} + 5\sqrt{3})$.

**Solution**

$$3\sqrt{2}(2\sqrt{5} + 5\sqrt{3}) = 3\sqrt{2} \cdot 2\sqrt{5} + 3\sqrt{2} \cdot 5\sqrt{3} \qquad \text{Distributive property}$$

$$= 3 \cdot 2 \cdot \sqrt{2}\sqrt{5} + 3 \cdot 5\sqrt{2}\sqrt{3} \qquad \text{Commutative property}$$

$$= 6\sqrt{10} + 15\sqrt{6}$$

Each item in the last line is in simplified form, so the problem is complete.

---

**E X A M P L E   4**    Multiply $(\sqrt{5} + 2)(\sqrt{5} + 7)$.

**Solution**    We multiply using the FOIL method that we used to multiply binomials:

$$(\sqrt{5} + 2)(\sqrt{5} + 7) = \underset{\text{F}}{\sqrt{5} \cdot \sqrt{5}} + \underset{\text{O}}{7\sqrt{5}} + \underset{\text{I}}{2\sqrt{5}} + \underset{\text{L}}{14}$$

$$= 5 + 9\sqrt{5} + 14$$

$$= 19 + 9\sqrt{5}$$

We must be careful not to try to simplify further by adding 19 and 9. We can add only radical expressions that have a common radical part; 19 and $9\sqrt{5}$ are not similar.

---

**E X A M P L E   5**    Multiply $(\sqrt{x} + 3)(\sqrt{x} - 7)$.

**Solution**    Remember, we are assuming that any variables that appear under a radical represent positive numbers.

$$(\sqrt{x} + 3)(\sqrt{x} - 7) = \underset{\text{F}}{\sqrt{x}\sqrt{x}} - \underset{\text{O}}{7\sqrt{x}} + \underset{\text{I}}{3\sqrt{x}} - \underset{\text{L}}{21}$$

$$= x - 4\sqrt{x} - 21$$

---

**E X A M P L E   6**    Expand and simplify $(\sqrt{3} - 2)^2$.

**Solution**    Multiplying $\sqrt{3} - 2$ times itself, we have

$$(\sqrt{3} - 2)^2 = (\sqrt{3} - 2)(\sqrt{3} - 2)$$

$$= \sqrt{3}\sqrt{3} - 2\sqrt{3} - 2\sqrt{3} + 4$$

$$= 3 - 4\sqrt{3} + 4$$

$$= 7 - 4\sqrt{3}$$

---

**E X A M P L E   7**    Multiply $(\sqrt{5} + \sqrt{2})(\sqrt{5} - \sqrt{2})$.

**Solution**    We can apply the formula $(x + y)(x - y) = x^2 - y^2$ to obtain

$$(\sqrt{5} + \sqrt{2})(\sqrt{5} - \sqrt{2}) = (\sqrt{5})^2 - (\sqrt{2})^2$$

$$= 5 - 2$$

$$= 3$$

We could also have multiplied the two expressions using the FOIL method. If we were to do so, the work would look like this:

$$(\sqrt{5} + \sqrt{2})(\sqrt{5} - \sqrt{2}) = \underset{F}{\sqrt{5}\,\sqrt{5}} - \underset{O}{\sqrt{2}\,\sqrt{5}} + \underset{I}{\sqrt{2}\,\sqrt{5}} - \underset{L}{\sqrt{2}\,\sqrt{2}}$$

$$= 5 - \sqrt{10} + \sqrt{10} - 2$$

$$= 5 - 2$$

$$= 3$$

In either case, the product is 3. Also, the expressions $\sqrt{5} + \sqrt{2}$ and $\sqrt{5} - \sqrt{2}$ are called *conjugates* of each other.

**EXAMPLE 8** Multiply $(\sqrt{a} + \sqrt{b})(\sqrt{a} - \sqrt{b})$.

**Solution** We can apply the formula $(x + y)(x - y) = x^2 - y^2$ to obtain

$$(\sqrt{a} + \sqrt{b})(\sqrt{a} - \sqrt{b}) = (\sqrt{a})^2 - (\sqrt{b})^2$$

$$= a - b$$

**EXAMPLE 9** Rationalize the denominator in the expression

$$\frac{\sqrt{3}}{\sqrt{3} - \sqrt{2}}$$

**Solution** To remove the two radicals in the denominator, we must multiply both the numerator and denominator by $\sqrt{3} + \sqrt{2}$. That way, when we multiply $\sqrt{3} - \sqrt{2}$ and $\sqrt{3} + \sqrt{2}$, we will obtain the difference of two squares in the denominator:

$$\frac{\sqrt{3}}{\sqrt{3} - \sqrt{2}} = \frac{\sqrt{3}}{\sqrt{3} - \sqrt{2}} \cdot \frac{\sqrt{3} + \sqrt{2}}{\sqrt{3} + \sqrt{2}}$$

$$= \frac{\sqrt{3}\,\sqrt{3} + \sqrt{3}\,\sqrt{2}}{(\sqrt{3})^2 - (\sqrt{2})^2}$$

$$= \frac{3 + \sqrt{6}}{3 - 2}$$

$$= \frac{3 + \sqrt{6}}{1}$$

$$= 3 + \sqrt{6}$$

**EXAMPLE 10** Rationalize the denominator in the expression $\dfrac{2}{5 - \sqrt{3}}$.

**Solution**   We use the same procedure as in Example 9. Multiply the numerator and denominator by the conjugate of the denominator, which is $5 + \sqrt{3}$:

$$\left(\frac{2}{5 - \sqrt{3}}\right)\left(\frac{5 + \sqrt{3}}{5 + \sqrt{3}}\right) = \frac{10 + 2\sqrt{3}}{5^2 - (\sqrt{3})^2}$$

$$= \frac{10 + 2\sqrt{3}}{25 - 3}$$

$$= \frac{10 + 2\sqrt{3}}{22}$$

The numerator and denominator of this last expression have a factor of 2 in common. We can reduce to lowest terms by dividing out the common factor 2. Continuing, we have

$$= \frac{2(5 + \sqrt{3})}{2 \cdot 11}$$

$$= \frac{5 + \sqrt{3}}{11}$$

The final expression is in simplified form.

**EXAMPLE   11**   Rationalize the denominator in the expression

$$\frac{\sqrt{2} + \sqrt{3}}{\sqrt{2} - \sqrt{3}}$$

**Solution**   We remove the two radicals in the denominator by multiplying both the numerator and denominator by the conjugate of $\sqrt{2} - \sqrt{3}$, which is $\sqrt{2} + \sqrt{3}$:

$$\frac{\sqrt{2} + \sqrt{3}}{\sqrt{2} - \sqrt{3}} = \left(\frac{\sqrt{2} + \sqrt{3}}{\sqrt{2} - \sqrt{3}}\right)\frac{(\sqrt{2} + \sqrt{3})}{(\sqrt{2} + \sqrt{3})}$$

$$= \frac{\sqrt{2}\,\sqrt{2} + \sqrt{2}\,\sqrt{3} + \sqrt{3}\,\sqrt{2} + \sqrt{3}\,\sqrt{3}}{(\sqrt{2})^2 - (\sqrt{3})^2}$$

$$= \frac{2 + \sqrt{6} + \sqrt{6} + 3}{2 - 3}$$

$$= \frac{5 + 2\sqrt{6}}{-1}$$

$$= -(5 + 2\sqrt{6}) \qquad \text{or} \qquad -5 - 2\sqrt{6}$$

## Getting Ready for Class

*After reading through the preceding section, respond in your own words and in complete sentences.*

**A.** Describe how you would use the commutative and associative properties to multiply $3\sqrt{5}$ and $2\sqrt{7}$.

**B.** Explain why $(\sqrt{5} + 7)^2$ is not the same as $5 + 49$.

**C.** Explain in words how you would rationalize the denominator in the expression $\dfrac{\sqrt{3}}{\sqrt{3} - \sqrt{2}}$.

**D.** What are conjugates?

# PROBLEM SET 8.5

Perform the following multiplications. All answers should be in simplified form for radical expressions.

**1.** $\sqrt{3}\,\sqrt{2}$

**2.** $\sqrt{5}\,\sqrt{6}$

**3.** $\sqrt{6}\,\sqrt{2}$

**4.** $\sqrt{6}\,\sqrt{3}$

**5.** $(2\sqrt{3})(5\sqrt{7})$

**6.** $(3\sqrt{2})(4\sqrt{5})$

**7.** $(4\sqrt{3})(2\sqrt{6})$

**8.** $(7\sqrt{6})(3\sqrt{2})$

**9.** $\sqrt{2}(\sqrt{3} - 1)$

**10.** $\sqrt{3}(\sqrt{5} + 2)$

**11.** $\sqrt{2}(\sqrt{3} + \sqrt{2})$

**12.** $\sqrt{5}(\sqrt{7} - \sqrt{5})$

**13.** $\sqrt{3}(2\sqrt{2} + \sqrt{3})$

**14.** $\sqrt{11}(3\sqrt{2} - \sqrt{11})$

**15.** $\sqrt{3}(2\sqrt{3} - \sqrt{5})$

**16.** $\sqrt{7}(\sqrt{14} - \sqrt{7})$

**17.** $2\sqrt{3}(\sqrt{2} + \sqrt{5})$

**18.** $3\sqrt{2}(\sqrt{3} + \sqrt{2})$

**19.** $(\sqrt{2} + 1)^2$

**20.** $(\sqrt{5} - 4)^2$

**21.** $(\sqrt{x} + 3)^2$

**22.** $(\sqrt{x} - 4)^2$

**23.** $(5 - \sqrt{2})^2$

**24.** $(2 + \sqrt{5})^2$

**25.** $\left(\sqrt{a} - \dfrac{1}{2}\right)^2$

**26.** $\left(\sqrt{a} + \dfrac{1}{2}\right)^2$

**27.** $(3 + \sqrt{7})^2$

**28.** $(3 - \sqrt{2})^2$

**29.** $(\sqrt{5} + 3)(\sqrt{5} + 2)$

**30.** $(\sqrt{7} + 4)(\sqrt{7} - 5)$

**31.** $(\sqrt{2} - 5)(\sqrt{2} + 6)$

**32.** $(\sqrt{3} + 8)(\sqrt{3} - 2)$

**33.** $\left(\sqrt{3} + \dfrac{1}{2}\right)\left(\sqrt{2} + \dfrac{1}{3}\right)$

**34.** $\left(\sqrt{5} - \dfrac{1}{4}\right)\left(\sqrt{3} + \dfrac{1}{5}\right)$

**35.** $(\sqrt{x} + 6)(\sqrt{x} - 6)$

**36.** $(\sqrt{x} + 7)(\sqrt{x} - 7)$

**37.** $\left(\sqrt{a} + \dfrac{1}{3}\right)\left(\sqrt{a} + \dfrac{2}{3}\right)$

**38.** $\left(\sqrt{a} + \dfrac{1}{4}\right)\left(\sqrt{a} + \dfrac{3}{4}\right)$

**39.** $(\sqrt{5} - 2)(\sqrt{5} + 2)$

**40.** $(\sqrt{6} - 3)(\sqrt{6} + 3)$

**41.** $(2\sqrt{7} + 3)(3\sqrt{7} - 4)$

**42.** $(3\sqrt{5} + 1)(4\sqrt{5} + 3)$

**43.** $(2\sqrt{x} + 4)(3\sqrt{x} + 2)$

**44.** $(3\sqrt{x} + 5)(4\sqrt{x} + 2)$

**45.** $(7\sqrt{a} + 2\sqrt{b})(7\sqrt{a} - 2\sqrt{b})$

**46.** $(3\sqrt{a} - 2\sqrt{b})(3\sqrt{a} + 2\sqrt{b})$

Rationalize the denominator. All answers should be expressed in simplified form.

**47.** $\dfrac{\sqrt{3}}{\sqrt{5} - \sqrt{2}}$

**48.** $\dfrac{\sqrt{2}}{\sqrt{6} + \sqrt{3}}$

**49.** $\dfrac{\sqrt{5}}{\sqrt{5} + \sqrt{2}}$

**50.** $\dfrac{\sqrt{7}}{\sqrt{7} - \sqrt{2}}$

**51.** $\dfrac{8}{3 - \sqrt{5}}$

**52.** $\dfrac{10}{5 + \sqrt{5}}$

**53.** $\dfrac{\sqrt{3} + \sqrt{2}}{\sqrt{3} - \sqrt{2}}$

**54.** $\dfrac{\sqrt{5} - \sqrt{2}}{\sqrt{5} + \sqrt{2}}$

**55.** $\dfrac{\sqrt{7} - \sqrt{3}}{\sqrt{7} + \sqrt{3}}$

**56.** $\dfrac{\sqrt{11} - \sqrt{6}}{\sqrt{11} + \sqrt{6}}$

**57.** $\dfrac{\sqrt{x} + 2}{\sqrt{x} - 2}$

**58.** $\dfrac{\sqrt{x} - 3}{\sqrt{x} + 3}$

**59.** $\dfrac{\sqrt{5} - \sqrt{2}}{\sqrt{5} + \sqrt{3}}$    **60.** $\dfrac{\sqrt{7} - \sqrt{3}}{\sqrt{5} + \sqrt{2}}$

## Applying the Concepts

**61. Comparing Expressions**  The following statement is false. Correct the right side to make the statement true.

$$2(3\sqrt{5}) = 6\sqrt{15}$$

**62. Comparing Expressions**  The following statement is false. Correct the right side to make the statement true.

$$5(2\sqrt{6}) = 10\sqrt{30}$$

**63. Comparing Expressions**  The following statement is false. Correct the right side to make the statement true.

$$(\sqrt{3} + 7)^2 = 3 + 49$$

**64. Comparing Expressions**  The following statement is false. Correct the right side to make the statement true.

$$(\sqrt{5} + \sqrt{2})^2 = 5 + 2$$

## Review Problems

The following problems review material we covered in Sections 6.6 and 7.7. Reviewing these problems will help you understand the next section.

Solve each equation. [6.6].

**65.** $x^2 + 5x - 6 = 0$    **66.** $x^2 + 5x + 6 = 0$
**67.** $x^2 - 3x = 0$    **68.** $x^2 + 5x = 0$

Solve each proportion. [7.7].

**69.** $\dfrac{x}{3} = \dfrac{27}{x}$    **70.** $\dfrac{x}{2} = \dfrac{8}{x}$

**71.** $\dfrac{x}{5} = \dfrac{3}{x + 2}$    **72.** $\dfrac{x}{2} = \dfrac{4}{x + 2}$

**73. Driving Distance**  Suppose you drive your car 375 miles in 15 hours. At this rate, how far will you drive in 20 hours?

**74. Map Reading**  If 2.5 inches on a map corresponds to an actual distance of 100 miles, then how many miles apart are two cities that are 3.5 inches apart on the map?

---

## 8.6    Equations Involving Radicals

In order to solve equations that contain one or more radical expressions, we need an additional property. From our work with exponents we know that if two quantities are equal, then so are the squares of those quantities. That is, for real numbers $a$ and $b$,

$$\text{if} \qquad a = b$$
$$\text{then} \qquad a^2 = b^2$$

The only problem with squaring both sides of an equation is that occasionally we will change a false statement into a true statement. Let's take the false statement $3 = -3$ as an example.

$$3 = -3 \qquad \text{A false statement}$$
$$(3)^2 = (-3)^2 \qquad \text{Square both sides.}$$
$$9 = 9 \qquad \text{A true statement}$$

We can avoid this problem by always checking our solutions if, at any time during the process of solving an equation, we have squared both sides of the equation. Here is how the property is stated.

### SQUARING PROPERTY OF EQUALITY

We can square both sides of an equation any time it is convenient to do so, as long as we check all solutions in the original equation.

**EXAMPLE  I**   Solve for $x$: $\sqrt{x + 1} = 7$

**Solution**   To solve this equation by our usual methods, we must first eliminate the radical sign. We can accomplish this by squaring both sides of the equation:

$$\sqrt{x + 1} = 7$$
$$(\sqrt{x + 1})^2 = 7^2 \qquad \text{Square both sides.}$$
$$x + 1 = 49$$
$$x = 48$$

To check our solution, we substitute $x = 48$ into the original equation:

$$\sqrt{48 + 1} \overset{?}{=} 7$$
$$\sqrt{49} = 7$$
$$7 = 7 \qquad \text{A true statement}$$

The solution checks.

**EXAMPLE  2**   Solve for $x$: $\sqrt{2x - 3} = -9$

**Solution**   We square both sides and proceed as in Example 1:

$$\sqrt{2x - 3} = -9$$
$$(\sqrt{2x - 3})^2 = (-9)^2 \qquad \text{Square both sides.}$$
$$2x - 3 = 81$$
$$2x = 84$$
$$x = 42$$

Checking our solution in the original equation, we have

$$\sqrt{2(42) - 3} \overset{?}{=} -9$$
$$\sqrt{84 - 3} = -9$$
$$\sqrt{81} = -9$$
$$9 = -9 \qquad \text{A false statement}$$

Our solution does not check because we end up with a false statement.

Squaring both sides of the equation has produced what is called an extraneous solution. This happens occasionally when we use the squaring property of equality.

We can always eliminate extraneous solutions by checking each solution in the original equation.

*Note:* As you can see, when we check $x = 42$ in the original equation, we find that it is not a solution to the equation. Actually, it was apparent from the beginning that the equation had no solution. That is, no matter what $x$ is, the equation

$$\sqrt{2x - 3} = -9$$

can never be true because the left side is a positive number (or zero) for any value of $x$, and the right side is always negative.

---

**EXAMPLE 3**   Solve for $a$: $\sqrt{3a - 2} + 3 = 5$

**Solution**   Before we can square both sides to eliminate the radical, we must isolate the radical on the left side of the equation. To do so, we add $-3$ to both sides:

$$\sqrt{3a - 2} + 3 = 5$$
$$\sqrt{3a - 2} = 2 \qquad \text{Add } -3 \text{ to both sides.}$$
$$(\sqrt{3a - 2})^2 = 2^2 \qquad \text{Square both sides.}$$
$$3a - 2 = 4$$
$$3a = 6$$
$$a = 2$$

Checking $a = 2$ in the original equation, we have

$$\sqrt{3 \cdot 2 - 2} + 3 \overset{?}{=} 5$$
$$\sqrt{4} + 3 = 5$$
$$5 = 5 \qquad \text{A true statement}$$

---

**EXAMPLE 4**   Solve for $x$: $\sqrt{x + 15} = x + 3$

**Solution**   We begin by squaring both sides:

$$(\sqrt{x + 15})^2 = (x + 3)^2 \qquad \text{Square both sides.}$$
$$x + 15 = x^2 + 6x + 9$$

We have a quadratic equation. We put it into standard form by adding $-x$ and $-15$ to both sides. Then we factor and solve as usual:

$$0 = x^2 + 5x - 6 \qquad\qquad \text{Standard form}$$
$$0 = (x + 6)(x - 1) \qquad\qquad \text{Factor.}$$
$$x + 6 = 0 \quad \text{or} \quad x - 1 = 0 \qquad \text{Set factors equal to 0.}$$
$$x = -6 \quad \text{or} \qquad x = 1$$

We check each solution in the original equation:

$$\text{Check:} \quad -6 \qquad\qquad \text{Check:} \quad 1$$

$$\sqrt{-6 + 15} \overset{?}{=} -6 + 3 \qquad \sqrt{1 + 15} \overset{?}{=} 1 + 3$$

$$\sqrt{9} = -3 \qquad\qquad \sqrt{16} = 4$$

$$3 = -3 \qquad\qquad\qquad 4 = 4$$

A false statement          A true statement

Since $x = -6$ does not check in the original equation, it cannot be a solution. The only solution is $x = 1$.

**E X A M P L E   5**    Graph $y = \sqrt{x}$ and $y = \sqrt[3]{x}$.

**Solution**   The graphs are shown in Figures 1 and 2. Notice that the graph of $y = \sqrt{x}$ appears in the first quadrant only, because in the equation $y = \sqrt{x}$, $x$ and $y$ cannot be negative.

The graph of $y = \sqrt[3]{x}$ appears in quadrants I and III, since the cube root of a positive number is also a positive number and the cube root of a negative number is a negative number. That is, when $x$ is positive, $y$ will be positive, and when $x$ is negative, $y$ will be negative.

The graphs of both equations will contain the origin, since $y = 0$ when $x = 0$ in both equations.

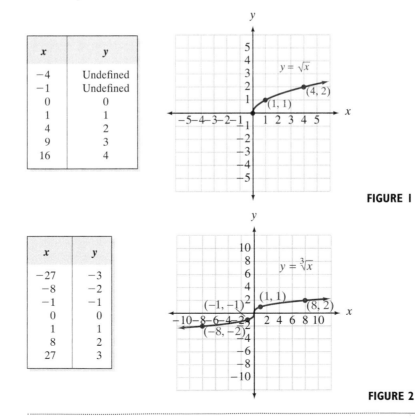

| $x$ | $y$ |
|-----|-----|
| $-4$ | Undefined |
| $-1$ | Undefined |
| $0$ | $0$ |
| $1$ | $1$ |
| $4$ | $2$ |
| $9$ | $3$ |
| $16$ | $4$ |

**FIGURE 1**

| $x$ | $y$ |
|-----|-----|
| $-27$ | $-3$ |
| $-8$ | $-2$ |
| $-1$ | $-1$ |
| $0$ | $0$ |
| $1$ | $1$ |
| $8$ | $2$ |
| $27$ | $3$ |

**FIGURE 2**

# Getting Ready for Class

*After reading through the preceding section, respond in your own words and in complete sentences.*

**A.** What is the squaring property of equality?

**B.** Under what conditions do we obtain extraneous solutions to equations that contain radical expressions?

**C.** Why do we check our solutions to equations when we have used the squaring property of equality?

**D.** Why are there no solutions to the equation $\sqrt{2x - 9} = -9$?

## PROBLEM SET 8.6

Solve each equation by applying the squaring property of equality. Be sure to check all solutions in the original equation.

**1.** $\sqrt{x + 1} = 2$

**2.** $\sqrt{x - 3} = 4$

**3.** $\sqrt{x + 5} = 7$

**4.** $\sqrt{x + 8} = 5$

**5.** $\sqrt{x - 9} = -6$

**6.** $\sqrt{x + 10} = -3$

**7.** $\sqrt{x - 5} = -4$

**8.** $\sqrt{x + 7} = -5$

**9.** $\sqrt{x - 8} = 0$

**10.** $\sqrt{x - 9} = 0$

**11.** $\sqrt{2x + 1} = 3$

**12.** $\sqrt{2x - 5} = 7$

**13.** $\sqrt{2x - 3} = -5$

**14.** $\sqrt{3x - 8} = -4$

**15.** $\sqrt{3x + 6} = 2$

**16.** $\sqrt{5x - 1} = 5$

**17.** $2\sqrt{x} = 10$

**18.** $3\sqrt{x} = 9$

**19.** $3\sqrt{a} = 6$

**20.** $2\sqrt{a} = 12$

**21.** $\sqrt{3x + 4} - 3 = 2$

**22.** $\sqrt{2x - 1} + 2 = 5$

**23.** $\sqrt{5y - 4} - 2 = 4$

**24.** $\sqrt{3y + 1} + 7 = 2$

**25.** $\sqrt{2x + 1} + 5 = 2$

**26.** $\sqrt{6x - 8} - 1 = 3$

**27.** $\sqrt{x + 3} = x - 3$

**28.** $\sqrt{x - 3} = x - 3$

**29.** $\sqrt{a + 2} = a + 2$

**30.** $\sqrt{a + 10} = a - 2$

**31.** $\sqrt{2x + 9} = x + 5$

**32.** $\sqrt{x + 6} = x + 4$

**33.** $\sqrt{y - 4} = y - 6$

**34.** $\sqrt{2y + 13} = y + 7$

**35. Pendulum Problem** The time (in seconds) it takes for the pendulum on a clock to swing through one complete cycle is given by the formula

$$T = \frac{11}{7}\sqrt{\frac{L}{2}}$$

1 sec

where $L$ is the length of the pendulum, in feet. The following table was constructed using this formula. Use the template to draw a line graph of the information in the table.

| Length $L$ (feet) | Time $T$ (seconds) |
|---|---|
| 1 | 1.11 |
| 2 | 1.57 |
| 3 | 1.92 |
| 4 | 2.22 |
| 5 | 2.48 |
| 6 | 2.72 |

**36. Lighthouse Problem**  The higher you are above the ground, the farther you can see. If your view is unobstructed, then the distance in miles that you can see from $h$ feet above the ground is given by the formula

$$d = \sqrt{\frac{3h}{2}}$$

The following table was constructed using this formula. Use the template to draw a line graph of the information in the table.

| Height $h$ (feet) | Distance $d$ (miles) |
|---|---|
| 10 | 3.9 |
| 50 | 8.7 |
| 90 | 11.6 |
| 130 | 14.0 |
| 170 | 16.0 |
| 190 | 16.9 |

Graph each equation. Begin by making a table using the given values of $x$. (In some cases you must use a calculator to approximate a square root.)

**37.** $y = \sqrt{x}, x = 0, 1, 2, 3, 4$

**38.** $y = \sqrt[3]{x}, x = -8, -4, -1, 0, 4, 8$

**39.** $y = 2\sqrt{x}, x = 0, 1, 4, 9$

**40.** $y = 3\sqrt[3]{x}, x = -8, -1, 0, 1, 8$

**41.** $y = \sqrt{x} + 2, x = 0, 1, 2, 4, 9$

**42.** $y = \sqrt[3]{x} + 3, x = -8, -1, 0, 1, 8$

## Applying the Concepts

**43. Number Problem**  The sum of a number and 2 is equal to the positive square root of 8 times the number. Find the number.

**44. Number Problem**  The sum of twice a number and 1 is equal to 3 times the positive square root of the number. Find the number.

**45. Number Problem**  The difference of a number and 3 is equal to twice the positive square root of the number. Find the number.

**46. Number Problem**  The difference of a number and 2 is equal to the positive square root of the number. Find the number.

**47. Pendulum Problem**  The number of seconds $T$ it takes the pendulum of a clock to swing through one complete cycle is given by the formula

$$T = \frac{11}{7}\sqrt{\frac{L}{2}}$$

where $L$ is the length, in feet, of the pendulum. Find how long the pendulum must be for one complete cycle to take 2 seconds by substituting 2 for $T$ in the formula and then solving for $L$.

**48. Pendulum Problem** How long must the pendulum on a clock be if one complete cycle is to take 1 second?

## Review Problems

The following problems review material we covered in Chapter 7.

**49.** Reduce to lowest terms $\dfrac{x^2 - x - 6}{x^2 - 9}$.

**50.** Divide using long division $\dfrac{x^2 - 2x + 6}{x - 4}$.

Perform the indicated operations.

**51.** $\dfrac{x^2 - 25}{x + 4} \cdot \dfrac{2x + 8}{x^2 - 9x + 20}$

**52.** $\dfrac{3x + 6}{x^2 + 4x + 3} \div \dfrac{x^2 + x - 2}{x^2 + 2x - 3}$

**53.** $\dfrac{x}{x^2 - 16} + \dfrac{4}{x^2 - 16}$

**54.** $\dfrac{2}{x^2 - 1} - \dfrac{5}{x^2 + 3x - 4}$

**55.** $\dfrac{1 - \dfrac{25}{x^2}}{1 - \dfrac{8}{x} + \dfrac{15}{x^2}}$

Solve each equation.

**56.** $\dfrac{x}{2} - \dfrac{5}{x} = -\dfrac{3}{2}$

**57.** $\dfrac{x}{x^2 - 9} - \dfrac{3}{x - 3} = \dfrac{1}{x + 3}$

**58. Speed of a Boat** A boat travels 30 miles up a river in the same amount of time it takes to travel 50 miles down the same river. If the current is 5 miles per hour, what is the speed of the boat in still water?

**59. Filling a Pool** A pool can be filled by an inlet pipe in 8 hours. The drain will empty the pool in 12 hours. How long will it take to fill the pool if both the inlet pipe and the drain are open?

**60. Mixture Problem** If 30 liters of a certain solution includes 2 liters of alcohol, how much alcohol is in 45 liters of the same solution?

**61.** $y$ varies directly with $x$. If $y = 8$ when $x$ is 12, find $y$ when $x$ is 36.

# CHAPTER 8    SUMMARY

**Examples**

**1.** The two square roots of 9 are 3 and $-3$:

$$\sqrt{9} = 3 \qquad \text{and} \qquad -\sqrt{9} = -3$$

## Roots [8.1]

Every positive real number $x$ has two square roots, one positive and one negative. The positive square root is written $\sqrt{x}$. The negative square root of $x$ is written $-\sqrt{x}$. In both cases the square root of $x$ is a number we square to get $x$.

The cube root of $x$ is written $\sqrt[3]{x}$ and is the number we cube to get $x$.

**2.**

## Notation [8.1]

In the expression $\sqrt[3]{8}$, 8 is called the *radicand*, 3 is the *index*, $\sqrt{\phantom{x}}$ is called the *radical sign*, and the whole expression $\sqrt[3]{8}$ is called the *radical*.

**3.** (a) $\sqrt{3} \cdot \sqrt{2} = \sqrt{3 \cdot 2} = \sqrt{6}$

(b) $\dfrac{\sqrt{12}}{\sqrt{3}} = \sqrt{\dfrac{12}{3}} = \sqrt{4} = 2$

(c) $\sqrt{5} \cdot \sqrt{5} = (\sqrt{5})^2 = 5$

## Properties of Radicals [8.2]

If $a$ and $b$ represent nonnegative real numbers, then

**1.** $\sqrt{a}\,\sqrt{b} = \sqrt{ab}$     The product of the square roots is the square root of the product.

**2.** $\dfrac{\sqrt{a}}{\sqrt{b}} = \sqrt{\dfrac{a}{b}}$   $(b \neq 0)$     The quotient of the square roots is the square root of the quotient.

**3.** $\sqrt{a} \cdot \sqrt{a} = (\sqrt{a})^2 = a$     This property shows that squaring and square roots are inverse operations.

**4.** Simplify $\sqrt{20}$ and $\sqrt{\dfrac{2}{3}}$.

$$\sqrt{20} = \sqrt{4 \cdot 5} = \sqrt{4}\,\sqrt{5} = 2\sqrt{5}$$

$$\sqrt{\dfrac{2}{3}} = \dfrac{\sqrt{2}}{\sqrt{3}} = \dfrac{\sqrt{2}}{\sqrt{3}} \cdot \dfrac{\sqrt{3}}{\sqrt{3}} = \dfrac{\sqrt{6}}{3}$$

## Simplified Form for Radicals [8.3]

A radical expression is in simplified form if:

**1.** There are no perfect squares that are factors of the quantity under the square root sign, no perfect cubes that are factors of the quantity under the cube root sign, and so on. We want as little as possible under the radical sign.

**2.** There are no fractions under the radical sign.

**3.** There are no radicals in the denominator.

**5.** (a) $5\sqrt{7} + 3\sqrt{7} = 8\sqrt{7}$

(b) $2\sqrt{18} - 3\sqrt{50}$
$= 2 \cdot 3\sqrt{2} - 3 \cdot 5\sqrt{2}$
$= 6\sqrt{2} - 15\sqrt{2}$
$= -9\sqrt{2}$

## Addition and Subtraction of Radical Expressions [8.4]

We add and subtract radical expressions by using the distributive property to combine terms that have the same radical parts. If the radicals are not in simplified form, we begin by writing them in simplified form and then combining similar terms, if possible.

**6.** (a) $\sqrt{3}(\sqrt{5} - \sqrt{3}) = \sqrt{15} - 3$

(b) $(\sqrt{7} + 3)(\sqrt{7} - 5)$
$= 7 - 5\sqrt{7} + 3\sqrt{7} - 15$
$= -8 - 2\sqrt{7}$

## Multiplication of Radical Expressions [8.5]

We multiply radical expressions by applying the distributive property or the FOIL method.

**7.** $\dfrac{7}{\sqrt{5} - \sqrt{3}}$

$= \dfrac{7}{\sqrt{5} - \sqrt{3}} \cdot \dfrac{\sqrt{5} + \sqrt{3}}{\sqrt{5} + \sqrt{3}}$

$= \dfrac{7\sqrt{5} + 7\sqrt{3}}{2}$

**8.** Solve $\sqrt{x - 3} = 2$.

$$(\sqrt{x - 3})^2 = 2^2$$

$$x - 3 = 4$$

$$x = 7$$

The solution checks in the original equation.

### Division of Radical Expressions [8.5]

To divide by an expression like $\sqrt{5} - \sqrt{3}$, we multiply the numerator and denominator by its conjugate, $\sqrt{5} + \sqrt{3}$. This process is also called rationalizing the denominator.

### Squaring Property of Equality [8.6]

We are free to square both sides of an equation whenever it is convenient, as long as we check all solutions in the original equation. We must check solutions because squaring both sides of an equation occasionally produces extraneous solutions.

## COMMON MISTAKES

**1.** A very common mistake with radicals is to think of $\sqrt{25}$ as representing both the positive and negative square roots of 25. The notation $\sqrt{25}$ stands for the *positive* square root of 25. If we want the negative square root of 25, we write $-\sqrt{25}$.

**2.** The most common mistake when working with radicals is to try to apply a property similar to Property 1 for radicals involving addition instead of multiplication. Here is an example:

$$\sqrt{16 + 9} = \sqrt{16} + \sqrt{9} \qquad \text{Mistake}$$

Although this example looks like it may be true, it isn't. If we carry it out further, the mistake becomes obvious:

$$\sqrt{16 + 9} \overset{?}{=} \sqrt{16} + \sqrt{9}$$

$$\sqrt{25} = 4 + 3$$

$$5 = 7 \qquad \text{False}$$

**3.** It is a mistake to try to simplify expressions like $2 + 3\sqrt{7}$. The 2 and 3 cannot be combined because the terms they appear in are not similar. Therefore, $2 + 3\sqrt{7} \neq 5\sqrt{7}$. The expression $2 + 3\sqrt{7}$ cannot be simplified further.

# CHAPTER 8 REVIEW

Numbers in brackets refer to sections of the text in which similar problems can be found.

Find the following roots. Assume all variables are positive. [8.1]

**1.** $\sqrt{25}$ **2.** $\sqrt{169}$ **3.** $\sqrt[3]{-1}$

**4.** $\sqrt[4]{625}$ **5.** $\sqrt{100x^2y^4}$ **6.** $\sqrt[3]{8a^3}$

Simplify. Assume all variables represent positive numbers. [8.2]

**7.** $\sqrt{24}$ **8.** $\sqrt{60x^2}$ **9.** $\sqrt{90x^3y^4}$

**10.** $-\sqrt{32}$ **11.** $3\sqrt{20x^3y}$

Simplify. Assume all variables represent positive numbers. [8.2]

**12.** $\sqrt{\dfrac{3}{49}}$ **13.** $\sqrt{\dfrac{8}{81}}$ **14.** $\sqrt{\dfrac{49}{64}}$

**15.** $\sqrt{\dfrac{49a^2b^2}{16}}$

Simplify. Assume all variables represent positive numbers. [8.2]

**16.** $\sqrt{\dfrac{80}{49}}$ **17.** $\sqrt{\dfrac{40a^2}{121}}$ **18.** $\dfrac{5\sqrt{84}}{7}$

**19.** $\dfrac{3\sqrt{120a^2b^2}}{\sqrt{25}}$ **20.** $\dfrac{-5\sqrt{20x^3y^2}}{\sqrt{144}}$

Write in simplest form. Assume all variables represent positive numbers. [8.3]

**21.** $\dfrac{2}{\sqrt{7}}$ **22.** $\sqrt{\dfrac{32}{5}}$ **23.** $\sqrt{\dfrac{5}{48}}$

**24.** $\dfrac{-3\sqrt{60}}{\sqrt{5}}$ **25.** $\sqrt{\dfrac{32ab^2}{3}}$ **26.** $\sqrt[3]{\dfrac{3}{4}}$

Write in simplest form. Assume all variables represent positive numbers. [8.5]

**27.** $\dfrac{3}{\sqrt{3}-4}$ **28.** $\dfrac{2}{3+\sqrt{7}}$ **29.** $\dfrac{3}{\sqrt{5}-\sqrt{2}}$

**30.** $\dfrac{\sqrt{5}}{\sqrt{3}-\sqrt{5}}$ **31.** $\dfrac{\sqrt{5}-\sqrt{2}}{\sqrt{5}+\sqrt{2}}$ **32.** $\dfrac{\sqrt{x}+3}{\sqrt{x}-3}$

Combine the following expressions. [8.4]

**33.** $3\sqrt{5}-7\sqrt{5}$ **34.** $3\sqrt{27}-5\sqrt{48}$

**35.** $-2\sqrt{45}-5\sqrt{80}+2\sqrt{20}$

**36.** $3\sqrt{50x^2}-x\sqrt{200}$ **37.** $\sqrt{40a^3b^2}-a\sqrt{90ab^2}$

Multiply. Write all answers in simplest form. [8.5]

**38.** $\sqrt{3}(\sqrt{3}+3)$ **39.** $4\sqrt{2}(\sqrt{3}+\sqrt{5})$

**40.** $(\sqrt{x}+7)(\sqrt{x}-7)$ **41.** $(2\sqrt{5}-4)(\sqrt{5}+3)$

**42.** $(\sqrt{x}+5)^2$

Solve each equation. [8.6]

**43.** $\sqrt{x-3}=3$ **44.** $\sqrt{3x-5}=4$

**45.** $5\sqrt{a}=20$ **46.** $\sqrt{3x-7}+6=2$

**47.** $\sqrt{2x+1}+10=8$ **48.** $\sqrt{7x+1}=x+1$

Find $x$ in each of the following right triangles.

**49.** **50.**

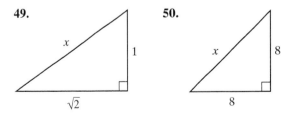

# CHAPTER 8 TEST

Find the following roots. [8.1]

**1.** $\sqrt{16}$            **2.** $-\sqrt{36}$

**3.** The square roots of 49    **4.** $\sqrt[3]{27}$

**5.** $\sqrt[3]{-8}$           **6.** $-\sqrt[4]{81}$

Put the following expressions into simplified form. [8.2, 8.3]

**7.** $\sqrt{75}$           **8.** $\sqrt{32}$

**9.** $\sqrt{\dfrac{2}{3}}$         **10.** $\dfrac{1}{\sqrt[3]{4}}$

**11.** $3\sqrt{50x^2}$       **12.** $\sqrt{\dfrac{12x^2y^3}{5}}$

Combine. [8.4]

**13.** $5\sqrt{12} - 2\sqrt{27}$     **14.** $2x\sqrt{18} + 5\sqrt{2x^2}$

Multiply. [8.5]

**15.** $\sqrt{3}(\sqrt{5} - 2)$      **16.** $(\sqrt{5} + 7)(\sqrt{5} - 8)$

**17.** $(\sqrt{x} + 6)(\sqrt{x} - 6)$     **18.** $(\sqrt{5} - \sqrt{3})^2$

Divide. (Rationalize the denominator.) [8.5]

**19.** $\dfrac{\sqrt{7} - \sqrt{3}}{\sqrt{7} + \sqrt{3}}$      **20.** $\dfrac{\sqrt{x}}{\sqrt{x} + 5}$

Solve the following equations. [8.6]

**21.** $\sqrt{2x + 1} + 2 = 7$     **22.** $\sqrt{3x + 1} + 6 = 2$

**23.** $\sqrt{2x - 3} = x - 3$

**24.** The difference of a number and 4 is equal to 3 times the positive square root of the number. Find the number.

**25.** Find $x$ in the following right triangle.

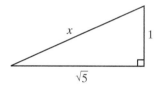

Simplify.

**1.** $\left(\dfrac{4}{5}\right)\left(\dfrac{5}{4}\right)$        **2.** $-(-3)$

**3.** $-\left|-\dfrac{1}{2}\right|$        **4.** $-40 + (-5)$

**5.** $\dfrac{4^2 - 8^2}{(4 - 8)^2}$        **6.** $-3(4)(-2)$

**7.** $4x - 7x$        **8.** $\dfrac{a^{10}}{a^2}$

**9.** $\dfrac{(b^6)^2(b^3)^4}{(b^{10})^3}$        **10.** $\dfrac{24a^{12}}{6a^3} + \dfrac{30a^{24}}{10a^{15}}$

**11.** $\left(\dfrac{1}{2}y + 2\right)\left(\dfrac{1}{2}y - 2\right)$    **12.** $5x(8x + 3)$

**13.** $\dfrac{\dfrac{1}{a + 6} + 3}{\dfrac{1}{a + 6} + 2}$

**14.** $\sqrt{120x^4y^3}$ (Assume $x$ and $y$ are positive.)

**15.** $\sqrt{\dfrac{90a^2}{169}}$ (Assume $a > 0$.)

**16.** $-\sqrt[4]{81}$

**17.** $\dfrac{7a}{a^2 - 3a - 54} + \dfrac{5}{a - 9}$

**18.** $\dfrac{3x}{3x - 4} - \dfrac{4}{3x - 4}$

Solve.

**19.** $3(5x - 1) = 6(2x + 3) - 21$

**20.** $5x - 1 = 9x + 9$

**21.** $x(3x + 2)(x - 4) = 0$

**22.** $1 + \dfrac{12}{x} = \dfrac{-35}{x^2}$    **23.** $\dfrac{4}{a} = \dfrac{a}{16}$

**24.** $\sqrt{x + 4} = 5$

Solve the following systems.

**25.** $x + y = 1$        **26.** $2x + 5y = 4$
    $x - y = 3$          $3x - 2y = -13$

**27.** $x - y = 5$
    $y = 3x - 1$

**28.** Solve the following system by graphing:
$$x - y = 5$$
$$y = -3x - 1$$

**29.** Graph on the number line: $x > -5$ and $x < 2$

Graph on a rectangular coordinate system.

**30.** $3x + 2y = 6$    **31.** $x = 1$    **32.** $y > 3x - 4$

**33.** For the equation $y = -x + 7$, find the $x$- and $y$-intercepts.

**34.** For the equation $y = -3x + 2$, find the slope and $y$-intercept.

**35.** Find the equation for the line with slope $-1$ that passes through point $(3, -3)$.

**36.** Find $y$ if the line through $(-2, -3)$ and $(2, y)$ has slope $-2$.

Factor completely.

**37.** $r^2 + r - 20$        **38.** $y^4 - 81$

**39.** $x^5 - x^4 - 30x^3$      **40.** $80x^3 - 5xy^2$

**41.** Simplify $\dfrac{(6 \times 10^5)(6 \times 10^{-3})}{9 \times 10^{-4}}$ and write the answer in scientific notation.

Rationalize the denominator.

**42.** $\dfrac{\sqrt{7} + \sqrt{3}}{\sqrt{7} - \sqrt{3}}$        **43.** $\dfrac{5}{\sqrt{3}}$

**44.** Find the value of the expression $2x + 9$ when $x = -2$.

**45.** Combine the expressions $5\sqrt{63x^2} - x\sqrt{28}$. (Assume $x > 0$.)

**46.** Divide $\dfrac{x^2 + 5x + 6}{x^2 - x - 6} \div \dfrac{x^2 + 6x + 9}{2x^2 - 5x - 3}$.

**47.** Reduce to lowest terms $\dfrac{x^2 - 9}{x^4 - 81}$.

**48.** **Geometry** A rectangle has a perimeter of 48 feet. If the length is twice the width, find the dimensions.

**49.** **Number Problem** One number is 5 more than twice another number. The sum of the numbers is 35. Find the numbers.

**50.** **Geometry** The length of a rectangle is one more than 5 times the width. The area is 48 inches$^2$. Find the dimensions.

# Quadratic Equations

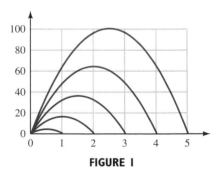

## INTRODUCTION

In the introduction to Chapter 6, we mentioned that hang time for a football depends only on the vertical velocity of the ball. It is also true that the maximum height attained by the ball is also dependent only on the initial vertical velocity. The height $h$ of the ball at time $t$ is always given by the equation

$$h = vt - 16t^2$$

where $v$ is the initial velocity of the ball. Table 1 shows the equations that give the height of the ball for different vertical velocities. Figure 1 shows the graph of each of those equations.

| TABLE 1 Height of a Football | |
| --- | --- |
| Initial Vertical Velocity (feet/second) | Height |
| 16 | $h = 16t - 16t^2$ |
| 32 | $h = 32t - 16t^2$ |
| 48 | $h = 48t - 16t^2$ |
| 64 | $h = 64t - 16t^2$ |
| 80 | $h = 80t - 16t^2$ |

**FIGURE 1**

## 9.1 More Quadratic Equations

Consider the equation $x^2 = 9$. Inspection shows that there are two solutions: $x = 3$ and $x = -3$, the two square roots of 9. Since every positive real number has two square roots, we can write the following property.

### SQUARE ROOT PROPERTY FOR EQUATIONS

For all positive real numbers $b$,

$$\text{If } a^2 = b, \text{ then } a = \sqrt{b} \text{ or } a = -\sqrt{b}.$$

*Notation*  A shorthand notation for

$$a = \sqrt{b} \quad \text{or} \quad a = -\sqrt{b}$$

is:

$$a = \pm \sqrt{b}$$

which is read "$a$ is plus or minus the square root of $b$."

We can use the square root property any time we feel it is helpful. We must make sure, however, that we include both the positive and the negative square roots.

**EXAMPLE 1**  Solve for $x$: $x^2 = 7$

**Solution**
$$x^2 = 7$$
$$x = \pm \sqrt{7} \qquad \text{Square root property}$$

The two solutions are $\sqrt{7}$ and $-\sqrt{7}$.

**Note:**  This method of solving quadratic equations is sometimes called *extraction of roots*.

**EXAMPLE 2**  Solve for $y$: $3y^2 = 60$

**Solution**  We begin by dividing both sides by 3 (which is the same as multiplying both sides by $\frac{1}{3}$):

$$3y^2 = 60$$
$$y^2 = 20 \qquad \text{Divide each side by 3.}$$
$$y = \pm\sqrt{20} \qquad \text{Square root property}$$
$$y = \pm 2\sqrt{5} \qquad \sqrt{20} = \sqrt{4 \cdot 5} = \sqrt{4}\,\sqrt{5} = 2\sqrt{5}$$

Our two solutions are $2\sqrt{5}$ and $-2\sqrt{5}$. Each of them will yield a true statement when used in place of the variable in the original equation $3y^2 = 60$.

**EXAMPLE 3**   Solve for $a$: $(a + 3)^2 = 16$

**Solution**   We proceed as we did in the previous example. The square root of $(a + 3)^2$ is $(a + 3)$:

$$(a + 3)^2 = 16$$

$$a + 3 = \pm 4$$

At this point we add $-3$ to both sides to get:

$$a = -3 \pm 4$$

which we can write as:

$$a = -3 + 4 \qquad \text{or} \qquad a = -3 - 4$$

$$a = 1 \qquad\qquad \text{or} \qquad a = -7$$

Our solutions are 1 and $-7$.

**EXAMPLE 4**   Solve for $x$: $(3x - 2)^2 = 25$

**Solution**

$$(3x - 2)^2 = 25$$

$$3x - 2 = \pm 5$$

Adding 2 to both sides, we have:

$$3x = 2 \pm 5$$

Dividing both sides by 3 gives us:

$$x = \frac{2 \pm 5}{3}$$

We separate the preceding equation into two separate statements:

$$x = \frac{2 + 5}{3} \qquad \text{or} \qquad x = \frac{2 - 5}{3}$$

$$x = \frac{7}{3} \qquad\qquad \text{or} \qquad x = \frac{-3}{3} = -1$$

**Note:**  We can solve the equation in Example 4 by factoring (as we did in Section 6.6) if we first expand $(3x - 2)^2$.

$$(3x - 2)^2 = 25$$

$$9x^2 - 12x + 4 = 25$$

$$9x^2 - 12x - 21 = 0$$

$$3(3x^2 - 4x - 7) = 0$$

$$3(3x - 7)(x + 1) = 0$$

$$x = \frac{7}{3} \quad \text{or} \quad x = -1$$

---

**E X A M P L E   5**   Solve for $y$: $(4y - 5)^2 = 6$

**Solution**

$$(4y - 5)^2 = 6$$

$$4y - 5 = \pm\sqrt{6}$$

$$4y = 5 \pm \sqrt{6} \qquad \text{Add 5 to both sides.}$$

$$y = \frac{5 \pm \sqrt{6}}{4} \qquad \text{Divide both sides by 4.}$$

Since $\sqrt{6}$ is irrational, we cannot simplify the expression further. The solution set is $\left\{ \dfrac{5 + \sqrt{6}}{4}, \dfrac{5 - \sqrt{6}}{4} \right\}$.

---

**E X A M P L E   6**   Solve for $x$: $(2x + 6)^2 = 8$

**Solution**

$$(2x + 6)^2 = 8$$

$$2x + 6 = \pm\sqrt{8}$$

$$2x + 6 = \pm 2\sqrt{2} \qquad \sqrt{8} = \sqrt{4 \cdot 2} = 2\sqrt{2}$$

$$2x = -6 \pm 2\sqrt{2} \qquad \text{Add } -6 \text{ to both sides.}$$

$$x = \frac{-6 \pm 2\sqrt{2}}{2} \qquad \text{Divide each side by 2.}$$

We can reduce the previous expression to lowest terms by factoring a 2 from each term in the numerator and then dividing that 2 by the 2 in the denominator. This is equivalent to dividing each term in the numerator by the 2 in the denominator. Here is what it looks like:

$$x = \frac{2(-3 \pm \sqrt{2})}{2} \qquad \text{Factor a 2 from each term in numerator.}$$

$$x = -3 \pm \sqrt{2} \qquad \text{Divide numerator and denominator by 2.}$$

The two solutions are $-3 + \sqrt{2}$ and $-3 - \sqrt{2}$.

We can check our two solutions in the original equation. Let's check our first solution, $-3 + \sqrt{2}$.

$$\text{When} \qquad\qquad x = -3 + \sqrt{2}$$

$$\text{the equation} \qquad (2x + 6)^2 = 8$$

becomes

$$[2(-3 + \sqrt{2}) + 6]^2 \overset{?}{=} 8$$

$$(-6 + 2\sqrt{2} + 6)^2 = 8$$

$$(2\sqrt{2})^2 = 8$$

$$4 \cdot 2 = 8$$

$$8 = 8 \qquad \text{A true statement}$$

The second solution, $-3 - \sqrt{2}$, checks also.

**EXAMPLE 7**    If an object is dropped from a height of $h$ feet, the amount of time in seconds it will take for the object to reach the ground (ignoring the resistance of air) is given by the formula

$$h = 16t^2$$

Solve this formula for $t$.

**Solution**    To solve for $t$, we apply the square root property:

$$h = 16t^2 \qquad \text{Original formula}$$

$$\pm\sqrt{h} = 4t \qquad \text{Square root property}$$

$$\pm\frac{\sqrt{h}}{4} = t \qquad \text{Divide each side by 4.}$$

Since $t$ represents the time it takes for the object to fall $h$ feet, $t$ will never be negative. Therefore, the formula that gives $t$ in terms of $h$ is:

$$t = \frac{\sqrt{h}}{4}$$

Whenever we are solving an application problem like this one and we obtain a result that includes the $\pm$ sign, we must ask ourselves if the result can actually be negative. If it cannot be, we delete the negative result and use only the positive result.

FACTS FROM
# Geometry

## Special Triangles

Two special triangles are shown on the next page. An *equilateral triangle* (Figure 1) is a triangle with three sides of equal length. If all three sides in a triangle have the same length, then the three interior angles in the triangle must also be equal. Since the sum of the interior angles in a triangle is always 180°, each of the three interior angles in any equilateral triangle must be 60°.

An *isosceles triangle* (Figure 2) is a triangle with two sides of equal length. Angles $A$ and $B$ in the isosceles triangle in Figure 2 are called the *base angles;* they are the angles opposite the two equal sides. In every isosceles triangle, the base angles are equal.

**Equilateral Triangle**                  **Isosceles Triangle**

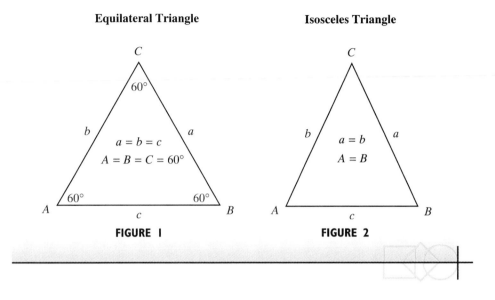

**FIGURE 1**                               **FIGURE 2**

**E X A M P L E  8**   The lengths of all three sides of an equilateral triangle are 8 centimeters. Find the height of the triangle.

**Solution**   Because the three sides of an equilateral triangle are equal, the height always divides the base into two equal line segments. Figure 3 illustrates this fact. We find the height by applying the Pythagorean theorem.

$$8^2 = x^2 + 4^2$$
$$64 = x^2 + 16$$
$$x^2 = 64 - 16$$
$$x^2 = 48$$
$$x = \sqrt{48}$$
$$x = \sqrt{16 \cdot 3}$$
$$x = 4\sqrt{3} \text{ cm}$$

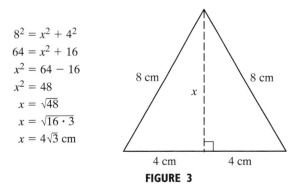

**FIGURE 3**

**Note:**  A calculator gives a decimal approximation of 6.9 centimeters for the height of the triangle in Example 8.

**E X A M P L E 9** The base of an isosceles triangle is 10 feet, and the length of the two equal sides is 7 feet, as shown in Figure 4. Find the height of the triangle.

10 ft

**FIGURE 4**

**Solution** Because two sides in every isosceles triangle are equal, the height that is drawn to the base always divides the base into two equal line segments, as shown in Figure 5. This fact allows us to form a right triangle in which two sides are known. We apply the Pythagorean theorem and solve for $x$.

$$7^2 = x^2 + 5^2$$
$$49 = x^2 + 25$$
$$x^2 = 49 - 25$$
$$x^2 = 24$$
$$x = \sqrt{24}$$
$$x = 2\sqrt{6} \text{ feet} = 4.9 \text{ feet to the nearest tenth}$$

**FIGURE 5**

## Getting Ready for Class

*After reading through the preceding section, respond in your own words and in complete sentences.*

**A.** What is the square root property for equations?

**B.** Describe how you would solve the equation $x^2 = 7$.

**C.** What does the symbol $\pm$ stand for?

**D.** What is an isosceles triangle?

## PROBLEM SET 9.1

Solve each of the following equations, using the methods learned in this section.

**1.** $x^2 = 9$    **2.** $x^2 = 16$    **3.** $a^2 = 25$

**4.** $a^2 = 36$    **5.** $y^2 = 8$    **6.** $y^2 = 75$

**7.** $2x^2 = 100$    **8.** $2x^2 = 54$    **9.** $3a^2 = 54$

**10.** $2a^2 = 64$

**12.** $(x - 3)^2 = 16$

**14.** $(x + 3)^2 = 64$

**16.** $(a - 4)^2 = 32$

**18.** $(y - 5)^2 = 27$

**11.** $(x + 2)^2 = 4$

**13.** $(x + 1)^2 = 25$

**15.** $(a - 5)^2 = 75$

**17.** $(y + 1)^2 = 50$

**19.** $(2x + 1)^2 = 25$

**20.** $(3x - 2)^2 = 16$

**21.** $(4a - 5)^2 = 36$

**22.** $(2a + 6)^2 = 64$

**23.** $(3y - 1)^2 = 12$

**24.** $(5y - 4)^2 = 12$

**25.** $(6x + 2)^2 = 27$

**26.** $(8x - 1)^2 = 20$

**27.** $(3x - 9)^2 = 27$

**28.** $(2x + 8)^2 = 32$

**29.** $(3x + 6)^2 = 45$

**30.** $(5x - 10)^2 = 75$

**31.** $(2y - 4)^2 = 8$

**32.** $(4y - 6)^2 = 48$

**33.** $\left(x - \dfrac{2}{3}\right)^2 = \dfrac{25}{9}$

**34.** $\left(x - \dfrac{3}{4}\right)^2 = \dfrac{49}{16}$

**35.** $\left(x + \dfrac{1}{2}\right)^2 = \dfrac{7}{4}$

**36.** $\left(x + \dfrac{1}{3}\right)^2 = \dfrac{5}{9}$

**37.** $\left(a - \dfrac{4}{5}\right)^2 = \dfrac{12}{25}$

**38.** $\left(a - \dfrac{3}{7}\right)^2 = \dfrac{18}{49}$

Since $a^2 + 2ab + b^2$ can be written as $(a + b)^2$, each of the following equations can be solved using our square root method. The first step is to write the trinomial on the left side of the equal sign as the square of a binomial. Solve each equation.

**39.** $x^2 + 10x + 25 = 7$

**40.** $x^2 + 6x + 9 = 11$

**41.** $x^2 - 2x + 1 = 9$

**42.** $x^2 + 8x + 16 = 25$

**43.** $x^2 + 12x + 36 = 8$

**44.** $x^2 - 4x + 4 = 12$

## Applying the Concepts

**45. Checking Solutions**   Check the solution

$$x = -1 + 5\sqrt{2}$$

in the equation $(x + 1)^2 = 50$.

**46. Checking Solutions**   Check the solution

$$x = -8 + 2\sqrt{6}$$

in the equation $(x + 8)^2 = 24$.

**47. Number Problem**   The square of the sum of a number and 3 is 16. Find the number. (There are two solutions.)

**48. Number Problem**   The square of the sum of twice a number and 3 is 25. Find the number. (There are two solutions.)

**49. Investing**   If you invest \$100 in an account with interest rate $r$ compounded annually, the amount of money $A$ in the account after 2 years is given by the formula

$$A = 100(1 + r)^2$$

Solve this formula for $r$.

**50. Investing**   If you invest $P$ dollars in an account with interest rate $r$ compounded annually, the amount of money $A$ in the account after 2 years is given by the formula

$$A = P(1 + r)^2$$

Solve this formula for $r$.

**51. Geometry**   The lengths of all three sides of an equilateral triangle are 10 feet. Find the height of the triangle.

**52. Geometry**   The lengths of all three sides of an equilateral triangle are 12 meters. Find the height of the triangle.

**53. Geometry**   The front of a tent forms an equilateral triangle with sides of 6 feet. Can a person 5 feet 8 inches tall stand up inside the tent?

**54. Geometry**   The front of a tent forms an equilateral triangle. The tent must be constructed so that a person 6 feet tall can stand up inside. Find the length of the three sides of the front of the tent.

**55. Geometry**   The base of an isosceles triangle is 8 feet, and the length of the two equal sides is 5 feet. Find the height of the triangle.

**56. Geometry**   An isosceles triangle has a base 8 feet long. If the length of the two equal sides is 6 feet, find the height of the triangle.

**Displacement**   The displacement, in cubic inches, of a car engine is given by the formula

$$d = \pi \cdot s \cdot c \cdot \left(\frac{1}{2} \cdot b\right)^2$$

where $s$ is the stroke and $b$ is the bore, as shown in the following figure, and $c$ is the number of cylinders. Calculate the bore for each of the following cars. Use 3.14 to approximate $\pi$.

**57. BMW M Coupe**   Six cylinders, 3.53 inches of stroke, and 192 cubic inches of displacement

**58. Chevrolet Corvette**   Eight cylinders, 3.62 inches of stroke, and 346 cubic inches of displacement

**59. Length of a Ramp**   A contractor is making a ramp to roll a wheelbarrow into the door of a building he is constructing. The ramp is 14 feet long, and the door is 6 feet above ground level. How far from the base of the building will the other end of the ramp be?

**60. Proportional Rectangles**   What proportions must a rectangle be so that when you cut it in half, the two new smaller rectangles have the same proportions as the original large rectangle? In the original large rectangle, let's say the shorter side is 1 unit and the longer side is $x$ units.

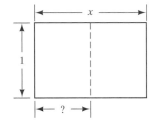

(a) What is the expression for the length of the short side of one of the smaller rectangles?

(b) Fill in this proportion with the correct expressions:

$$\frac{\text{Short side of small rectangle}}{\text{Long side of small rectangle}} = \frac{\text{short side of large rectangle}}{\text{long side of large rectangle}}$$

(c) Solve the resulting equation for $x$.

**61. Distance Up a Pyramid**   The Great Pyramid of Khufu (or Cheops) at Gizeh, Egypt, was one of the Seven Wonders of the Ancient World — and the only one that still exists. Before some of its outermost stones disintegrated or were stolen, its square base was 768 feet on each side, and it was 482 feet high ($BD = 768$ and $OC = 482$).

(a) How far is it if you walk straight up the middle of a face (along $AC$)?

(b) How far is it if you walk up an edge (along $BC$)?

(c) The volume of a pyramid is $\frac{1}{3}$ the height times the area of the base. What was the original volume of the Great Pyramid of Khufu?

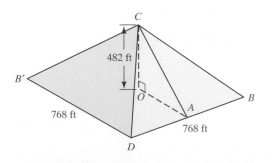

**62. Material for a Tent** A pyramidal tent 6 feet high with a square base 12 feet on each side is made of ripstop nylon with plasticized reinforcing along all the edges.

(a) Find the length of $AC$.

(b) Find the area of triangle $BCD$.

(c) What is the total area of all the ripstop nylon used for the tent (including the floor)?

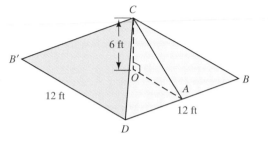

## Review Problems

The following problems review material we covered in Sections 5.6, 6.4, and 8.1. Reviewing the problems from Sections 5.6 and 6.4 will help you understand the material in the next section.

Multiply. [5.6]

**63.** $(x - 5)^2$        **64.** $(x + 5)^2$

Factor. [6.4]

**65.** $x^2 - 12x + 36$     **66.** $x^2 + 12x + 36$

**67.** $x^2 + 4x + 4$       **68.** $x^2 - 4x + 4$

Find the following roots. [8.1]

**69.** $\sqrt[3]{8}$          **70.** $\sqrt[3]{27}$

**71.** $\sqrt[4]{16}$        **72.** $\sqrt[4]{81}$

## 9.2 Completing the Square

In this section we will develop a method of solving quadratic equations that works whether or not the equation can be factored. Since we will be working with the individual terms of trinomials, we need some new definitions so that we can keep our vocabulary straight.

> **DEFINITION** In the trinomial $2x^2 + 3x + 4$, the first term, $2x^2$, is called the **quadratic term;** the middle term, $3x$, is called the **linear term;** and the last term, 4, is called the **constant term.**

Now consider the following list of perfect square trinomials and their corresponding binomial squares:

$$x^2 + \phantom{0}6x + \phantom{0}9 = (x + 3)^2$$
$$x^2 - \phantom{0}8x + 16 = (x - 4)^2$$
$$x^2 - 10x + 25 = (x - 5)^2$$
$$x^2 + 12x + 36 = (x + 6)^2$$

In each case the coefficient of $x^2$ is 1. A more important observation, however, stems from the relationship between the linear terms (middle terms) and the constant terms (last terms). Notice that the constant in each case is the square of half the coefficient of $x$ in the linear term. That is,

1. For the first trinomial, $x^2 + 6x + 9$, the last term, 9, is the square of half the coefficient of the middle term: $9 = (\frac{6}{2})^2$.
2. For the second trinomial, $x^2 - 8x + 16$, $16 = [\frac{1}{2}(-8)]^2$.
3. For the trinomial $x^2 - 10x + 25$, it also holds: $25 = [\frac{1}{2}(-10)]^2$.

Check and see that it also works for the final trinomial.

In summary, then, for every perfect square trinomial in which the coefficient of $x^2$ is 1, the final term is always the square of half the coefficient of the linear term. We can use this fact to build our own perfect square trinomials.

---

**EXAMPLES**   Write the correct final term to each of the following expressions so that each becomes a perfect square trinomial.

**1.** $x^2 - 2x$

   **Solution**   The coefficient of the linear term is $-2$. If we take the square of half of $-2$, we get 1. Adding the 1 as the final term, we have the perfect square trinomial:
$$x^2 - 2x + 1 = (x - 1)^2$$

**2.** $x^2 + 18x$

   **Solution**   Half of 18 is 9, the square of which is 81. If we add 81 at the end, we have:
$$x^2 + 18x + 81 = (x + 9)^2$$

**3.** $x^2 + 3x$

   **Solution**   Half of 3 is $\frac{3}{2}$, the square of which is $\frac{9}{4}$:
$$x^2 + 3x + \frac{9}{4} = \left(x + \frac{3}{2}\right)^2$$

---

We can use this procedure, along with the method developed in Section 9.1, to solve some quadratic equations.

---

**EXAMPLE   4**   Solve $x^2 - 6x + 5 = 0$ by completing the square.

**Solution**   We begin by adding $-5$ to both sides of the equation. We want just $x^2 - 6x$ on the left side so that we can add on our own final term to get a perfect square trinomial:

$$x^2 - 6x + 5 = 0$$
$$x^2 - 6x \qquad = -5 \qquad \text{Add } -5 \text{ to both sides.}$$

Now we can add 9 to both sides, and the left side will be a perfect square:

$$x^2 - 6x + 9 = -5 + 9$$
$$(x - 3)^2 = 4$$

The final line is in the form of the equations we solved in Section 9.1:

$$x - 3 = \pm 2$$

$$x = 3 \pm 2 \qquad \text{Add 3 to both sides.}$$

$$x = 3 + 2 \quad \text{or} \quad x = 3 - 2$$

$$x = 5 \quad \text{or} \quad x = 1$$

The two solutions are 5 and 1.

The preceding method of solution is called *completing the square.*

**Note:**  The equation in Example 4 can be solved quickly by factoring:

$$x^2 - 6x + 5 = 0$$

$$(x - 5)(x - 1) = 0$$

$$x - 5 = 0 \quad \text{or} \quad x - 1 = 0$$

$$x = 5 \quad \text{or} \quad x = 1$$

The reason we didn't solve it by factoring is that we want to practice completing the square on some simple equations.

**E X A M P L E   5**   Solve by completing the square: $2x^2 + 16x - 18 = 0$

**Solution**   We begin by moving the constant term to the other side:

$$2x^2 + 16x - 18 = 0$$

$$2x^2 + 16x \qquad = 18 \qquad \text{Add 18 to both sides.}$$

In order to complete the square, we must be sure the coefficient of $x^2$ is 1. To accomplish this, we divide both sides by 2:

$$\frac{2x^2}{2} + \frac{16x}{2} = \frac{18}{2}$$

$$x^2 + 8x = 9$$

We now complete the square by adding the square of half the coefficient of the linear term to both sides:

$$x^2 + 8x + \mathbf{16} = 9 + \mathbf{16} \qquad \text{Add } \mathbf{16} \text{ to both sides.}$$

$$(x + 4)^2 = 25$$

$$x + 4 = \pm 5 \qquad \text{Square root property}$$

$$x = -4 \pm 5 \qquad \text{Add } -4 \text{ to both sides.}$$

$$x = -4 + 5 \quad \text{or} \quad x = -4 - 5$$

$$x = 1 \quad \text{or} \quad x = -9$$

The solution set arrived at by completing the square is $\{1, -9\}$.

We will now summarize the preceding examples by listing the steps involved in solving quadratic equations by completing the square.

---

### STRATEGY FOR SOLVING A QUADRATIC EQUATION BY COMPLETING THE SQUARE

*Step 1:* Put the equation in the form $ax^2 + bx = c$. This usually involves moving only the constant term to the opposite side.

*Step 2:* Make sure the coefficient of the squared term is 1. If it is not 1, simply divide both sides by whatever it is.

*Step 3:* Add the square of half the coefficient of the linear term to both sides of the equation.

*Step 4:* Write the left-hand side of the equation as a binomial square and solve, using the methods developed in Section 9.1.

---

Here is one final example.

---

**E X A M P L E   6**    Solve for $y$: $3y^2 - 9y + 3 = 0$

**Solution**

$$3y^2 - 9y + 3 = 0$$

$$3y^2 - 9y \quad\quad = -3 \quad\quad\quad \text{Add } -3 \text{ to both sides.}$$

$$y^2 - 3y \quad\quad = -1 \quad\quad\quad \text{Divide by 3.}$$

$$y^2 - 3y + \frac{9}{4} = -1 + \frac{9}{4} \quad\quad\quad \text{Complete the square.}$$

$$\left(y - \frac{3}{2}\right)^2 = \frac{5}{4} \quad\quad\quad\quad -1 + \frac{9}{4} = -\frac{4}{4} + \frac{9}{4} = \frac{5}{4}$$

$$y - \frac{3}{2} = \pm\frac{\sqrt{5}}{2} \quad\quad\quad \text{Square root property}$$

$$y = \frac{3}{2} \pm \frac{\sqrt{5}}{2} \quad\quad \text{Add } \frac{3}{2} \text{ to both sides.}$$

$$y = \frac{3}{2} + \frac{\sqrt{5}}{2} \quad\quad \text{or} \quad\quad y = \frac{3}{2} - \frac{\sqrt{5}}{2}$$

$$y = \frac{3 + \sqrt{5}}{2} \quad\quad \text{or} \quad\quad y = \frac{3 - \sqrt{5}}{2}$$

The solutions are $\dfrac{3 + \sqrt{5}}{2}$ and $\dfrac{3 - \sqrt{5}}{2}$, which can be written in a shorter form as:

$$\frac{3 \pm \sqrt{5}}{2}$$

**Note:**  We can use a calculator to get decimal approximations to these solutions. If we use the approximation $\sqrt{5} \approx 2.236$, then

$$\frac{3 + \sqrt{5}}{2} \approx \frac{3 + 2.236}{2} = \frac{5.236}{2} = 2.618$$

$$\frac{3 - \sqrt{5}}{2} \approx \frac{3 - 2.236}{2} = \frac{0.764}{2} = 0.382$$

## Getting Ready for Class

*After reading through the preceding section, respond in your own words and in complete sentences.*

**A.** What kind of equation do we solve using the method of completing the square?

**B.** Explain in words how you would complete the square on $x^2 - 6x = 5$.

**C.** What is the first step in completing the square?

**D.** What is the linear term in a trinomial?

## PROBLEM SET 9.2

Give the correct final term for each of the following expressions to ensure that the resulting trinomial is a perfect square trinomial.

**1.** $x^2 + 6x$    **2.** $x^2 - 10x$    **3.** $x^2 + 2x$

**4.** $x^2 + 14x$    **5.** $y^2 - 8y$    **6.** $y^2 + 12y$

**7.** $y^2 - 2y$    **8.** $y^2 - 6y$    **9.** $x^2 + 16x$

**10.** $x^2 - 4x$    **11.** $a^2 - 3a$    **12.** $a^2 + 5a$

**13.** $x^2 - 7x$    **14.** $x^2 - 9x$    **15.** $y^2 + y$

**16.** $y^2 - y$    **17.** $x^2 - \frac{3}{2}x$    **18.** $x^2 + \frac{2}{3}x$

Solve each of the following equations by completing the square. Follow the steps given at the end of this section.

**19.** $x^2 + 4x = 12$    **20.** $x^2 - 2x = 8$

**21.** $x^2 - 6x = 16$    **22.** $x^2 + 12x = -27$

**23.** $a^2 + 2a = 3$    **24.** $a^2 - 8a = -7$

**25.** $x^2 - 10x = 0$    **26.** $x^2 + 4x = 0$

**27.** $y^2 + 2y - 15 = 0$    **28.** $y^2 - 10y - 11 = 0$

**29.** $x^2 + 4x - 3 = 0$    **30.** $x^2 + 6x + 5 = 0$

**31.** $x^2 - 4x = 4$    **32.** $x^2 + 4x = -1$

**33.** $a^2 = 7a + 8$    **34.** $a^2 = 3a + 1$

**35.** $4x^2 + 8x - 4 = 0$    **36.** $3x^2 + 12x + 6 = 0$

**37.** $2x^2 + 2x - 4 = 0$    **38.** $4x^2 + 4x - 3 = 0$

**39.** $4x^2 + 8x + 1 = 0$    **40.** $3x^2 + 6x + 2 = 0$

**41.** $2x^2 - 2x = 1$    **42.** $3x^2 - 3x = 1$

**43.** $4a^2 - 4a + 1 = 0$    **44.** $2a^2 + 4a + 1 = 0$

**45.** $3y^2 - 9y = 2$    **46.** $5y^2 - 10y = 4$

## Applying the Concepts

**47. Computer Screen**  An advertisement for a portable computer indicates it has a 14-inch viewing screen. This means that the diagonal of the screen measures 14 inches. If the ratio of the length to the width of the screen is 3 to 4, we can represent the length with $4x$ and the width with $3x$, and then use the Pythagorean theorem to solve for $x$.

Once we have $x$, the length will be $4x$, and the width will be $3x$. Find the length and width of this computer screen to the nearest tenth of an inch.

$3x$

14 in.

$4x$

**48. Television Screen** A 25-inch television has a rectangular screen on which the diagonal measures 25 inches. The ratio of the length to the width of the screen is 3 to 4. Let the length equal $4x$ and the width equal $3x$, and then use the Pythagorean theorem to solve for $x$. Then find the length and width of this television screen.

$4x$

25 in.

$3x$

**49. Finding Intercepts** The graph of $y = x^2 - 2x - 1$ is shown in the following figure. To find the intercepts for this graph we let $y = 0$, and then we solve the resulting equation, which is

$$0 = x^2 - 2x - 1$$

To the nearest tenth, where does this graph cross the $x$-axis?

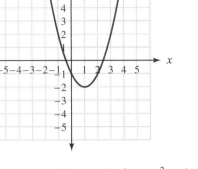

**50. Finding Intercepts** The graph of $y = x^2 - 4x + 2$ is shown in the following figure. To find the intercepts for this graph we let $y = 0$, and then we solve the resulting equation, which is

$$0 = x^2 - 4x + 2$$

To the nearest tenth, where does this graph cross the $x$-axis?

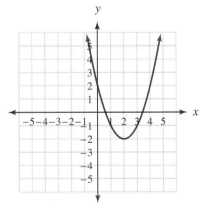

**51. Sum and Product of Solutions** The equation $x^2 + 4x - 3 = 0$ has two solutions, $-2 + \sqrt{7}$ and $-2 - \sqrt{7}$.
(a) Find the sum of these two solutions.
(b) Find the product of these two solutions.

**52. Sum and Product of Solutions** The equation $x^2 - 6x + 4 = 0$ has two solutions, $3 + \sqrt{5}$ and $3 - \sqrt{5}$.
(a) Find the sum of these two solutions.
(b) Find the product of these two solutions.

## Completing the Square, Visually

At the end of Problem Set 5.6, we gave some visual representations for $(a + b)^2$ and $(x + 5)^2$. The following diagram can be used to show why we add 16 to $x^2 + 8x$ in order to complete the square.

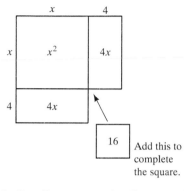

Add this to complete the square.

Draw a similar diagram to visualize completing the square on each of the following expressions.

**53.** $x^2 + 6x$

**54.** $x^2 + 10x$

**55.** $x^2 + 2x$

**56.** $x^2 + 4x$

## Review Problems

The following problems review material we covered in Sections 2.1 and 8.2. Reviewing the problems from Section 2.1 will help you with the next section.

Find the value of each expression if $a = 2$, $b = 4$, and $c = -3$. [2.1]

**57.** $2a$

**58.** $b^2$

**59.** $4ac$

**60.** $b^2 - 4ac$

**61.** $\sqrt{b^2 - 4ac}$

**62.** $-b + \sqrt{b^2 - 4ac}$

Put each expression in simplified form for radicals. [8.2]

**63.** $\sqrt{12}$

**64.** $\sqrt{50x^2}$

**65.** $\sqrt{20x^2 y^3}$

**66.** $3\sqrt{48x^4}$

**67.** $\sqrt{\dfrac{81}{25}}$

**68.** $\dfrac{6\sqrt{8x^2 y}}{\sqrt{9}}$

---

## 9.3   The Quadratic Formula

In this section we will derive the quadratic formula. It is one formula that you will use in almost all types of mathematics. We will first state the formula as a theorem and then prove it. The proof is based on the method of completing the square developed in the preceding section.

### THE QUADRATIC THEOREM

For any quadratic equation in the form $ax^2 + bx + c = 0$, where $a$, $b$, and $c$ are real numbers and $a \neq 0$, the two solutions are:

$$x = \frac{-b + \sqrt{b^2 - 4ac}}{2a} \quad \text{and} \quad x = \frac{-b - \sqrt{b^2 - 4ac}}{2a}$$

***Proof***  We will prove the theorem by completing the square on:

$$ax^2 + bx + c = 0$$

Adding $-c$ to both sides, we have:

$$ax^2 + bx = -c$$

To make the coefficient of $x^2$ one, we divide both sides by $a$:

$$\frac{ax^2}{a} + \frac{bx}{a} = -\frac{c}{a}$$

$$x^2 + \frac{b}{a}x = -\frac{c}{a}$$

Now, to complete the square, we add the square of half of $\dfrac{b}{a}$ to both sides:

$$x^2 + \frac{b}{a}x + \left(\frac{b}{2a}\right)^2 = -\frac{c}{a} + \left(\frac{b}{2a}\right)^2 \qquad \frac{1}{2} \text{ of } \frac{b}{a} \text{ is } \frac{b}{2a}.$$

Let's simplify the right side separately:

$$-\frac{c}{a} + \left(\frac{b}{2a}\right)^2 = -\frac{c}{a} + \frac{b^2}{4a^2}$$

The least common denominator is $4a^2$. We multiply the numerator and denominator of $-\dfrac{c}{a}$ by $4a$ to give it the common denominator. Then we combine numerators:

$$\frac{4a}{4a}\left(-\frac{c}{a}\right) + \frac{b^2}{4a^2} = -\frac{4ac}{4a^2} + \frac{b^2}{4a^2}$$

$$= \frac{-4ac + b^2}{4a^2}$$

$$= \frac{b^2 - 4ac}{4a^2}$$

Now, back to the equation. We use our simplified expression for the right side:

$$x^2 + \frac{b}{a}x + \left(\frac{b}{2a}\right)^2 = \frac{b^2 - 4ac}{4a^2}$$

$$\left(x + \frac{b}{2a}\right)^2 = \frac{b^2 - 4ac}{4a^2}$$

Applying the square root property, we have:

$$x + \frac{b}{2a} = \pm\frac{\sqrt{b^2 - 4ac}}{2a}$$

$$x = \frac{-b}{2a} \pm \frac{\sqrt{b^2 - 4ac}}{2a} \qquad \text{Add } \frac{-b}{2a} \text{ to both sides.}$$

$$x = \frac{-b \pm \sqrt{b^2 - 4ac}}{2a}$$

Our proof is now complete. What we have is this: If our equation is in the form $ax^2 + bx + c = 0$ (standard form), then the solution can always be found by using the quadratic formula:

$$x = \frac{-b \pm \sqrt{b^2 - 4ac}}{2a}$$

This formula is called the quadratic formula. You will see it many times if you continue taking math classes. By the time you are finished with this section and the problems in the problem set, you should have it memorized.

**EXAMPLE 1**   Solve $x^2 - 5x - 6 = 0$ by using the quadratic formula.

**Solution**   To use the quadratic formula, we must make sure the equation is in standard form; identify $a$, $b$, and $c$; substitute them into the formula, and work out the arithmetic.

For the equation $x^2 - 5x - 6 = 0$, $a = 1$, $b = -5$, and $c = -6$:

$$x = \frac{-b \pm \sqrt{b^2 - 4ac}}{2a} = \frac{-(-5) \pm \sqrt{(-5)^2 - 4(1)(-6)}}{2(1)}$$

$$= \frac{5 \pm \sqrt{49}}{2}$$

$$= \frac{5 \pm 7}{2}$$

$$x = \frac{5 + 7}{2} \quad \text{or} \quad x = \frac{5 - 7}{2}$$

$$x = \frac{12}{2} \quad\quad\quad\quad x = -\frac{2}{2}$$

$$x = 6 \quad\quad\quad\quad\quad x = -1$$

The two solutions are 6 and $-1$.

**Note:**  Whenever the solutions to our quadratic equations turn out to be rational numbers, as in Example 1, it means the original equation could have been solved by factoring. (We didn't solve the equation in Example 1 by factoring because we were trying to get some practice with the quadratic formula.)

**EXAMPLE 2**   Solve for $x$: $2x^2 = -4x + 3$.

**Solution**   Before we can identify $a$, $b$, and $c$, we must write the equation in standard form. To do so, we add $4x$ and $-3$ to each side of the equation:

$$2x^2 = -4x + 3$$

$$2x^2 + 4x - 3 = 0 \quad\quad\quad \text{Add } 4x \text{ and } -3 \text{ to each side.}$$

Now that the equation is in standard form, we see that $a = 2$, $b = 4$, and $c = -3$. Using the quadratic formula we have:

$$x = \frac{-b \pm \sqrt{b^2 - 4ac}}{2a}$$

$$= \frac{-4 \pm \sqrt{4^2 - 4(2)(-3)}}{2(2)}$$

$$= \frac{-4 \pm \sqrt{40}}{4}$$

$$= \frac{-4 \pm 2\sqrt{10}}{4}$$

We can reduce the final expression in the preceding equation to lowest terms by factoring 2 from the numerator and denominator and then dividing it out:

$$x = \frac{2(-2 \pm \sqrt{10})}{2 \cdot 2}$$

$$= \frac{-2 \pm \sqrt{10}}{2}$$

Our two solutions are $\dfrac{-2 + \sqrt{10}}{2}$ and $\dfrac{-2 - \sqrt{10}}{2}$.

---

**EXAMPLE 3**    Solve for $x$: $(x - 2)(x + 3) = 5$

**Solution**    We must put the equation into standard form before we can use the formula:

$$(x - 2)(x + 3) = 5$$

$$x^2 + x - 6 = 5 \qquad \text{Multiply out the left side.}$$

$$x^2 + x - 11 = 0 \qquad \text{Add } -5 \text{ to each side.}$$

Now, $a = 1$, $b = 1$, and $c = -11$; therefore:

$$x = \frac{-1 \pm \sqrt{1^2 - 4(1)(-11)}}{2(1)}$$

$$= \frac{-1 \pm \sqrt{45}}{2}$$

$$= \frac{-1 \pm 3\sqrt{5}}{2}$$

The solution set is $\left\{ \dfrac{-1 + 3\sqrt{5}}{2}, \dfrac{-1 - 3\sqrt{5}}{2} \right\}$.

**E X A M P L E   4**   Solve $x^2 - 6x = -7$.

**Solution**   We begin by writing the equation in standard form:

$$x^2 - 6x = -7$$

$$x^2 - 6x + 7 = 0 \qquad \text{Add 7 to each side.}$$

Using $a = 1$, $b = -6$, and $c = 7$ in the quadratic formula

$$x = \frac{-b \pm \sqrt{b^2 - 4ac}}{2a}$$

we have:

$$x = \frac{-(-6) \pm \sqrt{(-6)^2 - 4(1)(7)}}{2(1)}$$

$$= \frac{6 \pm \sqrt{36 - 28}}{2}$$

$$= \frac{6 \pm \sqrt{8}}{2}$$

$$= \frac{6 \pm 2\sqrt{2}}{2}$$

The two terms in the numerator have a 2 in common. We reduce to lowest terms by factoring the 2 from the numerator and then dividing numerator and denominator by 2:

$$x = \frac{2(3 \pm \sqrt{2})}{2}$$

$$= 3 \pm \sqrt{2}$$

The two solutions are $3 + \sqrt{2}$ and $3 - \sqrt{2}$. This time, let's check our solutions in the original equation $x^2 - 6x = -7$.

Checking $x = 3 + \sqrt{2}$, we have:

$$(3 + \sqrt{2})^2 - 6(3 + \sqrt{2}) \stackrel{?}{=} -7$$

$$9 + 6\sqrt{2} + 2 - 18 - 6\sqrt{2} = -7 \qquad \text{Multiply.}$$

$$11 - 18 + 6\sqrt{2} - 6\sqrt{2} = -7 \qquad \text{Add 9 and 2.}$$

$$-7 + 0 = -7 \qquad \text{Subtraction}$$

$$-7 = -7 \qquad \text{A true statement}$$

Checking $x = 3 - \sqrt{2}$, we have:

$$(3 - \sqrt{2})^2 - 6(3 - \sqrt{2}) \stackrel{?}{=} -7$$

$$9 - 6\sqrt{2} + 2 - 18 + 6\sqrt{2} = -7 \qquad \text{Multiply.}$$

$$11 - 18 - 6\sqrt{2} + 6\sqrt{2} = -7 \qquad \text{Add 9 and 2.}$$

$$-7 + 0 = -7 \qquad \text{Subtraction}$$

$$-7 = -7 \qquad \text{A true statement}$$

As you can see, both solutions yield true statements when used in place of the variable in the original equation.

## Getting Ready for Class

*After reading through the preceding section, respond in your own words and in complete sentences.*

**A.** What is the quadratic formula?

**B.** Under what circumstances should the quadratic formula be applied?

**C.** What is standard form for a quadratic equation?

**D.** What is the first step in solving a quadratic equation using the quadratic formula?

## PROBLEM SET 9.3

Solve the following equations by using the quadratic formula.

**1.** $x^2 + 3x + 2 = 0$    **2.** $x^2 - 5x + 6 = 0$

**3.** $x^2 + 5x + 6 = 0$    **4.** $x^2 - 7x - 8 = 0$

**5.** $x^2 + 6x + 9 = 0$    **6.** $x^2 - 10x + 25 = 0$

**7.** $x^2 + 6x + 7 = 0$    **8.** $x^2 - 4x - 1 = 0$

**9.** $2x^2 + 5x + 3 = 0$    **10.** $2x^2 + 3x - 20 = 0$

**11.** $4x^2 + 8x + 1 = 0$    **12.** $3x^2 + 6x + 2 = 0$

**13.** $x^2 - 2x + 1 = 0$    **14.** $x^2 + 2x - 3 = 0$

**15.** $x^2 - 5x = 7$    **16.** $2x^2 - 6x = 8$

**17.** $6x^2 - x - 2 = 0$    **18.** $6x^2 + 5x - 4 = 0$

**19.** $(x - 2)(x + 1) = 3$    **20.** $(x - 8)(x + 7) = 5$

**21.** $(2x - 3)(x + 2) = 1$    **22.** $(4x - 5)(x - 3) = 6$

**23.** $2x^2 - 3x = 5$    **24.** $3x^2 - 4x = 5$

**25.** $2x^2 = -6x + 7$    **26.** $5x^2 = -6x + 3$

**27.** $3x^2 = -4x + 2$    **28.** $3x^2 = 4x + 2$

**29.** $2x^2 - 5 = 2x$    **30.** $5x^2 + 1 = -10x$

**31.** Solve the equation $2x^3 + 3x^2 - 4x = 0$ by first factoring out the common factor $x$ and then using the quadratic formula. There are three solutions to this equation.

**32.** Solve the equation $5y^3 - 10y^2 + 4y = 0$ by first factoring out the common factor $y$ and then using the quadratic formula.

**33.** To apply the quadratic formula to the equation $3x^2 - 4x = 0$, you have to notice that $c = 0$. Solve the equation using the quadratic formula.

**34.** Solve the equation $9x^2 - 16 = 0$ using the quadratic formula. (Notice $b = 0$.)

**35.** Solve the following equation by first multiplying both sides by the LCD and then applying the quadratic formula to the result.

$$\frac{1}{2}x^2 - \frac{1}{2}x - \frac{1}{6} = 0$$

**36.** Solve the following equation by first multiplying both sides by the LCD and then applying the quadratic formula to the result.

$$\frac{1}{2}y^2 - y - \frac{3}{2} = 0$$

## Applying the Concepts

**37. Archery** Margaret shoots an arrow into the air. The equation for the height (in feet) of the tip of the arrow is $h = 8 + 64t - 16t^2$. To find the time at which the arrow is 56 feet above the ground we replace $h$ with 56 to obtain

$$56 = 8 + 64t - 16t^2$$

Solve this equation for $t$ to find the times at which the arrow is 56 feet above the ground.

6 ft                    $h$

8 ft

**38.** At the beginning of every football game, the referee flips a coin to see who will kick off. The equation that gives the height (in feet) of the coin tossed in the air is $h = 6 + 32t - 16t^2$. To find the times at which the coin is 18 feet above the ground we substitute 18 for $h$ in the equation, giving us

$$18 = 6 + 32t - 16t^2$$

Solve this equation for $t$ to find the times at which the coin is 18 feet above the ground.

## Review Problems

The following problems review the material we covered in Section 8.5. Reviewing these problems will help you with the material in the next section.

Multiply.

**39.** $(2\sqrt{3})(3\sqrt{5})$      **40.** $\sqrt{7}(\sqrt{7} + 4)$

**41.** $(\sqrt{6} + 2)(\sqrt{6} - 5)$      **42.** $(\sqrt{x} + 4)^2$

**43.** $(\sqrt{7} - \sqrt{2})(\sqrt{7} + \sqrt{2})$      **44.** $(\sqrt{a} + \sqrt{b})(\sqrt{a} - \sqrt{b})$

Rationalize the denominator.

**45.** $\dfrac{2}{3 + \sqrt{5}}$      **46.** $\dfrac{2 + \sqrt{3}}{2 - \sqrt{3}}$

---

## 9.4  Complex Numbers

To solve quadratic equations such as $x^2 = -4$, we need to introduce a new set of numbers. If we try to solve $x^2 = -4$ using real numbers, we always get no solution. There is no real number whose square is $-4$.

The new set of numbers is called *complex numbers* and is based on the following definition.

**DEFINITION**  The **number** $i$ is a number such that $i = \sqrt{-1}$.

The first thing we notice about this definition is that $i$ is not a real number. There are no real numbers that represent the square root of $-1$. The other observation we make about $i$ is that $i^2 = -1$. If $i = \sqrt{-1}$, then, squaring both sides, we must have $i^2 = -1$. The most common power of $i$ is $i^2$. Whenever we see $i^2$, we can write it as $-1$. We are now ready for a definition of complex numbers.

---

**DEFINITION**    A **complex number** is any number that can be put in the form $a + bi$, where $a$ and $b$ are real numbers and $i = \sqrt{-1}$.

---

**Note:**  The form $a + bi$ is called *standard form* for complex numbers. The definition in the box indicates that if a number can be written in the form $a + bi$, then it is a complex number.

*Example*  The following are complex numbers:

$$3 + 4i \qquad \frac{1}{2} - 6i \qquad 8 + i\sqrt{2} \qquad \frac{3}{4} - 2i\sqrt{5}$$

*Example*  The number $4i$ is a complex number because $4i = 0 + 4i$.

*Example*  The number $8$ is a complex number because $8 = 8 + 0i$.

From the previous example we can see that the real numbers are a subset of the complex numbers because any real number $x$ can be written as $x + 0i$.

## ADDITION AND SUBTRACTION OF COMPLEX NUMBERS

We add and subtract complex numbers according to the same procedure we used to add and subtract polynomials: We combine similar terms.

---

**EXAMPLE  1**    Combine $(3 + 4i) + (2 - 6i)$.

**Solution**

$$
\begin{aligned}
(3 + 4i) + (2 - 6i) &= (3 + 2) + (4i - 6i) && \text{Commutative and} \\
&&& \text{associative properties} \\
&= 5 + (-2i) && \text{Combine similar terms.} \\
&= 5 - 2i
\end{aligned}
$$

---

**EXAMPLE  2**    Combine $(2 - 5i) - (3 + 7i) + (2 - i)$.

**Solution**

$$(2 - 5i) - (3 + 7i) + (2 - i) = 2 - 5i - 3 - 7i + 2 - i$$

$$= (2 - 3 + 2) + (-5i - 7i - i)$$
$$= 1 - 13i$$

## MULTIPLICATION OF COMPLEX NUMBERS

Multiplication of complex numbers is very similar to multiplication of polynomials. We can simplify many answers by using the fact that $i^2 = -1$.

**EXAMPLE   3**   Multiply $4i(3 + 5i)$.

**Solution**   $4i(3 + 5i) = 4i(3) + 4i(5i)$      Distributive property

$$= 12i + 20i^2 \qquad \text{Multiplication}$$
$$= 12i + 20(-1) \qquad i^2 = -1$$
$$= -20 + 12i$$

**EXAMPLE   4**   Multiply $(3 + 2i)(4 - 3i)$.

**Solution**   $(3 + 2i)(4 - 3i)$

$$= 3 \cdot 4 + 3(-3i) + 2i(4) + 2i(-3i) \qquad \text{FOIL method}$$
$$= 12 - 9i + 8i - 6i^2$$
$$= 12 - 9i + 8i - 6(-1) \qquad\qquad i^2 = -1$$
$$= (12 + 6) + (-9i + 8i)$$
$$= 18 - i$$

## DIVISION OF COMPLEX NUMBERS

We divide complex numbers by applying the same process we used to rationalize denominators.

**EXAMPLE   5**   Divide $\dfrac{2i}{3 - 4i}$.

**Solution**   We multiply the numerator and denominator by the conjugate of the denominator, which is $3 + 4i$:

$$\left(\frac{2i}{3 - 4i}\right)\left(\frac{3 + 4i}{3 + 4i}\right) = \frac{6i + 8i^2}{9 - 16i^2}$$

$$= \frac{6i + 8(-1)}{9 - 16(-1)}$$

$$= \frac{-8 + 6i}{25}$$

**Note:** The conjugate of $a + bi$ is $a - bi$. When we multiply complex conjugates the result is always a real number because:

$$(a + bi)(a - bi) = a^2 - (bi)^2$$
$$= a^2 - b^2 i^2$$
$$= a^2 - b^2(-1)$$
$$= a^2 + b^2$$

which is a real number.

**EXAMPLE 6**   Divide $\dfrac{2 + i}{5 + 2i}$.

**Solution**   The conjugate of the denominator is $5 - 2i$:

$$\left(\frac{2 + i}{5 + 2i}\right)\left(\frac{5 - 2i}{5 - 2i}\right) = \frac{10 - 4i + 5i - 2i^2}{25 - 4i^2}$$

$$= \frac{10 - 4i + 5i - 2(-1)}{25 - 4(-1)} \qquad i^2 = -1$$

$$= \frac{12 + i}{29}$$

When we write our answer in standard form for complex numbers, we get:

$$\frac{12}{29} + \frac{1}{29}i$$

## Getting Ready for Class

*After reading through the preceding section, respond in your own words and in complete sentences.*

**A.** What is the number $i$?

**B.** What is a complex number?

**C.** What kind of number will always result when we multiply complex conjugates?

**D.** Explain how to divide complex numbers.

# PROBLEM SET 9.4

Combine the following complex numbers.

**1.** $(3 - 2i) + 3i$

**2.** $(5 - 4i) - 8i$

**3.** $(6 + 2i) - 10i$

**4.** $(8 - 10i) + 7i$

**5.** $(11 + 9i) - 9i$

**6.** $(12 + 2i) + 6i$

**7.** $(3 + 2i) + (6 - i)$

**8.** $(4 + 8i) - (7 + i)$

**9.** $(5 + 7i) - (6 + 8i)$

**10.** $(11 + 6i) - (3 + 6i)$

**11.** $(9 - i) + (2 - i)$      **12.** $(8 + 3i) - (8 - 3i)$

**13.** $(6 + i) - 4i - (2 - i)$

**14.** $(3 + 2i) - 5i - (5 + 4i)$

**15.** $(6 - 11i) + 3i + (2 + i)$

**16.** $(3 + 4i) - (5 + 7i) - (6 - i)$

**17.** $(2 + 3i) - (6 - 2i) + (3 - i)$

**18.** $(8 + 9i) + (5 - 6i) - (4 - 3i)$

Multiply the following complex numbers.

**19.** $3(2 - i)$          **20.** $4(5 + 3i)$

**21.** $2i(8 - 7i)$        **22.** $-3i(2 + 5i)$

**23.** $(2 + i)(4 - i)$    **24.** $(6 + 3i)(4 + 3i)$

**25.** $(2 + i)(3 - 5i)$   **26.** $(4 - i)(2 - i)$

**27.** $(3 + 5i)(3 - 5i)$  **28.** $(8 + 6i)(8 - 6i)$

**29.** $(2 + i)(2 - i)$    **30.** $(3 + i)(3 - i)$

Divide the following complex numbers.

**31.** $\dfrac{2}{3 - 2i}$          **32.** $\dfrac{3}{5 + 6i}$

**33.** $\dfrac{-3i}{2 + 3i}$        **34.** $\dfrac{4i}{3 + i}$

**35.** $\dfrac{6i}{3 - i}$          **36.** $\dfrac{-7i}{5 - 4i}$

**37.** $\dfrac{2 + i}{2 - i}$       **38.** $\dfrac{3 + 2i}{3 - 2i}$

**39.** $\dfrac{4 + 5i}{3 - 6i}$     **40.** $\dfrac{-2 + i}{5 + 6i}$

**41.** Use the FOIL method to multiply $(x + 3i)(x - 3i)$.

**42.** Use the FOIL method to multiply $(x + 5i)(x - 5i)$.

**43.** The opposite of $i$ is $-i$. The reciprocal of $i$ is $\dfrac{1}{i}$.

Multiply the numerator and denominator of $\dfrac{1}{i}$ by $i$ and simplify the result to see that the opposite of $i$ and the reciprocal of $i$ are the same number.

**44.** If $i^2 = -1$, what are $i^3$ and $i^4$? (*Hint:* $i^3 = i^2 \cdot i$.)

## Review Problems

The following problems review material we covered in Sections 9.1 and 8.3. Reviewing the problems from Section 9.1 will help you with the next section.

Solve each equation by taking the square root of each side. [9.1]

**45.** $(x - 3)^2 = 25$        **46.** $(x - 2)^2 = 9$

**47.** $(2x - 6)^2 = 16$      **48.** $(2x + 1)^2 = 49$

**49.** $(x + 3)^2 = 12$       **50.** $(x + 3)^2 = 8$

Put each expression into simplified form for radicals. [8.3]

**51.** $\sqrt{\dfrac{1}{2}}$          **52.** $\sqrt{\dfrac{5}{6}}$

**53.** $\sqrt{\dfrac{8x^2y^3}{3}}$    **54.** $\sqrt{\dfrac{45xy^4}{7}}$

**55.** $\sqrt[3]{\dfrac{1}{4}}$       **56.** $\sqrt[3]{\dfrac{2}{3}}$

---

## 9.5    Complex Solutions to Quadratic Equations

The quadratic formula tells us that solutions to equations of the form $ax^2 + bx + c = 0$ are always:

$$x = \frac{-b \pm \sqrt{b^2 - 4ac}}{2a}$$

The part of the quadratic formula under the radical sign is called the *discriminant*:

$$\text{Discriminant} = b^2 - 4ac$$

When the discriminant is negative, we have to deal with the square root of a negative number. We handle square roots of negative numbers by using the definition $i = \sqrt{-1}$. To illustrate, suppose we want to simplify an expression that contains $\sqrt{-9}$, which is not a real number. We begin by writing $\sqrt{-9}$ as $\sqrt{9(-1)}$. Then, we write this expression as the product of two separate radicals: $\sqrt{9}\,\sqrt{-1}$. Applying the definition $i = \sqrt{-1}$ to this last expression, we have

$$\sqrt{9}\,\sqrt{-1} = 3i$$

As you may recall from the previous section, the number $3i$ is called a complex number. Here are some further examples.

**EXAMPLE 1**   Write the following radicals as complex numbers:

(a) $\sqrt{-4} = \sqrt{4(-1)} = \sqrt{4}\,\sqrt{-1} = 2i$
(b) $\sqrt{-36} = \sqrt{36(-1)} = \sqrt{36}\,\sqrt{-1} = 6i$
(c) $\sqrt{-7} = \sqrt{7(-1)} = \sqrt{7}\,\sqrt{-1} = i\sqrt{7}$
(d) $\sqrt{-75} = \sqrt{75(-1)} = \sqrt{75}\,\sqrt{-1} = 5i\sqrt{3}$

In parts (c) and (d) of Example 1, we wrote $i$ before the radical because it is less confusing that way. If we put $i$ after the radical, it is sometimes mistaken for being under the radical.

Let's see how complex numbers relate to quadratic equations by looking at some examples of quadratic equations whose solutions are complex numbers.

**EXAMPLE 2**   Solve for $x$: $(x + 2)^2 = -9$

**Solution**   We can solve this equation by expanding the left side, putting the results into standard form, and then applying the quadratic formula. It is faster, however, simply to apply the square root property:

$$(x + 2)^2 = -9$$
$$x + 2 = \pm\sqrt{-9} \qquad \text{Square root property}$$
$$x + 2 = \pm 3i \qquad\qquad \sqrt{-9} = \sqrt{9}\,\sqrt{-1} = 3i$$
$$x = -2 \pm 3i \qquad \text{Add } -2 \text{ to both sides.}$$

The solution set contains two complex solutions. Notice that the two solutions are conjugates.

The solution set is $\{-2 + 3i, -2 - 3i\}$.

**EXAMPLE 3**   Solve for $x$: $\dfrac{1}{10}x^2 - \dfrac{1}{5}x = -\dfrac{1}{2}$

**Solution**   It will be easier to apply the quadratic formula if we clear the equation of fractions. Multiplying both sides of the equation by the LCD 10 gives us:

$$x^2 - 2x = -5$$

Next, we add 5 to both sides to put the equation into standard form:

$$x^2 - 2x + 5 = 0 \qquad \text{Add 5 to both sides.}$$

Applying the quadratic formula with $a = 1$, $b = -2$, and $c = 5$, we have:

$$x = \frac{-(-2) \pm \sqrt{(-2)^2 - 4(1)(5)}}{2(1)}$$

$$= \frac{2 \pm \sqrt{-16}}{2}$$

$$= \frac{2 \pm 4i}{2}$$

Dividing the numerator and denominator by 2, we have the two solutions:

$$x = 1 \pm 2i$$

The two solutions are $1 + 2i$ and $1 - 2i$.

**E X A M P L E   4**   Solve $(2x - 3)(2x - 1) = -4$.

**Solution**   We multiply the binomials on the left side and then add 4 to each side to write the equation in standard form. From there we identify $a$, $b$, and $c$ and apply the quadratic formula:

$$(2x - 3)(2x - 1) = -4$$

$$4x^2 - 8x + 3 = -4 \qquad \text{Multiply binomials on left side.}$$

$$4x^2 - 8x + 7 = 0 \qquad \text{Add 4 to each side.}$$

Placing $a = 4$, $b = -8$, and $c = 7$ in the quadratic formula we have:

$$x = \frac{-(-8) \pm \sqrt{(-8)^2 - 4(4)(7)}}{2(4)}$$

$$= \frac{8 \pm \sqrt{64 - 112}}{8}$$

$$= \frac{8 \pm \sqrt{-48}}{8}$$

$$= \frac{8 \pm 4i\sqrt{3}}{8} \qquad \sqrt{-48} = i\sqrt{48} = i\sqrt{16}\,\sqrt{3} = 4i\sqrt{3}$$

To reduce this final expression to lowest terms, we factor a 4 from the numerator and then divide the numerator and denominator by 4:

$$x = \frac{\cancel{4}(2 \pm i\sqrt{3})}{\cancel{4} \cdot 2}$$

$$= \frac{2 \pm i\sqrt{3}}{2}$$

**Note:** It would be a mistake to try to reduce this final expression further. Sometimes first-year algebra students will try to divide the 2 in the denominator into the 2 in the numerator, which is a mistake. Remember, when we reduce to lowest terms, we do so by dividing the numerator and denominator by any factors they have in common. In this case 2 is not a factor of the numerator. This expression is in lowest terms.

This completes our work with solving quadratic equations. We can solve any quadratic equation. Factoring is probably still the fastest method of solution, but again, factoring works only if the equation is factorable. Applying the quadratic formula always produces solutions, whether the equation is factorable or not.

# Getting Ready for Class

*After reading through the preceding section, respond in your own words and in complete sentences.*

**A.** Describe how you would write $\sqrt{-4}$ in terms of the number $i$.

**B.** When would the quadratic formula result in complex solutions?

**C.** How can we avoid working with fractions when using the quadratic formula?

**D.** Describe how you would simplify the expression $\dfrac{8 \pm 4i\sqrt{3}}{8}$.

# PROBLEM SET 9.5

Write the following radicals as complex numbers.

**1.** $\sqrt{-16}$     **2.** $\sqrt{-25}$     **3.** $\sqrt{-49}$
**4.** $\sqrt{-81}$     **5.** $\sqrt{-6}$     **6.** $\sqrt{-10}$
**7.** $\sqrt{-11}$     **8.** $\sqrt{-19}$     **9.** $\sqrt{-32}$
**10.** $\sqrt{-288}$     **11.** $\sqrt{-50}$     **12.** $\sqrt{-45}$
**13.** $\sqrt{-8}$     **14.** $\sqrt{-24}$     **15.** $\sqrt{-48}$
**16.** $\sqrt{-27}$

Solve the following quadratic equations. Use whatever method seems to fit the situation or is convenient for you.

**17.** $x^2 = 2x - 2$     **18.** $x^2 = 4x - 5$
**19.** $x^2 - 4x = -4$     **20.** $x^2 - 4x = 4$
**21.** $2x^2 + 5x = 12$     **22.** $2x^2 + 30 = 16x$
**23.** $(x - 2)^2 = -4$     **24.** $(x - 5)^2 = -25$
**25.** $\left(x + \dfrac{1}{2}\right)^2 = -\dfrac{9}{4}$     **26.** $\left(x - \dfrac{1}{4}\right)^2 = -\dfrac{1}{2}$

**27.** $\left(x - \dfrac{1}{2}\right)^2 = -\dfrac{27}{36}$     **28.** $\left(x + \dfrac{1}{2}\right)^2 = -\dfrac{32}{64}$

**29.** $x^2 + x + 1 = 0$     **30.** $x^2 - 3x + 4 = 0$
**31.** $x^2 - 5x + 6 = 0$     **32.** $x^2 + 2x + 2 = 0$

**33.** $\dfrac{1}{2}x^2 + \dfrac{1}{3}x + \dfrac{1}{6} = 0$

**34.** $\dfrac{1}{5}x^2 + \dfrac{1}{20}x + \dfrac{1}{4} = 0$

**35.** $\dfrac{1}{3}x^2 = -\dfrac{1}{2}x + \dfrac{1}{3}$

**36.** $\dfrac{1}{2}x^2 = -\dfrac{1}{3}x + \dfrac{1}{6}$

**37.** $(x + 2)(x - 3) = 5$
**38.** $(x - 1)(x + 1) = 6$
**39.** $(x - 5)(x - 3) = -10$
**40.** $(x - 2)(x - 4) = -5$

**41.** $(2x - 2)(x - 3) = 9$

**42.** $(x - 1)(2x + 6) = 9$

**43. Intercepts**   The graph of $y = x^2 - 4x + 5$ is shown in the following figure. As you can see, the graph does not cross the $x$-axis. If the graph did cross the $x$-axis, the $x$-intercepts would be solutions to the equation

$$0 = x^2 - 4x + 5$$

Solve this equation. The solutions will confirm the fact that the graph cannot cross the $x$-axis.

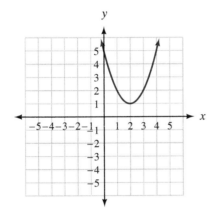

**44. Intercepts**   The graph of $y = x^2 + 2x + 5$ is shown in the following figure. As you can see, the graph does not cross the $x$-axis. If the graph did cross the $x$-axis, the $x$-intercepts would be solutions to the equation

$$0 = x^2 + 2x + 5$$

Solve this equation.

**45.** If one solution to a quadratic equation is $3 + 7i$, what do you think the other solution is?

**46.** If one solution to a quadratic equation is $4 - 2i$, what do you think the other solution is?

## Review Problems

The following problems review material we covered in Sections 3.2, 3.3, and 8.4. Reviewing the problems from Sections 3.2 and 3.3 will help you understand the next section.

Graph each line. [3.2, 3.3]

**47.** $y = x - 2$                **48.** $y = -x - 2$

**49.** $2x + 4y = 8$            **50.** $2x - 4y = 8$

Write each term in simplified form for radicals. Then combine similar terms. [8.4]

**51.** $3\sqrt{50} + 2\sqrt{32}$           **52.** $4\sqrt{18} + \sqrt{32} - \sqrt{2}$

**53.** $\sqrt{24} - \sqrt{54} - \sqrt{150}$   **54.** $\sqrt{72} - \sqrt{8} + \sqrt{50}$

**55.** $2\sqrt{27x^2} - x\sqrt{48}$      **56.** $5\sqrt{8x^3} + x\sqrt{50x}$

---

## 9.6    Graphing Parabolas

In this section we will graph equations of the form $y = ax^2 + bx + c$ and equations that can be put into this form. The graphs of this type of equation all have similar shapes.

We will begin this section by reviewing how we graphed the simplest quadratic equation, $y = x^2$, in Chapter 5. To get the idea of the shape of this graph, we need to find some ordered pairs that are solutions. We can do this by setting up the following table:

| $x$ | $y = x^2$ | $y$ |
|---|---|---|
|  |  |  |

We can choose any convenient numbers for $x$ and then use the equation $y = x^2$ to find the corresponding values for $y$. Let's use the values $-3, -2, -1, 0, 1, 2,$ and $3$ for $x$ and find corresponding values for $y$. Here is how the table looks when we let $x$ have these values:

| $x$ | $y = x^2$ | $y$ |
|---|---|---|
| $-3$ | $y = (-3)^2 = 9$ | 9 |
| $-2$ | $y = (-2)^2 = 4$ | 4 |
| $-1$ | $y = (-1)^2 = 1$ | 1 |
| 0 | $y = 0^2 = 0$ | 0 |
| 1 | $y = 1^2 = 1$ | 1 |
| 2 | $y = 2^2 = 4$ | 4 |
| 3 | $y = 3^2 = 9$ | 9 |

The table gives us the solutions $(-3, 9)$, $(-2, 4)$, $(-1, 1)$, $(0, 0)$, $(1, 1)$, $(2, 4)$, and $(3, 9)$ for the equation $y = x^2$. We plot each of the points on a rectangular coordinate system and draw a smooth curve between them, as shown in Figure 1.

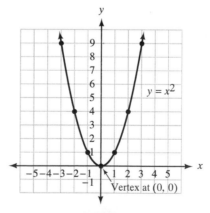

**FIGURE I**

This graph is called a *parabola*. All equations of the form $y = ax^2 + bx + c$ ($a \neq 0$) produce parabolas when graphed.

Note that the point $(0, 0)$ is called the vertex of the parabola in Figure 1. It is the lowest point on the graph. Although most of the parabolas in this section will open up, there are some parabolas that open downward. For those parabolas, the vertex is the highest point on the graph.

**E X A M P L E   1**   Graph the equation $y = x^2 - 3$.

**Solution**    We begin by making a table using convenient values for $x$:

| $x$ | $y = x^2 - 3$ | $y$ |
|---|---|---|
| $-2$ | $y = (-2)^2 - 3 = 4 - 3 = 1$ | $1$ |
| $-1$ | $y = (-1)^2 - 3 = 1 - 3 = -2$ | $-2$ |
| $0$ | $y = 0^2 - 3 = -3$ | $-3$ |
| $1$ | $y = 1^2 - 3 = 1 - 3 = -2$ | $-2$ |
| $2$ | $y = 2^2 - 3 = 4 - 3 = 1$ | $1$ |

The table gives us the ordered pairs $(-2, 1)$, $(-1, -2)$, $(0, -3)$, $(1, -2)$, and $(2, 1)$ as solutions to $y = x^2 - 3$. The graph is shown in Figure 2.

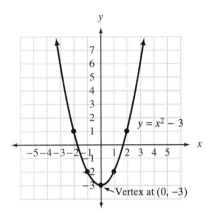

**FIGURE 2**

**E X A M P L E   2**   Graph $y = (x - 2)^2$.

**Solution**

| $x$ | $y = (x - 2)^2$ | $y$ |
|---|---|---|
| $-1$ | $y = (-1 - 2)^2 = (-3)^2 = 9$ | $9$ |
| $0$ | $y = (0 - 2)^2 = (-2)^2 = 4$ | $4$ |
| $1$ | $y = (1 - 2)^2 = (-1)^2 = 1$ | $1$ |
| $2$ | $y = (2 - 2)^2 = 0^2 = 0$ | $0$ |

We can continue the table if we feel more solutions will make the graph clearer.

| $3$ | $y = (3 - 2)^2 = 1^2 = 1$ | $1$ |
| $4$ | $y = (4 - 2)^2 = 2^2 = 4$ | $4$ |
| $5$ | $y = (5 - 2)^2 = 3^2 = 9$ | $9$ |

Putting the results of the table onto a coordinate system, we have the graph in Figure 3.

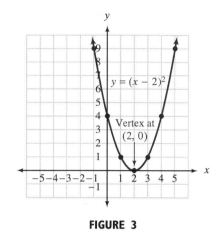

**FIGURE 3**

When graphing parabolas, it is sometimes easier to obtain the correct graph if you have some idea what the graph will look like before you start making your table.

If you look over the first two examples closely, you will see that the graphs in Figures 2 and 3 have the same shape as the graph of $y = x^2$. The difference is in the position of the vertex. For example, the graph of $y = x^2 - 3$, as shown in Figure 2, looks like the graph of $y = x^2$ with its vertex moved down three units vertically. Similarly, the graph of $y = (x - 2)^2$, shown in Figure 3, looks like the graph of $y = x^2$ with its vertex moved two units to the right.

Without showing the tables necessary to graph them, Figures 4–7 are four more parabolas and their corresponding equations. Can you see the correlation between the numbers in the equations and the position of the graphs on the coordinate systems?

**FIGURE 4**                             **FIGURE 5**

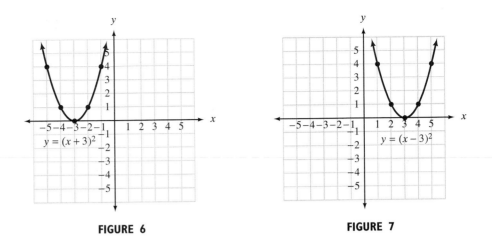

FIGURE 6                                 FIGURE 7

For our next example, we will graph a parabola in which the vertex has been moved both vertically and horizontally from the origin.

**E X A M P L E   3**    Graph $y = (x + 1)^2 - 3$.

**Solution**

| $x$ | $y = (x + 1)^2 - 3$ | $y$ |
|-----|---------------------|-----|
| $-4$ | $y = (-4 + 1)^2 - 3 = 9 - 3$ | $6$ |
| $-3$ | $y = (-3 + 1)^2 - 3 = 4 - 3$ | $1$ |
| $-1$ | $y = (-1 + 1)^2 - 3 = 0 - 3$ | $-3$ |
| $1$ | $y = (1 + 1)^2 - 3 = 4 - 3$ | $1$ |
| $2$ | $y = (2 + 1)^2 - 3 = 9 - 3$ | $6$ |

Graphing the results of the table, we have Figure 8.

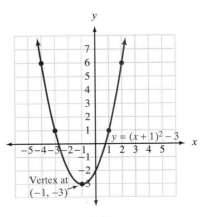

FIGURE 8

We summarize the information shown in the preceding examples and graphs with the following statement.

## GRAPHING PARABOLAS

The graph of each of the following equations is a parabola with the same basic shape as the graph of $y = x^2$.

| Equation | Vertex | To Obtain the Graph, Move the Graph of $y = x^2$ |
|---|---|---|
| $y = x^2 + k$ | $(0, k)$ | $k$ units vertically |
| $y = (x - h)^2$ | $(h, 0)$ | $h$ units horizontally |
| $y = (x - h)^2 + k$ | $(h, k)$ | $h$ units horizontally and $k$ units vertically |

Our last example for this section shows what we do if the equation of the parabola we want to graph is not in the form $y = (x - h)^2 + k$.

**EXAMPLE 4** Graph $y = x^2 - 6x + 5$.

**Solution** We could graph this equation by making a table of values of $x$ and $y$ as we have previously. However, there is an easier way! If we can write the equation in the form

$$y = (x - h)^2 + k$$

then, to graph it, we simply place the vertex at $(h, k)$ and draw a graph from there that has the same shape as $y = x^2$.

To write our equation in the form given in the preceding equation, we simply complete the square on the first two terms on the right side of the equation, $x^2 - 6x$. We do so by adding 9 to and subtracting 9 from the right side of the equation. This amounts to adding 0 to the equation, so we know we haven't changed its solutions. This is what it looks like:

$$y = (x^2 - 6x \quad) + 5$$
$$y = (x^2 - 6x + \mathbf{9}) + 5 - \mathbf{9}$$
$$y = (x - 3)^2 - 4$$

This final equation has the form we are looking for. Our graph will have the same shape as the graph of $y = x^2$. The vertex of our graph is at $(3, -4)$. The graph is shown in Figure 9.

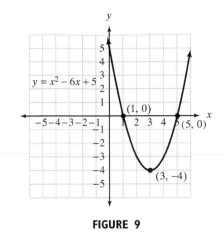

**FIGURE 9**

One final note: We can check to see that the *x*-intercepts in Figure 9 are correct by finding the *x*-intercepts from our original equation. Remember, the *x*-intercepts occur when *y* is 0. That is, to find the *x*-intercepts, we let $y = 0$ and solve for *x*:

$$0 = x^2 - 6x + 5$$
$$0 = (x - 5)(x - 1)$$
$$x = 5 \qquad \text{or} \qquad x = 1$$

 ## Getting Ready for Class

*After reading through the preceding section, respond in your own words and in complete sentences.*

**A.** What is a parabola?

**B.** What is the vertex of a parabola?

**C.** How do you find the *x*-intercepts of a parabola?

**D.** Describe the relationship between the graph of $y = x^2$ and the graph of $y = x^2 - 2$.

# PROBLEM SET 9.6

Graph each of the following equations.

**1.** $y = x^2 - 4$

**2.** $y = x^2 + 2$

**3.** $y = x^2 + 5$

**4.** $y = x^2 - 2$

**5.** $y = (x + 2)^2$

**6.** $y = (x + 5)^2$

**7.** $y = (x - 3)^2$

**8.** $y = (x - 2)^2$

**9.** $y = (x - 5)^2$

**10.** $y = (x + 3)^2$

**11.** $y = (x + 1)^2 - 2$

**12.** $y = (x - 1)^2 + 2$

**13.** $y = (x + 2)^2 - 3$

**14.** $y = (x - 2)^2 + 3$

**15.** $y = (x - 3)^2 + 2$

**16.** $y = (x + 4)^2 + 1$

Graph each of the following equations. Begin by completing the square on the first two terms as we did in Example 4 of this section.

**17.** $y = x^2 + 6x + 5$       **18.** $y = x^2 - 8x + 12$

**19.** $y = x^2 - 2x - 3$       **20.** $y = x^2 + 2x - 3$

The following equations have graphs that are also parabolas. However, the graphs of these equations will open downward; the vertex of each will be the highest point on the graph. Graph each equation by first making a table of ordered pairs, using the given values of $x$.

**21.** $y = 4 - x^2$       $x = -3, -2, -1, 0, 1, 2, 3$

**22.** $y = 3 - x^2$       $x = -3, -2, -1, 0, 1, 2, 3$

**23.** $y = -1 - x^2$       $x = -2, -1, 0, 1, 2$

**24.** $y = -2 - x^2$       $x = -2, -1, 0, 1, 2$

**25.** Graph the line $y = x + 2$ and the parabola $y = x^2$ on the same coordinate system. Name the points where the two graphs intersect.

**26.** Graph the line $y = x$ and the parabola $y = x^2 - 2$ on the same coordinate system. Name the points where the two graphs intersect.

**27.** Graph the parabola $y = 2x^2$ and the parabola $y = \frac{1}{2}x^2$ on the same coordinate system.

**28.** Graph the parabola $y = 3x^2$ and the parabola $y = \frac{1}{3}x^2$ on the same coordinate system.

## Applying the Concepts

**29. Comet Orbit**  If a comet approaches the Sun with *exactly* the right velocity and direction, its orbit around the sun will be a parabola. (Such a comet will only be seen once, and then leave the solar system forever.) A certain comet's orbit is given by $y = x^2 - 0.25$, where both $x$ and $y$ are measured in millions of miles, and the sun's position is at the origin $(0, 0)$.

 (a) Graph the comet's orbit between $x = -2$ and $x = 2$.

 (b) How far is the comet from the sun when the comet is at the vertex of its path?

 (c) How far is the comet from the Sun when the comet crosses the $x$-axis?

**30. Maximum Profit**  A company wants to know what price they should charge to make the maximum profit on their inexpensive digital watches. They have discovered that if they sell the watches for \$4, they can sell 2 million of them, but if they sell them for \$5, they can only sell 1 million.

 (a) Let $x$ be the number of *million* watches sold at $p$ dollars per watch. Assume that the relationship between $x$ and $p$ is linear and find an equation that gives that relationship.

 (b) Remember that revenue is price times number of items sold: $R = xp$. Solve the equation from part (a) for $x$ and substitute the result into the equation $R = xp$ in place of $x$.

 (c) Graph the equation from part (b) on a coordinate system where the horizontal axis is the $p$-axis and the vertical axis is the $R$-axis.

 (d) The highest point on the graph will give you the maximum revenue. What is the maximum revenue?

## Review Problems

The following problems review material we covered in Chapter 8.

Find each root.

**31.** $\sqrt{49}$       **32.** $\sqrt[3]{-8}$

Write in simplified form for radicals.

**33.** $\sqrt{50}$       **34.** $2\sqrt{18x^2y^3}$

**35.** $\sqrt{\dfrac{2}{5}}$       **36.** $\sqrt[3]{\dfrac{1}{2}}$

Perform the indicated operations.

**37.** $3\sqrt{12} + 5\sqrt{27}$       **38.** $\sqrt{3}(\sqrt{3} - 4)$

**39.** $(\sqrt{6} + 2)(\sqrt{6} - 5)$       **40.** $(\sqrt{x} + 3)^2$

Rationalize the denominator.

**41.** $\dfrac{8}{\sqrt{5} - \sqrt{3}}$       **42.** $\dfrac{\sqrt{5} - \sqrt{2}}{\sqrt{5} + \sqrt{2}}$

Solve for $x$.

**43.** $\sqrt{2x - 5} = 3$       **44.** $\sqrt{x + 15} = x + 3$

# CHAPTER 9 SUMMARY

**Examples**

**1.** $(x - 3)^2 = 25$

$$x - 3 = \pm 5$$

$$x = 3 \pm 5$$

$$x = -2 \quad \text{or} \quad x = 8$$

## Solving Quadratic Equations of the Form $(ax + b)^2 = c$ [9.1]

We can solve equations of the form $(ax + b)^2 = c$ by applying the square root property for equations, to write:

$$ax + b = \pm\sqrt{c}$$

**2.** $x^2 - 6x + 2 = 0$

$$x^2 - 6x \quad = -2$$

$$x^2 - 6x + 9 = -2 + 9$$

$$(x - 3)^2 = 7$$

$$x - 3 = \pm\sqrt{7}$$

$$x = 3 \pm \sqrt{7}$$

## Completing the Square [9.2]

To complete the square on a quadratic equation as a method of solution, we use the following steps:

> *Step 1:* Move the constant term to one side and the variable terms to the other. Then, divide each side by the coefficient of $x^2$ if it is other than 1.
>
> *Step 2:* Take the square of half the coefficient of the linear term and add it to both sides of the equation.
>
> *Step 3:* Write the left side as a binomial square and then apply the square root property.
>
> *Step 4:* Solve the resulting equation.

**3.** If $2x^2 + 3x - 4 = 0$

then $x = \dfrac{-3 \pm \sqrt{9 - 4(2)(-4)}}{2(2)}$

$= \dfrac{-3 \pm \sqrt{41}}{4}$

## The Quadratic Formula [9.3]

Any equation that is in the form $ax^2 + bx + c = 0$, where $a \neq 0$, has as its solutions:

$$x = \frac{-b \pm \sqrt{b^2 - 4ac}}{2a}$$

The expression under the square root sign, $b^2 - 4ac$, is known as the discriminant. When the discriminant is negative, the solutions are complex numbers.

**4.** The numbers $5, 3i, 2 + 4i$, and $7 - i$ are all complex numbers.

## Complex Numbers [9.4]

Any number that can be put in the form $a + bi$, where $i = \sqrt{-1}$, is called a complex number.

**5.** $(3 + 4i) + (6 - 7i)$

$= (3 + 6) + (4i - 7i)$

$= 9 - 3i$

## Addition and Subtraction of Complex Numbers [9.4]

We add (or subtract) complex numbers by using the same procedure we used to add (or subtract) polynomials: We combine similar terms.

**6.** $(2 + 3i)(3 - i)$

$= 6 - 2i + 9i - 3i^2$

$= 6 + 7i + 3$

$= 9 + 7i$

## Multiplication of Complex Numbers [9.4]

We multiply complex numbers in the same way we multiply binomials. The result, however, can be simplified further by substituting $-1$ for $i^2$ whenever it appears.

**7.** $\dfrac{3}{2 + 5i} = \dfrac{3}{2 + 5i} \cdot \dfrac{2 - 5i}{2 - 5i}$

$= \dfrac{6 - 15i}{29}$

**8.** Graph $y = x^2 - 2$.

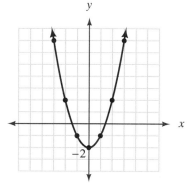

## Division of Complex Numbers [9.4]

Division with complex numbers is accomplished with the method for rationalizing the denominator that we developed while working with radical expressions. If the denominator has the form $a + bi$, we multiply both the numerator and the denominator by its conjugate, $a - bi$.

## Graphing Parabolas [9.6]

The graph of each of the following equations is a parabola with the same basic shape as the graph of $y = x^2$.

| Equation | Vertex | To Obtain the Graph, Move the Graph of $y = x^2$ |
|---|---|---|
| $y = x^2 + k$ | $(0, k)$ | $k$ units vertically |
| $y = (x - h)^2$ | $(h, 0)$ | $h$ units horizontally |
| $y = (x - h)^2 + k$ | $(h, k)$ | $h$ units horizontally and $k$ units vertically |

## COMMON MISTAKES

**1.** The most common mistake when working with complex numbers is to say $i = -1$. It does not; $i$ is the *square root* of $-1$, not $-1$ itself.

**2.** The most common mistake when working with the quadratic formula is to try to identify the constants $a$, $b$, and $c$ before putting the equation into standard form.

# CHAPTER 9  REVIEW

The numbers in brackets refer to the sections of the text in which similar problems can be found.

Solve each quadratic equation. [9.1, 9.5]

**1.** $a^2 = 32$

**2.** $a^2 = 60$

**3.** $2x^2 = 32$

**4.** $(x + 3)^2 = 36$

**5.** $(x - 2)^2 = 81$

**6.** $(3x + 2)^2 = 16$

**7.** $(2x + 5)^2 = 32$

**8.** $(3x - 4)^2 = 27$

**9.** $\left(x - \dfrac{2}{3}\right)^2 = -\dfrac{25}{9}$

Solve by completing the square. [9.2]

**10.** $x^2 + 8x = 4$

**11.** $x^2 - 8x = 4$

**12.** $x^2 - 4x - 7 = 0$

**13.** $x^2 + 4x + 3 = 0$

**14.** $a^2 = 9a + 3$

**15.** $a^2 = 5a + 6$

**16.** $2x^2 + 4x - 6 = 0$

**17.** $3x^2 - 6x - 2 = 0$

Solve by using the quadratic formula. [9.3, 9.5]

**18.** $x^2 + 7x + 12 = 0$

**19.** $x^2 - 8x + 16 = 0$

**20.** $x^2 + 5x + 7 = 0$

**21.** $2x^2 = -8x + 5$

**22.** $3x^2 = -3x - 4$

**23.** $\dfrac{1}{5}x^2 - \dfrac{1}{2}x = \dfrac{3}{10}$

**24.** $(2x + 1)(2x - 3) = -6$

Add and subtract the following complex numbers. [9.4]

**25.** $(4 - 3i) + 5i$

**26.** $(2 - 5i) - 3i$

**27.** $(5 + 6i) + (5 - i)$

**28.** $(2 + 5i) + (3 - 7i)$   **29.** $(3 - 2i) - (3 - i)$
**30.** $(5 - 7i) - (6 - 2i)$
**31.** $(3 + i) - 5i - (4 - i)$
**32.** $(2 - 3i) - (5 - 2i) + (5 - i)$

Multiply the following complex numbers. [9.4]

**33.** $2(3 - i)$   **34.** $-6(4 + 2i)$
**35.** $4i(6 - 5i)$   **36.** $(2 - i)(3 + i)$
**37.** $(3 - 4i)(5 + i)$   **38.** $(3 - i)^2$
**39.** $(4 + i)(4 - i)$   **40.** $(3 + 2i)(3 - 2i)$

Divide the following complex numbers. [9.4]

**41.** $\dfrac{i}{3 + i}$   **42.** $\dfrac{i}{2 - i}$

**43.** $\dfrac{5}{2 + 5i}$   **44.** $\dfrac{2i}{4 - i}$

**45.** $\dfrac{-3i}{3 - 2i}$   **46.** $\dfrac{3 + i}{3 - i}$

**47.** $\dfrac{4 - 5i}{4 + 5i}$   **48.** $\dfrac{2 + 3i}{3 - 5i}$

Write the following radicals as complex numbers. [9.5]

**49.** $\sqrt{-36}$   **50.** $\sqrt{-144}$
**51.** $\sqrt{-17}$   **52.** $\sqrt{-31}$
**53.** $\sqrt{-40}$   **54.** $\sqrt{-72}$
**55.** $\sqrt{-200}$   **56.** $\sqrt{-242}$

Graph each of the following equations. [9.6]

**57.** $y = x^2 + 2$   **58.** $y = (x - 2)^2$
**59.** $y = (x + 2)^2$   **60.** $y = (x + 3)^2 - 2$
**61.** $y = x^2 + 4x + 7$   **62.** $y = x^2 - 4x + 7$

## CHAPTER 9   TEST

Solve the following quadratic equations. [9.1, 9.3, 9.5]

**1.** $x^2 - 7x - 8 = 0$   **2.** $(x - 3)^2 = 12$
**3.** $\left(x - \dfrac{5}{2}\right)^2 = -\dfrac{75}{4}$   **4.** $\dfrac{1}{3}x^2 = \dfrac{1}{2}x - \dfrac{5}{6}$
**5.** $3x^2 = -2x + 1$   **6.** $(x + 2)(x - 1) = 6$
**7.** $9x^2 + 12x + 4 = 0$
**8.** Solve $x^2 - 6x - 6 = 0$ by completing the square. [9.2]

Write as complex numbers. [9.5]

**9.** $\sqrt{-9}$   **10.** $\sqrt{-121}$
**11.** $\sqrt{-72}$   **12.** $\sqrt{-18}$

Work the following problems involving complex numbers. [9.4]

**13.** $(3i + 1) + (2 + 5i)$   **14.** $(6 - 2i) - (7 - 4i)$
**15.** $(2 + i)(2 - i)$   **16.** $(3 + 2i)(1 + i)$
**17.** $\dfrac{i}{3 - i}$   **18.** $\dfrac{2 + i}{2 - i}$

Graph the following equations. [9.6]

**19.** $y = x^2 - 4$   **20.** $y = (x - 4)^2$
**21.** $y = (x + 3)^2 - 4$   **22.** $y = x^2 - 6x + 11$

Simplify.

**1.** $10 - 8 - 11$

**2.** $\dfrac{9(-6) - 10}{2(-8)}$

**3.** $-\dfrac{4}{5} \div \dfrac{8}{15}$

**4.** $\dfrac{x^4}{x^{-8}}$

**5.** $\dfrac{(3x^5)(20x^3)}{15x^{10}}$

**6.** $\dfrac{1 - \dfrac{16}{y^2}}{1 - \dfrac{4}{y} - \dfrac{32}{y^2}}$

**7.** $\sqrt{81}$

**8.** $\sqrt{121x^4y^2}$   (Assume $x$ and $y$ are positive.)

**9.** $\sqrt{\dfrac{200}{81}}$

**10.** $\dfrac{5}{\sqrt{7} + \sqrt{3}}$

**11.** $(3 + 3i) - 7i - (2 + 2i)$

**12.** $(3 + 5i) + (4 - 3i) - (6 - i)$

**13.** $(a - 2)(a^2 - 6a + 7)$

Solve each equation.

**14.** $\dfrac{2}{3}a = \dfrac{1}{6}a + 1$

**15.** $7 - 4(3x + 4) = -9x$

**16.** $3x^2 = 7x + 20$

**17.** $5x^2 = -15x$

**18.** $\dfrac{a}{a + 4} = \dfrac{7}{3}$

**19.** $\sqrt{6x - 2} = 3x - 5$

**20.** $(2x - 3)^2 = 49$

**21.** $\dfrac{1}{2}x^2 - \dfrac{1}{3}x = -\dfrac{1}{6}$

Graph the solution set on the number line.

**22.** $-3x < 9$

**23.** $-5 \le 2x - 1 \le 7$

**24.** $x < -5$ or $x > 2$

Graph on a rectangular coordinate system.

**25.** $x - y = 5$

**26.** $y = \dfrac{1}{3}x$

**27.** $y = -3x + 4$

**28.** $y = x^2 - 2$

**29.** $y = (x - 3)^2 - 2$

**30.** Solve the following system by graphing:
$x + y = 1$
$x - y = 3$

Solve each system.

**31.**   $x = y - 3$
$2x + 3y = 4$

**32.** $2x + 3y = 2$
$3x - y = 3$

Factor completely.

**33.** $x^2 - 5x - 24$

**34.** $12x^2 - 11xy - 15y^2$   **35.** $25 - y^2$

**36.** $25x^2 - 30xy + 9y^2$

**37.** Find the slope of the line that passes through the points $(3, -2)$ and $(1, 8)$.

**38.** Find the slope and $y$-intercept for the equation $8x - 4y = -12$.

**39.** Subtract $-2$ from 6.

**40.** What property justifies the statement $3(x + 2) = 3x + 6$?

**41.** Divide $15a^3b - 10a^2b^2 - 20ab^3$ by $5ab$.

**42.** Divide $\dfrac{3x^3 - 7x^2 + 14x - 10}{3x - 1}$.

**43.** Reduce $\dfrac{x^2 + 5x - 24}{x + 8}$ to lowest terms.

Find the differences.

**44.** $\dfrac{a - 3}{a - 7} - \dfrac{a + 10}{a - 7}$

**45.** $\dfrac{-1}{x^2 - 4} - \dfrac{-2}{x^2 - 4x - 12}$

**46.** Multiply $\dfrac{x^2 + 3x}{x^2 + 4x + 4} \cdot \dfrac{x^2 - 5x - 14}{x^2 + 6x + 9}$.

**47.** Combine the expressions $5\sqrt{200} + 9\sqrt{50}$.

**48.** Divide $\dfrac{3 - 2i}{5 - i}$.

**49.** Solve $3x - 8y = 24$ for $y$.

**50. Investing**   Randy invested $20,000 in two accounts. One paid 5% simple interest, and the other paid 7%. He earned $800 more in interest on the 7% account than he did on the 5% account. How much did he invest in each account?

# Appendix A

## Introduction to Functions

The ad shown in the margin appeared in the "Help Wanted" section of the local newspaper the day I was writing this part of the book. We can use the information in the ad to start an informal discussion of functions.

### AN INFORMAL LOOK AT FUNCTIONS

To begin with, suppose you have a job that pays $7.50 per hour and that you work anywhere from 0 to 40 hours per week. The amount of money you make in 1 week depends on the number of hours you work that week. In mathematics we say that your weekly earnings are a *function* of the number of hours you work. If we let the variable $x$ represent hours and the variable $y$ represent the money you make, then the relationship between $x$ and $y$ can be written as

$$y = 7.5x \qquad \text{for } 0 \leq x \leq 40$$

**EXAMPLE 1** Construct a table and graph for the function

$$y = 7.5x \qquad \text{for } 0 \leq x \leq 40$$

**Solution** Table 1 gives some of the paired data that satisfy the equation $y = 7.5x$. Figure 1 is the graph of the equation with the restriction $0 \leq x \leq 40$.

| TABLE 1 Weekly Wages | | |
|---|---|---|
| **Hours Worked** $x$ | **Rule** $y = 7.5x$ | **Pay** $y$ |
| 0 | $y = 7.5(0)$ | 0 |
| 10 | $y = 7.5(10)$ | 75 |
| 20 | $y = 7.5(20)$ | 150 |
| 30 | $y = 7.5(30)$ | 225 |
| 40 | $y = 7.5(40)$ | 300 |

*Ordered Pairs*

(0, 0)
(10, 75)
(20, 150)
(30, 225)
(40, 300)

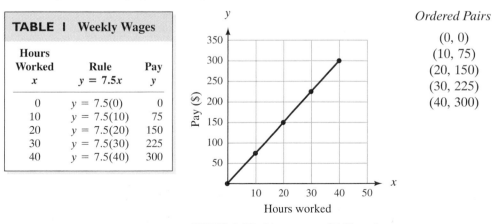

**FIGURE 1** Weekly wages at $7.50 per hour

The equation $y = 7.5x$ with the restriction $0 \leq x \leq 40$, Table 1, and Figure 1 are three ways to describe the same relationship. In all three, we *input* values of $x$ and then use the function rule to *output* values of $y$.

## DOMAIN AND RANGE OF A FUNCTION

We began this discussion by saying that the number of hours worked during the week was from 0 to 40, so these are the values that $x$ can assume. From the line graph in Figure 1, we see that the values of $y$ range from 0 to 300. We call the complete set of values that $x$ can assume the *domain* of the function. The values that are assigned to $y$ are called the *range* of the function.

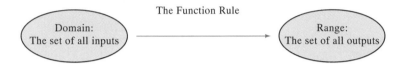

The Function Rule

Domain: The set of all inputs

Range: The set of all outputs

---

**EXAMPLE  2**    State the domain and range for the function

$$y = 7.5x \qquad 0 \leq x \leq 40$$

**Solution**    From the previous discussion we have

$$\text{Domain} = \{x \mid 0 \leq x \leq 40\}$$

$$\text{Range} = \{y \mid 0 \leq y \leq 300\}$$

---

## FUNCTION MAPS

Another way to visualize the relationship between $x$ and $y$ is with the diagram in Figure 2, which we call a *function map*.

Domain    Rule:   Multiply by 7.5    Range

10 → 75
20 → 150
$x$ → 7.5$x$

**FIGURE 2**  A function map

Although the diagram in Figure 2 does not show all the values that $x$ and $y$ can assume, it does give us a visual description of how $x$ and $y$ are related. It shows that values of $y$ in the range come from values of $x$ in the domain according to a specific rule (multiply by 7.5 each time).

## A FORMAL LOOK AT FUNCTIONS

What is apparent from the preceding discussion is that we are working with paired data. The solutions to the equation $y = 7.5x$ are pairs of numbers; the points on the line graph in Figure 1 come from paired data; and the diagram in Figure 2 pairs numbers in the domain with numbers in the range. We are now ready for the formal definition of a function.

> **DEFINITION**   A **function** is a rule that pairs each element in one set, called the **domain,** with exactly one element from a second set, called the **range.**

In other words, a function is a rule for which each input is paired with exactly one output.

## FUNCTIONS AS ORDERED PAIRS

The function rule $y = 7.5x$ from Example 1 produces ordered pairs of numbers $(x, y)$. The same thing happens with all functions: The function rule produces ordered pairs of numbers. We use this result to write an alternative definition for a function.

> **ALTERNATIVE DEFINITION**   A **function** is a set of ordered pairs in which no two different ordered pairs have the same first coordinate. The set of all first coordinates is called the **domain** of the function. The set of all second co-ordinates is called the **range** of the function.

The restriction on first coordinates in the alternative definition keeps us from assigning a number in the domain to more than one number in the range.

## A RELATIONSHIP THAT IS NOT A FUNCTION

You may be wondering if any sets of paired data fail to qualify as functions. The answer is yes, as the next example reveals.

**EXAMPLE   3**   Table 2 shows the prices of used Ford Mustangs that were listed in the local newspaper. The diagram in Figure 3 gives a visual representation of the data in Table 2. Why are these data not a function?

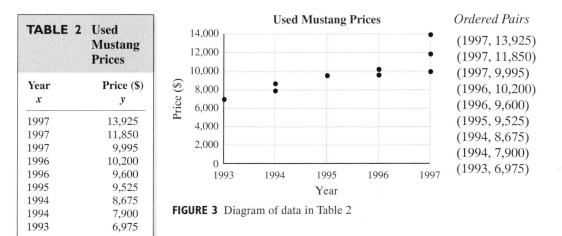

| TABLE 2 | Used Mustang Prices |
| --- | --- |
| Year<br>$x$ | Price ($)<br>$y$ |
| 1997 | 13,925 |
| 1997 | 11,850 |
| 1997 | 9,995 |
| 1996 | 10,200 |
| 1996 | 9,600 |
| 1995 | 9,525 |
| 1994 | 8,675 |
| 1994 | 7,900 |
| 1993 | 6,975 |

*Ordered Pairs*

(1997, 13,925)
(1997, 11,850)
(1997, 9,995)
(1996, 10,200)
(1996, 9,600)
(1995, 9,525)
(1994, 8,675)
(1994, 7,900)
(1993, 6,975)

**FIGURE 3** Diagram of data in Table 2

**Solution**    In Table 2, the year 1997 is paired with three different prices: $13,925, $11,850, and $9,995. That is enough to disqualify the data from belonging to a function. For a set of paired data to be considered a function, each number in the domain must be paired with exactly one number in the range.

Still, there is a relationship between the first coordinates and second coordinates in the used-car data. It is not a function relationship, but it is a relationship. To classify all relationships specified by ordered pairs, whether they are functions or not, we include the following two definitions.

> **DEFINITION**    A **relation** is a rule that pairs each element in one set, called the **domain,** with **one or more elements** from a second set, called the **range.**

> **ALTERNATIVE DEFINITION**    A **relation** is a set of ordered pairs. The set of all first coordinates is the **domain** of the relation. The set of all second coordinates is the **range** of the relation.

Here are some facts that will help clarify the distinction between relations and functions.

1. Any rule that assigns numbers from one set to numbers in another set is a relation. If that rule makes the assignment such that no input has more than one output, then it is also a function.
2. Any set of ordered pairs is a relation. If none of the first coordinates of those ordered pairs is repeated, the set of ordered pairs is also a function.
3. Every function is a relation.
4. Not every relation is a function.

## GRAPHING RELATIONS AND FUNCTIONS

To give ourselves a wider perspective on functions and relations, we consider some equations whose graphs are not straight lines.

**EXAMPLE    4**    Kendra is tossing a softball into the air with an underhand motion. The distance of the ball above her hand at any time is given by the function

$$h = 32t - 16t^2 \qquad \text{for } 0 \leq t \leq 2$$

where $h$ is the height of the ball in feet and $t$ is the time in seconds. Construct a table that gives the height of the ball at quarter-second intervals, starting with $t = 0$ and ending with $t = 2$. Construct a line graph from the table.

**Solution**    We construct Table 3, using the following values of $t$: $0, \frac{1}{4}, \frac{1}{2}, \frac{3}{4}, 1, \frac{5}{4}, \frac{3}{2}, \frac{7}{4}, 2$. The values of $h$ come from substituting these values of $t$ into the equation

$h = 32t - 16t^2$. (This equation comes from physics. If you take a physics class, you will learn how to derive this equation.) Then we construct the graph in Figure 4 from the table. The graph appears only in the first quadrant because neither $t$ nor $h$ can be negative.

Here is a summary of what we know about functions as it applies to this example: We input values of $t$ and output values of $h$ according to the function rule:

$$h = 32t - 16t^2 \qquad \text{for } 0 \le t \le 2$$

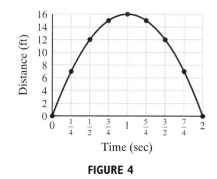

| TABLE 3   Tossing a Softball Into the Air | | |
| --- | --- | --- |
| Time (seconds) $t$ | Function Rule $h = 32t - 16t^2$ | Distance (feet) $h$ |
| 0 | $h = 32(0) - 16(0)^2 = 0 - 0 = 0$ | 0 |
| $\frac{1}{4}$ | $h = 32\left(\frac{1}{4}\right) - 16\left(\frac{1}{4}\right)^2 = 8 - 1 = 7$ | 7 |
| $\frac{1}{2}$ | $h = 32\left(\frac{1}{2}\right) - 16\left(\frac{1}{2}\right)^2 = 16 - 4 = 12$ | 12 |
| $\frac{3}{4}$ | $h = 32\left(\frac{3}{4}\right) - 16\left(\frac{3}{4}\right)^2 = 24 - 9 = 15$ | 15 |
| 1 | $h = 32(1) - 16(1)^2 = 32 - 16 = 16$ | 16 |
| $\frac{5}{4}$ | $h = 32\left(\frac{5}{4}\right) - 16\left(\frac{5}{4}\right)^2 = 40 - 25 = 15$ | 15 |
| $\frac{3}{2}$ | $h = 32\left(\frac{3}{2}\right) - 16\left(\frac{3}{2}\right)^2 = 48 - 36 = 12$ | 12 |
| $\frac{7}{4}$ | $h = 32\left(\frac{7}{4}\right) - 16\left(\frac{7}{4}\right)^2 = 56 - 49 = 7$ | 7 |
| 2 | $h = 32(2) - 16(2)^2 = 64 - 64 = 0$ | 0 |

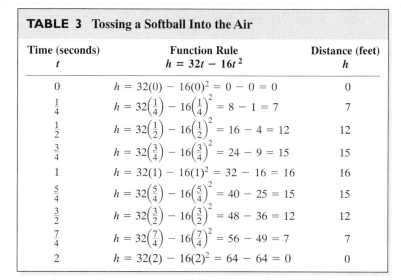

**FIGURE 4**

The domain is given by the inequality that follows the equation:

$$\text{Domain} = \{t \mid 0 \le t \le 2\}$$

The range is the set of all outputs that are possible by substituting the values of $t$ from the domain into the equation. From our table and graph, it seems that the range is:

$$\text{Range} = \{h \mid 0 \le h \le 16\}$$

**EXAMPLE 5**   Sketch the graph of $x = y^2$.

**Solution**   Without going into much detail, we graph the equation $x = y^2$ by finding a number of ordered pairs that satisfy the equation, plotting these points, and then drawing a smooth curve that connects them. A table of values for $x$ and $y$ that satisfy the equation follows, along with the graph of $x = y^2$ shown in Figure 5.

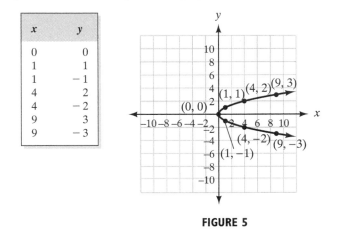

| $x$ | $y$ |
|-----|-----|
| 0 | 0 |
| 1 | 1 |
| 1 | $-1$ |
| 4 | 2 |
| 4 | $-2$ |
| 9 | 3 |
| 9 | $-3$ |

**FIGURE 5**

As you can see from looking at the table and the graph in Figure 5, several ordered pairs whose graphs lie on the curve have repeated first coordinates, for instance, (1, 1) and (1, −1), (4, 2) and (4, −2), as well as (9, 3) and (9, −3). The graph is therefore not the graph of a function.

## VERTICAL LINE TEST

Look back at the scatter diagram for used Mustang prices shown in Figure 3. Notice that some of the points on the diagram lie above and below each other along vertical lines. This is an indication that the data do not constitute a function. Two data points that lie on the same vertical line must have come from two ordered pairs with the same first coordinates.

Now, look at the graph shown in Figure 5. The reason this graph is the graph of a relation, but not of a function, is that some points on the graph have the same first coordinates, for example, the points (4, 2) and (4, −2). Furthermore, any time two points on a graph have the same first coordinates, those points must lie on a vertical line. [To convince yourself, connect the points (4, 2) and (4, −2) with a straight line. You will see that it must be a vertical line.] This allows us to write the following test that uses the graph to determine whether a relation is also a function.

### VERTICAL LINE TEST

If a vertical line crosses the graph of a relation in more than one place, the relation cannot be a function. If no vertical line can be found that crosses a graph in more than one place, then the graph is the graph of a function.

If we look back to the graph of $h = 32t - 16t^2$ as shown in Figure 4, we see that no vertical line can be found that crosses this graph in more than one place. The graph shown in Figure 4 is therefore the graph of a function.

**EXAMPLE 6**   Graph $y = |x|$. Use the graph to determine whether we have the graph of a function. State the domain and range.

**Solution**   We let $x$ take on values of $-4, -3, -2, -1, 0, 1, 2, 3,$ and 4. The corresponding values of $y$ are shown in the table. The graph is shown in Figure 6.

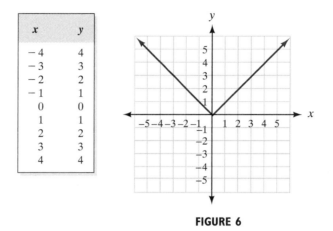

| $x$ | $y$ |
|-----|-----|
| $-4$ | 4 |
| $-3$ | 3 |
| $-2$ | 2 |
| $-1$ | 1 |
| 0 | 0 |
| 1 | 1 |
| 2 | 2 |
| 3 | 3 |
| 4 | 4 |

**FIGURE 6**

Since no vertical line can be found that crosses the graph in more than one place, $y = |x|$ is a function. The domain is all real numbers. The range is $\{y \mid y \geq 0\}$.

![notebook icon] **Getting Ready for Class**

*After reading through the preceding section, respond in your own words and in complete sentences.*

**A.** What is a function?

**B.** What is the vertical line test?

**C.** Is every line the graph of a function? Explain.

**D.** Which variable is usually associated with the domain of a function?

# PROBLEM SET A

1. Suppose you have a job that pays $8.50 per hour and you work anywhere from 10 to 40 hours per week.

   (a) Write an equation, with a restriction on the variable $x$, that gives the amount of money, $y$, you will earn for working $x$ hours in 1 week.

   (b) Use the function rule you have written in part (a) to complete Table 4.

   **TABLE 4  Weekly Wages**

   | Hours Worked $x$ | Function Rule | Gross Pay ($) $y$ |
   |---|---|---|
   | 10 | | |
   | 20 | | |
   | 30 | | |
   | 40 | | |

   (c) Use the template in Figure 7 to construct a line graph from the information in Table 4.

   **FIGURE 7** Template for line graph

   (d) State the domain and range of this function.

   (e) What is the minimum amount you can earn in a week with this job? What is the maximum amount?

2. The ad shown here was in the local newspaper. Suppose you are hired for the job described in the ad.

   (a) If $x$ is the number of hours you work per week and $y$ is your weekly gross pay, write the equation for $y$. (Be sure to include any restrictions on the variable $x$ that are given in the ad.)

   (b) Use the function rule you have written in part (a) to complete Table 5.  ·

   **TABLE 5  Weekly Wages**

   | Hours Worked $x$ | Function Rule | Gross Pay ($) $y$ |
   |---|---|---|
   | 15 | | |
   | 20 | | |
   | 25 | | |
   | 30 | | |

   **312 Help Wanted**

   ESPRESSO BAR OPERATOR

   Must be dependable, honest, service-oriented. Coffee exp desired. 15–30 hrs per wk. $5.25/hr. Start 5/31. Apply in person: Espresso Yourself, Central Coast Mall. Deadline 5/23.

   (c) Use the template in Figure 8 to construct a line graph from the information in Table 5.

   **FIGURE 8** Template for line graph

   (d) State the domain and range of this function.

   (e) What is the minimum amount you can earn in a week with this job? What is the maximum amount?

For each of the following relations, give the domain and range, and indicate which are also functions.

**3.** $\{(1, 3), (2, 5), (4, 1)\}$    **4.** $\{(3, 1), (5, 7), (2, 3)\}$

**5.** $\{(-1, 3), (1, 3), (2, -5)\}$

**6.** $\{(3, -4), (-1, 5), (3, 2)\}$

**7.** $\{(7, -1), (3, -1), (7, 4)\}$

**8.** $\{(5, -2), (3, -2), (5, -1)\}$

State whether each of the following graphs represents a function.

**9.**

**10.**

**11.**

**12.**

**13.**

**14.**

**15.**

**16.**

**17.**

**18.**

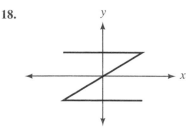

**19. Tossing a Coin**  Hali is tossing a quarter into the air with an underhand motion. The distance the quarter is above her hand at any time is given by the function

$$h = 16t - 16t^2 \qquad \text{for } 0 \le t \le 1$$

where $h$ is the height of the quarter in feet and $t$ is the time in seconds.

(a) Fill in the table.

| Time (sec) $t$ | Function Rule $h = 16t - 16t^2$ | Distance (ft) $h$ |
|---|---|---|
| 0 | | |
| 0.1 | | |
| 0.2 | | |
| 0.3 | | |
| 0.4 | | |
| 0.5 | | |
| 0.6 | | |
| 0.7 | | |
| 0.8 | | |
| 0.9 | | |
| 1 | | |

(b) State the domain and range of this function.
(c) Use the data from the table to graph the function.

**20. Intensity of Light**  The following formula gives the intensity of light that falls on a surface at various distances from a 100-watt light bulb:

$$I = \frac{120}{d^2} \qquad \text{for } d > 0$$

where $I$ is the intensity of light (in lumens per square foot) and $d$ is the distance (in feet) from the light bulb to the surface.

(a) Fill in the table.

| Distance (ft) $d$ | Function Rule | Intensity $I$ |
|---|---|---|
| 1 | | |
| 2 | | |
| 3 | | |
| 4 | | |
| 5 | | |
| 6 | | |

(b) Use the data from the table in part (a) to construct a graph of the function.

Determine the domain and range of the following functions. Assume the *entire* function is shown.

**21.**

**22.**

**24.**

**23.**

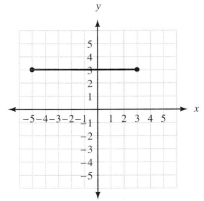

# Appendix B

## Function Notation

Let's return to the discussion that introduced us to functions. If a job pays $7.50 per hour for working from 0 to 40 hours a week, then the amount of money $y$ earned in 1 week is a function of the number of hours worked $x$. The exact relationship between $x$ and $y$ is written as follows:

$$y = 7.5x \qquad \text{for } 0 \le x \le 40$$

Since the amount of money earned $y$ depends on the number of hours worked $x$, we call $y$ the *dependent variable* and $x$ the *independent variable*. If we let $f$ represent all the ordered pairs produced by the equation, then we can write this:

$$f = \{(x, y) \mid y = 7.5x \text{ and } 0 \le x \le 40\}$$

Once we have named a function with a letter, we can use an alternative notation to represent the dependent variable $y$. The alternative notation for $y$ is $f(x)$. It is read "$f$ of $x$" and can be used instead of the variable $y$ when working with functions. The notation $y$ and the notation $f(x)$ are equivalent. That is,

$$y = 7.5x \Leftrightarrow f(x) = 7.5x$$

When we use the notation $f(x)$ we are using *function notation*. The benefit of using function notation is that we can write more information with fewer symbols than we can by using just the variable $y$. For example, asking how much money a person will make for working 20 hours is simply a matter of asking for $f(20)$. Without function notation, we would have to say "find the value of $y$ that corresponds to a value of $x = 20$." To illustrate further, using the variable $y$, we can say "$y$ is 150 when $x$ is 20." Using the notation $f(x)$, we simply say "$f(20) = 150$." Each expression indicates that you will earn $150 for working 20 hours.

**EXAMPLE 1** If $f(x) = 7.5x$, find $f(0), f(10),$ and $f(20)$.

**Solution** To find $f(0)$ we substitute 0 for $x$ in the expression $7.5x$ and simplify. We find $f(10)$ and $f(20)$ in a similar manner — by substitution.

$$\text{If} \qquad f(x) = 7.5x$$
$$\text{then} \qquad f(0) = 7.5(0) = 0$$
$$f(10) = 7.5(10) = 75$$
$$f(20) = 7.5(20) = 150$$

Input $x$

Function machine

Output $f(x)$

**Note:** Some students like to think of functions as machines. Values of $x$ are put into the machine, which transforms them into values of $f(x)$, which are then output by the machine.

If we changed the example in the discussion that opened this section so that the hourly wage was $6.50 per hour, we would have a new equation to work with:

$$y = 6.5x \qquad \text{for } 0 \le x \le 40$$

Suppose we name this new function with the letter $g$. Then:

$$g = \{(x, y) \mid y = 6.5x \text{ and } 0 \le x \le 40\}$$

and

$$g(x) = 6.5x$$

If we want to talk about both functions in the same discussion, having two different letters, $f$ and $g$, makes it easy to distinguish between them. For example, since $f(x) = 7.5x$ and $g(x) = 6.5x$, asking how much money a person makes for working 20 hours is simply a matter of asking for $f(20)$ or $g(20)$, avoiding any confusion over which hourly wage we are talking about.

The diagrams shown in Figure 1 further illustrate the similarities and differences between the two functions we have been discussing.

$$x \in \text{Domain} \quad \text{and} \quad f(x) \in \text{Range} \qquad\qquad x \in \text{Domain} \quad \text{and} \quad g(x) \in \text{Range}$$

**FIGURE 1** Function maps

## FUNCTION NOTATION AND GRAPHS

We can visualize the relationship between $x$ and $f(x)$ or $g(x)$ on the graphs of the two functions. Figure 2 shows the graph of $f(x) = 7.5x$ along with two additional line segments. The horizontal line segment corresponds to $x = 20$, and the vertical line segment corresponds to $f(20)$. Figure 3 shows the graph of $g(x) = 6.5x$ along with the horizontal line segment that corresponds to $x = 20$, and the vertical line segment that corresponds to $g(20)$. (Note that the domain in each case is restricted to $0 \le x \le 40$.)

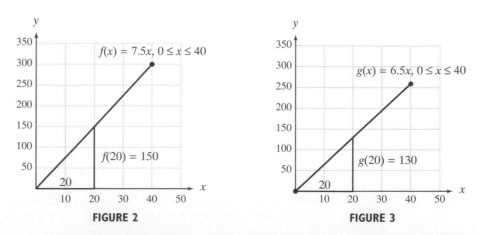

**FIGURE 2**                        **FIGURE 3**

## USING FUNCTION NOTATION

The remaining examples in this section show a variety of ways to use and interpret function notation.

**EXAMPLE 2**   If it takes Lorena $t$ minutes to run a mile, then her average speed $s(t)$ in miles per hour is given by the formula

$$s(t) = \frac{60}{t} \qquad \text{for } t > 0$$

Find $s(10)$ and $s(8)$, and then explain what they mean.

**Solution**   To find $s(10)$, we substitute 10 for $t$ in the equation and simplify:

$$s(\mathbf{10}) = \frac{60}{\mathbf{10}} = 6$$

*In words:*   When Lorena runs a mile in 10 minutes, her average speed is 6 miles per hour.

We calculate $s(8)$ by substituting 8 for $t$ in the equation. Doing so gives:

$$s(\mathbf{8}) = \frac{60}{\mathbf{8}} = 7.5$$

*In words:*   Running a mile in 8 minutes is running at a rate of 7.5 miles per hour.

**EXAMPLE 3**   A painting is purchased as an investment for $125. If its value increases continuously so that it doubles every 5 years, then its value is given by the function

$$V(t) = 125 \cdot 2^{t/5} \qquad \text{for } t \geq 0$$

where $t$ is the number of years since the painting was purchased and $V(t)$ is its value (in dollars) at time $t$. Find $V(5)$ and $V(10)$, and explain what they mean.

**Solution**   The expression $V(5)$ is the value of the painting when $t = 5$ (5 years after it is purchased). We calculate $V(5)$ by substituting 5 for $t$ in the equation $V(t) = 125 \cdot 2^{t/5}$. Here is our work:

$$V(\mathbf{5}) = 125 \cdot 2^{5/5} = 125 \cdot 2^1 = 125 \cdot 2 = 250$$

*In words:*   After 5 years, the painting is worth $250.

The expression $V(10)$ is the value of the painting after 10 years. To find this number, we substitute 10 for $t$ in the equation:

$$V(\mathbf{10}) = 125 \cdot 2^{10/5} = 125 \cdot 2^2 = 125 \cdot 4 = 500$$

*In words:*   The value of the painting 10 years after it is purchased is $500.

**EXAMPLE 4**   A balloon has the shape of a sphere with a radius of 3 inches. Use the following formulas to find the volume and surface area of the balloon.

$$V(r) = \frac{4}{3} \pi r^3 \qquad S(r) = 4 \pi r^2$$

**Solution**   As you can see, we have used function notation to write the two formulas for volume and surface area because each quantity is a function of the radius. To find these quantities when the radius is 3 inches, we evaluate $V(3)$ and $S(3)$:

$$V(\mathbf{3}) = \frac{4}{3}\pi\mathbf{3}^3 = \frac{4}{3}\pi27 = 36\pi \text{ cubic inches, or } 113 \text{ cubic inches } \text{ }^{\text{to the nearest}}_{\text{whole number}}$$

$$S(\mathbf{3}) = 4\pi\mathbf{3}^2 = 36\pi \text{ square inches, or } 113 \text{ square inches } \text{ }^{\text{to the nearest}}_{\text{whole number}}$$

The fact that $V(3) = 36\pi$ means that the ordered pair $(3, 36\pi)$ belongs to the function $V$. Likewise, the fact that $S(3) = 36\pi$ tells us that the ordered pair $(3, 36\pi)$ is a member of function $S$.

We can generalize the discussion at the end of Example 4 this way:

$$(a, b) \in f \quad \text{if and only if} \quad f(a) = b$$

**EXAMPLE 5**   If $f(x) = 3x^2 + 2x - 1$, find $f(0)$, $f(3)$, and $f(-2)$.

**Solution**   Since $f(x) = 3x^2 + 2x - 1$, we have

$$f(\mathbf{0}) = 3(\mathbf{0})^2 + 2(\mathbf{0}) - 1 \quad = \quad 0 + 0 - 1 = -1$$

$$f(\mathbf{3}) = 3(\mathbf{3})^2 + 2(\mathbf{3}) - 1 \quad = 27 + 6 - 1 = 32$$

$$f(\mathbf{-2}) = 3(\mathbf{-2})^2 + 2(\mathbf{-2}) - 1 = 12 - 4 - 1 = 7$$

In Example 5, the function $f$ is defined by the equation $f(x) = 3x^2 + 2x - 1$. We could just as easily have said $y = 3x^2 + 2x - 1$. That is, $y = f(x)$. Saying $f(-2) = 7$ is exactly the same as saying $y$ is 7 when $x$ is $-2$.

**EXAMPLE 6**   If $f(x) = 4x - 1$ and $g(x) = x^2 + 2$, then:

$$f(\mathbf{5}) = 4(\mathbf{5}) - 1 = 19 \qquad \text{and} \qquad g(\mathbf{5}) = \mathbf{5}^2 + 2 = 27$$

$$f(\mathbf{-2}) = 4(\mathbf{-2}) - 1 = -9 \qquad \text{and} \qquad g(\mathbf{-2}) = (\mathbf{-2})^2 + 2 = 6$$

$$f(\mathbf{0}) = 4(\mathbf{0}) - 1 = -1 \qquad \text{and} \qquad g(\mathbf{0}) = \mathbf{0}^2 + 2 = 2$$

$$f(z) = 4z - 1 \qquad \text{and} \qquad g(z) = z^2 + 2$$

$$f(a) = 4a - 1 \qquad \text{and} \qquad g(a) = a^2 + 2$$

**EXAMPLE 7**   If the function $f$ is given by:

$$f = \{(-2, 0), (3, -1), (2, 4), (7, 5)\}$$

then $f(-2) = 0$, $f(3) = -1$, $f(2) = 4$, and $f(7) = 5$.

**EXAMPLE 8** If $f(x) = 2x^2$ and $g(x) = 3x - 1$, find
(a) $f[g(2)]$; (b) $g[f(2)]$.

**Solution** The expression $f[g(2)]$ is read "$f$ of $g$ of 2."
(a) Since $g(2) = 3(2) - 1 = 5$,

$$f[g(2)] = f(5) = 2(5)^2 = 50$$

(b) Since $f(2) = 2(2)^2 = 8$,

$$g[f(2)] = g(8) = 3(8) - 1 = 23$$

 Getting Ready for Class

*After reading through the preceding section, respond in your own words and in complete sentences.*

**A.** Explain what you are calculating when you find $f(2)$ for a given function $f$.

**B.** If $s(t) = \dfrac{60}{t}$, how do you find $s(10)$?

**C.** If $f(2) = 3$ for a function $f$, what is the relationship between the numbers 2 and 3 and the graph of $f$?

## PROBLEM SET B

Let $f(x) = 2x - 5$ and $g(x) = x^2 + 3x + 4$. Evaluate the following.

**1.** $f(2)$      **2.** $f(3)$

**3.** $f(-3)$      **4.** $g(-2)$

**5.** $g(-1)$      **6.** $f(-4)$

**7.** $g(-3)$      **8.** $g(2)$

**9.** $g(4) + f(4)$      **10.** $f(2) - g(3)$

**11.** $f(3) - g(2)$      **12.** $g(-1) + f(-1)$

Let $f(x) = 3x^2 - 4x + 1$ and $g(x) = 2x - 1$. Evaluate the following.

**13.** $f(0)$      **14.** $g(0)$

**15.** $g(-4)$      **16.** $f(1)$

**17.** $f(-1)$      **18.** $g(-1)$

**19.** $g(10)$      **20.** $f(10)$

**21.** $f(3)$      **22.** $g(3)$

**23.** $g(\frac{1}{2})$      **24.** $g(\frac{1}{4})$

**25.** $f(a)$      **26.** $g(b)$

If $f = \{(1, 4), (-2, 0), (3, \frac{1}{2}), (\pi, 0)\}$ and $g = \{(1, 1), (-2, 2), (\frac{1}{2}, 0)\}$, find each of the following values of $f$ and $g$.

**27.** $f(1)$      **28.** $g(1)$      **29.** $g(\frac{1}{2})$

**30.** $f(3)$      **31.** $g(-2)$      **32.** $f(\pi)$

Let $f(x) = 2x^2 - 8$ and $g(x) = \frac{1}{2}x + 1$. Evaluate each of the following.

**33.** $f(0)$      **34.** $g(0)$      **35.** $g(-4)$

**36.** $f(1)$      **37.** $f(a)$      **38.** $g(z)$

**39.** $f(b)$     **40.** $g(t)$     **41.** $f[g(2)]$

**42.** $g[f(2)]$     **43.** $g[f(-1)]$     **44.** $f[g(-2)]$

**45.** $g[f(0)]$     **46.** $f[g(0)]$

**47.** Graph the function $f(x) = \frac{1}{2}x + 2$. Then draw and label the line segments that represent $x = 4$ and $f(4)$.

**48.** Graph the function $f(x) = -\frac{1}{2}x + 6$. Then draw and label the line segments that represent $x = 4$ and $f(4)$.

**49.** For the function $f(x) = \frac{1}{2}x + 2$, find the value of $x$ for which $f(x) = x$.

**50.** For the function $f(x) = -\frac{1}{2}x + 6$, find the value of $x$ for which $f(x) = x$.

**51.** Graph the function $f(x) = x^2$. Then draw and label the line segments that represent $x = 1$ and $f(1)$, $x = 2$ and $f(2)$, and, finally, $x = 3$ and $f(3)$.

**52.** Graph the function $f(x) = x^2 - 2$. Then draw and label the line segments that represent $x = 2$ and $f(2)$ and the line segments corresponding to $x = 3$ and $f(3)$.

## Applying the Concepts

**53. Investing in Art**   A painting is purchased as an investment for $150. If its value increases continuously so that it doubles every 3 years, then its value is given by the function

$$V(t) = 150 \cdot 2^{t/3} \qquad \text{for } t \geq 0$$

where $t$ is the number of years since the painting was purchased and $V(t)$ is its value (in dollars) at time $t$. Find $V(3)$ and $V(6)$, and then explain what they mean.

**54. Average Speed**   If it takes Minke $t$ minutes to run a mile, then her average speed $s(t)$, in miles per hour, is given by the formula

$$s(t) = \frac{60}{t} \qquad \text{for } t > 0$$

Find $s(4)$ and $s(5)$, and then explain what they mean.

**55. Dimensions of a Rectangle**   The length of a rectangle is 3 inches more than twice the width. Let $x$ represent the width of the rectangle and $P(x)$ represent the perimeter of the rectangle. Use function

notation to write the relationship between $x$ and $P(x)$, noting any restrictions on the variable $x$.

**56. Dimensions of a Rectangle**   The length of a rectangle is 3 inches more than twice the width. Let $x$ represent the width of the rectangle and $A(x)$ represent the area of the rectangle. Use function notation to write the relationship between $x$ and $A(x)$, noting any restrictions on the variable $x$.

**Area of a Circle**   The formula for the area $A$ of a circle with radius $r$ can be written with function notation as $A(r) = \pi r^2$.

**57.** Find $A(2)$, $A(5)$, and $A(10)$. (Use $\pi \approx 3.14$.)

**58.** Why doesn't it make sense to ask for $A(-10)$?

**59. Cost of a Phone Call**   Suppose a phone company charges 33¢ for the first minute and 24¢ for each additional minute to place a long-distance call between 5 P.M. and 11 P.M. If $x$ is the number of additional minutes and $f(x)$ is the cost of the call, then $f(x) = 0.24x + 0.33$.
(a) How much does it cost to talk for 10 minutes?
(b) What does $f(5)$ represent in this problem?
(c) If a call costs $1.29, how long was it?

**60. Cost of a Phone Call**   The same phone company mentioned in Problem 59 charges 52¢ for the first minute and 36¢ for each additional minute to place a long-distance call between 8 A.M. and 5 P.M.
(a) Let $g(x)$ be the total cost of a long-distance call between 8 A.M. and 5 P.M., and write an equation for $g(x)$.
(b) Find $g(5)$.
(c) Find the difference in price between a 10-minute call made between 8 A.M. and 5 P.M. and the same call made between 5 P.M. and 11 P.M.

**Straight-Line Depreciation**   Straight-line depreciation is an accounting method used to help spread the cost of new equipment over a number of years. It takes into account both the cost when new and the salvage value, which is the value of the equipment at the time it gets replaced.

**61. Value of a Copy Machine**   The function $V(t) = -3,300t + 18,000$, where $V$ is value and $t$ is time in years, can be used to find the value of a large copy machine during the first 5 years of use.
(a) What is the value of the copier after 3 years and 9 months?
(b) What is the salvage value of this copier?

(c) State the domain of this function.

(d) Sketch the graph of this function.

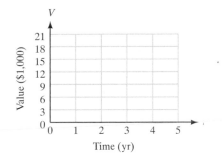

(e) What is the range of this function?

(f) After how many years will the copier be worth only $10,000?

62. **Value of a Forklift**    The function $V(t) = -16,500t + 125,000$, where $V$ is value and $t$ is

*(FPG International/Telegraph Colour Library)*

time in years, can be used to find the value of an electric forklift during the first 6 years of use.

(a) What is the value of the forklift after 2 years and 3 months?

(b) What is the salvage value of this forklift?

(c) State the domain of this function.

(d) Sketch the graph of this function.

*(FPG International/Telegraph Colour Library)*

(e) What is the range of this function?

(f) After how many years will the forklift be worth only $45,000?

# Appendix C

## Answers to Odd-Numbered Problems, Chapter Reviews, Chapter Tests, and Cumulative Reviews

### CHAPTER 1

#### PROBLEM SET 1.1

**1.** $x + 5 = 14$    **3.** $5y < 30$    **5.** $3y \leq y + 6$    **7.** $\dfrac{x}{3} = x + 2$    **9.** 9    **11.** 49    **13.** 8    **15.** 64

**17.** 16    **19.** 100    **21.** 121    **23.** 11    **25.** 16    **27.** 17    **29.** 42    **31.** 30    **33.** 30    **35.** 24
**37.** 80    **39.** 27    **41.** 35    **43.** 13    **45.** 4    **47.** 37    **49.** 37    **51.** 16    **53.** 16    **55.** 81
**57.** 41    **59.** 345    **61.** 2,345    **63.** 2    **65.** 148    **67.** 36    **69.** 36    **71.** 58    **73.** 62    **75.** 5
**77.** 10    **79.** 25    **81.** 10    **83.** (a) 600 mg   (b) 231 mg    **85.**

| Activity | Calories Burned in 1 Hour by a 150-Pound Person |
|---|---|
| Bicycling | 374 |
| Bowling | 265 |
| Handball | 680 |
| Jogging | 680 |
| Skiing | 544 |

**87.** $2(10 + 4)$    **89.** $3(50) - 14$    **91.** 10 cookies    **93.** 420 calories    **95.** Approximately 224 chips
**97.** 95 grams

#### PROBLEM SET 1.2

**1–7.**

**9.** $\dfrac{18}{24}$    **11.** $\dfrac{12}{24}$    **13.** $\dfrac{15}{24}$    **15.** $\dfrac{36}{60}$    **17.** $\dfrac{22}{60}$    **19.** $-10, \dfrac{1}{10}, 10$    **21.** $-\dfrac{3}{4}, \dfrac{4}{3}, \dfrac{3}{4}$    **23.** $-\dfrac{11}{2}, \dfrac{2}{11}, \dfrac{11}{2}$
**25.** $3, -\dfrac{1}{3}, 3$    **27.** $\dfrac{2}{5}, -\dfrac{5}{2}, \dfrac{2}{5}$    **29.** $-x, \dfrac{1}{x}, |x|$    **31.** $<$    **33.** $>$    **35.** $>$    **37.** $>$    **39.** $<$    **41.** $<$
**43.** 6    **45.** 22    **47.** 3    **49.** 7    **51.** 3    **53.** $\dfrac{8}{15}$    **55.** $\dfrac{3}{2}$    **57.** $\dfrac{5}{4}$    **59.** 1    **61.** 1    **63.** 1
**65.** $\dfrac{9}{16}$    **67.** $\dfrac{8}{27}$    **69.** $\dfrac{1}{10,000}$    **71.** $\dfrac{1}{9}$    **73.** $\dfrac{1}{25}$    **75.** 4 inches; 1 inch$^2$    **77.** 4.5 inches; 1.125 inches$^2$
**79.** 10.25 centimeters; 5 centimeters$^2$    **81.** $-8, -2$    **83.** $-64°F; -54°F$    **85.** $-15°F$

**87.**

| Month | Temperature (°F) |
|---|---|
| January | −36 |
| February | −30 |
| March | −14 |
| April | −2 |
| May | 19 |
| June | 22 |
| July | 35 |
| August | 30 |
| September | 19 |
| October | 15 |
| November | −11 |
| December | −26 |

**89.** −100 ft; −105 ft

**91.** Area = 93.5 inches$^2$: perimeter = 39 inches    **93.** 1,387 calories    **95.** 654 more calories

## PROBLEM SET 1.3

**1.** $3 + 5 = 8$; $3 + (-5) = -2$; $-3 + 5 = 2$; $-3 + (-5) = -8$
**3.** $15 + 20 = 35$; $15 + (-20) = -5$; $-15 + 20 = 5$; $-15 + (-20) = -35$    **5.** 3    **7.** −7    **9.** −14
**11.** −3    **13.** −25    **15.** −12    **17.** −19    **19.** −25    **21.** −8    **23.** −4    **25.** 6    **27.** 6    **29.** 8
**31.** −4    **33.** −14    **35.** −17    **37.** 4    **39.** 3    **41.** 15    **43.** −8    **45.** 12    **47.** 23, 28
**49.** 30, 35    **51.** 0, −5    **53.** −12, −18    **55.** −4, −8    **57.** Yes    **59.** $5 + 9 = 14$
**61.** $[-7 + (-5)] + 4 = -8$    **63.** $[-2 + (-3)] + 10 = 5$    **65.** 3    **67.** −3    **69.** $-12 + 4$
**71.** $10 + (-6) + (-8) = -4$    **73.** $-\$30 + \$40 = \$10$    **75.** \$6.50, \$6.75, \$7.00, \$7.25, \$7.50, \$7.75, \$8.00; yes
**77.** (a)

| Professions | Salaries |
|---|---|
| Engineering | $42,862 |
| Computer Science | $40,920 |
| Math/Statistics | $40,523 |
| Chemistry | $36,036 |
| Business Admin. | $34,831 |
| Accounting | $33,702 |
| Sales/Marketing | $33,252 |
| Teaching | $25,735 |

(b) $78,898

## PROBLEM SET 1.4

**1.** −3    **3.** −6    **5.** 0    **7.** −10    **9.** −16    **11.** −12    **13.** −7    **15.** 35    **17.** 0    **19.** −4
**21.** 4    **23.** −24    **25.** −28    **27.** 25    **29.** 4    **31.** 7    **33.** 17    **35.** 8    **37.** 4    **39.** 18
**41.** 10    **43.** 17    **45.** 1    **47.** 1    **49.** 27    **51.** −26    **53.** −2    **55.** 68    **57.** $-7 - 4 = -11$

**59.** $12 - (-8) = 20$    **61.** $-5 - (-7) = 2$    **63.** $[4 + (-5)] - 17 = -18$    **65.** $8 - 5 = 3$
**67.** $-8 - 5 = -13$    **69.** $8 - (-5) = 13$    **71.** 10    **73.** $-2$    **75.** $1,500 - 730$
**77.** $-\$35 + \$15 - \$20 = -\$40$    **79.** $\$98 - \$65 - \$53 = -\$20$
**81.** $4,500, $3,950, $3,400, $2,850, $2,300; yes    **83.** 769 feet    **85.** 439 feet    **87.** 2 seconds
**89.** (a)                                                (b) 84 million tons

| Year | Garbage (millions of tons) |
|------|----------------------------|
| 1960 | 88 |
| 1970 | 121 |
| 1980 | 152 |
| 1990 | 205 |
| 1997 | 217 |

**91.** (a)                    (b) 5 cents/minute    **93.** 35°    **95.** 60°

| Year | Cents/Minute |
|------|--------------|
| 1998 | 33 |
| 1999 | 28 |
| 2000 | 25 |
| 2001 | 23 |
| 2002 | 22 |
| 2003 | 20 |

## PROBLEM SET 1.5

**1.** Commutative    **3.** Multiplicative inverse    **5.** Commutative    **7.** Distributive    **9.** Commutative, associative
**11.** Commutative, associative    **13.** Commutative    **15.** Commutative, associative    **17.** Commutative
**19.** Additive inverse    **21.** $3x + 6$    **23.** $9a + 9b$    **25.** 0    **27.** 0    **29.** 10    **31.** $(4 + 2) + x = 6 + x$
**33.** $x + (2 + 7) = x + 9$    **35.** $(3 \cdot 5)x = 15x$    **37.** $(9 \cdot 6)y = 54y$    **39.** $\left(\frac{1}{2} \cdot 3\right) a = \frac{3}{2}a$    **41.** $\left(\frac{1}{3} \cdot 3\right) x = x$
**43.** $\left(\frac{1}{2} \cdot 2\right) y = y$    **45.** $\left(\frac{3}{4} \cdot \frac{4}{3}\right) x = x$    **47.** $\left(\frac{6}{5} \cdot \frac{5}{6}\right) a = a$    **49.** $8x + 16$    **51.** $8x - 16$    **53.** $4y + 4$
**55.** $18x + 15$    **57.** $6a + 14$    **59.** $54y - 72$    **61.** $\frac{3}{2}x - 3$    **63.** $x + 2$    **65.** $3x + 3y$    **67.** $8a - 8b$
**69.** $12x + 18y$    **71.** $12a - 8b$    **73.** $3x + 2y$    **75.** $4a + 25$    **77.** $6x + 12$    **79.** $14x + 38$    **81.** No
**83.** No, not commutative    **85.** $8 \div 4 \neq 4 \div 8$    **87.** $12(2,400 - 480) = \$23,040; 12(2,400) - 12(480) = \$23,040$
**89.** $P = 2(l + w)$

## PROBLEM SET 1.6

**1.** $-42$    **3.** $-21$    **5.** $-16$    **7.** 3    **9.** 121    **11.** 6    **13.** $-60$    **15.** 24    **17.** 49    **19.** $-27$
**21.** 6    **23.** 10    **25.** 9    **27.** 45    **29.** 14    **31.** $-2$    **33.** 216    **35.** $-2$    **37.** $-18$    **39.** 29
**41.** 38    **43.** $-5$    **45.** 37    **47.** 80    **49.** $-\frac{10}{21}$    **51.** $-4$    **53.** 1    **55.** $\frac{9}{16}$    **57.** $-\frac{8}{27}$    **59.** $-8x$
**61.** $42x$    **63.** $x$    **65.** $x$    **67.** $-4a - 8$    **69.** $-\frac{3}{2}x + 3$    **71.** $-6x + 8$    **73.** $-15x - 30$    **75.** $-25$
**77.** $2(-4x) = -8x$    **79.** $-26$    **81.** 8    **83.** $-80$    **85.** $\frac{1}{8}$    **87.** 1    **89.** $-24$    **91.** $60    **93.** 1°F
**95.** $500, $1,000, $2,000, $4,000, $8,000, $16,000; yes    **97.** 465 calories

## PROBLEM SET 1.7

**1.** $-2$    **3.** $-3$    **5.** $-\frac{1}{3}$    **7.** $3$    **9.** $\frac{1}{7}$    **11.** $0$    **13.** $9$    **15.** $-15$    **17.** $-36$    **19.** $-\frac{1}{4}$    **21.** $\frac{16}{15}$

**23.** $\frac{4}{3}$    **25.** $-\frac{8}{13}$    **27.** $-1$    **29.** $1$    **31.** $\frac{3}{5}$    **33.** $-\frac{5}{3}$    **35.** $-2$    **37.** $-3$    **39.** Undefined

**41.** Undefined    **43.** $5$    **45.** $-\frac{7}{3}$    **47.** $-1$    **49.** $-7$    **51.** $\frac{15}{17}$    **53.** $-\frac{32}{17}$    **55.** $\frac{1}{3}$    **57.** $1$    **59.** $1$

**61.** $-2$    **63.** $\frac{9}{7}$    **65.** $\frac{16}{11}$    **67.** $-1$    **69.** $3$    **71.** $-10$    **73.** $-3$    **75.** $-8$    **77.** $\$350$

**79.** Drops $3.5°$F each hour.

## PROBLEM SET 1.8

**1.** $0, 1$    **3.** $-3, -2.5, 0, 1, \frac{3}{2}$    **5.** All    **7.** $-10, -8, -2, 9$    **9.** $\pi$    **11.** True    **13.** False    **15.** False

**17.** True    **19.** Composite, $2^4 \cdot 3$    **21.** Prime    **23.** Composite, $3 \cdot 11 \cdot 31$    **25.** $2^4 \cdot 3^2$    **27.** $2 \cdot 19$

**29.** $3 \cdot 5 \cdot 7$    **31.** $2^2 \cdot 3^2 \cdot 5$    **33.** $5 \cdot 7 \cdot 11$    **35.** $11^2$    **37.** $2^2 \cdot 3 \cdot 5 \cdot 7$    **39.** $2^2 \cdot 5 \cdot 31$    **41.** $\frac{7}{11}$    **43.** $\frac{5}{7}$

**45.** $\frac{11}{13}$    **47.** $\frac{14}{15}$    **49.** $\frac{5}{9}$    **51.** $\frac{5}{8}$    **53.** $6^3 = (2 \cdot 3)^3 = 2^3 \cdot 3^3$    **55.** $9^4 \cdot 16^2 = (3^2)^4(2^4)^2 = 2^8 \cdot 3^8$

**57.** $3 \cdot 8 + 3 \cdot 7 + 3 \cdot 5 = 24 + 21 + 15 = 60 = 2^2 \cdot 3 \cdot 5$    **59.** Irrational numbers    **61.** $8, 21, 34$

## PROBLEM SET 1.9

**1.** $\frac{2}{3}$    **3.** $-\frac{1}{4}$    **5.** $\frac{1}{2}$    **7.** $\frac{x-1}{3}$    **9.** $\frac{3}{2}$    **11.** $\frac{x+6}{2}$    **13.** $-\frac{3}{5}$    **15.** $\frac{10}{a}$    **17.** $\frac{7}{8}$    **19.** $\frac{1}{10}$    **21.** $\frac{7}{9}$

**23.** $\frac{7}{3}$    **25.** $\frac{1}{4}$    **27.** $\frac{7}{6}$    **29.** $\frac{19}{24}$    **31.** $\frac{13}{60}$    **33.** $\frac{29}{35}$    **35.** $\frac{949}{1,260}$    **37.** $\frac{13}{420}$    **39.** $\frac{41}{24}$    **41.** $\frac{5}{4}$    **43.** $\frac{160}{63}$

**45.** $\frac{5}{8}$    **47.** $-\frac{2}{3}$    **49.** $\frac{7}{3}$    **51.** $\frac{1}{125}$

## CHAPTER 1 REVIEW

**1.** $-7 + (-10) = -17$    **2.** $(-7 + 4) + 5 = 2$    **3.** $(-3 + 12) + 5 = 14$    **4.** $4 - 9 = -5$

**5.** $9 - (-3) = 12$    **6.** $-7 - (-9) = 2$    **7.** $(-3)(-7) - 6 = 15$    **8.** $5(-6) + 10 = -20$

**9.** $2[(-8)(3x)] = -48x$    **10.** $\frac{-25}{-5} = 5$    **11.** $\frac{-40}{8} - 7 = -12$    **12.** $\frac{-45}{15} + 9 = 6$

**13–18.**

**19.** $12$    **20.** $3$    **21.** $\frac{4}{5}$    **22.** $\frac{7}{10}$    **23.** $1.8$    **24.** $-10$    **25.** $-6, \frac{1}{6}$    **26.** $-\frac{3}{10}, \frac{10}{3}$    **27.** $9, -\frac{1}{9}$

**28.** $\frac{12}{5}, -\frac{5}{12}$    **29.** $\frac{6}{35}$    **30.** $-5$    **31.** $-\frac{5}{4}$    **32.** $3$    **33.** $-38$    **34.** $0$    **35.** $-22$    **36.** $-25$

**37.** $1$    **38.** $-3$    **39.** $22$    **40.** $-4$    **41.** $-20$    **42.** $-15$    **43.** $14$    **44.** $12$    **45.** $28$    **46.** $-30$

**47.** $-12$    **48.** $-24$    **49.** $12$    **50.** $-4$    **51.** $-\frac{1}{4}$    **52.** $-\frac{2}{3}$    **53.** $23$    **54.** $47$    **55.** $-3$    **56.** $-35$

**57.** $32$    **58.** $2$    **59.** $30$    **60.** $-3$    **61.** $-98$    **62.** $70$    **63.** $2$    **64.** $\frac{17}{2}$    **65.** Undefined    **66.** $8$

**67.** $3$    **68.** Associative    **69.** Multiplicative identity    **70.** Commutative    **71.** Additive inverse

**72.** Multiplicative inverse    **73.** Additive identity    **74.** Commutative, associative    **75.** Distributive

**76.** $12 + x$    **77.** $28a$    **78.** $x$    **79.** $y$    **80.** $14x + 21$    **81.** $6a - 12$    **82.** $\frac{5}{2}x - 3$    **83.** $-\frac{3}{2}x + 3$

**84.** $-\frac{1}{3}, 0, 5, -4.5, \frac{2}{5}, -3$    **85.** $0, 5$    **86.** $\sqrt{7}, \pi$    **87.** $0, 5, -3$    **88.** $2 \cdot 3^2 \cdot 5$    **89.** $2^3 \cdot 3 \cdot 5 \cdot 7$    **90.** $\frac{173}{210}$

**91.** $\frac{109}{420}$    **92.** $-2$    **93.** $810$    **94.** $8$    **95.** $12$    **96.** $-1$    **97.** $\frac{1}{16}$

# CHAPTER 1 TEST

**1.** $x + 3 = 8$    **2.** $5y = 15$    **3.** 40    **4.** 16    **5.** Opposite: 4; reciprocal: $-\frac{1}{4}$; absolute value: 4
**6.** Opposite: $-\frac{3}{4}$; reciprocal: $\frac{4}{3}$; absolute value: $\frac{3}{4}$    **7.** $-4$    **8.** 17    **9.** $-12$    **10.** 0    **11.** c    **12.** e
**13.** d    **14.** a    **15.** $-21$    **16.** 64    **17.** $-2$    **18.** $-\frac{8}{27}$    **19.** 4    **20.** 204    **21.** 25    **22.** 52
**23.** 2    **24.** 2    **25.** $8 + 2x$    **26.** $10x$    **27.** $6x + 10$    **28.** $-2x + 1$    **29.** $1, -8$    **30.** $1, 1.5, \frac{3}{4}, -8$
**31.** $\sqrt{2}$    **32.** All of them    **33.** $2^4 \cdot 37$    **34.** $2^2 \cdot 5 \cdot 67$    **35.** $\frac{25}{42}$    **36.** $\frac{8}{x}$    **37.** $8 + (-3) = 5$
**38.** $-24 - 2 = -26$    **39.** $-5(-4) = 20$    **40.** $\frac{-24}{-2} = 12$    **41.** 12    **42.** $\frac{1}{2}$

# CHAPTER 2

## PROBLEM SET 2.1

**1.** $-3x$    **3.** $-a$    **5.** $12x$    **7.** $6a$    **9.** $6x - 3$    **11.** $7a + 5$    **13.** $5x - 5$    **15.** $4a + 2$
**17.** $-9x - 2$    **19.** $12a + 3$    **21.** $10x - 1$    **23.** $21y + 6$    **25.** $-6x + 8$    **27.** $-2a + 3$
**29.** $-4x + 26$    **31.** $4y - 16$    **33.** $-6x - 1$    **35.** $2x - 12$    **37.** $10a + 33$    **39.** $4x - 9$    **41.** $7y - 39$
**43.** $-19x - 14$    **45.** 5    **47.** $-9$    **49.** 4    **51.** 4    **53.** $-37$    **55.** $-41$    **57.** 64    **59.** 64
**61.** 144    **63.** 144    **65.** 3    **67.** 0    **69.** 15    **71.** 6    **73.** 5, 7, 9, 11    **75.** 2, 5, 10, 17
**77.** (a)

| $n$ | 1 | 2 | 3 | 4 |
|---|---|---|---|---|
| $3n$ | 3 | 6 | 9 | 12 |

(b)

| $n$ | 1 | 2 | 3 | 4 |
|---|---|---|---|---|
| $n^3$ | 1 | 8 | 27 | 64 |

**79.** (a) 42°F    (b) 28°F    (c) $-14$°F

**81.** (a) \$37.50    (b) \$40.00    (c) \$42.50    **83.** $0.71G$; \$887.50    **85.** $x - 5$; $-7$    **87.** $2(x + 10)$; 16
**89.** $\frac{10}{x}$; $-5$    **91.** $[3x + (-2)] - 5$; $-13$    **93.** $-\frac{7}{2}$    **95.** $\frac{51}{40}$

## PROBLEM SET 2.2

**1.** 11    **3.** 4    **5.** $-\frac{3}{4}$    **7.** $-5.8$    **9.** $-17$    **11.** $-\frac{1}{8}$    **13.** $-4$    **15.** $-3.6$    **17.** 1    **19.** $-\frac{7}{45}$
**21.** 3    **23.** $\frac{11}{8}$    **25.** 21    **27.** 7    **29.** 3.5    **31.** 22    **33.** $-6$    **35.** 0    **37.** $-2$    **39.** $-16$
**41.** $-3$    **43.** 10    **45.** $-12$    **47.** $-1$    **49.** 4    **51.** 2    **53.** $-5$    **55.** $-1$    **57.** $-3$    **59.** 8
**61.** $-8$    **63.** 2    **65.** 11    **67.** $x = 3, y = -1, z = 0$
**69.** (a) 6% of the light is reflected.    (b) 5% is reflected.    (c) 2% is absorbed.    (d) 75% is reflected.
**71.** 70°    **73.** 11    **75.** 2    **77.** $18x$    **79.** $x$    **81.** $y$    **83.** $x$    **85.** $a$

## PROBLEM SET 2.3

**1.** 2    **3.** 4    **5.** $-\frac{1}{2}$    **7.** $-2$    **9.** 3    **11.** 4    **13.** 0    **15.** 0    **17.** 6    **19.** $-50$    **21.** $\frac{3}{2}$
**23.** 12    **25.** $-3$    **27.** 32    **29.** $-8$    **31.** $\frac{1}{2}$    **33.** 4    **35.** 8    **37.** $-4$    **39.** 4    **41.** $-15$
**43.** $-\frac{1}{2}$    **45.** 3    **47.** 1    **49.** $\frac{1}{4}$    **51.** $-3$    **53.** 3    **55.** 2    **57.** $-\frac{3}{2}$    **59.** $-\frac{3}{2}$    **61.** 1    **63.** 1
**65.** $-2$    **67.** $-2$    **69.** 200 tickets    **71.** 2 three-pointers    **73.** (a) 9    (b) 15    (c) 18
**75.** $\frac{17}{3}$; $3x + 2 = 19$    **77.** 10; $2(x + 10) = 40$    **79.** $10x - 43$    **81.** $-3x - 13$    **83.** $-6y + 4$
**85.** $-5x + 7$

## PROBLEM SET 2.4

**1.** 3    **3.** −2    **5.** −1    **7.** 2    **9.** −4    **11.** −2    **13.** 0    **15.** 1    **17.** $\frac{1}{2}$    **19.** 7    **21.** 8
**23.** $-\frac{1}{3}$    **25.** $\frac{3}{4}$    **27.** 75    **29.** 2    **31.** 6    **33.** 8    **35.** 0    **37.** $\frac{3}{7}$    **39.** 1    **41.** $\frac{3}{2}$    **43.** 4
**45.** 1    **47.** $6x - 10$    **49.** $\frac{3}{2}x + 3$    **51.** $-x + 2$

## PROBLEM SET 2.5

**1.** 100 feet    **3.** 0    **5.** 2    **7.** 15    **9.** 10    **11.** 4    **13.** 2    **15.** $l = \dfrac{A}{w}$    **17.** $r = \dfrac{d}{t}$    **19.** $h = \dfrac{V}{lw}$
**21.** $P = \dfrac{nRT}{V}$    **23.** $a = P - b - c$    **25.** $x = 3y - 1$    **27.** $y = 3x + 6$    **29.** $y = -\frac{2}{3}x + 2$
**31.** $y = -2x + 4$    **33.** $y = \frac{5}{2}x - \frac{3}{2}$    **35.** $w = \dfrac{P - 2l}{2}$    **37.** $v = \dfrac{h - 16t^2}{t}$    **39.** $h = \dfrac{A - \pi r^2}{2\pi r}$
**41.** $y = -\frac{3}{2}x + 3$    **43.** $y = \frac{3}{7}x - 3$    **45.** $y = 2x + 8$    **47.** 60°; 150°    **49.** 45°; 135°    **51.** 10
**53.** 240    **55.** 25%    **57.** 35%    **59.** 64    **61.** 2,000
**63.** (a) $68,840,000 = 0.628T$    (b) 109,617,834 people    **65.** (a) 60%  (b) 40%    **67.** 100°C; yes
**69.** 20°C; yes    **71.** $C = \frac{5}{9}(F - 32)$    **73.** 60%    **75.** 26.5%
**77.**    **79.** (a) 7 meters    (b) $\frac{3}{2}$ or 1.5 inches

| Altitude (feet) | Temperature (°F) |
|---|---|
| 0 | 56 |
| 9,000 | 24.5 |
| 15,000 | 3.5 |
| 21,000 | −17.5 |
| 28,000 | −42 |
| 40,000 | −84 |

**81.** The sum of 4 and 1 is 5.    **83.** The difference between 6 and 2 is 4.    **85.** $2(6 + 3)$    **87.** $2 \cdot 5 + 3 = 13$

## PROBLEM SET 2.6

Along with the answers to the odd-numbered problems in this problem set and the next, we are including some of the equations used to solve the problems. Be sure that you try the problems on your own before looking here to see the correct equations.
**1.** 8    **3.** $2x + 4 = 14$; 5    **5.** −1    **7.** The two numbers are $x$ and $x + 2$; $x + (x + 2) = 8$; 3 and 5.
**9.** 6 and 14    **11.** Barney's age is $x$; Fred's age is $x + 4$; $(x − 5) + (x − 1) = 48$; Barney is 27; Fred is 31.
**13.** Lacy is 16; Jack is 32.    **15.** Patrick is 18; Pat is 38.    **17.** $l = 11$ inches; $w = 6$ inches
**19.** The length of a side is $x$; $4x = 48$; $x = 12$ meters.    **21.** $l = 17$ inches; $w = 10$ inches
**23.** If the number of nickels is $x$, then the number of dimes is $x + 9$; $5x + 10(x + 9) = 210$; 8 nickels, 17 dimes.
**25.** 15 dimes, 30 quarters    **27.** If she has $x$ nickels, then she has $x + 3$ dimes and $x + 5$ quarters: $5x + 10(x + 3) + 25(x + 5) = 435$; 7 nickels, 10 dimes, 12 quarters.    **29.** 4 is less than 10.    **31.** 9 is greater than or equal to −5.
**33.** $<$    **35.** $<$    **37.** 2    **39.** 12

## PROBLEM SET 2.7

**1.** $4,000 invested at 8%, $6,000 invested at 9%    **3.** If $x =$ the amount invested at 10%, then $x + 500$ is the amount invested at 12%; $0.10x + 0.12(x + 500) = 214$; $700 invested at 10%, $1,200 invested at 12%    **5.** $500 at 8%, $1,000 at 9%, $1,500 at 10%    **7.** 45°, 45°, 90°    **9.** 22.5°, 45°, 112.5°    **11.** 53°, 90°    **13.** Let $x =$ the number of minutes after the first minute; $41 + 32x = 521$; $x = 15$, so the call was 16 minutes.

**15.** Let $x$ = the number of hours past 35 hours; $35(12) + 18x = 492$; $x = 4$, so she worked 39 hours that week.
**17.** Let $x$ = the number of children's tickets; then $2x$ = the number of adult tickets; $4.5x + 6(2x) = 115.5$; $x = 7$; she sold 7 children's tickets and 14 adult tickets.    **19.** Jeff    **21.** $10.38    **23.** Yes    **25.** $54.00    **27.** Yes
**29.** $79,626    **31.** $-8$    **33.** $\frac{3}{4}$    **35.** 25    **37.** 0

## PROBLEM SET 2.8

**1.** $x < 12$          **3.** $a \leq 12$

**5.** $x > 13$          **7.** $y \geq 4$

**9.** $x > 9$           **11.** $x < 2$

**13.** $a \leq 5$       **15.** $x > 15$

**17.** $x < -3$         **19.** $x \leq 6$

**21.** $x \geq -50$     **23.** $y < -6$

**25.** $x < 6$          **27.** $y \geq -5$

**29.** $x < 3$          **31.** $x \leq 18$

**33.** $a < -20$        **35.** $y < 25$

**37.** $a \leq 3$       **39.** $x \geq \frac{15}{2}$

**41.** $x < -1$         **43.** $y \geq -2$

**45.** $x < -1$         **47.** $m \leq -6$

**49.** $x \leq -5$      **51.** $y < -\frac{3}{2}x + 3$    **53.** $y < \frac{2}{5}x - 2$    **55.** $y \leq \frac{3}{7}x + 3$

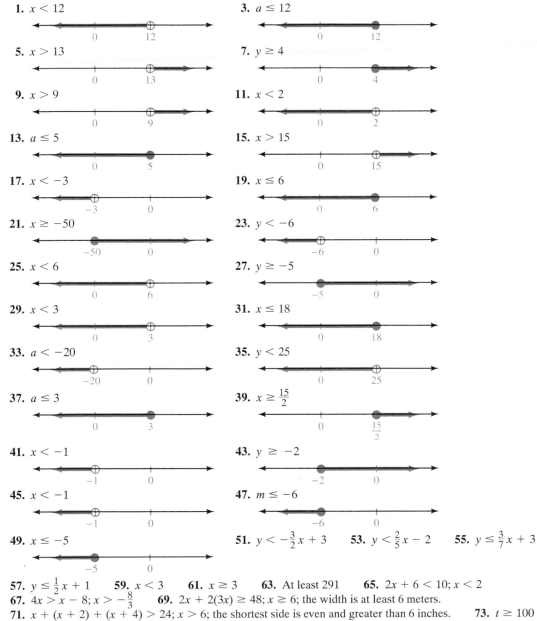

**57.** $y \leq \frac{1}{2}x + 1$    **59.** $x < 3$    **61.** $x \geq 3$    **63.** At least 291    **65.** $2x + 6 < 10$; $x < 2$
**67.** $4x > x - 8$; $x > -\frac{8}{3}$    **69.** $2x + 2(3x) \geq 48$; $x \geq 6$; the width is at least 6 meters.
**71.** $x + (x + 2) + (x + 4) > 24$; $x > 6$; the shortest side is even and greater than 6 inches.    **73.** $t \geq 100$
**75.** Lose money if they sell less than 200 tickets; make a profit if they sell more than 200 tickets    **77.** b    **79.** a
**81.** b and c

## PROBLEM SET 2.9

**41.** $-2 < x < 3$ **43.** $x \le -2$ or $x \ge 3$ **45.** (a) $2x + x > 10$; $x + 10 > 2x$; $2x + 10 > x$ (b) $\frac{10}{3} < x < 10$

**47.** (number line from 50 to 270)

**49.** $5 < x < 15$ **51.** $4 < x < 5$

**53.** The width is between 3 inches and $\frac{11}{2}$ inches. **55.** $-5$ **57.** $5$ **59.** $7$ **61.** $9$ **63.** $6$ **65.** $2x - 3$

**67.** $-3, 0, 2$

## CHAPTER 2 REVIEW

**1.** $-3x$ **2.** $-2x - 3$ **3.** $4a - 7$ **4.** $-2a + 3$ **5.** $-6y$ **6.** $-6x + 1$ **7.** $19$ **8.** $-11$

**9.** $-18$ **10.** $-13$ **11.** $8$ **12.** $6$ **13.** $-8$ **14.** $\frac{15}{14}$ **15.** $2$ **16.** $-5$ **17.** $-5$ **18.** $0$

**19.** $12$ **20.** $-8$ **21.** $-1$ **22.** $1$ **23.** $5$ **24.** $\frac{5}{2}$ **25.** $0$ **26.** $0$ **27.** $10$ **28.** $2$ **29.** $2$

**30.** $-\frac{8}{3}$ **31.** $0$ **32.** $-5$ **33.** $0$ **34.** $-4$ **35.** $-8$ **36.** $4$ **37.** $y = \frac{2}{5}x - 2$ **38.** $y = \frac{5}{2}x - 5$

**39.** $h = \frac{V}{\pi r^2}$ **40.** $w = \frac{P - 2l}{2}$ **41.** $206.4$ **42.** $9\%$ **43.** $11$ **44.** 5 meters; 25 meters

**45.** \$500 at 9%; \$800 at 10% **46.** 10 nickels; 5 dimes **47.** $x > -2$ **48.** $x < 2$ **49.** $a \ge 6$

**50.** $a < -15$ **51.** (number line) **52.** (number line)

**53.** (number line) **54.** (number line)

**55.** (number line) **56.** (number line)

## CHAPTER 2 TEST

**1.** $-4x + 5$ **2.** $3a - 4$ **3.** $-3y - 12$ **4.** $11x + 28$ **5.** $22$ **6.** $25$ **7.** $6$ **8.** $\frac{4}{3}$ **9.** $2$

**10.** $-\frac{1}{2}$ **11.** $-3$ **12.** $70$ **13.** $-3$ **14.** $-1$ **15.** $5.7$ **16.** $2,000$ **17.** $3$ **18.** $\frac{28}{3}$ inches

**19.** $y = -\frac{2}{5}x + 4$    **20.** $v = \dfrac{h - x - 16t^2}{t}$    **21.** Rick is 20; Dave is 40.

**22.** Width is 10 inches; length is 20 inches.    **23.** 8 quarters, 15 dimes    **24.** $800 at 7%; $1,400 at 9%

**25.** $x < 1$    **26.** $a < -4$

**27.** $x \le -3$    **28.** $m \ge -2$

**29.**    **30.**

## CHAPTERS 1 & 2 CUMULATIVE REVIEW

**1.** 30    **2.** $-7$    **3.** $-14$    **4.** $-12$    **5.** $2x$    **6.** 8    **7.** $-\frac{8}{27}$    **8.** $-72x$    **9.** $-\frac{4}{5}$    **10.** $\frac{8}{35}$

**11.** $-\frac{8}{3}$    **12.** 0    **13.** $\frac{1}{4}$    **14.** $\frac{8}{9}$    **15.** $-\frac{1}{15}$    **16.** $11x + 7$    **17.** 4    **18.** $-16$    **19.** $-50$    **20.** 5

**21.** $\frac{3}{2}$    **22.** 1    **23.** $\frac{1}{3}$    **24.** 12    **25.** 48    **26.** $c = P - a - b$    **27.** $y = -\frac{3}{4}x + 3$, or $\dfrac{-3x + 12}{4}$

**28.** $x \le 6$    **29.** $x > 3$

**30.** $x < 2$ or $x > 3$    **31.** $-3 < x < 4$

**32.** $x - 5 = 12$    **33.** $\frac{2}{3}, -\frac{3}{2}, \frac{2}{3}$    **34.** $P = 12$ inches, $A = 9$ inches$^2$    **35.** $-8$    **36.** Additive inverse
**37.** $2x - 1$    **38.** $-5, -3, -1.7, 2.3, \frac{12}{7}$    **39.** $\frac{3}{4}$    **40.** 11    **41.** 25    **42.** 15    **43.** 12    **44.** She lost $9.
**45.** $48°$ and $90°$    **46.** Width = 5 inches, length = 9 inches    **47.** $65°$    **48.** 4 quarters and 12 dimes
**49.** $400 at 5%; $600 at 6%    **50.** 18 hours

## CHAPTER 3

## PROBLEM SET 3.1

**1–17.**

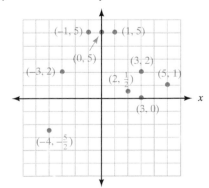

**19.** $(-4, 4)$    **21.** $(-4, 2)$    **23.** $(-3, 0)$    **25.** $(2, -2)$
**27.** $(-5, -5)$

**29.** Yes **31.** No **33.** Yes **35.** No

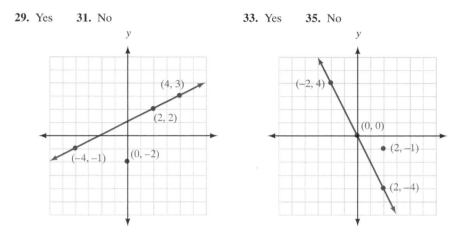

**37.** Yes **39.** No **41.** No **43.** No

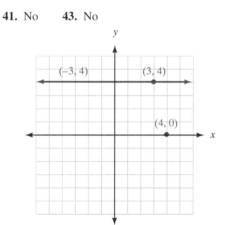

**45.** Every point on this line has a $y$-coordinate of $-3$.

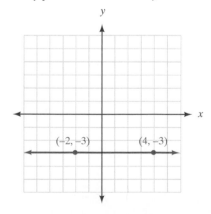

**47.** Along the $y$-axis

**49.** Any three:

(0, 0), (5, 40), (10, 80),
(15, 120), (20, 160), (25, 200),
(30, 240), (35, 280), (40, 320)

**51.**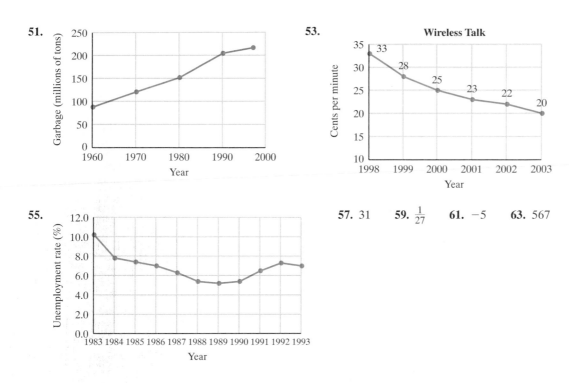

**53.**

**55.**

**57.** 31   **59.** $\frac{1}{27}$   **61.** $-5$   **63.** 567

**65.** 15

# PROBLEM SET 3.2

**1.** $(0, 6), (3, 0), (6, -6)$   **3.** $(0, 3), (4, 0), (-4, 6)$   **5.** $(1, 1), (\frac{3}{4}, 0)(5, 17)$   **7.** $(2, 13), (1, 6), (0, -1)$
**9.** $(-5, 4), (-5, -3), (-5, 0)$   **11.**

| x | y |
|---|---|
| 1 | 3 |
| -3 | -9 |
| 4 | 12 |
| 6 | 18 |

**13.**

| x | y |
|---|---|
| 0 | 0 |
| $-\frac{1}{2}$ | -2 |
| -3 | -12 |
| 3 | 12 |

**15.**

| x | y |
|---|---|
| 2 | 3 |
| 3 | 2 |
| 5 | 0 |
| 9 | -4 |

**17.**

| x | y |
|---|---|
| 2 | 0 |
| 3 | 2 |
| 1 | -2 |
| -3 | -10 |

**19.**

| x | y |
|---|---|
| 0 | -1 |
| -1 | -7 |
| -3 | -19 |
| $\frac{3}{2}$ | 8 |

**21.** $(0, -2)$   **23.** $(1, 5), (0, -2), (-2, -16)$   **25.** $(1, 6), (-2, -12), (0, 0)$

**27.** $(2, -2)$   **29.** $(3, 0), (3, -3)$   **31.** 12 inches

**33.**

| Minutes | Cost |
|---------|---------|
| 0 | $ 3.00 |
| 10 | $ 4.00 |
| 20 | $ 5.00 |
| 30 | $ 6.00 |
| 40 | $ 7.00 |
| 50 | $ 8.00 |
| 60 | $ 9.00 |
| 70 | $10.00 |
| 80 | $11.00 |
| 90 | $12.00 |
| 100 | $13.00 |

**35.**

| Hours | Cost |
|-------|------|
| 0 | $10 |
| 1 | $13 |
| 2 | $16 |
| 3 | $19 |
| 4 | $22 |
| 5 | $25 |
| 6 | $28 |
| 7 | $31 |
| 8 | $34 |
| 9 | $37 |
| 10 | $40 |

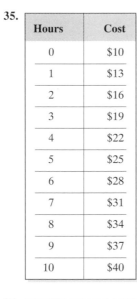

**37.**

| Bags Collected | Cost |
|----------------|---------|
| 5 | $20.50 |
| 7 | $23.50 |
| 9 | $26.50 |
| 11 | $29.50 |
| 13 | $32.50 |

**39.**

| Gallons per Month | Monthly Cost ($) |
|---|---|
| 15 | 23.50 |
| 20 | 29.00 |
| 25 | 34.50 |
| 30 | 40.00 |
| 35 | 45.50 |

**41.** −3    **43.** 2    **45.** 0    **47.** $y = -5x + 4$    **49.** $y = \frac{3}{2}x - 3$

## PROBLEM SET 3.3

**1.**

**3.**

**5.**

**7.**

**9.**

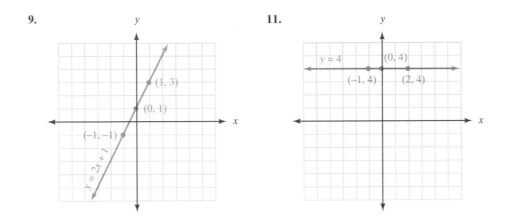

(1, 3)

(0, 1)

(−1, −1)

$y = 2x + 1$

**11.**

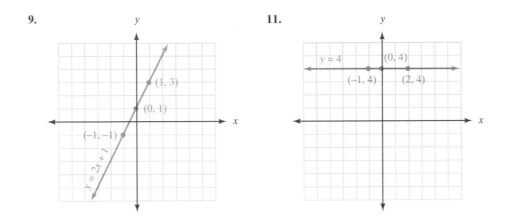

$y = 4$ (0, 4)

(−1, 4) (2, 4)

**13.**

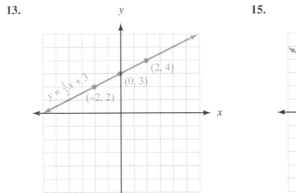

(2, 4)

(0, 3)

$y = \frac{1}{3}x + 3$

(−2, 2)

**15.**

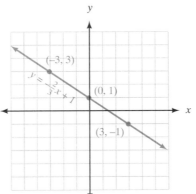

(−3, 3)

$y = -\frac{2}{3}x + 1$

(0, 1)

(3, −1)

**17.**

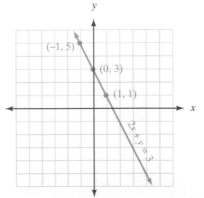

(−1, 5)

(0, 3)

(1, 1)

$2x + y = 3$

**19.**

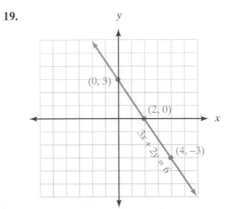

(0, 3)

(2, 0)

$3x + 2y = 6$

(4, −3)

**21.**

**23.**

**25.**

**27.**

**29.**

**31.**

**33.**

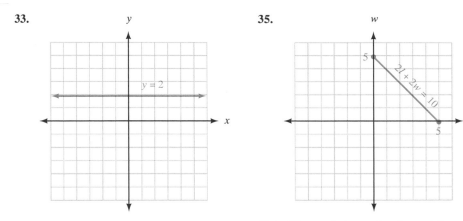

**35.**

Note: The graph does not go beyond the x- and y-axes because *l* and *w* are the length and width of a rectangle and therefore cannot be negative.

**37.**

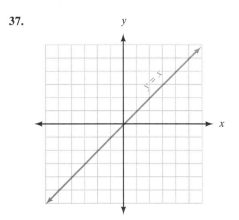

**39.**

| x | y |
|---|---|
| −4 | −3 |
| −2 | −2 |
| 0 | −1 |
| 2 | 0 |
| 6 | 2 |

**41.**

**43.**

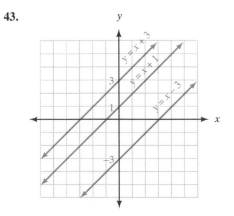

**45.**

| Distance From Airport (miles) | Height (miles) |
|---|---|
| 0 | 0 |
| 15 | 1.1 |
| 30 | 2.2 |
| 45 | 3.3 |
| 60 | 4.4 |
| 75 | 5.5 |
| 90 | 6.6 |

**47.**

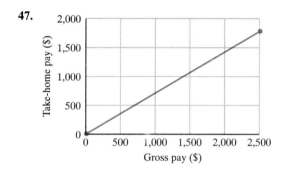

**49.** 5    **51.** −2    **53.** 3    **55.** −6

# PROBLEM SET 3.4

**1.**

**3.**

**5.**

**7.**

**9.**

**11.**

**13.**

**15.**

**17.**

**19.**

**21.**

**23.**

**25.**

**27.**

**29.**

**31.**

**33.**

**35.**

**37.**

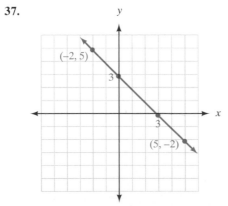

**39.**

| x | y |
|----|----|
| −2 | 1 |
| 0 | −1 |
| −1 | 0 |
| 1 | −2 |

**41.**

**43.**

**45.**

**47.** $x \leq -4$     **49.** $x \geq 3$     **51.** $x > -4$

## PROBLEM SET 3.5

**1.**

**3.**

**5.**

**7.**

**9.**

**11.**

**13.**

**15.**

**17.**

**19.**

**21.**

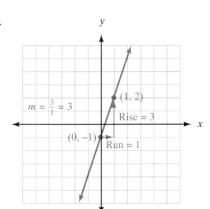

**23.** Slope $= 3$, $y$-intercept $= 2$     **25.** Slope $= 2$, $y$-intercept $= -2$

**27.**

**29.**

**31.**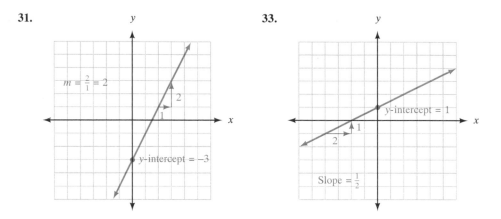

**33.**

**35.** 6 **37.** Slopes: $A$, 3.3; $B$, 3.1; $C$, 5.3; $D$, 1.7 **39.** Slopes: $A$, $-5$; $B$, $-2$; $C$, $-2$ **41.** $-15$ **43.** 27
**45.** $-27$ **47.** 13

## PROBLEM SET 3.6

**1.** $y = \frac{2}{3}x + 1$ **3.** $y = \frac{3}{2}x - 1$ **5.** $y = -\frac{2}{5}x + 3$ **7.** $y = 2x - 4$ **9.** $y = -3x + 2$

**11.** $m = 2, b = 4$ **13.** $m = -3, b = 3$

**15.** $m = -\frac{3}{2}, b = 3$ **17.** $m = \frac{4}{5}, b = -4$

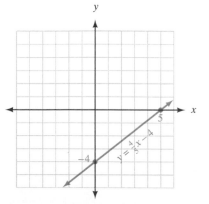

**19.** $m = -\frac{2}{5}, b = -2$        **21.** $y = 2x - 1$

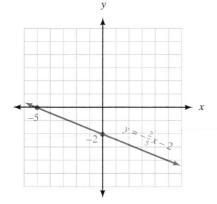

**23.** $y = -\frac{1}{2}x - 1$    **25.** $y = \frac{3}{2}x - 6$    **27.** $y = -3x + 1$    **29.** $y = x + 2$    **31.** $y = x - 2$

**33.** $y = 2x - 3$    **35.** $y = \frac{4}{3}x + 2$    **37.** $y = -\frac{2}{3}x - 3$    **39.** $y = -x - 1$    **41.** $m = 3, b = 3, y = 3x + 3$

**43.** $m = \frac{1}{4}, b = -1; y = \frac{1}{4}x - 1$    **45.** $y = -\frac{2}{3}x + 2$    **47.** $y = -\frac{5}{2}x - 5$    **49.** $x = 3$    **51.** $y = 3$

**53.** (a) \$75.00    (b) \$300.00    (c) $m = -4$    (d) $y = -4x + 300$    **55.** 75    **57.** 400

**59.**

**61.**

**63.**

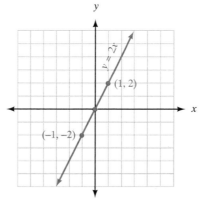

## PROBLEM SET 3.7

**1.**

**3.**

**5.**

**7.**

**9.**

**11.**

**13.**

**15.**

**17.**

**19.**

**21.**

**23.**

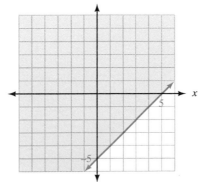

**25.**

**27.**

**29.** $-6x + 11$

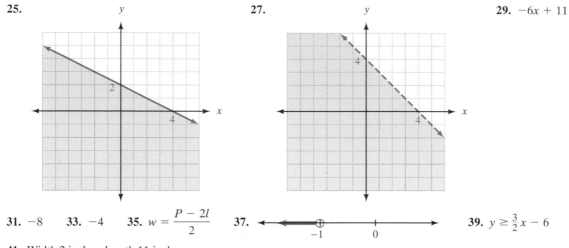

**31.** $-8$   **33.** $-4$   **35.** $w = \dfrac{P - 2l}{2}$   **37.**

**39.** $y \geq \frac{3}{2}x - 6$

**41.** Width 2 inches, length 11 inches

## CHAPTER 3 REVIEW

**1.** $(4, -6), (0, 6), (1, 3), (2, 0)$   **2.** $(5, -2), (0, -4), (15, 2), (10, 0)$   **3.** $(4, 2), (2, -2), (\frac{9}{2}, 3)$

**4.** $(2, 13), (-\frac{3}{5}, 0), (-\frac{6}{5}, -3)$   **5.** $(2, -3), (-1, -3), (-3, -3)$   **6.** $(6, 5), (6, 0), (6, -1)$   **7.** $(2, -\frac{3}{2})$

**8.** $(-\frac{8}{3}, -1), (-3, -2)$   **9–14.**

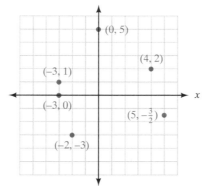

**15.** $(-2, 0), (0, -2), (1, -3)$

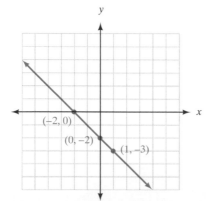

**16.** $(-1, -3), (1, 3), (0, 0)$

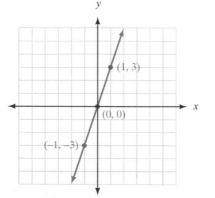

**17.** $(1, 1), (0, -1), (-1, -3)$

**18.**

**19.**

**20.**

**21.**

**22.**

**23.** x-intercept = 2　　**24.** x-intercept = 12　　**25.** x-intercept = 3　　**26.** x-intercept = 2
　　y-intercept = −6　　　　y-intercept = −4　　　　y-intercept = −3　　　　y-intercept = −6

**27.** $m = 2$　　**28.** $m = -1$　　**29.** $m = 2$　　**30.** $x = 6$　　**31.** $y = -2x + 2$

**32.** $y = \frac{1}{2}x + 1$　　**33.** $y = -\frac{3}{4}x + \frac{1}{4}$　　**34.** $y = 3x + 2$　　**35.** $y = -x + 6$　　**36.** $y = -\frac{1}{3}x + \frac{3}{4}$

**37.** $m = 4, b = -1$　　**38.** $m = -2, b = -5$　　**39.** $m = -2, b = 3$　　**40.** $m = -\frac{5}{2}, b = 4$

**41.**

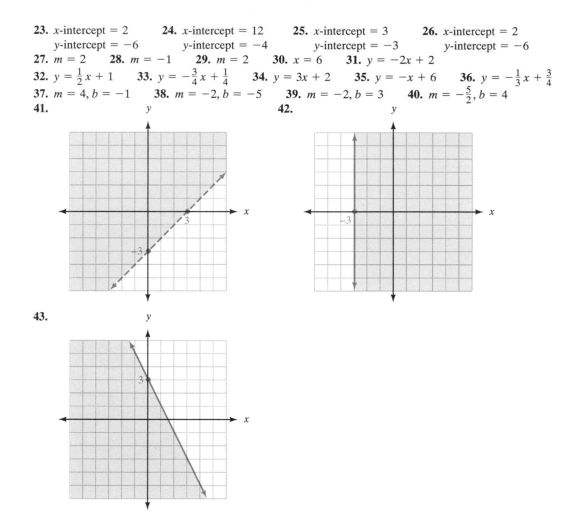

**42.**

**43.**

## CHAPTER 3 TEST

**1.** $(0, -2), (5, 0), (10, 2), \left(-\frac{5}{2}, -3\right)$　　**2.** $(2, 5), (0, -3)$　　**3.**

**4.**

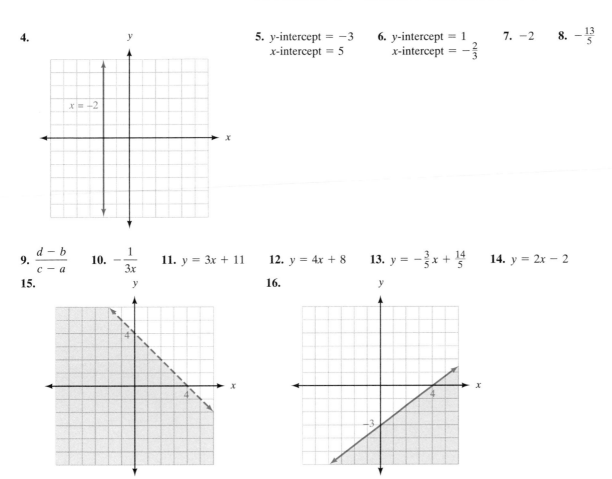

**5.** $y$-intercept $= -3$
$x$-intercept $= 5$

**6.** $y$-intercept $= 1$
$x$-intercept $= -\frac{2}{3}$

**7.** $-2$

**8.** $-\frac{13}{5}$

**9.** $\dfrac{d - b}{c - a}$    **10.** $-\dfrac{1}{3x}$    **11.** $y = 3x + 11$    **12.** $y = 4x + 8$    **13.** $y = -\frac{3}{5}x + \frac{14}{5}$    **14.** $y = 2x - 2$

**15.**

**16.**

# CHAPTERS 1–3 CUMULATIVE REVIEW

**1.** $-5$    **2.** $22$    **3.** $18$    **4.** $40$    **5.** $55$    **6.** Undefined    **7.** $\frac{1}{35}$    **8.** $40x$    **9.** $a - 3$    **10.** $4$

**11.** $2$    **12.** $2$    **13.** $-4$    **14.** $2$    **15.** $150$    **16.** $x < -7$

**17.** $x \leq -2$

**18.**

**19.**

**20.**

**21.**

**22.**

**23.** (4, 7) no; (1, 2) yes

**24.** $y$-intercept = 2

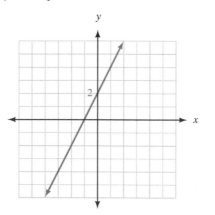

**25.** $x$-intercept $= 5$
    $y$-intercept $= 2$

**26.**

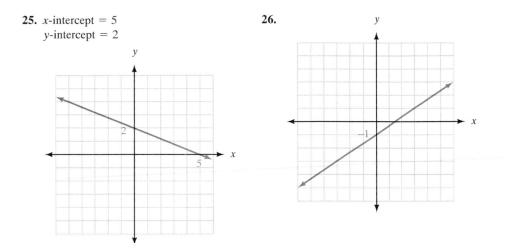

**27.** $\frac{4}{3}$    **28.** $y = 3x + 1$    **29.** $-\frac{2}{3}$    **30.** $y = \frac{1}{2}x + 4$

**31.** $y = 2x - 1$    **32.** $y = \frac{4}{3}x - 4$    **33.**

| $x$ | $y$ |
|-----|-----|
| 4 | 3 |
| 0 | −3 |

**34.** $(0, 2), (4, \frac{2}{5})$    **35.** $x + 7 = 4$

**36.** $15 - 2(11) = -7$    **37.** $\frac{1}{36}$    **38.** $-2, \frac{1}{2}, 2$    **39.** $-17$    **40.** $-3, 0, 5$    **41.** $\frac{5}{9}$    **42.** $\frac{13}{5}$    **43.** 9

**44.** $-3$    **45.** 75%    **46.** 14    **47.** Width is $\frac{17}{3}$ centimeters; length is $\frac{49}{3}$ centimeters.    **48.** 3 ounces

**49.** \$1,200 at 6% and \$2,100 at 8%    **50.** d

## CHAPTER 4

## PROBLEM SET 4.1

**1.**

**3.**

**5.**

**7.**

**9.**

**11.**

**13.**

**15.**

**17.**

**19.**

**21.**

**23.**

**25.** No solution

**27.** Infinitely many solutions

**29.** 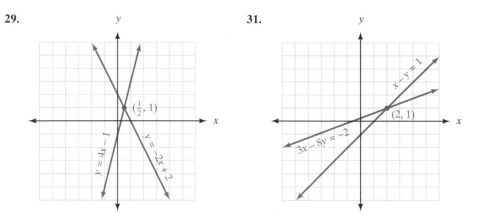 **31.**

**33.** (a) 25 hours  (b) Gigi's  (c) Marcy's    **35.** (a) 4 hours  (b) Computer Service  (c) ICM World

**37.** 1    **39.** $-\frac{3}{5}$    **41.**

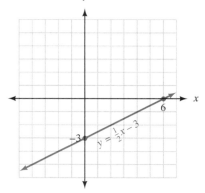

## PROBLEM SET 4.2

**1.** $(2, 1)$    **3.** $(3, 7)$    **5.** $(2, -5)$    **7.** $(-1, 0)$    **9.** Lines coincide.    **11.** $(4, 8)$    **13.** $(\frac{1}{5}, 1)$    **15.** $(1, 0)$

**17.** $(-1, -2)$    **19.** $(-5, \frac{3}{4})$    **21.** $(-4, 5)$    **23.** $(-3, -10)$    **25.** $(3, 2)$    **27.** $(5, \frac{1}{3})$    **29.** $(-2, \frac{2}{3})$

**31.** $(2, 2)$    **33.** Lines are parallel.    **35.** $(1, 1)$    **37.** Lines are parallel.    **39.** $(7, 5)$    **41.** $(10, 12)$

**43.** $y = \frac{1}{2}x - 2, m = \frac{1}{2}, b = 2$    **45.** $y = 3x$    **47.** $y = x - 2$

## PROBLEM SET 4.3

**1.** $(4, 7)$    **3.** $(3, 17)$    **5.** $(\frac{3}{2}, 2)$    **7.** $(2, 4)$    **9.** $(0, 4)$    **11.** $(-1, 3)$    **13.** $(1, 1)$    **15.** $(2, -3)$

**17.** $(-2, \frac{3}{5})$    **19.** $(-3, 5)$    **21.** Lines are parallel.    **23.** $(3, 1)$    **25.** $(\frac{1}{2}, \frac{3}{4})$    **27.** $(2, 6)$    **29.** $(4, 4)$

**31.** $(5, -2)$    **33.** $(18, 10)$    **35.** Lines coincide.    **37.** $(10, 12)$

**39.** (a) 1,000 miles  (b) Car  (c) Truck  (d) Miles $\geq 0$    **41.** 40 gallons    **43.** $l = 9$ meters; $w = 3$ meters

**45.** 12 nickels, 15 dimes    **47.** 480    **49.** $800 at 8%, $1,600 at 10%

## PROBLEM SET 4.4

**1.** $x + y = 25$    The two numbers    **3.** 3 and 12    **5.** $x - y = 5$    The two numbers    **7.** 6 and 29

$y = x + 5$    are 10 and 15.    $x = 2y + 1$    are 9 and 4.

**9.** Let $x$ = the amount invested at 6% and $y$ = the amount invested at 8%.    **11.** $2,000 at 6%, $8,000 at 5%

$x + y = 20,000$        He had $9,000 at 8%

$0.06x + 0.08y = 1,380$        and $11,000 at 6%.

**13.** 6 nickels, 8 quarters    **15.** Let $x$ = the number of dimes and $y$ = the number of quarters.

$x + y = 21$        He has 12 dimes

$0.10x + 0.25y = 3.45$        and 9 quarters.

**17.** Let $x$ = the number of liters of 50% solution and $y$ = the number of liters of 20% solution.

$x + y = 18$            6 liters of 50% solution

$0.50x + 0.20y = 0.30(18)$        12 liters of 20% solution

**19.** 10 gallons of 10% solution, 20 gallons of 7% solution    **21.** 20 adults, 50 kids    **23.** 16 feet wide, 32 feet long

**25.** 33 $5 chips, 12 $25 chips    **27.** 50 at $11, 100 at $20    **29.** 2, 3, 4

**31.**                              **33.** 2    **35.** $y = \frac{1}{2}x + 2$    **37.** $y = 2x + 1$

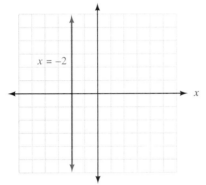

## CHAPTER 4 REVIEW

**1.** $(4, -2)$    **2.** $(-3, 2)$    **3.** $(3, -2)$    **4.** $(2, 0)$    **5.** $(2, 1)$    **6.** $(-1, -2)$    **7.** $(1, -3)$    **8.** $(-2, 5)$
**9.** Lines coincide.    **10.** $(2, -2)$    **11.** $(1, 1)$    **12.** Lines are parallel.    **13.** $(-2, -3)$    **14.** $(5, -1)$
**15.** $(-2, 7)$    **16.** $(-4, -2)$    **17.** $(-1, 4)$    **18.** $(2, -6)$    **19.** Lines are parallel.    **20.** $(-3, -10)$
**21.** $(1, -2)$    **22.** Lines coincide.    **23.** 10, 8    **24.** 24, 8    **25.** $4,000 at 4%, $8,000 at 5%
**26.** $3,000 at 6%, $11,000 at 8%    **27.** 10 dimes, 7 nickels    **28.** 9 dimes, 6 quarters
**29.** 40 liters of 10% solution, 10 liters of 20% solution    **30.** 20 liters of 25% solution, 20 liters of 15% solution

## CHAPTER 4 TEST

**1.**            **2.**

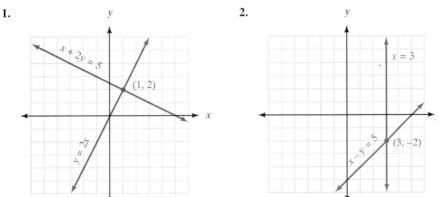

**3.** $(-3, -4)$  **4.** $(5, -3)$  **5.** $(2, -2)$  **6.** Lines coincide.  **7.** $(2, 7)$  **8.** $(4, 3)$  **9.** $(2, 9)$
**10.** $(-5, -1)$  **11.** $5, 7$  **12.** $3, 12$  **13.** \$6,000  **14.** 7 nickels, 5 quarters

# CHAPTERS 1–4 CUMULATIVE REVIEW

**1.** 17  **2.** 48  **3.** 36  **4.** $2x - 2$  **5.** 32  **6.** 16  **7.** $\frac{5}{4}$  **8.** $-10a + 29$  **9.** $-8$  **10.** 0
**11.** 7  **12.** $-6$  **13.** $x \leq -9$

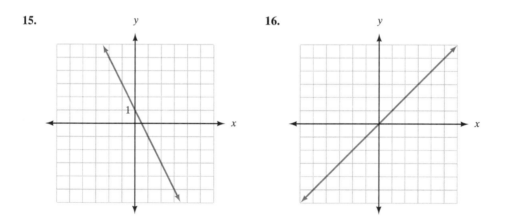

**14.** $x \geq 12$

**15.**

**16.**

**17.**

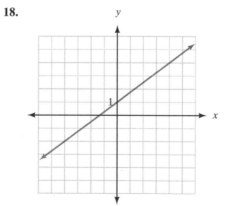

**18.**

**19.**  **20.** $(1, 0)$

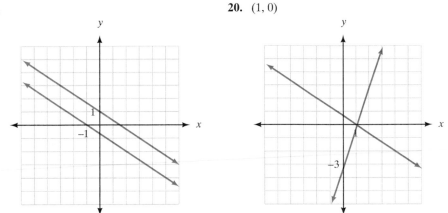

**21.** Lines coincide.    **22.** $(1, -3)$    **23.** $(2, 3)$    **24.** $(6, 7)$    **25.** $(9, 3)$    **26.** $(4, -1)$    **27.** $(4, 0)$
**28.** $(3, -1)$    **29.** $(-1, -3)$    **30.** Lines coincide.    **31.** Commutative property of addition    **32.** 32
**33.** $-5$    **34.** $2^2 \cdot 3^2 \cdot 5$    **35.** 25%    **36.** 15.5    **37.** 13    **38.** $-21$    **39.** $-13$    **40.** $(-2, -4)$
**41.** $5 - (-8) = 13$    **42.**                         **43.** $(4, 0), (0, -3)$    **44.** $\frac{1}{3}$    **45.** $\frac{5}{4}$    **46.** $-\frac{1}{12}$

| $x$ | $y$ |
|-----|-----|
| 2 | $-\frac{4}{5}$ |
| 5 | 1 |

**47.** $y = \frac{2}{3}x + 3$    **48.** $y = 2x - 7$    **49.** $y = \frac{3}{2}x - 3$    **50.** 8 nickels, 7 dimes

## CHAPTER 5

### PROBLEM SET 5.1

**1.** Base 4; exponent 2; 16    **3.** Base 0.3; exponent 2; 0.09    **5.** Base 4; exponent 3; 64
**7.** Base $-5$; exponent 2; 25    **9.** Base 2; exponent 3; $-8$    **11.** Base 3; exponent 4; 81    **13.** Base $\frac{2}{3}$; exponent 2; $\frac{4}{9}$
**15.** Base $\frac{1}{2}$; exponent 4; $\frac{1}{16}$    **17.** (a)                         (b) Either *larger* or *greater* will work.

| Number $x$ | Square $x^2$ |
|------------|--------------|
| 1 | 1 |
| 2 | 4 |
| 3 | 9 |
| 4 | 16 |
| 5 | 25 |
| 6 | 36 |
| 7 | 49 |

**19.** $x^9$    **21.** $y^{30}$    **23.** $2^{12}$    **25.** $x^{28}$    **27.** $x^{10}$    **29.** $5^{12}$    **31.** $y^9$    **33.** $2^{50}$    **35.** $a^{3x}$    **37.** $b^{xy}$
**39.** $16x^2$    **41.** $32y^5$    **43.** $81x^4$    **45.** $0.25a^2b^2$    **47.** $64x^3y^3z^3$    **49.** $8x^{12}$    **51.** $16a^6$    **53.** $x^{14}$

**55.** $a^{11}$    **57.** $128x^7$    **59.** $432x^{10}$    **61.** $16x^4y^6$    **63.** $\frac{8}{27}a^{12}b^{15}$

**65.**

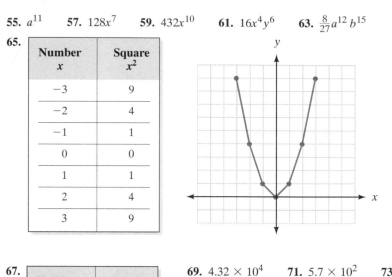

| Number $x$ | Square $x^2$ |
|---|---|
| $-3$ | 9 |
| $-2$ | 4 |
| $-1$ | 1 |
| 0 | 0 |
| 1 | 1 |
| 2 | 4 |
| 3 | 9 |

**67.**

| Number $x$ | Square $x^2$ |
|---|---|
| $-2.5$ | 6.25 |
| $-1.5$ | 2.25 |
| $-0.5$ | 0.25 |
| 0 | 0 |
| 0.5 | 0.25 |
| 1.5 | 2.25 |
| 2.5 | 6.25 |

**69.** $4.32 \times 10^4$    **71.** $5.7 \times 10^2$    **73.** $2.38 \times 10^5$    **75.** 2,490    **77.** 352

**79.** 28,000    **81.** 27 inches$^3$    **83.** 15.6 inches$^3$    **85.** 36 inches$^3$
**87.** Yes, the dimensions could be 2 feet $\times$ 3 feet $\times$ 7 feet.    **89.** $6.5 \times 10^8$ seconds    **91.** $740,000
**93.** 180,000    **95.** 219 inches$^3$    **97.** 182 inches$^3$
**99.** (a) Length of Stage 1 is $\frac{4}{3}$ feet.    (b) Length of Stage 2 is $\left(\frac{4}{3}\right)^2 = \frac{16}{9}$ feet.

(c) Length of Stage 3 is $\left(\frac{4}{3}\right)^3 = \frac{64}{27}$ feet.    (d) Length of Stage 10 is $\left(\frac{4}{3}\right)^{10}$ feet.
**101.** $-3$    **103.** 11    **105.** $-5$    **107.** 5

## PROBLEM SET 5.2

**1.** $\frac{1}{9}$    **3.** $\frac{1}{36}$    **5.** $\frac{1}{64}$    **7.** $\frac{1}{125}$    **9.** $\frac{2}{x^3}$    **11.** $\frac{1}{8x^3}$    **13.** $\frac{1}{25y^2}$    **15.** $\frac{1}{100}$

**17.**

| Number $x$ | Square $x^2$ | Power of 2 $2^x$ |
|---|---|---|
| −3 | 9 | $\frac{1}{8}$ |
| −2 | 4 | $\frac{1}{4}$ |
| −1 | 1 | $\frac{1}{2}$ |
| 0 | 0 | 1 |
| 1 | 1 | 2 |
| 2 | 4 | 4 |
| 3 | 9 | 8 |

**19.** $\frac{1}{25}$    **21.** $x^6$    **23.** 64    **25.** $8x^3$    **27.** $6^{10}$

**29.** $\frac{1}{6^{10}}$    **31.** $\frac{1}{2^8}$    **33.** $2^8$    **35.** $27x^3$    **37.** $81x^4y^4$    **39.** 1    **41.** $2a^2b$    **43.** $\frac{1}{49y^6}$    **45.** $\frac{1}{x^8}$

**47.** $\frac{1}{y^3}$    **49.** $x^2$    **51.** $a^6$    **53.** $\frac{1}{y^9}$    **55.** $y^{40}$    **57.** $\frac{1}{x}$    **59.** $x^9$    **61.** $a^{16}$    **63.** $\frac{1}{a^4}$

**65.**

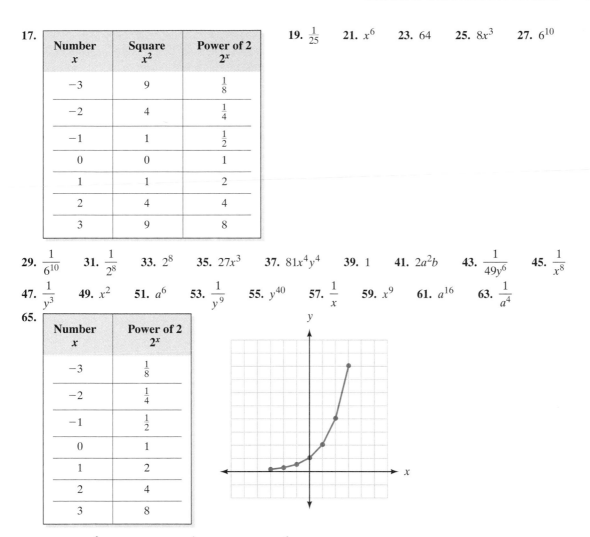

| Number $x$ | Power of 2 $2^x$ |
|---|---|
| −3 | $\frac{1}{8}$ |
| −2 | $\frac{1}{4}$ |
| −1 | $\frac{1}{2}$ |
| 0 | 1 |
| 1 | 2 |
| 2 | 4 |
| 3 | 8 |

**67.** $4.8 \times 10^{-3}$    **69.** $2.5 \times 10^1$    **71.** $9 \times 10^{-6}$

**73.**

| Expanded Form | Scientific Notation $n \times 10^r$ |
|---|---|
| 0.000357 | $3.57 \times 10^{-4}$ |
| 0.00357 | $3.57 \times 10^{-3}$ |
| 0.0357 | $3.57 \times 10^{-2}$ |
| 0.357 | $3.57 \times 10^{-1}$ |
| 3.57 | $3.57 \times 10^0$ |
| 35.7 | $3.57 \times 10^1$ |
| 357 | $3.57 \times 10^2$ |
| 3,570 | $3.57 \times 10^3$ |
| 35,700 | $3.57 \times 10^4$ |

**75.** 0.00423    **77.** 0.00008    **79.** 4.2    **81.** 0.002 seconds

**83.** $6 \times 10^{-3}$ inches   **85.** $2.5 \times 10^4$   **87.** $2.35 \times 10^5$   **89.** $8.2 \times 10^{-4}$   **91.** 100 inches²; 400 inches²; 4
**93.** $x^2$; $4x^2$; 4   **95.** 216 inches³; 1,728 inches³; 8   **97.** $x^3$; $8x^3$; 8   **99.** $7x$   **101.** $2a$   **103.** $10y$

## PROBLEM SET 5.3

**1.** $12x^7$   **3.** $-16y^{11}$   **5.** $32x^2$   **7.** $200a^6$   **9.** $-24a^3b^3$   **11.** $24x^6y^8$   **13.** $3x$   **15.** $\dfrac{6}{y^3}$   **17.** $\dfrac{1}{2a}$

**19.** $-\dfrac{3a}{b^2}$   **21.** $\dfrac{x^2}{9z^2}$   **23.**   **25.** $6 \times 10^8$

| $a$ | $b$ | $ab$ | $\dfrac{a}{b}$ | $\dfrac{b}{a}$ |
|---|---|---|---|---|
| 10 | $5x$ | $50x$ | $\dfrac{2}{x}$ | $\dfrac{x}{2}$ |
| $20x^3$ | $6x^2$ | $120x^5$ | $\dfrac{10x}{3}$ | $\dfrac{3}{10x}$ |
| $25x^5$ | $5x^4$ | $125x^9$ | $5x$ | $\dfrac{1}{5x}$ |
| $3x^{-2}$ | $3x^2$ | 9 | $\dfrac{1}{x^4}$ | $x^4$ |
| $-2y^4$ | $8y^7$ | $-16y^{11}$ | $-\dfrac{1}{4y^3}$ | $-4y^3$ |

**27.** $1.75 \times 10^{-1}$   **29.** $1.21 \times 10^{-6}$   **31.** $4.2 \times 10^3$   **33.** $3 \times 10^{10}$   **35.** $5 \times 10^{-3}$   **37.** $8x^2$
**39.** $-11x^5$   **41.** 0   **43.** $4x^3$   **45.** $31ab^2$   **47.**   **49.** $4x^3$

| $a$ | $b$ | $ab$ | $a + b$ |
|---|---|---|---|
| $5x$ | $3x$ | $15x^2$ | $8x$ |
| $4x^2$ | $2x^2$ | $8x^4$ | $6x^2$ |
| $3x^3$ | $6x^3$ | $18x^6$ | $9x^3$ |
| $2x^4$ | $-3x^4$ | $-6x^8$ | $-x^4$ |
| $x^5$ | $7x^5$ | $7x^{10}$ | $8x^5$ |

**51.** $\dfrac{1}{b^2}$   **53.** $\dfrac{6y^{10}}{x^4}$   **55.** $2 \times 10^6$   **57.** $1 \times 10^1$   **59.** $4.2 \times 10^{-6}$   **61.** $9x^3$   **63.** $-20a^2$   **65.** $6x^5y^2$
**67.** 2   **69.** 4   **71.** $(4 + 5)^2 = 9^2 = 81$; $4^2 + 5^2 = 16 + 25 = 41$
**73.** $(3 + 4)^2 = 7^2 = 49$; $3^2 + 2(3)(4) + 4^2 = 49$   **75.** $P = 6x$; $A = 2x^2$   **77.** $V = 8x^2$ inches³
**79.** (a) $4.7 \times 10^3$ feet³   (b) $3.1 \times 10^4$ boxes   **81.** $(-4, 16), (-2, 4), (-1, 1), (0, 0), (1, 1), (2, 4), (4, 16)$

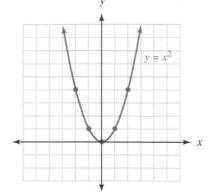

**83.** $(-3, 18), (-2, 8), (-1, 2), (0, 0), (1, 2), (2, 8), (3, 18)$

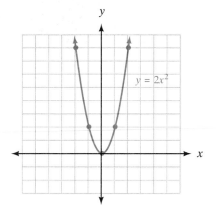

**85.** $(-4, 8), (-2, 2), (-1, \frac{1}{2}), (0, 0), (1, \frac{1}{2}), (2, 2), (4, 8)$

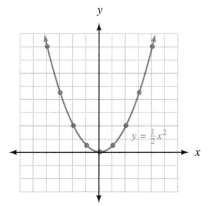

**87.** $(-4, 4), (-2, 1), (-1, \frac{1}{4}), (0, 0), (1, \frac{1}{4}), (2, 1), (4, 4)$

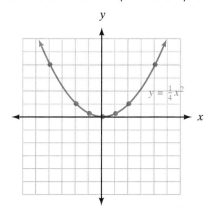

**89.** 9 **91.** 0

**93.**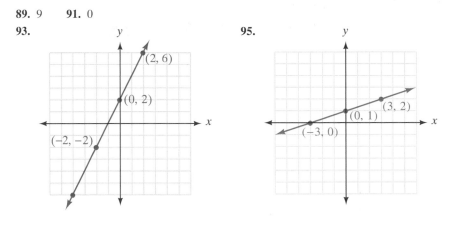

**95.**

## PROBLEM SET 5.4

**1.** Trinomial, 3 **3.** Trinomial, 3 **5.** Binomial, 1 **7.** Binomial, 2 **9.** Monomial, 2 **11.** Monomial, 0
**13.** $5x^2 + 5x + 9$ **15.** $5a^2 - 9a + 7$ **17.** $x^2 + 6x + 8$ **19.** $6x^2 - 13x + 5$ **21.** $x^2 - 9$
**23.** $3y^2 - 11y + 10$ **25.** $6x^3 + 5x^2 - 4x + 3$ **27.** $2x^2 - x + 1$ **29.** $2a^2 - 2a - 2$
**31.** $-\frac{1}{9}x^3 - \frac{2}{3}x^2 - \frac{5}{2}x + \frac{7}{4}$ **33.** $-4y^2 + 15y - 22$ **35.** $x^2 - 33x + 63$ **37.** $8y^2 + 4y + 26$
**39.** $75x^2 - 150x - 75$ **41.** $12x + 2$ **43.** 4 **45.** 25 **47.** 16 **49.** $18\pi \text{ inches}^3$ **51.** $-15x^2$ **53.**
$6x^3$ **55.** $6x^4$

## PROBLEM SET 5.5

**1.** $6x^2 + 2x$ **3.** $6x^4 - 4x^3 + 2x^2$ **5.** $2a^3b - 2a^2b^2 + 2ab$ **7.** $3y^4 + 9y^3 + 12y^2$
**9.** $8x^5y^2 + 12x^4y^3 + 32x^2y^4$ **11.** $x^2 + 7x + 12$ **13.** $x^2 + 7x + 6$ **15.** $x^2 + 2x + \frac{3}{4}$ **17.** $a^2 + 2a - 15$
**19.** $xy + bx - ay - ab$ **21.** $x^2 - 36$ **23.** $y^2 - \frac{25}{36}$ **25.** $2x^2 - 11x + 12$ **27.** $2a^2 + 3a - 2$
**29.** $6x^2 - 19x + 10$ **31.** $2ax + 8x + 3a + 12$ **33.** $25x^2 - 16$ **35.** $2x^2 + \frac{5}{2}x - \frac{3}{4}$ **37.** $3 - 10a + 8a^2$
**39.**

| | $x$ | 3 |
|---|---|---|
| $x$ | $x^2$ | $3x$ |
| 2 | $2x$ | 6 |

$(x + 2)(x + 3) = x^2 + 2x + 3x + 6$
$= x^2 + 5x + 6$

**41.**

| | $x$ | $x$ | 2 |
|---|---|---|---|
| $x$ | $x^2$ | $x^2$ | $2x$ |
| 1 | $x$ | $x$ | 2 |

$(x + 1)(2x + 2) = 2x^2 + 4x + 2$

**43.** $a^3 - 6a^2 + 11a - 6$

**45.** $x^3 + 8$ **47.** $2x^3 + 17x^2 + 26x + 9$ **49.** $5x^4 - 13x^3 + 20x^2 + 7x + 5$ **51.** $2x^4 + x^2 - 15$
**53.** $6a^6 + 15a^4 + 4a^2 + 10$ **55.** $x^3 + 12x^2 + 47x + 60$ **57.** $A = x(2x + 5) = 2x^2 + 5x$
**59.** $A = x(x + 1) = x^2 + x$ **61.** $R = (1,200 - 100p)p = 1,200p - 100p^2$
**63.** $R = (1,700 - 100p)p = 1,700p - 100p^2$

**65.**

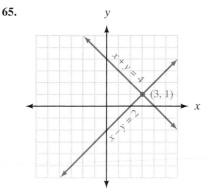

**67.** $(1, -1)$    **69.** $(-\frac{1}{2}, -\frac{1}{2})$    **71.** 6 dimes, 5 quarters

## PROBLEM SET 5.6

**1.** $x^2 - 4x + 4$    **3.** $a^2 + 6a + 9$    **5.** $x^2 - 10x + 25$    **7.** $a^2 - a + \frac{1}{4}$    **9.** $x^2 + 20x + 100$
**11.** $a^2 + 1.6a + 0.64$    **13.** $4x^2 - 4x + 1$    **15.** $16a^2 + 40a + 25$    **17.** $9x^2 - 12x + 4$
**19.** $9a^2 + 30ab + 25b^2$    **21.** $16x^2 - 40xy + 25y^2$    **23.** $49m^2 + 28mn + 4n^2$    **25.** $36x^2 - 120xy + 100y^2$
**27.** $x^4 + 10x^2 + 25$    **29.** $a^4 + 2a^2 + 1$    **31.**

| $x$ | $(x + 3)^2$ | $x^2 + 9$ | $x^2 + 6x + 9$ |
|---|---|---|---|
| 1 | 16 | 10 | 16 |
| 2 | 25 | 13 | 25 |
| 3 | 36 | 18 | 36 |
| 4 | 49 | 25 | 49 |

**35.** $a^2 - 25$    **37.** $y^2 - 1$    **39.** $81 - x^2$

**33.**

| $a$ | $b$ | $(a + b)^2$ | $a^2 + b^2$ | $a^2 + ab + b^2$ | $a^2 + 2ab + b^2$ |
|---|---|---|---|---|---|
| 1 | 1 | 4 | 2 | 3 | 4 |
| 3 | 5 | 64 | 34 | 49 | 64 |
| 3 | 4 | 49 | 25 | 37 | 49 |
| 4 | 5 | 81 | 41 | 61 | 81 |

**41.** $4x^2 - 25$    **43.** $16x^2 - \frac{1}{9}$    **45.** $4a^2 - 49$    **47.** $36 - 49x^2$    **49.** $x^4 - 9$    **51.** $a^4 - 16$
**53.** $25y^8 - 64$    **55.** $2x^2 - 34$    **57.** $-12x^2 + 20x + 8$    **59.** $a^2 + 4a + 6$    **61.** $8x^3 + 36x^2 + 54x + 27$
**63.** $(50 - 1)(50 + 1) = 2{,}500 - 1 = 2{,}499$    **65.** Both equal 25.    **67.** $x^2 + (x + 1)^2 = 2x^2 + 2x + 1$
**69.** $x^2 + (x + 1)^2 + (x + 2)^2 = 3x^2 + 6x + 5$    **71.** $a^2 + ab + ba + b^2 = a^2 + 2ab + b^2$

**73.**

| | $x$ | $x$ | $1$ |
|---|---|---|---|
| $x$ | $x^2$ | $x^2$ | $x$ |
| $x$ | $x^2$ | $x^2$ | $x$ |
| $1$ | $x$ | $x$ | $1$ |

$(2x + 1)^2 = 4x^2 + 4x + 1$    **75.** $5x$    **77.** $\dfrac{a^4}{2b^2}$

**79.**    **81.**

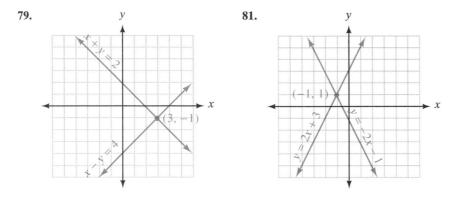

## PROBLEM SET 5.7

**1.** $x - 2$    **3.** $3 - 2x^2$    **5.** $5xy - 2y$    **7.** $7x^4 - 6x^3 + 5x^2$    **9.** $10x^4 - 5x^2 + 1$    **11.** $-4a + 2$

**13.** $-8a^4 - 12a^3$    **15.** $-4b - 5a$    **17.** $-6a^2b + 3ab^2 - 7b^3$    **19.** $-\dfrac{a}{2} - b - \dfrac{b^2}{2a}$    **21.** $3x + 4y$

**23.** $-y + 3$    **25.** $5y - 4$    **27.** $xy - x^2y^2$    **29.** $-1 + xy$    **31.** $-a + 1$    **33.** $x^2 - 3xy + y^2$

**35.** $2 - 3b + 5b^2$    **37.** $-2xy + 1$    **39.** $xy - \dfrac{1}{2}$    **41.** $\dfrac{1}{4x} - \dfrac{1}{2a} + \dfrac{3}{4}$    **43.** $\dfrac{4x^2}{3} + \dfrac{2}{3x} + \dfrac{1}{x^2}$

**45.** $3a^{3m} - 9a^m$    **47.** $2x^{4m} - 5x^{2m} + 7$    **49.** $3x^2 - x + 6$    **51.** $4$    **53.** $x + 5$    **55.** Both equal 7.

**57.** $\dfrac{3(10) + 8}{2} = 19;\ 3(10) + 4 = 34$    **59.** $(7, -1)$    **61.** $(2, 3)$    **63.** $(1, 1)$    **65.** Lines coincide.

## PROBLEM SET 5.8

**1.** $x - 2$    **3.** $a + 4$    **5.** $x - 3$    **7.** $x + 3$    **9.** $a - 5$    **11.** $x + 2 + \dfrac{2}{x + 3}$    **13.** $a - 2 + \dfrac{12}{a + 5}$

**15.** $x + 4 + \dfrac{9}{x - 2}$    **17.** $x + 4 + \dfrac{-10}{x + 1}$    **19.** $a + 1 + \dfrac{-1}{a + 2}$    **21.** $x - 3 + \dfrac{17}{2x + 4}$

**23.** $3a - 2 + \dfrac{7}{2a + 3}$    **25.** $2a^2 - a - 3$    **27.** $x^2 - x + 5$    **29.** $x^2 + x + 1$    **31.** $x^2 + 2x + 4$

**33.** $5, 20$    **35.** $800$ at 8%, $400 at 9%    **37.** Eight \$5 bills, twelve \$10 bills

**39.** 10 gallons of 20%, 6 gallons of 60%

## CHAPTER 5 REVIEW

**1.** $-1$    **2.** $-64$    **3.** $\dfrac{9}{49}$    **4.** $y^{12}$    **5.** $x^{30}$    **6.** $x^{35}$    **7.** $2^{24}$    **8.** $27y^3$    **9.** $-8x^3y^3z^3$    **10.** $\dfrac{1}{49}$

**11.** $\dfrac{4}{x^5}$    **12.** $\dfrac{1}{27y^3}$    **13.** $a^6$    **14.** $\dfrac{1}{x^4}$    **15.** $x^{15}$    **16.** $\dfrac{1}{x^5}$    **17.** $1$    **18.** $-3$    **19.** $9x^6y^4$

**20.** $64a^{22}b^{20}$    **21.** $\dfrac{-1}{27x^3y^6}$    **22.** $b$    **23.** $\dfrac{1}{x^{35}}$    **24.** $5x^7$    **25.** $\dfrac{6y^7}{x^5}$    **26.** $4a^6$    **27.** $2x^5y^2$

**28.** $6.4 \times 10^7$    **29.** $2.3 \times 10^8$    **30.** $8 \times 10^3$    **31.** $8a^2 - 12a - 3$    **32.** $-12x^2 + 5x - 14$

**33.** $-4x^2 - 6x$    **34.** $14$    **35.** $12x^2 - 21x$    **36.** $24x^5y^2 - 40x^4y^3 + 32x^3y^4$    **37.** $a^3 + 6a^2 + a - 4$

**38.** $x^3 + 125$    **39.** $6x^2 - 29x + 35$    **40.** $25y^2 - \dfrac{1}{25}$    **41.** $a^4 - 9$    **42.** $a^2 - 10a + 25$

**43.** $9x^2 + 24x + 16$    **44.** $y^4 + 6y^2 + 9$    **45.** $-2b - 4a$    **46.** $-5x^4y^3 + 4x^2y^2 + 2x$    **47.** $x + 9$

**48.** $2x + 5$    **49.** $x^2 - 4x + 16$    **50.** $x - 3 + \dfrac{16}{3x + 2}$    **51.** $x^2 - 4x + 5 + \dfrac{5}{2x + 1}$    **52.** $V = 3x^3$    **53.** Yes

**54.** (a) Surface area $= 50.24$ ft$^2$    (b) Volume $= 33.5$ ft$^3$    **55.** The trampoline has a radius of 6 feet.

## CHAPTER 5 TEST

**1.** 81    **2.** $\frac{9}{16}$    **3.** $72x^{18}$    **4.** $\frac{1}{9}$    **5.** 1    **6.** $a^2$    **7.** $x$    **8.** $2.78 \times 10^{-2}$    **9.** 243,000    **10.** $\frac{y^2}{2x^4}$

**11.** $3a$    **12.** $10x^5$    **13.** $9 \times 10^9$    **14.** $8x^2 + 2x + 2$    **15.** $3x^2 + 4x + 6$    **16.** $3x - 4$    **17.** 10

**18.** $6a^4 - 10a^3 + 8a^2$    **19.** $x^2 + \frac{5}{6}x + \frac{1}{6}$    **20.** $8x^2 + 2x - 15$    **21.** $x^3 - 27$    **22.** $x^2 + 10x + 25$

**23.** $9a^2 - 12ab + 4b^2$    **24.** $9x^2 - 16y^2$    **25.** $a^4 - 9$    **26.** $2x^2 + 3x - 1$    **27.** $4x + 3 + \dfrac{4}{2x - 3}$

**28.** $3x^2 + 9x + 25 + \dfrac{76}{x - 3}$    **29.** $15.625$ centimeters$^3$    **30.** $x^3$

## CHAPTERS 1–5 CUMULATIVE REVIEW

**1.** $\frac{3}{4}$    **2.** 24    **3.** 105    **4.** Undefined    **5.** $9a + 15$    **6.** 1    **7.** $-12$    **8.** 63    **9.** 45    **10.** $y^{16}$

**11.** $16y^4$    **12.** $108a^{18}b^{29}$    **13.** $\dfrac{1,152y^{13}}{x^3}$    **14.** $5 \times 10^{-5}$    **15.** $25x^2 - 10x + 1$    **16.** $-13x + 10$

**17.** $x^3 - 1$    **18.** 4    **19.** $\frac{5}{2}$    **20.** 4    **21.** $-2$    **22.**

**23.**

**24.**

**25.**    **26.** $2x + 7 + \dfrac{11}{2x - 3}$    **27.** $3 - \dfrac{2}{x^3} + \dfrac{4}{x^4}$

**28.**

**29.**

**30.**

**31.**

**32.**                          **33.**

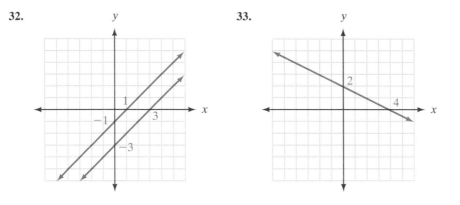

**34.** (3, 1)      **35.** Lines are parallel.      **36.** (−2, 1)      **37.** (3, 2)      **38.** (4, 5)      **39.** (5, 1)      **40.** −17

**41.** 21      **42.** $h = \dfrac{2A}{b}$      **43.** $-\sqrt{2},\ \pi$      **44.** 3      **45.** $(4, 0), \left(\frac{16}{3}, 1\right)$      **46.** $x$-intercept −2; $y$-intercept 4

**47.** $x$-intercept: 4; $y$-intercept: 2      **48.** $\frac{2}{5}$      **49.** $y = \frac{4}{3}x + \frac{20}{3}$

**50.** 5 liters of 40% solution and 15 liters of 16% solution

# CHAPTER 6

## PROBLEM SET 6.1

**1.** $5(3x + 5)$      **3.** $3(2a + 3)$      **5.** $4(x − 2y)$      **7.** $3(x^2 − 2x − 3)$      **9.** $3(a^2 − a − 20)$
**11.** $4(6y^2 − 13y + 6)$      **13.** $x^2(9 − 8x)$      **15.** $13a^2(1 − 2a)$      **17.** $7xy(3x − 4y)$      **19.** $11ab^2(2a − 1)$
**21.** $7x(x^2 + 3x − 4)$      **23.** $11(11y^4 − x^4)$      **25.** $25x^2(4x^2 − 2x + 1)$      **27.** $8(a^2 + 2b^2 + 4c^2)$
**29.** $4ab(a − 4b + 8ab)$      **31.** $11a^2b^2(11a − 2b + 3ab)$      **33.** $12x^2y^3(1 − 6x^3 − 3x^2y)$      **35.** $(x + 3)(y + 5)$
**37.** $(x + 2)(y + 6)$      **39.** $(a − 3)(b + 7)$      **41.** $(a − b)(x + y)$      **43.** $(2x − 5)(a + 3)$      **45.** $(b − 2)(3x − 4)$
**47.** $(x + 2)(x + a)$      **49.** $(x − b)(x − a)$      **51.** $(x + y)(a + b + c)$      **53.** $(3x + 2)(2x + 3)$
**55.** $(10x − 1)(2x + 5)$      **57.** $(4x + 5)(5x + 1)$      **59.** $(x + 2)(x^2 + 3)$      **61.** $(3x − 2)(2x^2 + 5)$      **63.** 6
**65.** $3(4x^2 + 2x + 1)$      **67.** $A = 1,000(1 + r)$; when $r = 0.12, A = 1,000(1.12) = \$1,120.$
**69.** (a) $1,000,000(1 + r)$      (b) 1,300,000 bacteria      **71.** (a) $A = 7,000(1 − r)$      (b) 5,390 names in the year 2000
**73.** $x^2 − 5x − 14$      **75.** $x^2 − x − 6$      **77.** $x^3 + 27$      **79.** $2x^3 + 9x^2 − 2x − 3$

## PROBLEM SET 6.2

**1.** $(x + 3)(x + 4)$      **3.** $(x + 1)(x + 2)$      **5.** $(a + 3)(a + 7)$      **7.** $(x − 2)(x − 5)$      **9.** $(y − 3)(y − 7)$
**11.** $(x − 4)(x + 3)$      **13.** $(y + 4)(y − 3)$      **15.** $(x + 7)(x − 2)$      **17.** $(r − 9)(r + 1)$      **19.** $(x − 6)(x + 5)$
**21.** $(a + 7)(a + 8)$      **23.** $(y + 6)(y − 7)$      **25.** $(x + 6)(x + 7)$      **27.** $2(x + 1)(x + 2)$      **29.** $3(a + 4)(a − 5)$
**31.** $100(x − 2)(x − 3)$      **33.** $100(p − 5)(p − 8)$      **35.** $x^2(x + 3)(x − 4)$      **37.** $2r(r + 5)(r − 3)$
**39.** $2y^2(y + 1)(y − 4)$      **41.** $x^3(x + 2)(x + 2)$      **43.** $3y^2(y + 1)(y − 5)$      **45.** $4x^2(x − 4)(x − 9)$
**47.** $(x + 2y)(x + 3y)$      **49.** $(x − 4y)(x − 5y)$      **51.** $(a + 4b)(a − 2b)$      **53.** $(a − 5b)(a − 5b)$
**55.** $(a + 5b)(a + 5b)$      **57.** $(x − 6a)(x + 8a)$      **59.** $(x + 4b)(x − 9b)$      **61.** $(x^2 − 3)(x^2 − 2)$
**63.** $(x − 100)(x + 20)$      **65.** $(x − \frac{1}{2})(x − \frac{1}{2})$      **67.** $(x + 0.2)(x + 0.4)$      **69.** $(x + 16)$      **71.** $4x^2 − x − 3$
**73.** $6a^2 + 13a + 2$      **75.** $6a^2 + 7a + 2$      **77.** $6a^2 + 8a + 2$      **79.** $2x^2 + 7x − 11$      **81.** $3x + 8$
**83.** $3x^2 + 4x − 5$

## PROBLEM SET 6.3

**1.** $(2x + 1)(x + 3)$    **3.** $(2a - 3)(a + 1)$    **5.** $(3x + 5)(x - 1)$    **7.** $(3y + 1)(y - 5)$    **9.** $(2x + 3)(3x + 2)$
**11.** $(2x - 3y)(2x - 3y)$    **13.** $(4y + 1)(y - 3)$    **15.** $(4x - 5)(5x - 4)$    **17.** $(10a - b)(2a + 5b)$
**19.** $(4x - 5)(5x + 1)$    **21.** $(6m - 1)(2m + 3)$    **23.** $(4x + 5)(5x + 3)$    **25.** $(3a - 4b)(4a - 3b)$
**27.** $(3x - 7y)(x + 2y)$    **29.** $(2x + 5)(7x - 3)$    **31.** $(3x - 5)(2x - 11)$    **33.** $(5t - 19)(3t - 2)$
**35.** $2(2x + 3)(x - 1)$    **37.** $2(4a - 3)(3a - 4)$    **39.** $x(5x - 4)(2x - 3)$    **41.** $x^2(3x + 2)(2x - 5)$
**43.** $2a(5a + 2)(a - 1)$    **45.** $3x(5x + 1)(x - 7)$    **47.** $5y(7y + 2)(y - 2)$    **49.** $a^2(5a + 1)(3a - 1)$
**51.** $3y(4x + 5)(2x - 3)$    **53.** $2y(2x - y)(3x - 7y)$    **55.** Both equal 25.    **57.** $4x^2 - 9$    **59.** $x^4 - 81$
**61.** $h = 2(4 - t)(1 + 8t)$    **63.** (a) $x(9 - 2x)(11 - 2x)$    (b) 9 inches by 11 inches    **65.** $x^2 - 9$
**67.** $36a^2 - 1$    **69.** $x^2 + 8x + 16$    **71.** $4x^2 + 12x + 9$

| Time $t$ (seconds) | Height $h$ (feet) |
|---|---|
| 0 | 8 |
| 1 | 54 |
| 2 | 68 |
| 3 | 50 |
| 4 | 0 |

## PROBLEM SET 6.4

**1.** $(x + 3)(x - 3)$    **3.** $(a + 6)(a - 6)$    **5.** $(x + 7)(x - 7)$    **7.** $4(a + 2)(a - 2)$    **9.** Cannot be factored
**11.** $(5x + 13)(5x - 13)$    **13.** $(3a + 4b)(3a - 4b)$    **15.** $(3 + m)(3 - m)$    **17.** $(5 + 2x)(5 - 2x)$
**19.** $2(x + 3)(x - 3)$    **21.** $32(a + 2)(a - 2)$    **23.** $2y(2x + 3)(2x - 3)$    **25.** $(a^2 + b^2)(a + b)(a - b)$
**27.** $(4m^2 + 9)(2m + 3)(2m - 3)$    **29.** $3xy(x + 5y)(x - 5y)$    **31.** $(x - 1)^2$    **33.** $(x + 1)^2$    **35.** $(a - 5)^2$
**37.** $(y + 2)^2$    **39.** $(x - 2)^2$    **41.** $(m - 6)^2$    **43.** $(2a + 3)^2$    **45.** $(7x - 1)^2$    **47.** $(3y - 5)^2$
**49.** $(x + 5y)^2$    **51.** $(3a + b)^2$    **53.** $3(a + 3)^2$    **55.** $2(x + 5y)^2$    **57.** $5x(x + 3y)^2$
**59.** $(x + 3 + y)(x + 3 - y)$    **61.** $(x + y + 3)(x + y - 3)$    **63.** 14    **65.** 25
**67.** (a) $x^2 - 16$    (b) $(x + 4)(x - 4)$    (c)

**69.** $a^2 - b^2 = (a - b)(a + b)$    **71.** $x - 2 + \dfrac{2}{x - 3}$    **73.** $3x - 2 + \dfrac{9}{2x + 3}$

## PROBLEM SET 6.5

**1.** $(x + 9)(x - 9)$    **3.** $(x + 5)(x - 3)$    **5.** $(x + 3)^2$    **7.** $(y - 5)^2$    **9.** $2ab(a^2 + 3a + 1)$
**11.** Cannot be factored    **13.** $3(2a + 5)(2a - 5)$    **15.** $(3x - 2y)^2$    **17.** $4x(x^2 + 4y^2)$    **19.** $2y(y + 5)^2$
**21.** $a^4(a^2 + 4b^2)$    **23.** $(x + 4)(y + 3)$    **25.** $(x^2 + 4)(x + 2)(x - 2)$    **27.** $(x + 2)(y - 5)$    **29.** $5(a + b)^2$
**31.** Cannot be factored    **33.** $3(x + 2y)(x + 3y)$    **35.** $(2x + 19)(x - 2)$    **37.** $100(x - 2)(x - 1)$
**39.** $(x + 8)(x - 8)$    **41.** $(x + a)(x + 3)$    **43.** $a^5(7a + 3)(7a - 3)$    **45.** Cannot be factored
**47.** $a(5a + 1)(5a + 3)$    **49.** $(x + y)(a - b)$    **51.** $3a^2b(4a + 1)(4a - 1)$    **53.** $5x^2(2x + 3)(2x - 3)$

**55.** $(3x + 41y)(x - 2y)$ **57.** $2x^3(2x - 3)(4x - 5)$ **59.** $(2x + 3)(x + a)$ **61.** $(y^2 + 1)(y + 1)(y - 1)$
**63.** $3x^2y^2(2x + 3y)^2$ **65.** 5 **67.** $-\frac{3}{2}$ **69.** $-\frac{3}{4}$ **71.** $x^{15}$ **73.** $72x^{18}$ **75.** $5.76 \times 10^4$

## PROBLEM SET 6.6

**1.** $-2, 1$ **3.** $4, 5$ **5.** $0, -1, 3$ **7.** $-\frac{2}{3}, -\frac{3}{2}$ **9.** $0, -\frac{4}{3}, \frac{4}{3}$ **11.** $0, -\frac{1}{3}, -\frac{3}{5}$ **13.** $-1, -2$ **15.** $4, 5$
**17.** $6, -4$ **19.** $2, 3$ **21.** $-3$ **23.** $4, -4$ **25.** $\frac{3}{2}, -4$ **27.** $-\frac{2}{3}$ **29.** 5 **31.** $4, -\frac{5}{2}$ **33.** $\frac{5}{3}, -4$
**35.** $\frac{7}{2}, -\frac{7}{2}$ **37.** $0, -6$ **39.** $0, 3$ **41.** $0, 4$ **43.** $0, 5$ **45.** $2, 5$ **47.** $\frac{1}{2}, -\frac{4}{3}$ **49.** $4, -\frac{5}{2}$
**51.** $8, -10$ **53.** $5, 8$ **55.** $6, 8$ **57.** $-4$ **59.** $5, 8$ **61.** $6, -8$ **63.** $0, -\frac{3}{2}, -4$ **65.** $0, 3, -\frac{5}{2}$
**67.** $0, \frac{1}{2}, -\frac{5}{2}$ **69.** $0, \frac{3}{5}, -\frac{3}{2}$ **71.** $-3, -2, 2$ **73.** $-4, -1, 4$ **75.** Bicycle, \$75; suit, \$15
**77.** House, \$2,400; lot, \$600 **79.** $\frac{1}{8}$ **81.** $x^8$ **83.** $x^{18}$ **85.** $5.6 \times 10^{-3}$

## PROBLEM SET 6.7

**1.** Two consecutive even integers are $x$ and $x + 2$; $x(x + 2) = 80$; 8, 10 and $-10, -8$ **3.** 9, 11 and $-11, -9$
**5.** $x(x + 2) = 5(x + x + 2) - 10$; 8, 10 and 0, 2 **7.** 8, 6
**9.** The numbers are $x$ and $5x + 2$; $x(5x + 2) = 24$; 2, 12 and $-\frac{12}{5}, -10$ **11.** 5, 20 and 0, 0
**13.** Let $x =$ the width; $x(x + 1) = 12$; width 3 inches, length 4 inches
**15.** Let $x =$ the base; $\frac{1}{2}(x)(2x) = 9$; base 3 inches **17.** $x^2 + (x + 2)^2 = 10^2$; 6 inches and 8 inches **19.** 12 meters
**21.** $1,400 = 400 + 700x - 100x^2$; 200 or 500 **23.** 200 videotapes or 800 videotapes
**25.** $R = xp = (1,200 - 100p)p = 3,200$; \$4 or \$8 **27.** \$7 or \$10 **29.** (a) 5 feet (b) 12 feet
**31.** (a) 25 seconds later (b)

| t (sec) | h (feet) |
|---------|----------|
| 0 | 100 |
| 5 | 1,680 |
| 10 | 2,460 |
| 15 | 2,440 |
| 20 | 1,620 |
| 25 | 0 |

**33.** $200x^{24}$ **35.** $x^7$ **37.** $8 \times 10^1$ **39.** $10ab^2$ **41.** $6x^4 + 6x^3 - 2x^2$ **43.** $9y^2 - 30y + 25$
**45.** $4a^4 - 49$

## CHAPTER 6 REVIEW

**1.** $10(x - 2)$ **2.** $x^2(4x - 9)$ **3.** $5(x - y)$ **4.** $x(7x^2 + 2)$ **5.** $4(2x + 1)$ **6.** $2(x^2 + 7x + 3)$
**7.** $8(3y^2 - 5y + 6)$ **8.** $15xy^2(2y - 3x^2)$ **9.** $7(7a^3 - 2b^3)$ **10.** $6ab(b + 3a^2b^2 - 4a)$ **11.** $(x + a)(y + b)$
**12.** $(x - 5)(y + 4)$ **13.** $(2x - 3)(y + 5)$ **14.** $(5x - 4a)(x - 2b)$ **15.** $(y + 7)(y + 2)$
**16.** $(w + 5)(w + 10)$ **17.** $(a - 6)(a - 8)$ **18.** $(r - 6)(r - 12)$ **19.** $(y + 9)(y + 11)$ **20.** $(y + 6)(y + 2)$
**21.** $(2x + 3)(x + 5)$ **22.** $(2y - 5)(2y - 1)$ **23.** $(5y + 6)(y + 1)$ **24.** $(5a - 3)(4a - 3)$
**25.** $(2r + 3t)(3r - 2t)$ **26.** $(2x - 7)(5x + 3)$ **27.** $(n + 9)(n - 9)$ **28.** $(2y + 3)(2y - 3)$
**29.** Cannot be factored. **30.** $(6y + 11x)(6y - 11x)$ **31.** $(8a + 11b)(8a - 11b)$ **32.** $(8 + 3m)(8 - 3m)$
**33.** $(y + 10)^2$ **34.** $(m - 8)^2$ **35.** $(8t + 1)^2$ **36.** $(4n - 3)^2$ **37.** $(2r - 3t)^2$ **38.** $(3m + 5n)^2$
**39.** $2(x + 4)(x + 6)$ **40.** $a(a - 3)(a - 7)$ **41.** $3m(m + 1)(m - 7)$ **42.** $5y^2(y - 2)(y + 4)$

**43.** $2(2x + 1)(2x + 3)$    **44.** $a(3a + 1)(a - 5)$    **45.** $2m(2m - 3)(5m - 1)$    **46.** $5y(2x - 3y)(3x - y)$
**47.** $4(x + 5)^2$    **48.** $x(2x + 3)^2$    **49.** $5(x + 3)(x - 3)$    **50.** $3x(2x + 3y)(2x - 3y)$    **51.** $3ab(2a + b)(a + 5b)$
**52.** $x^3(x + 1)(x - 1)$    **53.** $y^4(4y^2 + 9)$    **54.** $4x^3(x + 2y)(3x - y)$    **55.** $5a^2b(3a - b)(2a + 3b)$
**56.** $3ab^2(2a - b)(3a + 2b)$    **57.** $-2, 5$    **58.** $-\frac{5}{2}, \frac{5}{2}$    **59.** $2, -5$    **60.** $7, -7$    **61.** $0, 9$    **62.** $-\frac{3}{2}, -\frac{2}{3}$
**63.** $0, -\frac{5}{3}, \frac{2}{3}$    **64.** $10, 12$ and $-12, -10$    **65.** $10, 11$ and $-11, -10$    **66.** $5, 7,$ and $-1, 1$    **67.** $5, 15$
**68.** $6, 11$ and $-\frac{11}{2}, -12$    **69.** 2 inches

# CHAPTER 6 TEST

**1.** $5(x - 2)$    **2.** $9xy(2x - 1 - 4y)$    **3.** $(x + 2a)(x - 3b)$    **4.** $(x - 7)(y + 4)$    **5.** $(x - 2)(x - 3)$
**6.** $(x - 3)(x + 2)$    **7.** $(a + 4)(a - 4)$    **8.** Cannot be factored    **9.** $(x^2 + 9)(x + 3)(x - 3)$
**10.** $3(3x - 5y)(3x + 5y)$    **11.** $(x + 5)(x + 3)(x - 3)$    **12.** $(x - b)(x + 5)$    **13.** $2(2a + 1)(a + 5)$
**14.** $3(m - 3)(m + 2)$    **15.** $(2y - 1)(3y + 5)$    **16.** $2x(2x + 1)(3x - 5)$    **17.** $-3, -4$    **18.** $2$    **19.** $-6, 6$
**20.** $5, -4$    **21.** $5, 6$    **22.** $0, 4, -4$    **23.** $3, -\frac{5}{2}$    **24.** $0, 1, -\frac{1}{3}$    **25.** $4, 16$    **26.** $3, 5,$ and $-3, -1$
**27.** 3 feet, 14 feet    **28.** 5 meters and 12 meters    **29.** 200 or 300 items    **30.** \$3 or \$6

# CHAPTERS 1–6 CUMULATIVE REVIEW

**1.** $-9$    **2.** $-6$    **3.** $29$    **4.** $9$    **5.** $9$    **6.** $-64$    **7.** $2$    **8.** $5a - 5$    **9.** $-8a - 3$    **10.** $x^{40}$
**11.** $1$    **12.** $x^4$    **13.** $4x^4y^6$    **14.** $15x^2 + 14x - 8$    **15.** $-6$    **16.** $10$    **17.** $-21$    **18.** $-2$
**19.** $0, 7, \frac{7}{2}$    **20.** $-\frac{9}{4}, \frac{9}{4}$    **21.** $x < 4$    **22.** $x \le 2$ or $x \ge 7$    **23.**

**24.**

**25.**

**26.** $x$-intercept $= 3$; $y$-intercept $= -10$    **27.** $\frac{7}{9}$    **28.** $y = \frac{2}{3}x + 8$    **29.** $y = -\frac{2}{5}x - \frac{2}{3}$

**30.** Slope $= \frac{3}{4}$, $y$-intercept $= 4$ **31.** **32.** (2, 5) **33.** (2, −4)

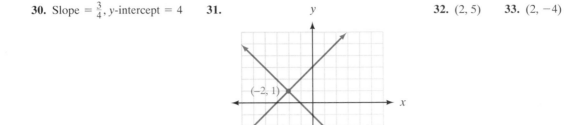

**34.** (1, 2) **35.** (1,500, 3,500) **36.** (−2, 1) **37.** $(n − 9)(n + 4)$ **38.** $(2x + 5y)(7x − 2y)$
**39.** $(4 + a)(4 − a)$ **40.** $(7x − 1)^2$ **41.** $5y(3x − y)^2$ **42.** $3x(3x + y)(2x − y)$ **43.** $(y + 5)(3x − 2)$
**44.** Opposite is 2; reciprocal is $−\frac{1}{2}$. **45.** Commutative **46.** $1.05 \times 10^9$ **47.** $−4x^3y^2 + 2xy − 3$
**48.** $x^2 + 3x + 9$ **49.** 34 inches, 38 inches **50.** 6 burgers, 8 fries

# CHAPTER 7

## PROBLEM SET 7.1

**1.** $\frac{1}{x − 2}, x \neq 2$ **3.** $\frac{1}{a + 3}, a \neq −3, 3$ **5.** $\frac{1}{x − 5}, x \neq −5, 5$ **7.** $\frac{(x + 2)(x − 2)}{2}$ **9.** $\frac{2(x − 5)}{3(x − 2)}, x \neq 2$

**11.** 2 **13.** $\frac{5(x − 1)}{4}$ **15.** $\frac{1}{x − 3}$ **17.** $\frac{1}{x + 3}$ **19.** $\frac{1}{a − 5}$ **21.** $\frac{1}{3x + 2}$ **23.** $\frac{x + 5}{x + 2}$

**25.** $\frac{2m(m + 2)}{m − 2}$ **27.** $\frac{x − 1}{x − 4}$ **29.** $\frac{2(2x − 3)}{2x + 3}$ **31.** $\frac{2x − 3}{2x + 3}$ **33.** $\frac{1}{(x^2 + 9)(x − 3)}$ **35.** $\frac{3x − 5}{(x^2 + 4)(x − 2)}$

**37.** $\frac{2x(7x + 6)}{2x + 1}$ **39.** $\frac{x + 2}{x + 5}$ **41.** $\frac{x + a}{x + b}$ **43.** $\frac{x + a}{x + 3}$ **45.** $\frac{4}{3}$ **47.** $\frac{4}{5}$ **49.** $\frac{8}{1}$

**51.**

| Checks Written $x$ | Total Cost $2.00 + 0.15x$ | Cost per Check $\frac{2.00 + 0.15x}{x}$ |
|---|---|---|
| 0 | $2.00 | Undefined |
| 5 | $2.75 | $0.55 |
| 10 | $3.50 | $0.35 |
| 15 | $4.25 | $0.28 |
| 20 | $5.00 | $0.25 |

**53.** 40.7 miles/hour **55.** 39.25 feet/minute

**57.** 12.95 feet/second
**59.** Level ground: 10 minutes/mile or 0.1 mile/minute; downhill: $\frac{20}{3}$ minutes/mile or $\frac{3}{20}$ mile/minute

**61.** 48 miles/gallon    **63.** $5 + 4 = 9$    **65.**

| $x$ | $\dfrac{x - 3}{3 - x}$ |
|---|---|
| $-2$ | $-1$ |
| $-1$ | $-1$ |
| $0$ | $-1$ |
| $1$ | $-1$ |
| $2$ | $-1$ |

The entries are all $-1$ because the numerator and denominator are opposites.

Or, $\dfrac{x - 3}{3 - x} = \dfrac{-1(x - 3)}{3 - x} = -1.$

**67.**

| $x$ | $\dfrac{x - 5}{x^2 - 25}$ | $\dfrac{1}{x + 5}$ |
|---|---|---|
| $0$ | $\frac{1}{5}$ | $\frac{1}{5}$ |
| $2$ | $\frac{1}{3}$ | $\frac{1}{3}$ |
| $-2$ | $\frac{1}{7}$ | $\frac{1}{7}$ |
| $5$ | Undefined | Undefined |
| $-5$ | Undefined | $\frac{1}{10}$ |

**69.** 0    **71.** $16a^2b^2$    **73.** $19x^4 + 21x^2 - 42$

**75.** $7a^4b^4 + 9b^3 - 11a^3$

## PROBLEM SET 7.2

**1.** 2    **3.** $\dfrac{x}{2}$    **5.** $\dfrac{3}{2}$    **7.** $\dfrac{1}{2(x - 3)}$    **9.** $\dfrac{4a(a + 5)}{7(a + 4)}$    **11.** $\dfrac{y - 2}{4}$    **13.** $\dfrac{2(x + 4)}{x - 2}$    **15.** $\dfrac{x + 3}{(x - 3)(x + 1)}$

**17.** 1    **19.** $\dfrac{y - 5}{(y + 2)(y - 2)}$    **21.** $\dfrac{x + 5}{x - 5}$    **23.** 1    **25.** $\dfrac{a + 3}{a + 4}$    **27.** $\dfrac{2y - 3}{y - 6}$    **29.** $\dfrac{x - 1}{x - 2}$    **31.** $\dfrac{3}{2}$

**33.** $2(x - 3)$    **35.** $2a$    **37.** $(x + 2)(x + 1)$    **39.** $-2x(x - 5)$    **41.** $\dfrac{2(x + 5)}{x(y + 5)}$    **43.** $\dfrac{2x}{x - y}$    **45.** $\dfrac{1}{x - 2}$

**47.** $\dfrac{1}{5}$    **49.** $\dfrac{1}{100}$    **51.** 2.7 miles    **53.** 742 miles/hour    **55.** 0.45 mile/hour    **57.** 8.8 miles/hour

**59.** Level ground: 6 miles/hour; downhill: 9 mile/hour    **61.** 3    **63.** $\dfrac{11}{4}$    **65.** $\dfrac{11}{35}$    **67.** $9x^2$    **69.** $6ab^2$

## PROBLEM SET 7.3

**1.** $\dfrac{7}{x}$    **3.** $\dfrac{4}{a}$    **5.** 1    **7.** $y + 1$    **9.** $x + 2$    **11.** $x - 2$    **13.** $\dfrac{6}{x + 6}$    **15.** $\dfrac{(y + 2)(y - 2)}{2y}$

**17.** $\dfrac{3 + 2a}{6}$    **19.** $\dfrac{7x + 3}{4(x + 1)}$    **21.** $\dfrac{1}{5}$    **23.** $\dfrac{1}{3}$    **25.** $\dfrac{3}{x - 2}$    **27.** $\dfrac{4}{a + 3}$    **29.** $\dfrac{2x - 20}{(x + 5)(x - 5)}$

**31.** $\dfrac{x + 2}{x + 3}$    **33.** $\dfrac{a + 1}{a + 2}$    **35.** $\dfrac{1}{(x + 3)(x + 4)}$    **37.** $\dfrac{y}{(y + 5)(y + 4)}$    **39.** $\dfrac{3(x - 1)}{(x + 4)(x + 1)}$    **41.** $\dfrac{1}{3}$

**43.**

| Number $x$ | Reciprocal $\frac{1}{x}$ | Sum $1 + \frac{1}{x}$ | Sum $\frac{x+1}{x}$ |
|---|---|---|---|
| 1 | 1 | 2 | 2 |
| 2 | $\frac{1}{2}$ | $\frac{3}{2}$ | $\frac{3}{2}$ |
| 3 | $\frac{1}{3}$ | $\frac{4}{3}$ | $\frac{4}{3}$ |
| 4 | $\frac{1}{4}$ | $\frac{5}{4}$ | $\frac{5}{4}$ |

**45.**

| $x$ | $x + \frac{4}{x}$ | $\frac{x^2 + 4}{x}$ | $x + 4$ |
|---|---|---|---|
| 1 | 5 | 5 | 5 |
| 2 | 4 | 4 | 6 |
| 3 | $\frac{13}{3}$ | $\frac{13}{3}$ | 7 |
| 4 | 5 | 5 | 8 |

**47.** $\dfrac{x+3}{x+2}$ **49.** $\dfrac{x+2}{x+3}$ **51.** $x + \dfrac{2}{x} = \dfrac{x^2+2}{x}$ **53.** $\dfrac{1}{x} + \dfrac{1}{2x} = \dfrac{3}{2x}$ **55.** 3 **57.** $-2$ **59.** $\frac{1}{5}$
**61.** $-2, -3$ **63.** $3, -2$ **65.** $0, 5$

## PROBLEM SET 7.4

**1.** $-3$ **3.** 20 **5.** $-1$ **7.** 5 **9.** $-2$ **11.** 4 **13.** $3, 5$ **15.** $-8, 1$ **17.** 2 **19.** 1 **21.** 8
**23.** 5 **25.** Possible solution 2, which does not check; $\varnothing$ **27.** 3
**29.** Possible solution $-3$, which does not check; $\varnothing$ **31.** 0 **33.** Possible solutions 2 and 3, but only 2 checks; 2
**35.** $-4$ **37.** $-1$ **39.** $-6, -7$ **41.** Possible solutions $-3$ and $-1$, but only $-1$ checks; $-1$ **43.** $1, 4$
**45.** 7 **47.** Length 13 inches, width 4 inches **49.** $6, 8$ or $-8, -6$ **51.** 6 inches, 8 inches

## PROBLEM SET 7.5

**1.** $\frac{1}{4}, \frac{3}{4}$ **3.** $x + \dfrac{1}{x} = \dfrac{13}{6}; \frac{2}{3}$ and $\frac{3}{2}$ **5.** $-2$ **7.** $\dfrac{1}{x} + \dfrac{1}{x+2} = \dfrac{5}{12}; 4$ and 6
**9.** Let $x =$ the speed of the boat in still water.

| | $d$ | $r$ | $t$ |
|---|---|---|---|
| Upstream | 26 | $x - 3$ | $\dfrac{26}{x-3}$ |
| Downstream | 38 | $x + 3$ | $\dfrac{38}{x+3}$ |

The equation is $\dfrac{26}{x-3} = \dfrac{38}{x+3}; x = 16$ miles per hour.
**11.** 300 miles/hour **13.** 170 miles/hour, 190 miles/hour **15.** 9 miles/hour **17.** 8 miles/hour
**19.** $\dfrac{1}{12} - \dfrac{1}{15} = \dfrac{1}{x}; 60$ hours **21.** $\frac{60}{11}$ minutes **23.** 12 minutes

**25.**

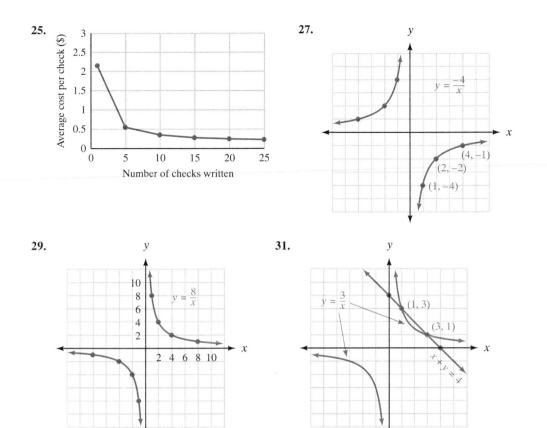

**27.**

**29.**

**31.**

**33.** $5ab(3a^2b^2 - 4a - 7b)$    **35.** $(x - 6)(x + 2)$    **37.** $(x^2 + 4)(x + 2)(x - 2)$    **39.** $5x(x - 6)(x + 1)$

**41.** $0, 6$    **43.** $-10, 8$    **45.** 9 inches and 12 inches

## PROBLEM SET 7.6

**1.** $6$    **3.** $\frac{1}{6}$    **5.** $xy^2$    **7.** $\frac{xy}{2}$    **9.** $\frac{y}{x}$    **11.** $\frac{a + 1}{a - 1}$    **13.** $\frac{x + 1}{2(x - 3)}$    **15.** $a - 3$    **17.** $\frac{y + 3}{y + 2}$

**19.** $x + y$    **21.** $\frac{a + 1}{a}$    **23.** $\frac{1}{x}$    **25.** $\frac{2a + 3}{3a + 4}$    **27.** $\frac{x - 1}{x + 2}$    **29.** $\frac{x + 4}{x + 1}$    **31.** $\frac{7}{3}, \frac{17}{7}, \frac{41}{17}$

**33.**

| Number $x$ | Reciprocal $\dfrac{1}{x}$ | Quotient $\dfrac{x}{\frac{1}{x}}$ | Square $x^2$ |
|:---:|:---:|:---:|:---:|
| 1 | 1 | 1 | 1 |
| 2 | $\frac{1}{2}$ | 4 | 4 |
| 3 | $\frac{1}{3}$ | 9 | 9 |
| 4 | $\frac{1}{4}$ | 16 | 16 |

**35.**

| Number | Reciprocal | Sum | Quotient |
|---|---|---|---|
| $x$ | $\dfrac{1}{x}$ | $1 + \dfrac{1}{x}$ | $\dfrac{1 + \dfrac{1}{x}}{\dfrac{1}{x}}$ |
| 1 | 1 | 2 | 2 |
| 2 | $\dfrac{1}{2}$ | $\dfrac{3}{2}$ | 3 |
| 3 | $\dfrac{1}{3}$ | $\dfrac{4}{3}$ | 4 |
| 4 | $\dfrac{1}{4}$ | $\dfrac{5}{4}$ | 5 |

**37.** $x < 1$     **39.** $x \geq -7$     **41.** $x < 6$

**43.** $x \leq 2$

## Problem Set 7.7

**1.** 1   **3.** 10   **5.** 40   **7.** $\frac{5}{4}$   **9.** $\frac{7}{3}$   **11.** $3, -4$   **13.** $4, -4$   **15.** $2, -4$   **17.** $6, -1$   **19.** 15 hits

**21.** 21 milliliters   **23.** 45.5 grams   **25.** 14.7 inches   **27.** 343 miles   **29.** $\dfrac{x + 2}{x + 3}$   **31.** $\dfrac{2(x + 5)}{x - 4}$

**33.** $\dfrac{1}{x - 4}$

## Problem Set 7.8

**1.** 8   **3.** 130   **5.** 5   **7.** $-3$   **9.** 3   **11.** 243   **13.** 2   **15.** $\frac{1}{2}$   **17.** 1   **19.** 5   **21.** 25

**23.** 9   **25.** 84 pounds   **27.** $\frac{735}{2}$, or 367.5   **29.** \$235.50   **31.** 96 pounds   **33.** 12 amps   **35.** $(2, -1)$

**37.** $(4, 3)$   **39.** $(1, 1)$   **41.** $(5, 2)$

## Chapter 7 Review

**1.** $\dfrac{1}{2(x - 2)}, x \neq 2$   **2.** $\dfrac{1}{a - 6}, a \neq -6, 6$   **3.** $\dfrac{2x - 1}{x + 3}, x \neq -3$   **4.** $\dfrac{1}{x + 4}, x \neq 4$   **5.** $\dfrac{3x - 2}{2x - 3}, x \neq \frac{3}{2}, -6, 0$

**6.** $\dfrac{1}{(x - 2)(x^2 + 4)}, x \neq \pm 2$   **7.** $x - 2, x \neq -7$   **8.** $a + 8, a \neq -8$   **9.** $\dfrac{y + b}{y + 5}, x \neq -a, y \neq -5$   **10.** $\dfrac{x}{2}$

**11.** $\dfrac{x + 2}{x - 4}$   **12.** $(a - 6)^2$   **13.** $\dfrac{x - 1}{x + 2}$   **14.** 1   **15.** $x - 9$   **16.** $\dfrac{13}{a + 8}$   **17.** $\dfrac{x^2 + 5x + 45}{x(x + 9)}$

**18.** $\dfrac{4x + 5}{4(x + 5)}$   **19.** $\dfrac{1}{(x + 6)(x + 2)}$   **20.** $\dfrac{a - 6}{(a + 5)(a + 3)}$   **21.** 4   **22.** 9   **23.** 1, 6

**24.** Possible solution 2, which does not check; $\varnothing$   **25.** 5   **26.** 3 and $\frac{7}{3}$   **27.** 15 miles/hour

**28.** 84 hours   **29.** $\frac{1}{2}$   **30.** $\dfrac{y + 3}{y + 7}$   **31.** $\dfrac{4a - 7}{a - 1}$   **32.** $\frac{2}{5}$   **33.** $\frac{2}{9}$   **34.** 12   **35.** $-6, 6$

**36.** $-6, 8$   **37.** $-35$   **38.** $\frac{1}{2}$

## Chapter 7 Test

**1.** $\dfrac{x + 4}{x - 4}$   **2.** $\dfrac{2}{a + 2}$   **3.** $\dfrac{y + 7}{x + a}$   **4.** 3   **5.** $(x - 7)(x - 1)$   **6.** $\dfrac{x + 4}{3x - 4}$   **7.** $(x - 3)(x + 2)$

**8.** $\dfrac{-3}{x - 2}$   **9.** $\dfrac{2x + 3}{(x + 3)(x - 3)}$   **10.** $\dfrac{3x}{(x + 1)(x - 2)}$   **11.** $\frac{11}{5}$   **12.** 1   **13.** 6   **14.** 15 miles/hour

**15.** 60 hours    **16.** $\frac{1}{2}$ and $\frac{1}{3}$    **17.** 132    **18.** $\dfrac{x+1}{x-1}$    **19.** $\dfrac{x+4}{x+2}$    **20.** 100    **21.** 2

# CHAPTERS 1–7 CUMULATIVE REVIEW

**1.** $-3$    **2.** $-6$    **3.** $-4$    **4.** $-3$    **5.** $-4x-4$    **6.** $4-x$    **7.** $\frac{1}{81}$    **8.** $\dfrac{1}{x^4}$    **9.** 6    **10.** 1

**11.** $-2a^3 - 10a^2 - 5a + 13$    **12.** $a^4 - 49$    **13.** $x - 7$    **14.** $\dfrac{5x+3}{5x+15}$    **15.** $\dfrac{x-2}{4(x+2)}$    **16.** $\frac{11}{4}$

**17.** $\frac{19}{12}$    **18.** $-2$    **19.** $-\frac{3}{7}, \frac{3}{7}$    **20.** $0, \frac{7}{2}, 2$    **21.** 6    **22.** No solution    **23.** $-3, 6$    **24.** $(3, -1)$

**25.** $(\frac{1}{3}, -\frac{1}{2})$    **26.** $(17, 5)$    **27.** $(6, -12)$

**28.**    **29.**

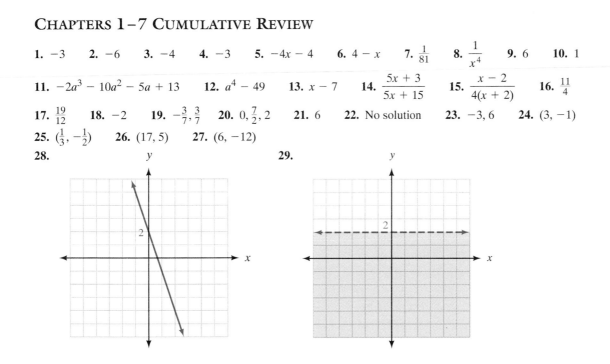

**30.** $a \geq 6$    **31.** $(x+a)(y+5)$    **32.** $(a+7)(a-5)$    **33.** $(4y-3)(5y-3)$    **34.** $(2r-3t)(2r+3t)$

**35.** $(4x+9y)^2$    **36.**    **37.** $(1, -3)$

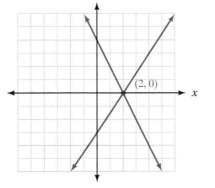

**38.** $x$-intercept $= 5$; $y$-intercept $= 2$    **39.** Slope $= 3$; $y$-intercept $= -4$

**40.** $y = -5x - 1$    **41.** $y = -\frac{2}{5}x - \frac{9}{5}$    **42.** $-14$    **43.** Associative property    **44.** $-3, 0, \frac{3}{4}, 1.5$

**45.** $\dfrac{x+1}{4(x+2)}$    **46.** $x - 7$    **47.** $\dfrac{2x+3}{3x+2}$    **48.** $l = \dfrac{P-2w}{2}$    **49.** 29, 11    **50.** \$7,500 at 6%, \$10,500 at 8%

# CHAPTER 8

## PROBLEM SET 8.1

**1.** 3 **3.** −3 **5.** Not a real number **7.** −12 **9.** 25 **11.** Not a real number **13.** −8 **15.** −10
**17.** 35 **19.** 1 **21.** −2 **23.** −5 **25.** −1 **27.** −3 **29.** −2 **31.** $x$ **33.** $3x$ **35.** $xy$
**37.** $a + b$ **39.** $7xy$ **41.** $x$ **43.** $2x$ **45.** $x^2$ **47.** $6a^3$ **49.** $5a^4b^2$ **51.** $x^2$ **53.** $3a^4$ **55.** $x^2$
**57.** 7 **59.** 5 **61.** 17 **63.** 13 **65.** 6, −1 **67.** 3, −1 **69.** $x + 3$

**71.**

| Length $L$ (feet) | Time $T$ (seconds) |
|---|---|
| 1 | 1.11 |
| 2 | 1.57 |
| 3 | 1.92 |
| 4 | 2.22 |
| 5 | 2.48 |
| 6 | 2.72 |

**73.** $r = \dfrac{\sqrt{65} - \sqrt{50}}{\sqrt{50}} \approx 0.14 = 14\%$ **75.** Approximately 0.095, or 9.5%

**77.** 5 **79.** $\sqrt{125} \approx 11.2$ **81.** 30 feet **83.** 13 feet **85.** $c = 5$ centimeters, $d = 13$ centimeters **87.** $x - 4$
**89.** $\dfrac{2}{a - 2}$ **91.** $\dfrac{2x + 1}{x}$ **93.** $\dfrac{x + 2}{x + a}$

## PROBLEM SET 8.2

**1.** $2\sqrt{2}$ **3.** $2\sqrt{3}$ **5.** $2\sqrt[3]{3}$ **7.** $5x\sqrt{2}$ **9.** $3ab\sqrt{5}$ **11.** $3x\sqrt[3]{2}$ **13.** $4x^2\sqrt{2}$ **15.** $20\sqrt{5}$ **17.** $x\sqrt{7x}$
**19.** $2x^2\sqrt[3]{x}$ **21.** $6a^2\sqrt[3]{a^2}$ **23.** $4a\sqrt{5a}$ **25.** $15y\sqrt{2x}$ **27.** $14x\sqrt{3y}$ **29.** $\dfrac{4}{5}$ **31.** $\dfrac{2}{3}$ **33.** $\dfrac{2}{3}$ **35.** $\dfrac{2}{3}$
**37.** $2x$ **39.** $3ab$ **41.** $\dfrac{3x}{2y}$ **43.** $\dfrac{5\sqrt{2}}{3}$ **45.** $\sqrt{3}$ **47.** $\dfrac{8\sqrt{2}}{7}$ **49.** $\dfrac{12\sqrt{2x}}{5}$ **51.** $\dfrac{3a\sqrt{6}}{5}$ **53.** $\dfrac{15\sqrt{2}}{2}$
**55.** $\dfrac{14y\sqrt{7}}{3}$ **57.** $5ab\sqrt{2}$ **59.** $6x\sqrt{2y}$ **61.**

| $x$ | $\sqrt{x}$ | $2\sqrt{x}$ | $\sqrt{4x}$ |
|---|---|---|---|
| 1 | 1 | 2 | 2 |
| 2 | 1.414 | 2.828 | 2.828 |
| 3 | 1.732 | 3.464 | 3.464 |
| 4 | 2 | 4 | 4 |

**63.**

| $x$ | $\sqrt{x}$ | $3\sqrt{x}$ | $\sqrt{9x}$ |
|---|---|---|---|
| 1 | 1 | 3 | 3 |
| 2 | 1.414 | 4.243 | 4.243 |
| 3 | 1.732 | 5.196 | 5.196 |
| 4 | 2 | 6 | 6 |

**65.** $\dfrac{5}{4}$ seconds **67.** $\dfrac{2(x + 5)}{x(x + 1)}$ **69.** $\dfrac{x + 1}{x - 1}$

**71.** $(x + 6)(x + 3)$

## PROBLEM SET 8.3

**1.** $\dfrac{\sqrt{2}}{2}$    **3.** $\dfrac{\sqrt{3}}{3}$    **5.** $\dfrac{\sqrt{10}}{5}$    **7.** $\dfrac{\sqrt{6}}{2}$    **9.** $\dfrac{2\sqrt{15}}{3}$    **11.** $\dfrac{\sqrt{30}}{2}$    **13.** 2    **15.** $\sqrt{7}$    **17.** $\sqrt{5}$    **19.** $2\sqrt{5}$

**21.** $2\sqrt{3}$    **23.** $\dfrac{\sqrt{7}}{2}$    **25.** $xy\sqrt{2}$    **27.** $\dfrac{x\sqrt{15y}}{3}$    **29.** $\dfrac{4a^2\sqrt{5}}{5}$    **31.** $\dfrac{6a^2\sqrt{10a}}{5}$    **33.** $\dfrac{2xy\sqrt{15y}}{3}$

**35.** $\dfrac{4xy\sqrt{5y}}{2}$    **37.** $\dfrac{18ab\sqrt{6b}}{5}$    **39.** $18x\sqrt{x}$    **41.** $\dfrac{\sqrt[3]{4}}{2}$    **43.** $\dfrac{\sqrt[3]{3}}{3}$    **45.** $\dfrac{\sqrt[3]{12}}{2}$

**47.**

| $x$ | $\sqrt{x}$ | $\dfrac{1}{\sqrt{x}}$ | $\dfrac{\sqrt{x}}{x}$ |
|---|---|---|---|
| 1 | 1 | 1 | 1 |
| 2 | 1.414 | 0.707 | 0.707 |
| 3 | 1.732 | 0.577 | 0.577 |
| 4 | 2 | 0.5 | 0.5 |
| 5 | 2.236 | 0.447 | 0.447 |
| 6 | 2.449 | 0.408 | 0.408 |

**49.**

| $x$ | $\sqrt{x^2}$ | $\sqrt{x^3}$ | $x\sqrt{x}$ |
|---|---|---|---|
| 1 | 1 | 1 | 1 |
| 2 | 2 | 2.828 | 2.828 |
| 3 | 3 | 5.196 | 5.196 |
| 4 | 4 | 8 | 8 |
| 5 | 5 | 11.180 | 11.180 |
| 6 | 6 | 14.697 | 14.697 |

**51.** 6 miles

**53.**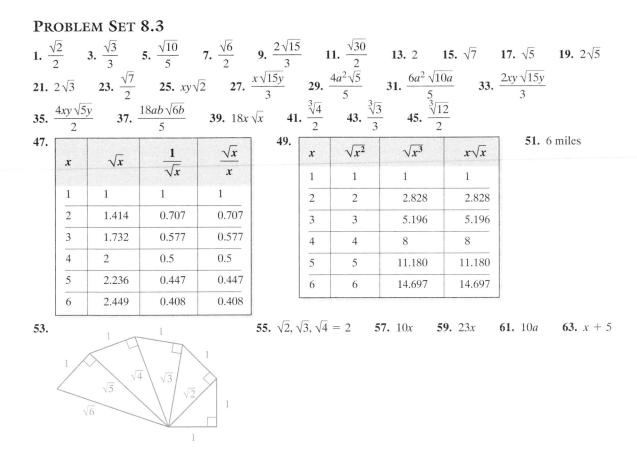

**55.** $\sqrt{2}, \sqrt{3}, \sqrt{4} = 2$    **57.** $10x$    **59.** $23x$    **61.** $10a$    **63.** $x + 5$

**65.** $\dfrac{5a + 6}{15}$    **67.** $\dfrac{1}{(a + 2)(a + 3)}$

## PROBLEM SET 8.4

**1.** $7\sqrt{2}$    **3.** $2\sqrt{5}$    **5.** $7\sqrt{3}$    **7.** $\dfrac{11}{56}\sqrt{5}$    **9.** $13\sqrt{13}$    **11.** $6\sqrt{10}$    **13.** $6\sqrt{5}$    **15.** $4\sqrt{2}$    **17.** 0

**19.** $-30\sqrt{3}$    **21.** $-6\sqrt{3}$    **23.** 0    **25.** $\dfrac{3}{2}\sqrt{2} + \dfrac{3}{2}\sqrt{3}$    **27.** 0    **29.** $\dfrac{23}{4}\sqrt{2}$    **31.** $-3\sqrt{7}$    **33.** $35\sqrt{3}$

**35.** $-6\sqrt{2} - 6\sqrt{3}$    **37.** $2x\sqrt{x}$    **39.** $4a\sqrt{3}$    **41.** $15x\sqrt{2x}$    **43.** $13x\sqrt{3xy}$    **45.** $-b\sqrt{5a}$

**47.** $27x\sqrt{2x} - 8x\sqrt{3x}$    **49.** $23x\sqrt{2y}$    **51.** $\dfrac{4 - \sqrt{6}}{3}$    **53.** $3 + \sqrt{2}$    **55.** $\dfrac{-2 + \sqrt{2}}{2}$

**57.**

| $x$ | $\sqrt{x^2 + 9}$ | $x + 3$ |
|---|---|---|
| 1 | 3.162 | 4 |
| 2 | 3.606 | 5 |
| 3 | 4.243 | 6 |
| 4 | 5 | 7 |
| 5 | 5.831 | 8 |
| 6 | 6.708 | 9 |

**59.**

| $x$ | $\sqrt{x + 3}$ | $\sqrt{x} + \sqrt{3}$ |
|---|---|---|
| 1 | 2 | 2.732 |
| 2 | 2.236 | 3.146 |
| 3 | 2.450 | 3.464 |
| 4 | 2.646 | 3.732 |
| 5 | 2.828 | 3.968 |
| 6 | 3 | 4.182 |

**61.** $9\sqrt{3}$    **63.** $9x^2 + 6xy + y^2$

**65.** $9x^2 - 16y^2$    **67.** 9    **69.** 2, 3    **71.** Possible solutions are 2 and 4; only 2 checks.

## PROBLEM SET 8.5

**1.** $\sqrt{6}$    **3.** $2\sqrt{3}$    **5.** $10\sqrt{21}$    **7.** $24\sqrt{2}$    **9.** $\sqrt{6} - \sqrt{2}$    **11.** $\sqrt{6} + 2$    **13.** $2\sqrt{6} + 3$    **15.** $6 - \sqrt{15}$
**17.** $2\sqrt{6} + 2\sqrt{15}$    **19.** $3 + 2\sqrt{2}$    **21.** $x + 6\sqrt{x} + 9$    **23.** $27 - 10\sqrt{2}$    **25.** $a - \sqrt{a} + \frac{1}{4}$    **27.** $16 + 6\sqrt{7}$
**29.** $11 + 5\sqrt{5}$    **31.** $-28 + \sqrt{2}$    **33.** $\sqrt{6} + \frac{1}{3}\sqrt{3} + \frac{1}{2}\sqrt{2} + \frac{1}{6}$    **35.** $x - 36$    **37.** $a + \sqrt{a} + \frac{2}{9}$    **39.** 1
**41.** $30 + \sqrt{7}$    **43.** $6x + 16\sqrt{x} + 8$    **45.** $49a - 4b$    **47.** $\dfrac{\sqrt{15} + \sqrt{6}}{3}$    **49.** $\dfrac{5 - \sqrt{10}}{3}$    **51.** $6 + 2\sqrt{5}$
**53.** $5 + 2\sqrt{6}$    **55.** $\dfrac{5 - \sqrt{21}}{2}$    **57.** $\dfrac{x + 4\sqrt{x} + 4}{x - 4}$    **59.** $\dfrac{5 - \sqrt{15} - \sqrt{10} + \sqrt{6}}{2}$    **61.** $6\sqrt{5}$    **63.** $52 + 14\sqrt{3}$
**65.** $-6, 1$    **67.** 0, 3    **69.** $-9, 9$    **71.** $3, -5$    **73.** 500 miles

## PROBLEM SET 8.6

**1.** 3    **3.** 44    **5.** $\varnothing$    **7.** $\varnothing$    **9.** 8    **11.** 4    **13.** $\varnothing$    **15.** $-\frac{2}{3}$    **17.** 25    **19.** 4    **21.** 7    **23.** 8
**25.** $\varnothing$    **27.** Possible solutions are 1 and 6; only 6 checks.    **29.** $-1, -2$    **31.** $-4$
**33.** Possible solutions are 5 and 8; only 8 checks.

**35.**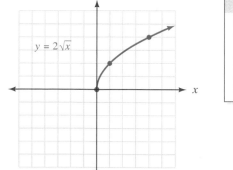

**37.**

| x | y |
|---|---|
| 0 | 0 |
| 1 | 1 |
| 2 | 1.4 |
| 3 | 1.7 |
| 4 | 2 |

**39.**

| x | y |
|---|---|
| 0 | 0 |
| 1 | 2 |
| 4 | 4 |
| 9 | 6 |

$y = \sqrt{x}$

$y = 2\sqrt{x}$

**41.**

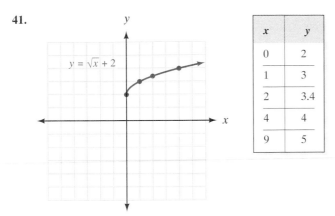

| $x$ | $y$ |
|---|---|
| 0 | 2 |
| 1 | 3 |
| 2 | 3.4 |
| 4 | 4 |
| 9 | 5 |

**43.** $x + 2 = \sqrt{8x}; x = 2$    **45.** $x - 3 = 2\sqrt{x}$; possible solutions are 1 and 9; only 9 checks.    **47.** $\dfrac{392}{121} \approx 3.2$ feet

**49.** $\dfrac{x + 2}{x + 3}$    **51.** $\dfrac{2(x + 5)}{x - 4}$    **53.** $\dfrac{1}{x - 4}$    **55** $\dfrac{x + 5}{x - 3}$    **57.** $-2$    **59.** 24 hours    **61.** 24

## CHAPTER 8 REVIEW

**1.** 5    **2.** 13    **3.** $-1$    **4.** 5    **5.** $10xy^2$    **6.** $2a$    **7.** $2\sqrt{6}$    **8.** $2x\sqrt{15}$    **9.** $3xy^2\sqrt{10x}$    **10.** $-4\sqrt{2}$
**11.** $6x\sqrt{5xy}$    **12.** $\dfrac{\sqrt{3}}{7}$    **13.** $\dfrac{2\sqrt{2}}{9}$    **14.** $\dfrac{7}{8}$    **15.** $\dfrac{7ab}{4}$    **16.** $\dfrac{4\sqrt{5}}{7}$    **17.** $\dfrac{2a\sqrt{10}}{11}$    **18.** $\dfrac{10\sqrt{21}}{7}$
**19.** $\dfrac{6ab\sqrt{30}}{5}$    **20.** $\dfrac{-5xy\sqrt{5x}}{6}$    **21.** $\dfrac{2\sqrt{7}}{7}$    **22.** $\dfrac{4\sqrt{10}}{5}$    **23.** $\dfrac{\sqrt{15}}{12}$    **24.** $-6\sqrt{3}$    **25.** $\dfrac{4b\sqrt{6a}}{3}$    **26.** $\dfrac{\sqrt[3]{6}}{2}$
**27.** $\dfrac{-3\sqrt{3} - 12}{13}$    **28.** $3 - \sqrt{7}$    **29.** $\sqrt{5} + \sqrt{2}$    **30.** $\dfrac{-\sqrt{15} - 5}{2}$    **31.** $\dfrac{7 - 2\sqrt{10}}{3}$    **32.** $\dfrac{x + 6\sqrt{x} + 9}{x - 9}$
**33.** $-4\sqrt{5}$    **34.** $-11\sqrt{3}$    **35.** $-22\sqrt{5}$    **36.** $5x\sqrt{2}$    **37.** $-ab\sqrt{10a}$    **38.** $3 + 3\sqrt{3}$    **39.** $4\sqrt{6} + 4\sqrt{10}$
**40.** $x - 49$    **41.** $2\sqrt{5} - 2$    **42.** $x + 10\sqrt{x} + 25$    **43.** 12    **44.** 7    **45.** 16    **46.** No solution
**47.** No solution    **48.** 0, 5    **49.** $\sqrt{3}$    **50.** $8\sqrt{2}$

## CHAPTER 8 TEST

**1.** 4    **2.** $-6$    **3.** 7, $-7$    **4.** 3    **5.** $-2$    **6.** $-3$    **7.** $5\sqrt{3}$    **8.** $4\sqrt{2}$    **9.** $\dfrac{\sqrt{6}}{3}$    **10.** $\dfrac{\sqrt[3]{2}}{2}$

**11.** $15x\sqrt{2}$    **12.** $\dfrac{2xy\sqrt{15y}}{5}$    **13.** $4\sqrt{3}$    **14.** $11x\sqrt{2}$    **15.** $\sqrt{15} - 2\sqrt{3}$    **16.** $-51 - \sqrt{5}$    **17.** $x - 36$

**18.** $8 - 2\sqrt{15}$    **19.** $\dfrac{5 - \sqrt{21}}{2}$    **20.** $\dfrac{x - 5\sqrt{x}}{x - 25}$    **21.** 12    **22.** $\varnothing$
**23.** Possible solutions are 2 and 6; only 6 checks.    **24.** Possible solutions are 1 and 16; only 16 checks.    **25.** $\sqrt{6}$

## CHAPTERS 1–8 CUMULATIVE REVIEW

**1.** 1    **2.** 3    **3.** $-\dfrac{1}{2}$    **4.** $-45$    **5.** $-3$    **6.** 24    **7.** $-3x$    **8.** $a^8$    **9.** $\dfrac{1}{b^6}$    **10.** $7a^9$

**11.** $\dfrac{1}{4}y^2 - 4$    **12.** $40x^2 + 15x$    **13.** $\dfrac{3a + 19}{2a + 13}$    **14.** $2x^2y\sqrt{30y}$    **15.** $\dfrac{3a\sqrt{10}}{13}$    **16.** $-3$

**17.** $\dfrac{12a + 30}{(a - 9)(a + 6)}$    **18.** 1    **19.** 0    **20.** $-\dfrac{5}{2}$    **21.** $0, -\dfrac{2}{3}, 4$    **22.** $-7, -5$    **23.** $-8, 8$    **24.** 21

**25.** $(2, -1)$    **26.** $(-3, 2)$    **27.** $(-2, -7)$

**28.**    **29.**

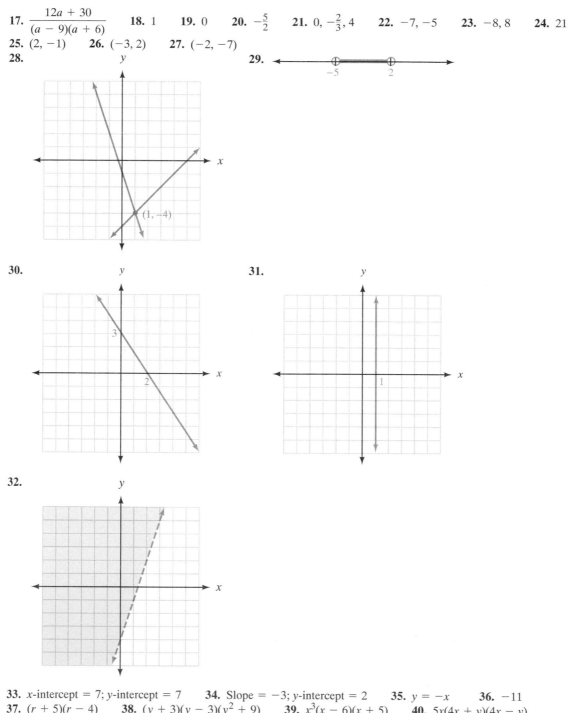

$(1, -4)$

$-5 \qquad 2$

**30.**    **31.**

**32.**

**33.** $x$-intercept $= 7$; $y$-intercept $= 7$    **34.** Slope $= -3$; $y$-intercept $= 2$    **35.** $y = -x$    **36.** $-11$

**37.** $(r + 5)(r - 4)$    **38.** $(y + 3)(y - 3)(y^2 + 9)$    **39.** $x^3(x - 6)(x + 5)$    **40.** $5x(4x + y)(4x - y)$

**41.** $4 \times 10^6$    **42.** $\dfrac{5 + \sqrt{21}}{2}$    **43.** $\dfrac{5\sqrt{3}}{3}$    **44.** 5    **45.** $13x\sqrt{7}$    **46.** $\dfrac{2x + 1}{x + 3}$    **47.** $\dfrac{1}{x^2 + 9}$

**48.** Width $= 8$ feet; length $= 16$ feet    **49.** 10, 25    **50.** Width $= 3$ inches; length $= 16$ inches

## CHAPTER 9

### PROBLEM SET 9.1

**1.** $\pm 3$ **3.** $\pm 5$ **5.** $\pm 2\sqrt{2}$ **7.** $\pm 5\sqrt{2}$ **9.** $\pm 3\sqrt{2}$ **11.** $0, -4$ **13.** $4, -6$ **15.** $5 \pm 5\sqrt{3}$

**17.** $-1 \pm 5\sqrt{2}$ **19.** $2, -3$ **21.** $\frac{11}{4}, -\frac{1}{4}$ **23.** $\frac{1 \pm 2\sqrt{3}}{3}$ **25.** $\frac{-2 \pm 3\sqrt{3}}{6}$ **27.** $3 \pm \sqrt{3}$ **29.** $-2 \pm \sqrt{5}$

**31.** $2 \pm \sqrt{2}$ **33.** $\frac{7}{3}, -1$ **35.** $\frac{-1 \pm \sqrt{7}}{2}$ **37.** $\frac{4 \pm 2\sqrt{3}}{5}$ **39.** $-5 \pm \sqrt{7}$ **41.** $4, -2$ **43.** $-6 \pm 2\sqrt{2}$

**45.** $(-1 + 5\sqrt{2} + 1)^2 = (5\sqrt{2})^2 = 50$ **47.** $-7, 1$ **49.** $r = -1 \approx \frac{\sqrt{A}}{10}$ **51.** $5\sqrt{3}$ feet

**53.** No, tent height is $3\sqrt{3}$ feet $\approx 5.20$ feet, which is less than 5 feet 8 inches ($5\frac{8}{12}$ feet $+ 5.67$ feet). **55.** 3 feet
**57.** 3.4 inches **59.** $x = 4\sqrt{10}$ feet $\approx 12.65$ feet $\approx 12$ feet 8 inches
**61.** (a) $AC = \sqrt{384^2 + 482^2} \approx 616$ feet (b) $BC = \sqrt{482^2 + (384\sqrt{2})^2} \approx 726$ feet (c) 94,765,056 cubic feet
**63.** $x^2 - 10x + 25$ **65.** $(x - 6)^2$ **67.** $(x + 2)^2$ **69.** 2 **71.** 2

### PROBLEM SET 9.2

**1.** 9 **3.** 1 **5.** 16 **7.** 1 **9.** 64 **11.** $\frac{9}{4}$ **13.** $\frac{49}{4}$ **15.** $\frac{1}{4}$ **17.** $\frac{9}{16}$ **19.** $-6, 2$ **21.** $-2, 8$

**23.** $-3, 1$ **25.** $0, 10$ **27.** $-5, 3$ **29.** $-2 \pm \sqrt{7}$ **31.** $2 \pm 2\sqrt{2}$ **33.** $8, -1$ **35.** $-1 \pm \sqrt{2}$

**37.** $-2, 1$ **39.** $\frac{-2 \pm \sqrt{3}}{2}$ **41.** $\frac{1 \pm \sqrt{3}}{2}$ **43.** $\frac{1}{2}$ **45.** $\frac{9 \pm \sqrt{105}}{6}$

**47.** Length $= 11.2$ inches, width $= 8.4$ inches **49.** $-0.4$ and $2.4$ **51.** (a) $-4$ (b) $-3$
**53.** **55.** **57.** 4 **59.** $-24$ **61.** $2\sqrt{10}$ **63.** $2\sqrt{3}$

**65.** $2xy\sqrt{5y}$ **67.** $\frac{9}{5}$

### PROBLEM SET 9.3

**1.** $-1, -2$ **3.** $-2, -3$ **5.** $-3$ **7.** $-3 \pm \sqrt{2}$ **9.** $-1, -\frac{3}{2}$ **11.** $\frac{-2 \pm \sqrt{3}}{2}$ **13.** 1 **15.** $\frac{5 \pm \sqrt{53}}{2}$

**17.** $\frac{2}{3}, -\frac{1}{2}$ **19.** $\frac{1 \pm \sqrt{21}}{2}$ **21.** $\frac{-1 \pm \sqrt{57}}{4}$ **23.** $\frac{5}{2}, -1$ **25.** $\frac{-3 \pm \sqrt{23}}{2}$ **27.** $\frac{-2 \pm \sqrt{10}}{3}$

**29.** $\frac{1 \pm \sqrt{11}}{2}$ **31.** $0, \frac{-3 \pm \sqrt{41}}{4}$ **33.** $0, \frac{4}{3}$ **35.** $\frac{3 \pm \sqrt{21}}{6}$ **37.** 1 second and 3 seconds **39.** $6\sqrt{15}$

**41.** $-4 - 3\sqrt{6}$ **43.** 5 **45.** $\frac{3 - \sqrt{5}}{2}$

## PROBLEM SET 9.4

**1.** $3 + i$ **3.** $6 - 8i$ **5.** $11$ **7.** $9 + i$ **9.** $-1 - i$ **11.** $11 - 2i$ **13.** $4 - 2i$ **15.** $8 - 7i$

**17.** $-1 + 4i$ **19.** $6 - 3i$ **21.** $14 + 16i$ **23.** $9 + 2i$ **25.** $11 - 7i$ **27.** $34$ **29.** $5$ **31.** $\dfrac{6 + 4i}{13}$

**33.** $\dfrac{-9 - 6i}{13}$ **35.** $\dfrac{-3 + 9i}{5}$ **37.** $\dfrac{3 + 4i}{5}$ **39.** $\dfrac{-6 + 13i}{15}$ **41.** $x^2 + 9$ **43.** $\dfrac{1}{i} \cdot \dfrac{i}{i} = \dfrac{i}{i^2} = \dfrac{i}{-1} = -i$

**45.** $8, -2$ **47.** $5, 1$ **49.** $-3 \pm 2\sqrt{3}$ **51.** $\dfrac{\sqrt{2}}{2}$ **53.** $\dfrac{2xy\sqrt{6y}}{3}$ **55.** $\dfrac{\sqrt[3]{2}}{2}$

## PROBLEM SET 9.5

**1.** $4i$ **3.** $7i$ **5.** $i\sqrt{6}$ **7.** $i\sqrt{11}$ **9.** $4i\sqrt{2}$ **11.** $5i\sqrt{2}$ **13.** $2i\sqrt{2}$ **15.** $4i\sqrt{3}$ **17.** $1 \pm i$ **19.** $2$

**21.** $\dfrac{3}{2}, -4$ **23.** $2 \pm 2i$ **25.** $\dfrac{-1 \pm 3i}{2}$ **27.** $\dfrac{1 \pm i\sqrt{3}}{2}$ **29.** $\dfrac{-1 \pm i\sqrt{3}}{2}$ **31.** $2, 3$ **33.** $\dfrac{-1 \pm i\sqrt{2}}{3}$

**35.** $\dfrac{1}{2}, -2$ **37.** $\dfrac{1 \pm 3\sqrt{5}}{2}$ **39.** $4 \pm 3i$ **41.** $\dfrac{4 \pm \sqrt{22}}{2}$ **43.** $2 + i$ and $2 - i$ **45.** $3 - 7i$

**47.**

**49.**

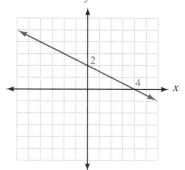

**51.** $23\sqrt{2}$ **53.** $-6\sqrt{6}$ **55.** $2x\sqrt{3}$

## PROBLEM SET 9.6

**1.**

**3.**

**5.**

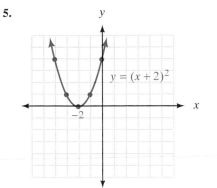

$y = (x + 2)^2$

$-2$

**7.**

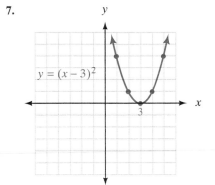

$y = (x - 3)^2$

$3$

**9.**

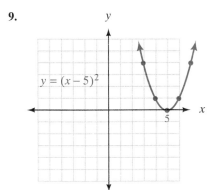

$y = (x - 5)^2$

$5$

**11.**

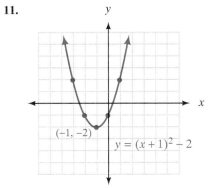

$(-1, -2)$

$y = (x + 1)^2 - 2$

**13.**

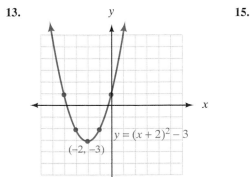

$y = (x + 2)^2 - 3$

$(-2, -3)$

**15.**

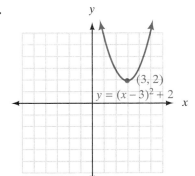

$(3, 2)$

$y = (x - 3)^2 + 2$

**17.**

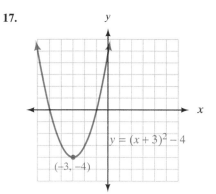

$y = (x + 3)^2 - 4$

$(-3, -4)$

**19.**

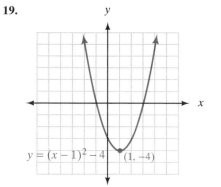

$y = (x - 1)^2 - 4$   $(1, -4)$

**21.**

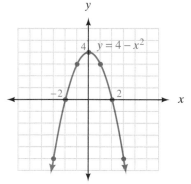

$y = 4 - x^2$

| x | y |
|---|---|
| −3 | −5 |
| −2 | 0 |
| −1 | 3 |
| 0 | 4 |
| 1 | 3 |
| 2 | 0 |
| 3 | −5 |

**23.**

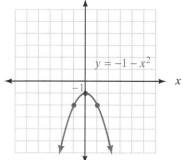

$y = -1 - x^2$

| x | y |
|---|---|
| −2 | −5 |
| −1 | −2 |
| 0 | 1 |
| 1 | −2 |
| 2 | −5 |

**25.**

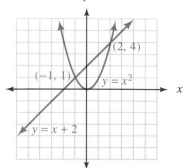

$(2, 4)$

$(-1, 1)$   $y = x^2$

$y = x + 2$

**27.**

$y = 2x^2$   $y = \frac{1}{2}x^2$

**29.** (a)

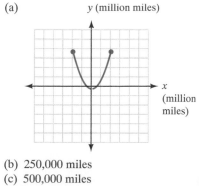

y (million miles)

x (million miles)

(b) 250,000 miles
(c) 500,000 miles

**31.** 7    **33.** $5\sqrt{2}$    **35.** $\dfrac{\sqrt{10}}{5}$    **37.** $21\sqrt{3}$    **39.** $-4 - 3\sqrt{6}$    **41.** $4\sqrt{5} + 4\sqrt{3}$    **43.** 7

## CHAPTER 9 REVIEW

**1.** $\pm 4\sqrt{2}$    **2.** $\pm 2\sqrt{15}$    **3.** $\pm 4$    **4.** $-9, 3$    **5.** $-7, 11$    **6.** $-2, \frac{2}{3}$    **7.** $\dfrac{-5 \pm 4\sqrt{2}}{2}$    **8.** $\dfrac{4 \pm 3\sqrt{3}}{3}$

**9.** $\frac{2}{3} \pm \frac{5}{3}i$    **10.** $-4 \pm 2\sqrt{5}$    **11.** $4 \pm 2\sqrt{5}$    **12.** $2 \pm \sqrt{11}$    **13.** $-3, -1$    **14.** $\dfrac{9 \pm \sqrt{93}}{2}$    **15.** $-1, 6$

**16.** $-3, 1$    **17.** $\dfrac{3 \pm \sqrt{15}}{3}$    **18.** $-4, -3$    **19.** 4    **20.** $\dfrac{-5 \pm i\sqrt{3}}{2}$    **21.** $\dfrac{-4 \pm \sqrt{26}}{2}$    **22.** $\dfrac{-3 \pm i\sqrt{39}}{6}$

**23.** $-\frac{1}{2}, 3$    **24.** $\dfrac{1 \pm i\sqrt{2}}{2}$    **25.** $4 + 2i$    **26.** $2 - 8i$    **27.** $10 + 5i$    **28.** $5 - 2i$    **29.** $-i$

**30.** $-1 - 5i$    **31.** $-1 - 3i$    **32.** $2 - 2i$    **33.** $6 - 2i$    **34.** $-24 - 12i$    **35.** $20 + 24i$    **36.** $7 - i$

**37.** $19 - 17i$    **38.** $8 - 6i$    **39.** 17    **40.** 13    **41.** $\dfrac{1 + 3i}{10}$    **42.** $\dfrac{-1 + 2i}{5}$    **43.** $\dfrac{10 - 25i}{29}$

**44.** $\dfrac{-2 + 8i}{17}$    **45.** $\dfrac{6 - 9i}{13}$    **46.** $\dfrac{4 + 3i}{5}$    **47.** $\dfrac{-9 - 40i}{41}$    **48.** $\dfrac{-9 + 19i}{34}$    **49.** $6i$    **50.** $12i$

**51.** $i\sqrt{17}$    **52.** $i\sqrt{31}$    **53.** $2i\sqrt{10}$    **54.** $6i\sqrt{2}$    **55.** $10i\sqrt{2}$    **56.** $11i\sqrt{2}$

**57.**

**58.**

**59.**

**60.**

**61.**

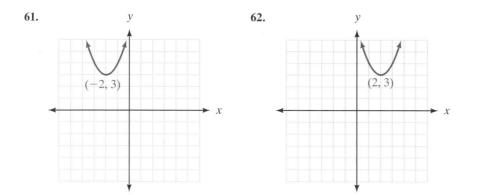

$(-2, 3)$

**62.**

$(2, 3)$

## CHAPTER 9 TEST

**1.** $-1, 8$    **2.** $3 \pm 2\sqrt{3}$    **3.** $\dfrac{5 \pm 5i\sqrt{3}}{2}$    **4.** $\dfrac{3 \pm i\sqrt{31}}{4}$    **5.** $\dfrac{1}{3}, -1$    **6.** $\dfrac{-1 \pm \sqrt{33}}{3}$    **7.** $-\dfrac{2}{3}$

**8.** $3 \pm \sqrt{15}$    **9.** $3i$    **10.** $11i$    **11.** $6i\sqrt{2}$    **12.** $3i\sqrt{2}$    **13.** $3 + 8i$    **14.** $-1 + 2i$    **15.** $5$

**16.** $1 + 5i$    **17.** $\dfrac{-1 + 3i}{10}$    **18.** $\dfrac{3 + 4i}{5}$

**19.**

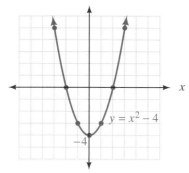

$y = x^2 - 4$
$-4$

**20.**

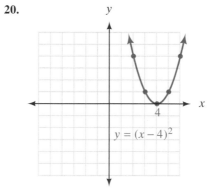

$4$
$y = (x - 4)^2$

**21.**

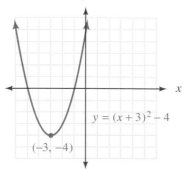

$y = (x + 3)^2 - 4$
$(-3, -4)$

**22.**

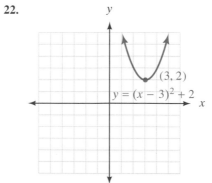

$(3, 2)$
$y = (x - 3)^2 + 2$

## CHAPTERS 1–9 CUMULATIVE REVIEW

**1.** $-9$   **2.** $4$   **3.** $-\frac{3}{2}$   **4.** $x^{12}$   **5.** $\dfrac{4}{x^2}$   **6.** $\dfrac{y-4}{y-8}$   **7.** $9$   **8.** $11x^2y$   **9.** $\dfrac{10\sqrt{2}}{9}$

**10.** $\dfrac{5\sqrt{7}-5\sqrt{3}}{4}$   **11.** $1-6i$   **12.** $1+3i$   **13.** $a^3-8a^2+19a-14$   **14.** $2$   **15.** $-3$   **16.** $-\frac{5}{3},4$

**17.** $-3,0$   **18.** $-7$   **19.** Possible solutions 3 and 1; only 3 checks.   **20.** $5,-2$   **21.** $\dfrac{1\pm i\sqrt{2}}{3}$

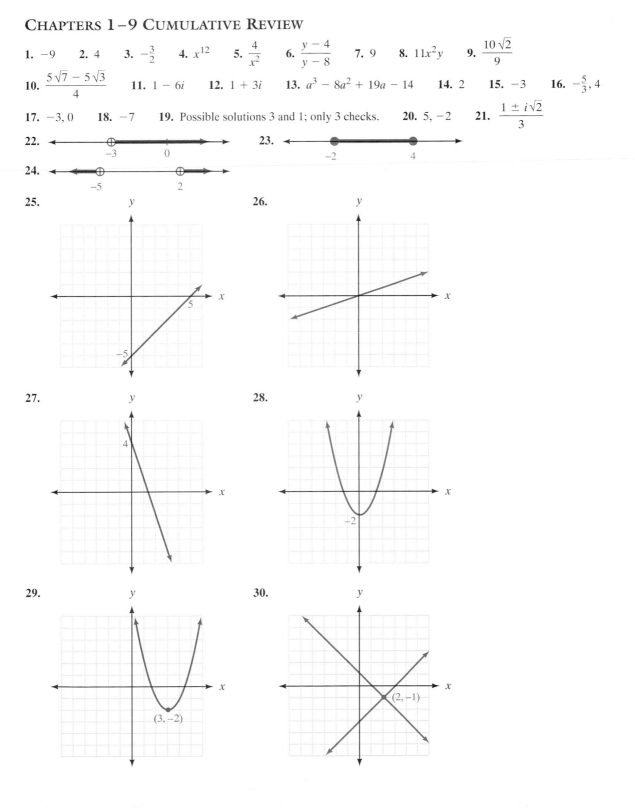

**22.** number line: open circle at $-3$, shaded to the right, with $0$ marked

**23.** number line: closed circles at $-2$ and $4$, shaded between

**24.** number line: open circles at $-5$ and $2$, shaded to the left of each

**25.** graph

**26.** graph

**27.** graph

**28.** graph

**29.** graph, $(3,-2)$

**30.** graph, $(2,-1)$

**31.** $(-1, 2)$   **32.** $(1, 0)$   **33.** $(x - 8)(x + 3)$   **34.** $(3x - 5y)(4x + 3y)$   **35.** $(5 + y)(5 - y)$
**36.** $(5x - 3y)^2$   **37.** $-5$   **38.** Slope $= 2$; $y$-intercept $= 3$   **39.** $8$   **40.** Distributive property
**41.** $3a^2 - 2ab - 4b^2$   **42.** $x^2 - 2x + 4 - \dfrac{6}{3x - 1}$   **43.** $x - 3$   **44.** $\dfrac{-13}{a - 7}$   **45.** $\dfrac{1}{(x - 2)(x - 6)}$
**46.** $\dfrac{x(x - 7)}{(x + 2)(x + 3)}$   **47.** $95\sqrt{2}$   **48.** $\dfrac{17 - 7i}{26}$   **49.** $y = \frac{3}{8}x - 3$   **50.** \$15,000 at 7%, \$5,000 at 5%

## APPENDIX A

## PROBLEM SET A

**1.** (a) $y = 8.5x$ for $10 \le x \le 40$

(b)

| Hours Worked $x$ | Function Rule $y = 8.5x$ | Gross Pay (\$) $y$ |
|---|---|---|
| 10 | $y = 8.5(10) = 85$ | 85 |
| 20 | $y = 8.5(20) = 170$ | 170 |
| 30 | $y = 8.5(30) = 255$ | 255 |
| 40 | $y = 8.5(40) = 340$ | 340 |

(c)

(d) Domain $= \{x \mid 10 \le x \le 40\}$; range $= \{y \mid 85 \le y \le 340\}$   (e) Minimum $= \$85$; maximum $= \$340$
**3.** Domain $= \{1, 2, 4\}$; range $= \{3, 5, 1\}$; a function   **5.** Domain $= \{-1, 1, 2\}$; range $= \{3, -5\}$; a function
**7.** Domain $= \{7, 3\}$; range $= \{-1, 4\}$; not a function   **9.** Yes   **11.** No   **13.** No   **15.** Yes   **17.** Yes

**19.** (a)

| Time (sec) $t$ | Function Rule $h = 16t - 16t^2$ | Distance (ft) $h$ |
|---|---|---|
| 0 | $h = 16(0) - 16(0)^2$ | 0 |
| 0.1 | $h = 16(0.1) - 16(0.1)^2$ | 1.44 |
| 0.2 | $h = 16(0.2) - 16(0.2)^2$ | 2.56 |
| 0.3 | $h = 16(0.3) - 16(0.3)^2$ | 3.36 |
| 0.4 | $h = 16(0.4) - 16(0.4)^2$ | 3.84 |
| 0.5 | $h = 16(0.5) - 16(0.5)^2$ | 4 |
| 0.6 | $h = 16(0.6) - 16(0.6)^2$ | 3.84 |
| 0.7 | $h = 16(0.7) - 16(0.7)^2$ | 3.36 |
| 0.8 | $h = 16(0.8) - 16(0.8)^2$ | 2.56 |
| 0.9 | $h = 16(0.9) - 16(0.9)^2$ | 1.44 |
| 1 | $h = 16(1) - 16(1)^2$ | 0 |

(b) Domain $= \{t \mid 0 \le t \le 1\}$;
range $= \{h \mid 0 \le h \le 4\}$

(c)

**21.** Domain $= \{x \mid -5 \le x \le 5\}$; range $= \{y \mid 0 \le y \le 5\}$   **23.** Domain $= \{x \mid -5 \le x \le 3\}$; range $= \{y \mid y = 3\}$

## APPENDIX B

### PROBLEM SET B

**1.** −1    **3.** −11    **5.** 2    **7.** 4    **9.** 35    **11.** −13    **13.** 1    **15.** −9    **17.** 8    **19.** 19    **21.** 16

**23.** 0    **25.** $3a^2 - 4a + 1$    **27.** 4    **29.** 0    **31.** 2    **33.** −8    **35.** −1    **37.** $2a^2 - 8$    **39.** $2b^2 - 8$

**41.** 0    **43.** −2    **45.** −3    **47.**    **49.** $x = 4$

**51.**

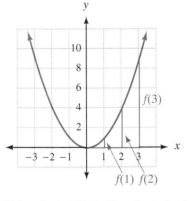

**53.** $V(3) = 300$, the painting is worth \$300 in 3 years; $V(6) = 600$, the painting is worth \$600 in 6 years.

**55.** $P(x) = 2x + 2(2x + 3) = 6x + 6$, where $x > 0$

**57.** $A(2) = 3.14(4) = 12.56$; $A(5) = 3.14(25) = 78.5$; $A(10) = 3.14(100) = 314$

**59.** (a) \$2.49    (b) \$1.53 for a 6-minute call    (c) 5 minutes    **61.** (a) \$5,625    (b) \$1,500

(c) Domain $= \{t \,|\, 0 \le t \le 5\}$    (d)

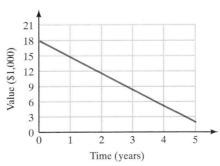

(e) Range $= \{V(t) \,|\, 1,500 \le V(t) \le 18,000\}$    (f) 2.42 years

# Index